T0202467

Graduate Texts in Physics

For further volumes:
www.springer.com/series/8431

Graduate Texts in Physics

Graduate Texts in Physics publishes core learning/teaching material for graduate- and advanced-level undergraduate courses on topics of current and emerging fields within physics, both pure and applied. These textbooks serve students at the MS- or PhD-level and their instructors as comprehensive sources of principles, definitions, derivations, experiments and applications (as relevant) for their mastery and teaching, respectively. International in scope and relevance, the textbooks correspond to course syllabi sufficiently to serve as required reading. Their didactic style, comprehensiveness and coverage of fundamental material also make them suitable as introductions or references for scientists entering, or requiring timely knowledge of, a research field.

Anwar Kamal

Particle Physics

Anwar Kamal *(deceased)*
Murphy, TX, USA

ISSN 1868-4513 ISSN 1868-4521 (electronic)
Graduate Texts in Physics
ISBN 978-3-662-52357-5 ISBN 978-3-642-38661-9 (eBook)
DOI 10.1007/978-3-642-38661-9
Springer Heidelberg New York Dordrecht London

Softcover reprint of the hardcover 1st edition 2014

Printed on acid-free paper

Springer is part of Springer Science+Business Media (www.springer.com)

Dedicated to my parents

Preface

This is an introductory textbook of particle physics for upper undergraduate students. The book is based on lectures given at British, American and Indian Universities over several years. The idea of writing a textbook on this subject was mooted some forty years ago. But it was never easy to write on a subject like particle physics which started expanding explosively in sixties and seventies. All this happened due to the development of high energy accelerators and sophisticated detecting systems on the experimental side, paralleled by new concepts on the theoretical side. And it became very difficult to keep pace with the rapidly changing scenario. Now, when the dust is settled down and the underlined physics is understood in a big way, although quite a bit is still obscure, there appears a justification in expounding the basic ideas of particle physics in the form of a textbook. It is attempted to survey the major developments in particle physics during the past 100 years. In Rutherford's time and early 1950s, only a few Elementary particles were known and the existence of neutrino was taken for granted. At that time, the entire knowledge of elementary particles and cosmic rays was so scanty that it could be accommodated in a single chapter. But, a few decades ago, particle physics branched off from nuclear physics and became virtually an independent discipline. No text book would be balanced until equal weightage is given to these two disciplines. There is also a complementary textbook on nuclear physics available by the same author.

The development of the subject is so fascinating that we were inclined to present the historical facts in the chronological order. The prerequisites for the use of this book are the elements of quantum mechanics comprising Schrodinger's equation and applications, Born's approximation, the golden rule, differential equations and Vector Calculus.

Basic concepts are explained with line diagrams wherever required. An attempt is made to strike a balance between theory and experiment. Theoretical predictions are compared with latest observations to show agreement or discrepancies with the theory. The subject matter is developed in each chapter with necessary mathematical details. Feynman diagrams are used extensively to explain the fundamental interactions.

The subject matter in various chapters is so much intimately connected that the logical sequential presentation of various topics becomes a vexing problem. For example, from the point of view of introducing quarks, the logical sequence would be strong, electromagnetic, weak and electroweak interactions, but from the point of view of introducing Feynman's diagrams, the desirable sequence would be electromagnetic, weak, electroweak and strong interactions, which is why one finds variance in sequences for both Nuclear and Particle physics in various textbooks. The only remedy is to make cross references to the chapters which were studied before and to those in which the relevant material is anticipated.

At the end of each chapter, a set of questions is given. A large number of worked examples are given. A comparable number of unworked problems with answers is delivered in the book to test the understanding of the student. The examples and problems are not necessarily of plug-in type but are given to explain the underlying physics. Various useful appendices are provided at the end of the book.

Murphy, TX Anwar Kamal

Note: These two volumes are the last books by my father Dr. Ahmad Kamal, the work he had conceived as his dream project and indeed his scientific masterpiece. Unfortunately, he passed away before he could see his manuscript in print. While we have tried our best to bring the publishing process to as satisfactory conclusion as possible, we regret any errors you may discover, in particular, that some of the references could not be as completely specifically cited as would otherwise be the case. We trust that these errors however do not compromise the quality or standard of the content of the text.

Suraiya Kamal
Daughter of Dr. Ahmad Kamal

Acknowledgements

I am grateful to God for helping us to complete the dream project of Dr. Ahmad Kamal after his demise, which seemed very difficult and even impossible at times.

I would like to thank Springer-Verlag, in particular Dr. Claus Asheron, Mr. Donatas Akmanavičius, Ms. Adelheid Duhm and Ms. Elke Sauer for their constant encouragement, patience, cooperation, and for bringing the book to its current form.

This project would not be complete without the constant support and encouragement of Mrs. Maryam Kamal, wife of Dr. Ahmad Kamal. Her determination kept us all moving to get these books done. My sincere thanks is due to the family, friends and well-wishers of the author who helped and prayed for the completion of these books.

Suraiya Kamal
Daughter of Dr. Ahmad Kamal

Contents

Chapter 1
Nuclear Radiation Detectors

1.1 Classification of Detectors

Experiments in Nuclear and Particle Physics depend upon the detection of primary radiation/particle and that of the product particles if any. The detection is made possible by the interaction of nuclear radiation with atomic electrons directly or indirectly. We may conveniently classify the detectors into two classes (i) Electrical (ii) Optical, as shown in Table 1.1.

The same detector may be used to study different types of radiation according to different phenomena. Thus, G.M. counters register all sorts of charged particles through ionization effects. Scintillation counters detect gamma rays by photoelectric effect, Compton scattering or pair production depending on gamma ray energy. Cerenkov counters detect a charged particle moving with speed exceeding that of light in a medium. Neutrons of high energy can be detected indirectly by the ionization caused by the recoil protons, and slow neutrons through the alpha particles produced in boron or by the fission of U-235 nuclei.

Instruments which provide one type of information accurately may give another type much less accurately. Thus, for example, semi-conductor detectors used to determine the energy of alpha particles yield precise determination of total energy released in an event but may furnish only a moderately accurate time resolution. Another example is the Cerenkov counter which furnishes accurate determination of the time at which an event occurs (within 1 nsec) but gives small spatial resolution. Photographic emulsions used to record the tracks of particles provide excellent spatial resolution but fail to distinguish between the old events and the new ones.

The detection efficiency of an instrument is of great consideration in an investigation. The detection efficiency which is the probability of detection when the particle crosses it, varies widely. In most of the direct ionizing radiation, it is nearly 1. In the detection of neutrons, it may vary from a few percent at high energy to nearly 1 for low energy. Similarly, gamma-ray counter efficiency is a function of energy.

A. Kamal, *Particle Physics*, Graduate Texts in Physics,
DOI 10.1007/978-3-642-38661-9_1, © Springer-Verlag Berlin Heidelberg 2014

Table 1.1 Classification of detectors

Type	Detectors
Electric	Ionization Chamber
	Proportional Counter
	Geiger-Muller Counter
	Semi-conductor detector
	Neutron Detector
	Scintillation Counter
	Cerenkov Counter
Optical	Photographic Emulsion
	Expansion Cloud Chamber
	Diffusion Cloud Chamber
	Bubble Chamber
	Spark Chamber

The electric detectors necessarily carry auxiliary electronic equipment like pulse shapers, amplifiers, discriminators, coincidence circuits, scalars etc. Again, computers are invariably used to analyze more than hundred thousand pictures in a typical bubble chamber experiment.

After an event is detected, most of the instruments lose their sensitivity for certain time called "Dead time". In order that the counter efficiency be high, it is important that the dead time be smaller than the mean time interval between successive events. If r is the counting rate and td is the dead time then for high efficiency, the condition $rtd \ll 1$, must be satisfied.

According to Hofstadter a perfect detector might have the following characteristics

i. 100 percent detection efficiency
ii. high-speed counting and timing ability
iii. good energy resolution
iv. linearity of response
v. application to virtually to all types of particles and radiations
vi. large dynamic range
vii. virtually no limit to the highest energy detectable
viii. reasonably large solid angles of acceptance
ix. discrimination between types of particles
x. directional information
xi. low background, and
xii. picturization of the event.

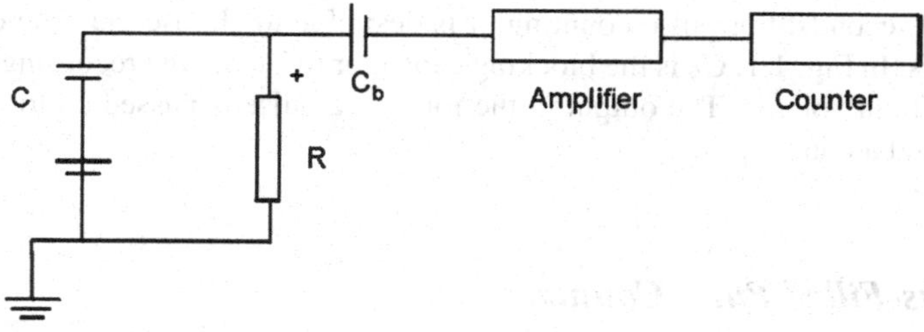

Fig. 1.1 A pulse counting circuit

1.2 Ionization Chambers

1.2.1 Ionization Chambers

An ionization chamber is a device which measures the amount of ionization created by charged particles passing through a volume of gas enclosed in a vessel. If an electric field be maintained in a gas by a pair of electrodes, the positive and negative ions will drift apart inducing charges on the electrodes. In their traversal the ions may undergo recombination processes, and the charge collected by the electrodes alone will result in the ionization current measured in the external circuit. When every ion is collected, with no loss due to recombination, the maximum current is obtained called saturation current which will be proportional to the intensity of radiation.

1.2.2 Pulse Chambers

When an ionization chamber is exposed to a highly intense radiation the rate of ionizing events will be large. The ionization currents measured in the external circuit will be the average values, fluctuating about the mean value due to random fluctuations of events. For low intensity radiation it is often desirable to count primary individual events or pulses rather than register average values of current or voltage. In this case it is necessary to use short ion-collection and recovery time.

Applying Kirchhoff's law to the RC circuit (Fig. 1.1) it is found that the signal voltage developed across R is given by

$$V = \left(\frac{Q}{C}\right) e^{-\frac{t}{RC}} \tag{1.1}$$

where C is the capacitance of the chamber. Equation (1.1) shows that the polarity of the pulse will be opposite to that of the applied voltage. The peak amplitude of the signal is determined by the ratio Q/C and the rate of decay by the product RC.

The product RC is called the time constant. At time $t = RC$, the pulse amplitude will be reduced to $1/e = 0.37$ of its peak value. If R is in ohms and C in farads then t is in seconds. For pulse counting, it is desirable to choose the time constant 10^{-4} or less. In Fig. 1.1, C_b is the blocking capacitor to isolate the recording circuits from D.C. high voltage. The output in the form of a pulse is passed on to the A.C. amplifier and a counter.

1.2.3 Gas-Filled Pulse Counter

Most of the gas-filled pulse counters take the advantage of large multiplication of ion pairs, through the occurrence of secondary ionization. This is called gas multiplication. The ions are accelerated towards the charged electrodes until re-combination takes place or lose their energy through a collision. At atmospheric pressure, under moderate electric fields, the mean free path between successive collisions is short and the energy acquired by an ion between collisions is small and the collisions are mainly elastic. However, if the gas pressure is reduced or if the applied voltage is substantially increased, an ion can pick up appreciable energy and the occurrence of inelastic collisions becomes a distinct possibility. The inelastic collisions result in the molecular excitation or ionization. An additional ion pair is caused by ionization. The secondary ions produce tertiary ions and so on in the subsequent collisions, thus producing enormously large number of ion pairs. This kind of multiplication process is called avalanche ionization. If a single primary pair finally results in A ion pairs then the gas multiplication is A. In an ionization chamber A will be unity as no secondaries are formed and may be as large as 10^{10} in a Geiger-Muller tube.

1.2.3.1 Geometry of the Chamber

For the plane parallel electrodes field strength E is given by

$$E = \frac{V}{d} \tag{1.2}$$

Now, for gas amplification in a mean free path an ion must pick up enough energy to ionize a neutral molecule. For typical ionization potentials which range from 10–25 eV, the field strength required is of the order of 10^6 volts cm^{-1}. Such field strengths are not practical.

The mean-free-path can be increased by reducing the pressure is determined by making a compromise between these conflicting requirements. Even with the best choice of gas pressure, the electric field requirement is much excessive for plane geometry. However, other type of geometry for the electrodes, in particular that of a co-axial cylinder, permits the use of gas at moderate pressure, a large field strength without employing very high voltages.

Consider an outer cylinder as the cathode with a coaxial wire of smaller diameter as the anode. Figure 1.2 shows the cross-section of the cylindrical configuration.

The cathode has radius b and the anode radius a. Consider a cylindrical surface of radius r in between.

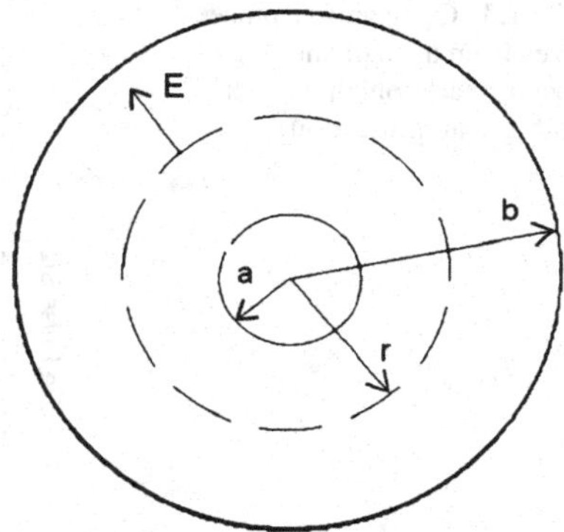

Fig. 1.2 Cross section of a counter with cylindrical geometry

The normal component of electric field acts radially outward and is constant. Applying Gauss theorem to the surface,

$$\oint E_n ds = 2\pi r E = \frac{q}{\varepsilon} \tag{1.3}$$

where $E_n = E =$ constant. Integration is done only on the cylindrical surface since the normal component at the closed surfaces is zero.

The potential difference between the two electrodes is given by

$$V = \int_a^b E_n dr = \frac{q}{2\pi\varepsilon}\int_a^b \frac{dr}{r} = \frac{q}{2\pi\varepsilon}\ln\frac{b}{a} \tag{1.4}$$

Combining (1.3) and (1.4),

$$E = \frac{V}{r\ln(\frac{b}{a})} \tag{1.5}$$

For given values of a and b, the most intense field will be close to the surface of the wire. Equation (1.5) shows that for a given voltage large values of E can be obtained for small diameter anode wire. Structural strength requirements limit the minimum diameter to 0.08 mm.

1.2.4 Variation of Pulse Sizes

We now consider the pulse sizes produced by a radiation source of constant intensity. At low voltages, the counter behaves as an ionization chamber (Region I, Fig. 1.3). This is a recombination region. As the applied voltage is increased, saturation region II is reached. The larger size pulse has been taken as 10^3 times greater than the smaller size pulse for illustration. With further increase of voltage secondary ioniza-

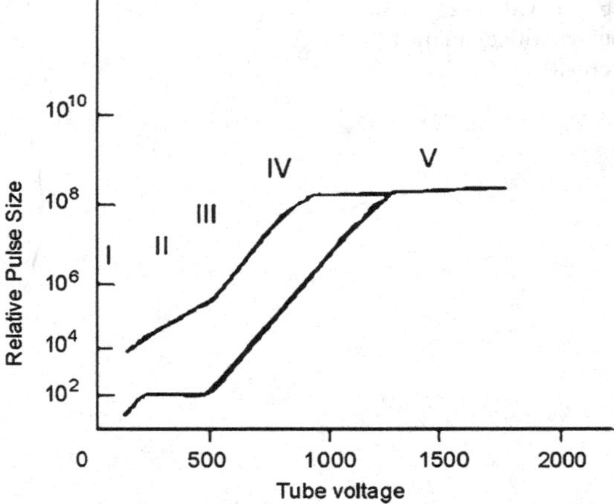

Fig. 1.3 Comparative pulse sizes from a small and large size primary ionizing event with gas amplification

tion ensues. Secondary ions will be produced when the integral of (1.4) taken over one mean-free-path equals the ionization potential of the counter gas. Secondary ionization will occur over the mean-free-path that lies just outside the central wire since the electric field is maximum close to the central wire. Primary ions formed outside of this mean-free-path experience a weaker field. They will be accelerated towards the respective electrodes but will not acquire enough energy to generate secondaries.

With the initiation of secondary ionization, each pulse will be enhanced compared to that of the primary ions alone by a factor which depends on the voltage but not upon the number of primary ions The size of each pulse is proportional to the size of the primary initiating ionization by a constant factor. This is the proportional region III, Fig. 1.3. With further increase of voltage, secondary ionization may be produced beyond one mean-free-path from the central wire. The gas amplification will increase but the proportionality factor will be maintained until a gas amplification factor of about 10^4 is reached.

As the voltage is further increased, secondary ions will be formed farther and farther from the wire until gas amplifications of 10^5–10^7 are reached. This is the region of limited proportionality (Region IV). Here the proportionality becomes weaker and weaker until it is lost. The two curves merge into one.

With further increase of voltage larger avalanche results until in the Geiger region (region V), as many as 10^{10} secondary ion pairs may be produced from a single ion pair. Here the size of the output pulse is saturated. In this region each pulse of ionization spreads axially along the entire length of the wire and radially it extends outward over several mean-free-paths.

Beyond Geiger's plateau there is a voltage breakdown and continuous discharge takes place.

1.3 Proportional Counter

1.3.1 Construction

Gas proportional counters usually have spherical or semi-hemispherical geometry. They are used when extreme values of gas amplification A are not required. A typical value of A may be 10^3. In the hemispherical type the anode is in the form of a small loop. The electric field changes rapidly in the vicinity of the anode as in the case of a cylindrical counter. The voltage-pulse size relations are also similar.

In the proportional region, with a typical gas amplification of 10^3, the pulse generated by alpha might contain 10^8 ions while for beta particles or gamma rays the corresponding figure is about 10^5. Even with these amplifications the pulses are not large enough to be detected. A pre-amplifier, with little gain serves the purpose of producing an output signal across a relatively low impedance. This is followed by a linear amplifier with voltage gain 200–500 and is used to amplify the pulse preserving the relative size. A discriminator allows only those pulses which are above a certain level. In this manner, all the small pulses due to beta particles and gamma rays are rejected and only relatively larger pulses due to alpha particles are counted. Similarly fission pulses can be counted against heavy background of alpha particles. Figure 1.4 is a representative diagram of a spherical chamber with 4π geometry. Window absorption can be introduced directly into the sensitive volume of the chamber. The source is placed on a thin film to minimize back scattering and self absorption. Filling gas at suitable pressure is flushed into the chamber to displace the air. Such a chamber is called windowless gas-flow counter. For the filling gas, pure Argon, though highly dense, is not suitable as the meta-stable states which are formed upon de-excitation after a relativity long time lead to spurious discharges. However, 90 percent argon and 10 percent methane mixture or a 96 percent Helium and 4 percent isobutane are commonly used. A voltage of 1500–4000 volts is generally used.

The advantage of the 4π windowless counter, Fig. 1.4, is that all the particles are counted and is therefore ideal for the absolute decay rate determinations of radioactive substances. Proportional counters are very useful for high counting rates, because the negative ions have to move only a few mean-free-paths to be collected in an intense electric field.

1.4 Geiger Muller Counter

1.4.1 Construction and Characteristics

Geiger Muller Counter (GM counter) is one of the old devices used to detect charged particles, regardless of their energy and identity. Nevertheless, it is by no means outdated. Figure 1.5 shows the GM tube along with the block diagram of the electronic circuit.

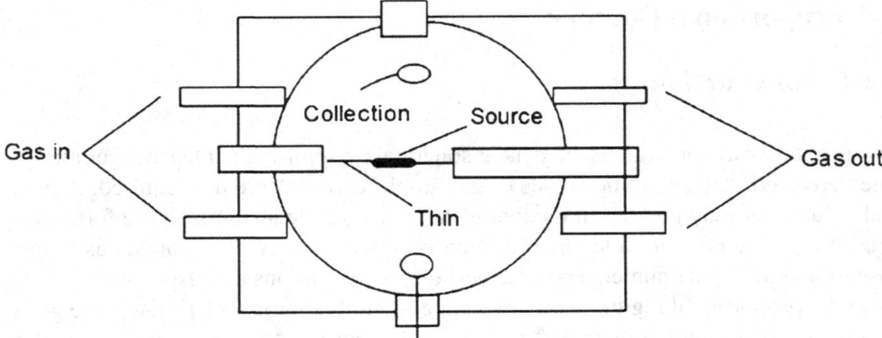

Fig. 1.4 Representative diagram of a spherical chamber with 4π geometry

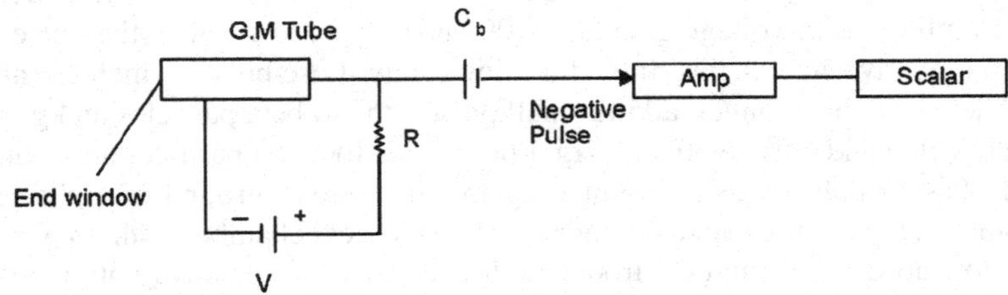

Fig. 1.5 A GM tube circuit when GM tube is operated, a negative pulse is developed for the output

The GM tube is usually of a cylindrical geometry, the outer hollow cylinder of diameter 1–10 cm and length 2–10 times as great, forms the cathode and the inner central wire of small radius, concentric with outer cylinder forms the anode. Cathode surface is coated with a substance to provide a large work function. The central wire is usually made of tungsten, tantalum or stainless steel with a smooth surface and of desired diameter. End-windows as thin as 1 mg cm^{-2} made of cleaved mica or thin glass permit low energy beta particles to be counted. A high voltage of 800 V to 1100 V is generally used. The blocking capacitor C_b (Fig. 1.5) isolates the electronic circuit from the high voltage that is applied but permits the pulses developed across the resistor R to get through, and recorded by the electronic circuit.

Let a GM tube in such a circuit be exposed to radiation of constant intensity. There is a threshold A as the circuits do not respond below it, Fig. 1.6. The threshold generally corresponds to the region of limited proportionality, where the gas amplification is of the order of 10^6–10^7. With the increase of voltage above threshold, the count rate rises sharply through B, as the small primary events also get counted. The point C is the beginning of the Geiger region and extends up to D, the flat part of the curve CD is called plateau. In the region CD where every event regardless of its original size is counted. Actually a plateau has a small upward slope. GM tubes with plateau length 200 volts and a change of counting rate of only 1–2 percent for a change 100 volts are available. The length as well as slope of plateau depend

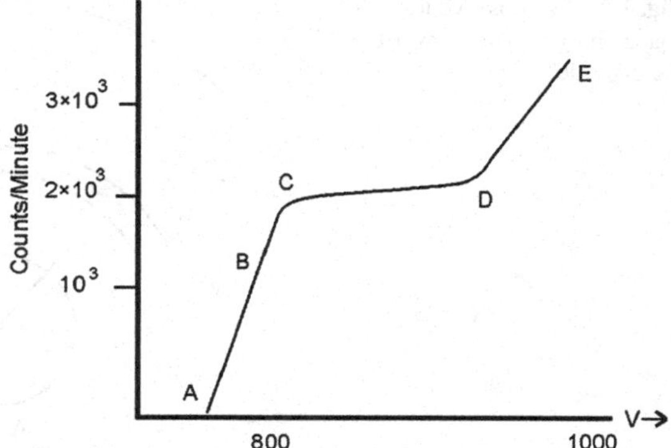

Fig. 1.6 Typical variation of counting rate as a function of applied voltage for a GM tube

on several factors, such as the filling gas, the condition of central wire and cathode surface.

When the voltage is further raised, beyond the end of plateau, multiple discharge called continuous discharge occurs, in the region E. Continuous discharge may be triggered by a single pulse. The continuous discharge rate will be independent of any radiation-initiated event and will depend only on the circuit elements. GM tubes are not operated in this region. If operated, the life would be shortened or the tube may even be damaged beyond repair.

1.4.2 Pulse Formation and Decay

The initial ionization pulse which consists of equal number of positive ions and electrons is formed as a thin cylindrical sheath just outside the central wire. Immediately, the ions start separating, the electrons moving toward the central wire and the positive ions outward toward the cathode. Because of their high mobility, the electrons are collected at the wire in a short time of the order 10^{-5} sec, while the positive ion cloud (space charge) will be still close to the wire.

Let the space charge move outward to a distance r_0, Fig. 1.7, after some time. Let there be q^- charges per unit length on the wire and q^+ charges per unit length in the space charge. The field intensity E_1 inside the sheath will be different from E_2 outside.

From (1.3),

$$E_1 = \frac{q^-}{2\pi \varepsilon r} \quad \text{(inside)} \tag{1.6}$$

$$E_2 = \frac{q^- + q^+}{2\pi \varepsilon r} \quad \text{(outside)} \tag{1.7}$$

Assuming that the applied voltage is constant, the voltage developed is given by the integrals of the field strengths.

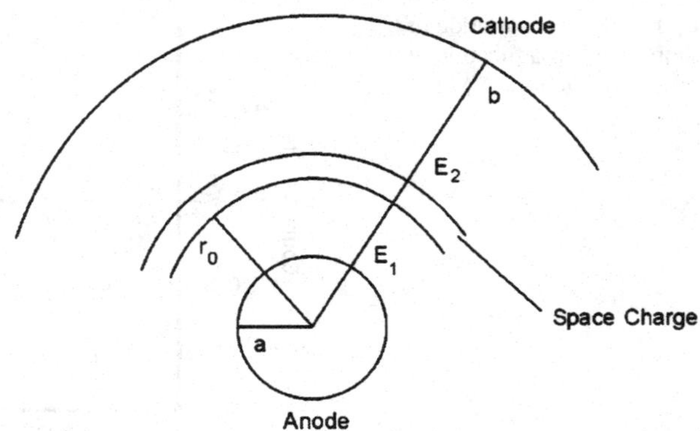

Fig. 1.7 The positive ion space charge moves toward the cathode

$$\begin{aligned} V &= \int_a^{r_0} \frac{q^- dr}{2\pi\varepsilon r} + \int_{r_0}^{b} \frac{(q^- + q^+)dr}{2\pi\varepsilon r} \\ &= \frac{q^-}{2\pi\varepsilon} \ln\frac{r_0}{a} + \frac{(q^- + q^+)}{2\pi\varepsilon} \ln\frac{b}{r_0} \\ &= \frac{q^-}{2\pi\varepsilon} \ln\frac{b}{a} + \frac{q^+}{2\pi\varepsilon} \ln\frac{b}{r_0} \end{aligned}$$

The charge on the central wire will be

$$q^- = \frac{2\pi\varepsilon}{\ln(b/a)}\left[V - \frac{q^+}{2\pi\varepsilon}\ln\frac{b}{r_0}\right] \tag{1.8}$$

Equation (1.8) shows that both q^- and E_1 in the avalanche region are reduced during the pulse. As the positive ion sheath moves toward the cathode, r_0 increases, the negative term in (1.8) decreases and the field strength is restored to its original value in the state prior to ionization.

A typical time in which the positive-ion cloud reaches the cathode is of the order of 10^{-4} sec. When a positive ion comes within about 10^{-7} cm from the cathode, it will capture an electron from the cathode surface to become a neutral molecule. In this process the molecule will get into an excitation state. As a result of de-excitation, photons will be emitted, some of them in the ultra violet region. These photons upon hitting nearby cathode surface will liberate electrons by photoelectric effect. The photo electrons thus liberated will now be accelerated toward the anode. Upon reaching the vicinity of the wire, they would have acquired sufficient energy to initiate a secondary avalanche. Thus a single discharge will eventually lead to a series of discharges at a rate determined by the parameters of the counter and the circuit.

1.4.3 Quenching the Discharge

From the discussion at the end of Sect. 1.4.2, it is obvious that there must be a mechanism for stopping the occurrence of repetitive discharges. This mechanism is called

Quenching. In practice this is accomplished by making the counter self- quenching on adding a small percentage of a quenching gas to the main gas. The main gas is usually argon and the quenching gas can be either organic such as alcohol, xylene or isobutane or halogen gases like Cl_2 or Br_2.

Consider a GM counter filled with a mixture of 90 % argon and 10 % of ethyl alcohol at 10 cm of mercury. In the first phase of avalanche the space-charge will contain both types of positive ions. Now the ionization energy of argon atom is 15.7 eV while for ethyl alcohol molecule, it is 11.3 eV. In an argon-alcohol collision, it is energetically possible for the argon to become neutral, while alcohol molecule is ionized. On the other hand, it is not energetically possible for an alcohol molecule to become neutral while argon atom is ionized. Each ion will make some thousand collisions while the space charge is on its way to the cathode. Before the space charge reaches the cathode, it now almost entirely consists of alcohol ions.

On reaching close to the cathode surface, the alcohol ion picks up an electron and the excitation energy is given by

$$E^* = I - \varphi \tag{1.9}$$

where I is the ionization potential of alcohol and φ is the work function of the cathode material. If a cathode surface of large work function be chosen, such as colloidal ($\varphi = 5$ eV) then less excitation energy will be available for the alcohol molecule. The de-excitation via molecular dissociation is strongly favored compared to radiation. The photon energies involved in the dissociation process are too small to initiate a second discharge. In the event of de-excitation by radiation, most of the resulting photons will have energy much below the work function of the cathode, and the photo electron emission would be energetically impossible. In each collision the energy transfer from argon to alcohol results in the emission of photons of energy $15.7 - 11.3 = 4.4$ eV, a value much below the work function of the selected cathode surface. Most of the photons emitted will be reabsorbed in the broad bands in the ultraviolet. Occasionally, an argon ion may escape and may extract an electron from the cathode. By (1.9) the excitation energy will be in excess of 4.4 eV compared to that in alcohol. This increases the probability for photon emission and hence photo-electric emission from the cathode surface. As the voltage increases the transit time decreases and further the ion density in the original plasma will be greater. This leads to occasional spurious discharges. The upward slope of the plateau is attributed to this phenomenon. With further increase in voltage, the occurrence of spurious discharger becomes a dominant phenomenon (Fig. 1.8). One must avoid working the GM tube in this region otherwise, the life of the tube may be shortened or it may be irreparably damaged.

When organic quenching gases are used some of the molecules are dissociated in each discharge. In a typical counter there may be 10^{20} molecules of alcohol. In each pulse 10^{10} molecules may be dissociated. Theoretically, 10^{10} pulses can be counted. But much before theoretical limit is reached, the counter starts behaving erratically. The life of such counters is therefore finite.

When halogens like Cl_2 are used as a quenching gas, the dissociated molecules of Cl_2 by recombination process become neutral again, and are therefore available

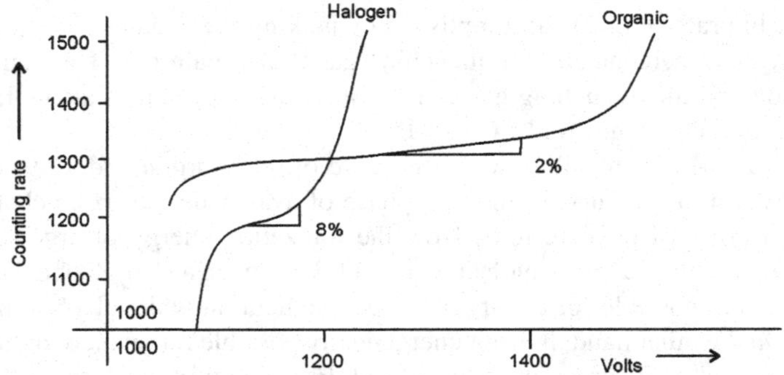

Fig. 1.8 Typical plateau slopes for halogen and organic quenched GM counters

for subsequent discharges. Theoretically, such counters have infinite life time. The plateau, however, is inferior than that for organic counters. Typical plateau length is limited to 150 V, with a slope of 8 % per 100 V length. These values are to be compared with 300 V length and 2 % slope for organic counters, Fig. 1.8.

1.4.4 Scaling Circuits

The pulses from a GM counter are negative, rising to maximum amplitude of about 1 volt in a time of the order of 10^{-5} sec and decaying in about 10^{-3} sec. It is preferable to amplify the pulses before feeding them to a counting system. As the mechanical register cannot respond to fast occurring pulses they are passed through a scalar. A binary scalar consists of several bi-stable circuits in tandem. At each stage a single pulse is transmitted for every two pulses received. At each stage the counts are cut down by a factor 2. Thus, the scaling factors of 2, 4, 8, 16, 32, 64, … will be available. A binary scalar can be converted into a decade scalar. With the use of a suitable feedback circuit, a scale of 16 can be reset to zero at the tenth pulse. A timer is automatically actuated at the moment the push-button is on, and the counting is stopped after a fixed duration. The counting rate is then simply found out.

1.5 Resolving Time

It has been pointed out that the space charge reduces the gas amplification to the extent that the secondary ionization temporarily becomes impossible. During this interval of time no event can be recorded. This is called **Dead Time**. As the space charge moves out toward cathode, its effect on the electric field in the immediate neighborhood of the central wire gradually decreases until the original voltage is

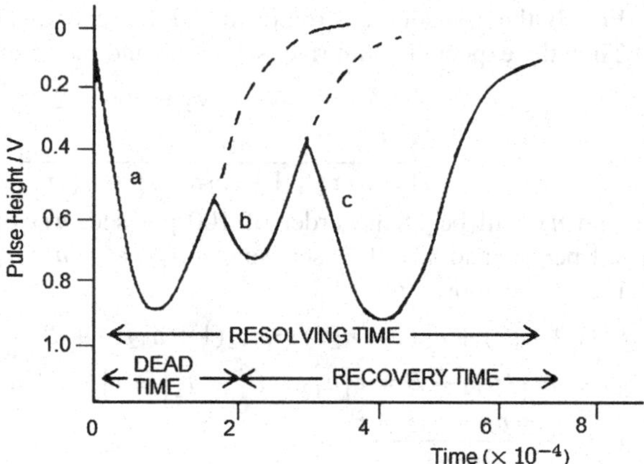

Fig. 1.9 Dead time, Recovery time and Resolving time of a GM tube and counting circuit

restored. Suppose a pulse of minimum amplitude of 0.5 volts is required to get it registered. Consider three pulses a, b, and c, Fig. 1.9, caused by three particles.

Pulse a starts at time $t = 0$ and has full amplitude of 0.9 V at 1.0×10^{-4} sec, and will certainly be counted. Due to arrival of a second particle, pulse b starts at time 1.8×10^{-4} sec and builds up to 0.1 V at maximum amplitude and will not be counted. At $t = 3.0 \times 10^{-4}$ sec, the pulse c starts due to the third particle, but has maximum amplitude of 0.5 V at $t = 4.0 \times 10^{-4}$ sec, barely sufficient to actuate the counting system. The time τ between just recordable pulses is known as the **Resolving Time**. It is obvious that because of finite resolution time, some of the random pulses will be lost.

$$\text{Resolving time} = \text{Dead time} + \text{Recovery time}$$

Let the true counting rate be N, i.e. without losses due to finite resolution, and n observed counting rate. The insensitive time will be $n\tau$ per unit time and the number of counts missed will be $Nn\tau$ which is also equal to $N - n$. Hence

$$N - n = Nn\tau \quad \text{or}$$

$$N = \frac{n}{1 - n\tau} \tag{1.10}$$

1.5.1 Determination of Resolving Time: Double Source Method

This method involves four counts using two radioactive sources. First, the background rate B per sec is found. One of the sources is then positioned in such a way that the resolving time losses are substantial. Let the expected count rate be $N_1 + B$ and that observed $n_1 + B$. With the first source still in position the second one is introduced and adjusted until the count rate is approximately doubled.

Then the expected count rate is $N_1 + N_2 + B$ and the observed count rate is $n_{12} + B$.

Finally the first source is removed and the second is left in position.
Then the expected count rate is $N_2 + B$ and the observed count rate is $n_2 + B$.

$$N_1 + N_2 = N_{12} + B \tag{1.11}$$

$$\frac{n_1}{1 - n_1\tau} + \frac{n_2}{1 - n_2\tau} = \frac{n_{12}}{1 - n_{12}\tau} + \frac{B}{1 - B\tau} \tag{1.12}$$

n_1 and n_2 will be of the order of 100 per sec, n_{12} of the order of 200 per sec, $B \cong 1$ per sec and $\tau \cong 10^{-4}$ sec, so that $n_1\tau \ll 1$, $n_2\tau \ll 1$, $n_{12}\tau \ll 1$, $B\tau \ll 1$, and (1.12) can be approximated to

$$n_1(1 + n_1\tau) + n_2(1 + n_2\tau) = n_{12}(1 + n_{12}\tau) + B$$

$$n_1 + n_2 - n_{12} - B = \tau\left[n_{12}^2 - n_1^2 - n_2^2\right] \cong \tau\left[(n_1 + n_2)^2 - n_1^2 - n_2^2\right] \quad \text{or}$$

$$\tau = \frac{n_1 + n_2 - n_{12} - B}{2n_1 n_2} \tag{1.13}$$

1.6 Resolving Time of a Coincidence Circuit

As the pulses arriving at the input of the coincidence circuit will have finite width, two independent (not coincident) signals may give rise to a pulse in the output (spurious coincidence or chance coincidence or accidental coincidence) if their relative delay is smaller than a certain time τ, the resolving time. Their number per unit time C_c depends on the single counting rates m_1 and m_2 of the two counters. The probability that a certain pulse from the first counter is accompanied by a pulse from the second counter within $\pm\tau$ is $p \cong m_1 m_2(2\tau)$. The number of chance coincidences is therefore

$$C_c = 2\tau m_1 m_2 \tag{1.14}$$

1.6.1 Determination of the Resolving Time of the Coincidence Circuit

Two independent sources are used and counting is made with two counters. They are well shielded from each other so that no genuine coincidence is registered. The observed coincidences are then only due to the finite resolving time and to the background produced by cosmic rays. Cosmic rays flux is maximum from the top, and if the counters are placed above each other, a single cosmic ray particle can penetrate both and produce a coincidence. On the other hand if they are placed horizontally, they can be triggered only by at least two shower particles which come from the same source. These are rather rare events. Thus the cosmic background B_c produced by the cosmic rays depends strongly on the relative position of the counters. If coincidence counting rate m_c is measured for different counting rate for different source strengths, the coincidence rate is given by

$$m_c = B_c + 2\tau m_1 m_2 \tag{1.15}$$

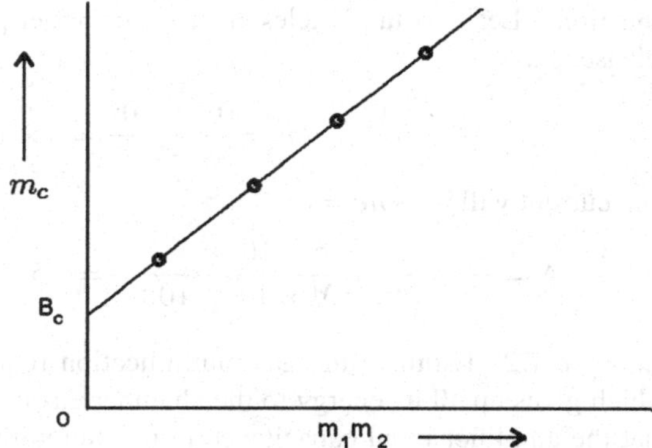

Fig. 1.10 Determination of resolving time. Plot of m_c against m_1m_2

Assuming that τ is constant, the plot of m_c vs. m_1m_2 is a straight line intercepting at the background value B_c and having a slope equal to 2τ, Fig. 1.10.

1.6.1.1 Limitation of the Useful Source Strength by the Resolving Time

It is desirable that the number of chance coincidences be smaller than the number of true coincidences. As an example, consider the beta emission from a certain isotope followed by gamma emission. If D is the source strength, beta and gamma counting rates are

$$m_\beta = DP_\beta \tag{1.16}$$

$$m_\gamma = DP_\gamma \tag{1.17}$$

where P_β and P_γ are the probabilities of detecting a given beta particle in the beta counter and the gamma counter, respectively. The probability of detecting a beta ray and at the same time in coincidence a gamma ray is the product of the two probabilities. The true coincidence is then

$$m_{\beta\gamma} = DP_\beta P_\gamma \tag{1.18}$$

On the other hand, the chance coincidence is

$$C_c = 2\tau m_\beta m_\gamma = 2\tau D^2 P_\beta P_\gamma \tag{1.19}$$

The condition $C_c < m_{\beta\gamma}$ becomes

$$2\tau D < 1 \tag{1.20}$$

For the resolving time of 1 μsec, the source strength must be smaller than 5×10^5 decays/sec i.e. smaller than 13.5 μci.

Example 1.1 An ionization chamber is used with an electrometer capable of measuring 5×10^{-11} ampere to assay a source of 0.7 MeV beta particles. Assuming saturation conditions and that all the particles are stopped within the chamber, calculate the rate at which the beta particles must enter the chamber to just produce a measurable response. Given the ionization potential for the gas atoms is 35 eV.

Solution Let N beta particles enter the chamber per second. Number of ion pairs released,

$$n = \frac{0.7 \times 10^6}{35} = 2 \times 10^4$$

The current will be $Nne = i$

$$N = \frac{i}{ne} = \frac{5 \times 10^{-11}}{2 \times 10^4 \times 1.6 \times 10^{-19}} = 1.56 \times 10^4 \text{ beta particles/sec}$$

Example 1.2 Estimate the gas multiplication required to count a 1 MeV proton which gives up all its energy to the chamber gas in a proportional counter. Assume that the amplifier input capacitance in parallel with the counter is 1×10^{-9} F and that its input sensitivity is 1 mV. Energy required to produce one ion pair is 35 eV.

Solution

$$V = \frac{MNq}{C}$$

$$M = \frac{CV}{qN} = \frac{10^{-9} \times 10^{-3}}{1.6 \times 10^{-19} \times (10^6/35)} = 219$$

Example 1.3 A proportional counter is used with an electrometer capable of measuring 5×10^{-11} Ampere to assay a source of 0.7 MeV beta particles. The beta particles produce on an average 60 ion pairs with a gas multiplication factor of 6×10^4. What rate of particle incidence will be required to produce an average current of 5×10^{-11} ampere?

Solution Let N beta particles be incident per second.
Current $i = NnMe$, where n is the number of ion pairs

$$N = \frac{i}{nMe} = \frac{5 \times 10^{-11}}{60 \times 6 \times 10^4 \times 1.6 \times 10^{-19}} = 87$$

Example 1.4 Calculate the pulse height obtained from a proportional counter when a 10 keV electron gives up all its energy to the gas. The gas multiplication factor of the proportional counter is 800, capacitance of the circuit is 20×10^{-12} F and energy required to produce an ion pair is 32 eV.

Solution Number of primary ion pairs produced by a 10 keV electron,

$$N = 10^4/32$$

$$\text{Charge collected } Q = Mne$$

where M is the gas multiplication and e the elementary charge is equal to 1.6×10^{-19} Coulomb.

Pulse height obtained is

$$V = \frac{Q}{C} = \frac{MNe}{C} = \frac{800}{20 \times 10^{-12}} \times \left(\frac{10^4}{32}\right) \times 1.6 \times 10^{-19}$$

$$\begin{aligned} V &= 2 \times 10^{-3} \text{ volt} \\ &= 2 \text{ mV} \end{aligned}$$

Example 1.5 Calculate the resolving time of a GM counter from the following observations by the double source method: Background count 60 cpm, first source in position 8220 cpm, both sources in position 16,860 cpm, second source in position 9360 cpm.

Solution

$$\begin{aligned} B &= 60 \text{ cpm} = 1 \text{ cps} \\ n_1 + B &= 8220 \text{ cpm} = 137 \text{ cps} \\ n_{12} + B &= 16{,}860 \text{ cpm} = 281 \text{ cps} \\ n_2 + B &= 9{,}360 \text{ cpm} = 156 \text{ cps} \\ \tau &= \frac{n_1 + n_2 - n_{12} - B}{2n_1 n_2} \\ &= \frac{(137-1) + (156-1) - (281-1) - 1}{2(137-1)(156-1)} \\ &= 240.8 \times 10^{-6} \text{ sec} \\ &= 241 \text{ µsec} \end{aligned}$$

Example 1.6 A counting rate of 16,200 counts/min is indicated by a GM tube having a dead time of 250 µsec. Calculate the counting rate that would be observed in the absence of dead time loss.

Solution

$$N = \frac{n}{1 - n\tau} = \frac{16200}{1 - \frac{16200}{60} \times 250 \times 10^{-6}} = 17{,}373 \text{ counts/min}$$

Example 1.7 The plateau of a G.M. counter working at 1 kV has a slope of 2.5 % count rate per 100 V. By how much can the working voltage be allowed to vary if the count rate is to be limited to 0.1 %?

Solution

$$\text{Slope} = \frac{n_2 - n_1}{n_{av}} \times \frac{100}{V_2 - V_1} \times 100 \text{ percent per 100 V}$$

$$\text{Given } \frac{n_2 - n_1}{n_{av}} = \frac{0.1}{100}$$

$$\text{Slope} = \frac{0.1}{100} \times \frac{100}{\Delta V} \times 100 = 2.5$$

or

$$\Delta V = \frac{10}{2.5} = 4 \text{ volts}$$

The voltage should not vary more than ±2 volts from the operating voltage of 1 kV.

Example 1.8 An organic-quenched G.M. tube has the following characteristics.

Working voltage 1000 V
Diameter of anode 0.2 mm
Diameter of cathode 2.0 cm
Maximum life time 10^9 counts

What is the maximum radial field in the tube and how long will it last if used for 15 hr per week of 6000 counts per minute?

Solution

$$E = \frac{V}{r \ln(b/a)}$$

Field will be maximum close to the anode.

$$r = 0.1 \text{ mm} = 1 \times 10^{-4} \text{ m}$$

$$E_{\max} = \frac{1000}{10^{-4} \times \ln(20/0.2)} = 2.17 \times 10^6 \text{ V/m}$$

The lifetime of G.M. tube in years is given by dividing total number of counts possible by counts per year.

$$t = \frac{10^9}{52 \times 15 \times 6000 \times 60} = 3.56 \text{ years.}$$

Example 1.9 A G.M. tube with a cathode and anode of 2 cm and 0.12 mm radii, respectively is filled with Argon gas to 10 cm Hg pressure. If the tube has 1.2 kV applied across it, estimate the distance from the anode, at which electron gains just enough energy in one mean free path to ionize Argon. Ionization potential of Argon is 15.7 eV and mean free path in Argon is 2×10^{-4} cm at 76 cm Hg pressure.

Solution Since the mean free path is inversely proportional to pressure, M.F.P at 10 cm pressure will be $2 \times 10^{-4} \times (76/10)$ or 1.52×10^{-3} cm.

The electric field at a distance = M.F.P, should be such that it acquires just sufficient energy to ionize Argon (15.7 eV). The required value of E is

$$E = \frac{15.7}{1.52 \times 10^{-3}} = 1.03 \times 10^4 \text{ volt/cm}$$

$$r = \frac{V}{E \ln(b/a)} = \frac{1200}{1.03 \times 10^4 \ln(20/0.12)}$$

$$= 0.0227 \text{ cm} \quad \text{or}$$

$$= 0.227 \text{ mm}$$

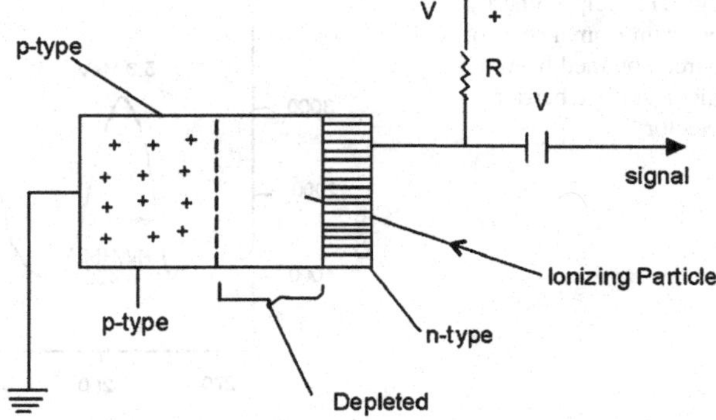

Fig. 1.11 A typical silicon depletion layer detector. A negative pulse appears across *R*

1.7 Semi-conductor Particle Detectors

1.7.1 Resolution

In gases incident charged particles produce negatively charged electrons and positively charged ions, where as in semi-conductors they produce free e^- and positively charged "holes". Second difference is that in silicon a charged particle will produce one electron-hole pair for each 3.5 eV absorbed. This is an order of magnitude smaller than the corresponding value in a gas (~35 eV). This leads to the production of ten times the number of carrier pairs, and therefore smaller relative straggling for the recorded particles. Thus, more accurate energy measurements can be made. A 5 MeV particle stopping in silicon releases 1.43×10^6 electron-hole pairs with a S.D. $= \pm 1.2 \times 10^3$ or relative S.D. $(\sigma/m) = 0.08\ \%$. Limiting fluctuation for gas ionization chamber is $\pm 0.23\ \%$ and is approximately 0.8 % for scintillation detector. Theoretical resolution is not achieved because of electronic noise in the semiconductor detector and its associated pre-amplifier and for inefficient charge collection. Thus, semiconductor detectors are preferred where energy resolution is important.

1.7.2 Reversed Biased p-n Junction Particle Detector

Figure 1.11 is a diagrammatic representation of a silicon depletion-layer detector. The p-type silicon contains an excess of acceptor impurities so that there are more holes than free electrons and is reversed biased. It occupies the main body of the detector. The n-type silicon is made extremely thin, usually about 0.1 μm. It is heavily doped with a donor and this region contains high concentration of free electrons and practically no free holes. The thickness of the depletion layer ranges from 10 μm to 5 mm, depending upon the detector construction and the value of the applied voltage. Since the n-type region is so thin that the p-n junction may be considered to be on the surface for all practical purposes. Hence this type of detector is also called surface barrier detector. A typical surface barrier detector for alpha-particle

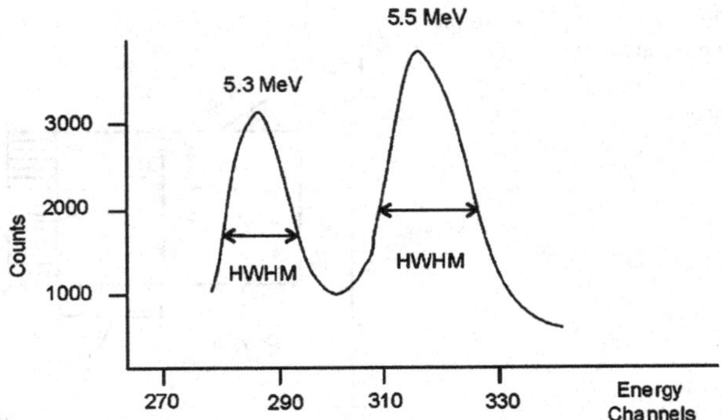

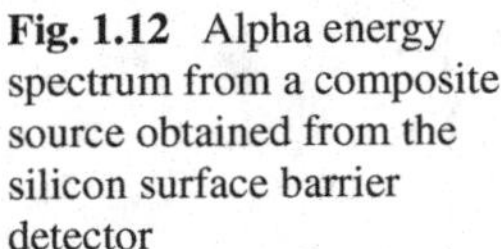
Fig. 1.12 Alpha energy spectrum from a composite source obtained from the silicon surface barrier detector

measurements may have a depletion layer of 100 μm with an applied potential of 100 volts. Sensitive areas of 2–3 cm^2 or larger are available. When a reverse bias voltage is applied as in Fig. 1.11, the depletion layer contains neither free electrons nor holes. Under these conditions, current flow through the device is severely limited. A charged particle or gamma ray that passes through the depletion layer of the junction, generates electron-hole pairs which drift to the opposite sides of the junction under the action of a reverse bias potential and result in a fast voltage pulse ($\sim 10^{-8}$ sec). An upper limit for the carrier velocities is $\sim 10^7$ cm/sec, and this together with the depletion layer thickness, sets a lower limit for the charge collection time. The magnitude of the pulse is proportional to energy lost by the incident particle crossing the thin junction region.

From the point of view of noise, the field effect transistor (FET) in the first stage of the amplifier is the best electronic device and it limits the resolution by itself to approximately 60 eV. The energy spectrum of radiation can be obtained using a pulse height analyzer, as in the case of scintillation counter.

Very thin transmission detectors give a signal proportional to dE/dx, while total absorption of the particle gives E. The combination of the two signals permits the identity of particles generating them. Figure 1.12 shows the alpha energy spectrum of a composite source consisting of ^{210}Po and ^{238}Pu obtained from a silicon barrier layer detector and a pulse height analyzer. The half-width at half maximum (HWHM) is measured on the higher side of the peaks to minimize the effects of finite source thickness.

Figure 1.13 shows the gamma-ray spectrum from ^{60}Co observed by a germanium detector. The photo peaks due to 1.17 MeV and 1.33 MeV are indicated. The Compton shoulders as well as the sharp drops in the curve at channel No: 680 and No: 790 which are Compton edges corresponding to maximum transfer of energy to electrons are also shown.

Figure 1.14 shows the mass separation for the fragments produced in the collisions of 5 GeV protons with the nuclei of Uranium. Each product particle is passed through a thin silicon detector for dE/dx measurement and stopped in the second detector for E measurement. The two signals are combined in coincidence to permit the product nuclei to be identified. The mass spectrum also shows the first observation of the short lived isotopes ^{11}Li, ^{14}B and ^{15}B.

Fig. 1.13 Gamma-ray spectrum due to 1.17 MeV and 1.33 MeV from ^{60}Co [1]

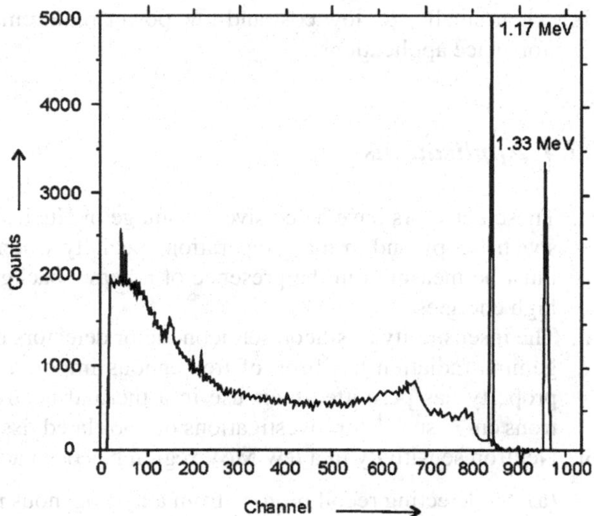

Fig. 1.14 The mass spectrum of isotopes produced in fragmentation of Uranium nuclei by 5 GeV protons obtained by a detector telescope and particle identifier [1]

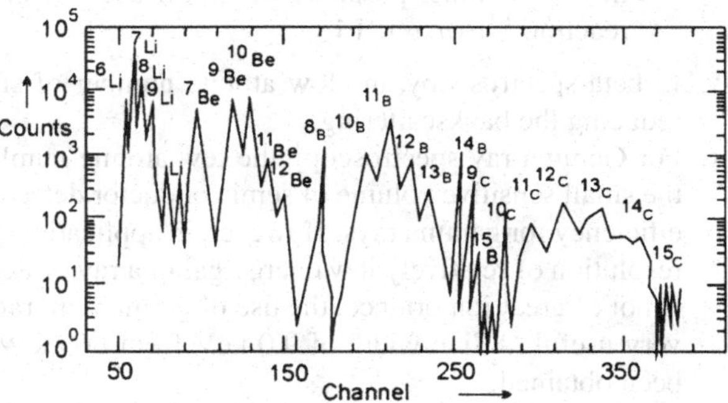

1.7.3 Advantages

i. These detectors have excellent energy resolution.
ii. Their energy response is linear over a wide range of particles and energies.
iii. The pulse rise time in these detectors is short, being of the order of 10^{-8} sec. These features make them suitable for Nuclear Spectroscopy, which includes nuclear product separation, Neutron spectroscopy, Beta spectroscopy, Gamma-ray spectroscopy and Fission studies.
iv. The output is insensitive to counting rate.
v. They involve substantially windowless operation. This means that low energy particles can easily enter the sensitive region.
vi. They have variable sensitive thickness.
vii. Unlike scintillation counters, they are insensitive to magnetic fields.
viii. They involve low temperature operation.

ix. Their small size, low cost and low power requirements make them amenable for space applications.

1.7.4 Applications

i. These detectors have a decisive advantage in Nuclear reaction charged particle spectroscopy and in mass separation specially when heavy energetic particles must be measured in the presence of a heavy background of light particles of high energies.
ii. The insensitivity of silicon semiconductor detectors to low energy neutrons and gamma-radiation has been of tremendous importance in fission studies. This property has permitted their use in a thermal-neutron flux of about 10^9 neutrons cm^{-2} sec^{-1} for investigations of correlated fission kinetic energies.
iii. Neutron sensitivity in a few MeV region has been achieved in two ways

    (a) by detecting recoil protons from a hydrogenous radiator and
    (b) by detecting charged particles produced in suitable neutron induced reactions. The other possibility is to diffuse ^{10}B inside silicon and exploit the reaction $^{10}B(n, \alpha)\,^7Li$.

iv. In beta-spectroscopy, the low atomic number of silicon has the advantage of reducing the backscattering.
v. For Gamma-ray spectroscopy the low atomic number of silicon together with the small sensitive volume of semi conductor detector results in a low detection efficiency for gamma rays. However, in applications which require good energy resolution of relatively low energy gamma rays where high detection efficiency is not of great importance, the use of germanium rather than silicon has proved very useful. A line width of 9.0 keV from 660 keV gamma-rays of ^{137}Cs has been obtained.
vi. Semiconductor detectors can detect minimum ionizing particles since the rate of loss of energy of a singly charged minimum ionizing particles in silicon is $\cong$400 keV/mm and it is possible to use devices with several mm sensitive thickness.
vii. Properties of semiconductors like energy requirement of electron-hole pairs in silicon and germanium, the electron multiplication phenomenon in germanium and silicon, and photo-conductive processes in germanium, carrier life times, carrier diffusion lengths and carrier capture cross section have been determined from the well defined amount of charge liberated by monoenergetic beams of protons and of α's of different energies.

1.7.5 Disadvantages

i. Long range particles are not stopped.
ii. These devices have short operating life time

iii. The output signal is small
iv. The devices can not be operated at high temperatures
v. The counting behavior changes due to radiation damages.
vi. There is a change of characteristics with different ambient for some of the detectors.

Example 1.10 A depletion-layer detector has an electrical capacitance determined by the thickness of the insulating dielectric. Estimate the capacitance of a silicon detector with the following characteristics: area 1.6 cm^2, dielectric constant 11, depletion layer 45 μm. What potential will be developed across this capacitance by the absorption of a 4.5 MeV alpha particle which produces one ion pair for each 3.5 eV dissipated?

Solution

$$C = \frac{\varepsilon AK}{d} = \frac{8.8 \times 10^{-12} \times 1.6 \times 10^{-4} \times 11}{45 \times 10^{-6}} = 344 \times 10^{-12}\ \text{F}$$

$$q = \frac{4.5 \times 10^{6} \times 1.6 \times 10^{-19}}{3.5} = 2.06 \times 10^{-13}\ \text{coulomb}$$

$$V = \frac{q}{C} = \frac{2.06 \times 10^{-13}}{344 \times 10^{-12}} = 6 \times 10^{-4}\ \text{volts}$$

1.8 Neutron Detectors

1.8.1 Principle

Neutrons being neutral cannot be detected directly. They can be detected only by means of secondary charged particles which are released in their nuclear reactions or scattering which they undergo in passing through matter. The secondary particles must be produced at energies which can produce ionization and which can be detected. The secondary charged particles could be protons recoiling in the collision of neutrons with hydrogen nuclei or they may be the products of nuclear reactions induced by neutrons with nuclei of suitable targets. One or more charged particles are emitted. Exothermic reactions (n, p), (n, α), (n, f) and (n, γ) are suitable as they have large cross sections for slow neutrons ($\sigma \propto 1/v$). A typical example is the reaction

$$^{10}\text{B} + n \rightarrow {}^{7}\text{Li} + \alpha \tag{1.21}$$

Alternatively, the secondary particles could be the radiations from radioactive product nuclei formed as the result of neutron capture. The detection of neutrons is dictated by the requirement that measurable ionization effects be produced.

1.8.2 Desired Characteristics of Neutron Detectors

i. The detection efficiency must be high.
ii. Neutron capture cross section must be large, so also the Q-value of nuclear reaction.
iii. Other competitive reactions must have negligible rates.
iv. In the time-of-flight method, detector must have high detection efficiency and fast rise time of pulses. Organic scintillators serving both as proton target and as a proton detector are commonly used for such applications.

Slow neutrons on collision with hydrogen nuclei will not produce protons which would generate enough ionization. Also, fast neutrons often do not have high capture cross-section to induce nuclear reactions or to produce radioactive nuclei. Consequently, detection can be classified to a large extent in terms of the energy of the neutrons.

1.8.3 Slow Neutron Detectors

1.8.3.1 Foil Activation Method

Numerous nuclides have a large activation cross section for (n, γ) reaction at low neutron energy. In this method, a thin foil of the given material is exposed to neutrons of known intensity and the induced radioactivity is determined. Consider, for example, the absorption of thermal neutrons by ^{27}Al to form 2.3 min ^{28}Al according to

$$^{27}\text{Al} + n \rightarrow {}^{28}\text{Al} \rightarrow {}^{28}\text{Si} + \beta^- + \overline{\nu} \tag{1.22}$$

If the thickness of the foil is small so that radiation intensity may be assumed constant across the material, then considering 1 cm area of the foil that is exposed

$$\frac{\Delta I}{\Delta x} = I\Sigma_\alpha \tag{1.23}$$

where I is the initial beam intensity, Σ_α is the macroscopic absorption cross section. But for each neutron lost from the beam, one atom of ^{28}Al is formed and since the volume of aluminum is $\Delta x \times 1^2$ or Δx cm^3,

$$\text{Interactions/cm}^3\text{/sec} = I\Sigma_\alpha \tag{1.24}$$

As soon as ^{28}Al is formed, it will start decaying at a rate determined by its decay constant λ. If Q is the number of atoms of ^{28}Al present at any time t,

$$\text{(Net rate of change of } Q) = \text{(Production rate)} - \text{(Decay rate)}$$

$$\frac{dQ}{dt} = I\Sigma_\alpha - \lambda Q \tag{1.25}$$

$$t = \int dt = \int \frac{dQ}{I\Sigma_\alpha - \lambda Q} + C \tag{1.26}$$

where C is the constant of integration.

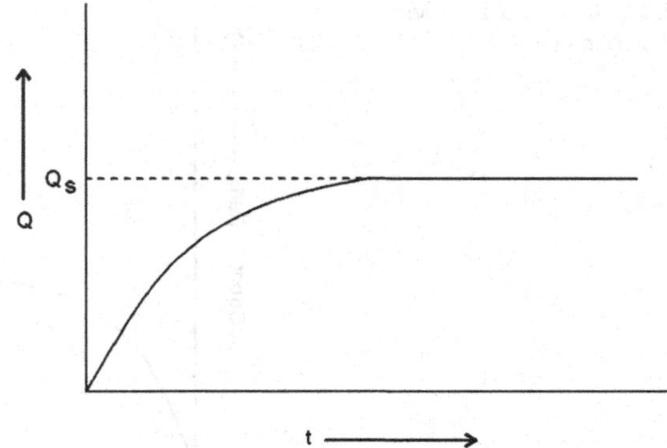

Fig. 1.15 Number of ^{28}Al atoms as a function of time

Solving (1.26) with the condition that $Q = 0$ at $t = 0$, we easily find,

$$Q = \frac{I\Sigma_\alpha}{\lambda}\left(1 - e^{-\lambda t}\right) \tag{1.27}$$

After a time much longer than say 5 times $1/\lambda$, the exponential term tends to zero, and

$$Q \to Q_s = \frac{I\Sigma_a}{\lambda} \tag{1.28}$$

Q_s represents the saturation value, Fig. 1.15. At this time and beyond, $dQ/dt = 0$ as is evident by substituting (1.28) in (1.25), i.e. Rate of production = Rate of decay.

After this state of affairs is reached, the foil is removed and after the lapse of known time t_w (called waiting time), the radioactivity is measured. Then

$$Q = Q_s e^{-t_w\lambda} \tag{1.29}$$

and

$$\left|\frac{dQ}{dt}\right| = \lambda Q = \lambda Q_s e^{-t_w\lambda} = I\Sigma_a e^{-t_w\lambda} \tag{1.30}$$

knowing the value of Σ_α, intensity of neutrons I can be determined.

1.8.3.2 Boron Detectors

BF$_3$ (Boron-Trifluoride) Proportional Counters

A counter or an ionization chamber filled with BF$_3$ gas can detect neutrons through reaction (1.21). The gas serves the purpose of both target material as well as the medium in which charged product nuclei may be arrested. BF$_3$ counters like other proportional counters are fabricated in cylindrical geometry. The outer tube acts as a cathode and the central wire as the anode. Operating voltage is usually 2000–2500 V, Fig. 1.16. Typical values of the gas multiplication are 100 to 500. BF$_3$ counter can discriminate thermal neutron induced pulses against those induced by gamma rays.

If a thermal neutron is absorbed by ^{10}B nucleus, about 2.3 MeV appears as kinetic energy of α and the recoiling nucleus ^{7}Li. If the counter is operated in the

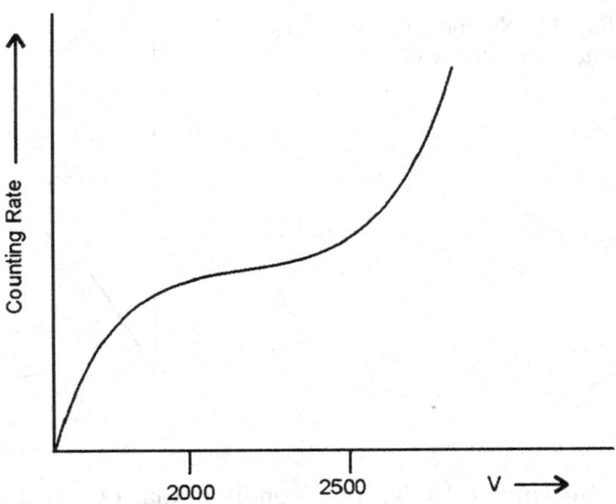

Fig. 1.16 Characteristics of BF_3 counter

proportional region, it is possible to count the large pulses against a heavy background of smaller pulses due to γ rays. Assume a speed distribution n_v such that there are no neutrons outside the $1/v$ limit for boron, i.e. $E > 1$ keV. The product $n_v dv$ is the number of neutrons/unit volume with speeds between v and $v + dv$. If N_0 is the number of boron nuclei in the counter there is a contribution dR to the detector counting rate of

$$dR = \sigma N_0 n_v v dv \tag{1.31}$$

The total detector counting rate is

$$R = \int \sigma N_0 n_v v dv = \sigma v N_0 \int n_v dv \tag{1.32}$$

as (σv) is independent of v. Also, the last integral in (1.32) represents n, the neutron density regardless of speed

$$R = N_0 n(\sigma v) \tag{1.33}$$

A thin boron detector then gives an output proportional to the neutron density. This is true either for a collimated beam or an isotropic flux.

Boron Lined Proportional Counters

The counter is lined with a boron compound, and filled with a suitable gas. Only one of the particles deposits energy in the counter gas. These counters show less satisfactory plateau.

1.8.3.3 ^{3}He Proportional Counters

These counters use ^{3}He $(n, p)^3$H reaction. The absorption cross section is large ($\sigma = 5327$ barns) for thermal neutrons and the Q-value of 0.77 MeV is favourable.

Further Helium is a good filling gas for proportional counters. These features make ^{3}He $(n, p)^3$H reaction quite attractive both for neutron counting and spectrometry.

1.8.4 Fission Detectors

Fission reactions induced by slow and thermal neutrons in the nuclei of ^{235}U, ^{233}U and ^{239}Pu are quite suitable for neutron detection. When the fissile elements are used in the proportional counters, because of large energy that is released (170 MeV) they provide a good discrimination against alphas from natural radioactivity.

1.8.5 Fast Neutron Detectors

The activation cross section of most nuclides is quite small for neutrons of energy greater than 1 MeV so that the activation method or the use of neutron induced nuclear reaction detectors is ruled out. However, the observation of recoil protons in n–p scattering furnishes a method for neutron detection. First the scattering cross sections for n–p scattering are reasonably large, being 4 barns at 1 MeV neutron energy and 1 barn at 10 MeV energy. An organic scintillation counter is a suitable detector. Under the assumption that n–p scattering is isotropic in the C.M. system, it can be shown that the recoil protons in the Lab system have a uniform distribution.

$$\frac{d\sigma}{d\omega^*} = \frac{\sigma}{4\pi} \tag{1.34}$$

$$\frac{d\sigma}{dE_p} = \frac{d\sigma}{d\omega^*} \cdot \frac{d\omega^*}{dE_p} = \frac{\sigma}{4\pi} \cdot \frac{2\pi \sin\phi^* d\phi^*}{dE_p} \tag{1.35}$$

where quantities with * refer to CMS. But

$$\phi^* = 2\phi \quad \text{and} \quad d\phi^* = 2d\phi \tag{1.36}$$

If E_0 is the initial neutron energy in the Lab system,

$$P_0^2 = 2ME_0$$

The momentum of the recoil proton P_p is given by

$$\begin{aligned} P_p &= P_0 \cos\phi \\ P_p^2 &= P_0^2 \cos^2\phi \\ E_p &= E_0 \cos^2\phi \\ dE_p &= 2E_0 \cos\phi \sin\phi d\phi \end{aligned} \tag{1.37}$$

where we have ignored the negative sign in (1.37) as E_p decreases when ϕ increases.

Using (1.36) and (1.37) in (1.35) and simplifying

$$\frac{d\sigma}{dE_p} = \frac{\sigma}{E_0} = \text{const.} \tag{1.38}$$

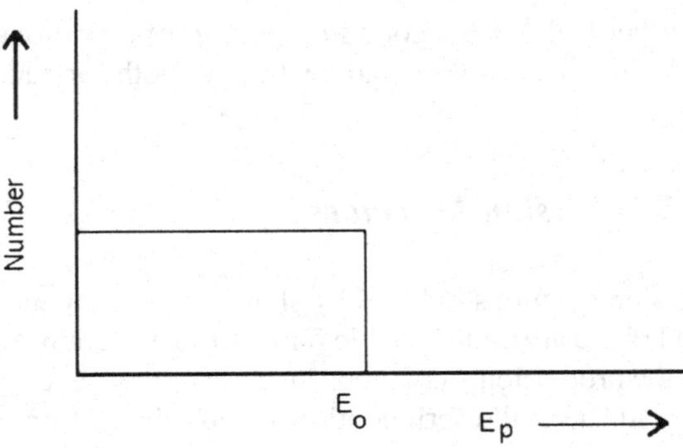

Fig. 1.17 Energy distribution of recoil protons

Equation (1.38) shows that the recoil energy of protons is uniform, Fig. 1.17, the proton energy ranging from 0 to E_0 for $\phi = 90°$ to $\phi = 0$. The number of proton recoils per sec in the energy.

Interval E and $E + dE$ is

$$d\sigma = \frac{\sigma}{E_0} N_0 \varphi dE_p \tag{1.39}$$

where N_0 = total number of atoms in the chamber and φ is the flux of neutrons. Knowing $d\sigma/dE_p$ and σ, the flux φ can be determined.

1.8.6 Transmission Experiments

Neutrons disappear from collimated beams like "One shot" process similar to gamma rays, the intensity decreasing exponentially with distance in the medium. The absorption coefficient μ can be determined. This is simply the macroscopic cross section.

Example 1.11 ^{60}Co is produced from natural cobalt in a reactor with a thermal neutron flux density of $2 \times 10^{12} n$ cm^{-2} sec^{-1}. Determine the maximum specific activity. Given $\sigma_{act} = 20$ barns.

Solution $Q_{\max} = \frac{\varphi \Sigma_{act}}{\lambda} = \frac{\varphi \sigma_a}{\lambda} \cdot \frac{N_{(av)}}{A}$

$$\left|\frac{dQ}{dt}\right| = Q_{\max}\lambda = \frac{\varphi \sigma_{act} N_{(av)}}{A} = \frac{2 \times 10^{12} \times 20 \times 10^{-24} \times 6 \times 10^{23}}{60}$$

$$= 4 \times 10^{11} \text{ D.P. s}$$

$$= \frac{4 \times 10^{11}}{3.7 \times 10^{10}} \text{ Ci}$$

$$= 11 \text{ Ci}$$

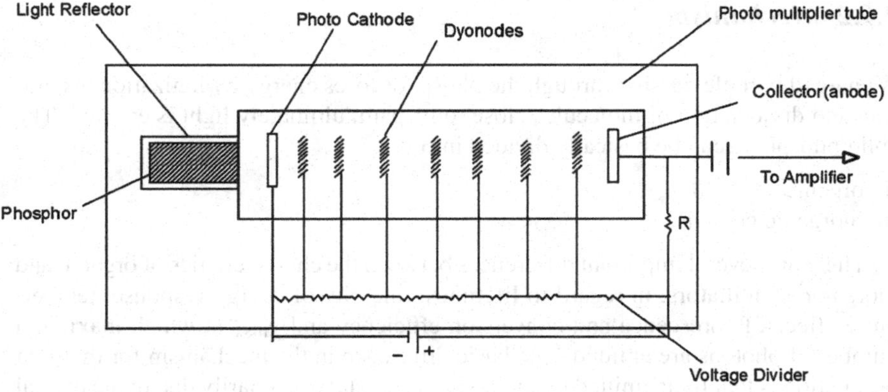

Fig. 1.18 Block diagram of a scintillation counter

Example 1.12 Natural cobalt is irradiated in a reactor with a thermal neutron flux density of $4 \times 10^{12} n\ \text{cm}^{-2}\ \text{sec}^{-1}$. How long an irradiation will be required to reach 25 % of the maximum activity? Given $T_{1/2} = 5.3$ years.

Solution $|\frac{dQ}{dt}| = Q\lambda = \varphi\Sigma_{act}(1 - e^{-\lambda t})$

$$\frac{Q}{Q_s} = \frac{25}{100} = \left(1 - e^{-0.693t/5.3}\right)$$

Solving for t, we find $t = 2.2$ years.

1.9 Scintillation Counter

1.9.1 Construction

Figure 1.18 is a representative diagram of a scintillation counter. It consists of a photomultiplier tube to which is fixed a scintillator. A high voltage (of the order of a kV) is applied between the photo-cathode and the anode. The dynodes incorporated in the tube produce electron multiplication and by the use of a voltage divider provide progressively larger voltage between cathode and anode. Scintillators exist in several forms, crystals (organic or inorganic), liquids, plastic solids and gases. The scintillation phenomenon depends on the fact that suitable "flours" give off pulses of light when traversed by a charged particle. This light is directed on to a photomultiplier cathode where it ejects electrons by photo-electric effect. These electrons are multiplied in the dynode structure of the tube. In each stage the number of secondary electrons is multiplied which are finally collected at the anode and recorded as a pulse by suitable circuits. The phosphor is in optical contact with the tube and is protected from external light. A reflector surrounding the phosphor enhances light falling on the photo-cathode for higher efficiency.

1.9.2 Mechanism

A charged particle passing through the phosphor loses energy by ionization, excitation and dissociation of molecules close to its path, ultimately light is emitted. The solid phosphor scan be basically divided into

i. organic
ii. inorganic crystals.

There are several important differences between the characteristics of organic and inorganic scintillators, in regard to lifetimes, linearity of energy response, temperature effects, fluorescence and conversion efficiency and ν_{max} at which maximum number of photons are emitted. The basic difference in the mechanism for the light production is that light emitted by an inorganic crystal is primarily due to the crystal structure, where as organic substances exhibit luminescence by virtue of molecular properties.

1.9.3 Desirable Characteristics of Luminescent Materials

i. The phosphor must have high efficiency for conversion of incident energy of radiation or particles into that of the emitted luminescence. In the case of inorganic phosphor material a small percentage impurity is essential while for organic phosphors, material must be pure.
ii. The spectrum of the emitted light must closely match the spectral response of the cathode of the photomultiplier used (in the blue, violet or u.v. region).
iii. The luminescent material must be transparent to their own luminescence radiation.
iv. The material used must be a large optically homogeneous mass, either as a single crystal without defects or in solution, solid or liquid, moulded or machined to any convenient shape.
v. The phosphor must have a high stopping power for the radiation to be detected.
vi. The rise and decay of luminescence during and after excitation should occur in a short time.
vii. The phosphor must be stable against vacuum conditions and under prolonged irradiation.
viii. The refractive index μ of the crystal should not be too high, otherwise light will not be able to come out easily due to internal reflections.

1.9.4 Organic Scintillators

i. In organic scintillators, the main mechanism is believed to be that of collisions which are responsible for the energy transfer from the molecules, either by excitation transfer or by a dipole resonance interaction

ii. The energy response, i.e. light output as a function of energy loss of particle, is not quite linear. Light output is very much dependent upon the nature of the particle.
iii. Organic scintillators have lifetimes of the order of 10^{-9} to 10^{-8} sec.
iv. The phenomenon of phosphorescence is absent.
v. The fluorescent conversion efficiency is generally smaller than the inorganic phosphors, conversion efficiency of anthracene is about half that of sodium iodide.
vi. Suitable organic scintillators are anthracene, ($\lambda_{max} = 4400$ A), diphenylacetyline, terphenyl naphthalene and stilbene. Among plastics, anthracene in polystyrene is found to be useful. Terephenyl in toluene and terphenyl in xylene are good liquid organic scintillators.
vii. Organic phosphors are extensively used for fast neutron detection. The interactions of fast neutrons with hydrogen produce fast recoil protons which can be detected with high efficiency in large crystals.

1.9.5 Inorganic Scintillators

i. In inorganic crystals, for example alkali halides notably NaI with thallium impurity, charged particles may raise electrons into the conduction bands or into excitation levels. The electron and the hole left move rapidly through out the crystal as an exciton until captured by the imperfection, giving up the energy in the form of vibrational transfer or until captured by an impurity. The impurity gets excited and acts as a scintillator. In the case of halides, thallium is added as an impurity to the extent of 0.1 to 0.2 percent. These crystals are highly transparent to their own radiation.
ii. The light output from inorganic crystals like NaI is very nearly proportional to energy loss down to about 1 MeV for protons and about 15 MeV for α's.
iii. The life times of inorganic scintillators are generally longer than those of organic scintillators, being of the order of 10^{-6} sec.
iv. The phenomenon of phosphorescence which is delayed emission of photons can in certain cases cause generation of secondary pulses which are indistinguishable from the primary pulses. As sodium iodide is deliquescent, it must be protected from moisture; nevertheless it is the most widely used inorganic phosphor. Large size crystals up to several inches in diameter and length are available. It has a high density and contains high Z atoms of Iodine and is an efficient detector for gamma rays as the absorption cross section for the three important processes, photoelectric, Compton and pair-production vary as $Z^{4.5}$, Z and Z^2 respectively. Of the other inorganic phosphors, zinc sulphide is useful for alpha-particle detection and lithium iodide for neutron detection, the relevant nuclear reaction being

$$^{6}\text{Li} + n \rightarrow {}^{3}\text{H} + {}^{4}\text{He} + 4.8\ \text{MeV} \tag{1.40}$$

v. Important inorganic scintillators are, sodium iodide (thallium activated, λ_{max} being 4100 A), cesium iodide (cooled to 77 K), zinc sulphide (copper activated) and lithium iodide (europium activated).
vi. Some of the inorganic phosphors have a high value of the refractive index (∼2). Difficulty is experienced in getting light out of them.

1.9.6 Photo-Multiplier (PM) Characteristics

The desired characteristics of Photo-multiplier (PM) are:

i. PM must have a photo cathode of large cathode area with an end-window.
ii. PM must be of a high efficiency for converting photons into photo-electrons. A peak of conversion efficiency in the blue or ultra violet region is desirable.
iii. PM must provide a high gain.
iv. It must provide a good signal-to-noise ratio. In the absence of light the output from a photomultiplier consists of numerous pulses (noise) of various sizes, principally due to thermal emission of electrons from the photo-cathode. This constitutes the so-called dark current which depends on the photo-cathode material. It can be reduced by cooling the cathode. Popular photomultipliers are 56 AVP (Philips), 6810 A, 7264 (RCA), 6292 Du Mont. The number of dyonodes varies normally from 10 to 15.

1.9.7 Light Collection

A crystal scintillation counter is normally placed in a metal container. When hygroscopic alkali halides are used, they are protected against moisture by sealing the container. Good optical contact is made between the surface of phosphor and the end face of PM with a layer of clear vacuum grease and by placing a good reflector in optical contact with other crystal surfaces so that light which would otherwise escape will be returned to the photomultiplier with improved efficiency. A highly polished foil or a diffuse reflector such as magnesium oxide is used as a specular reflector. When the arrangement is such that the PM can not be in direct contact with the scintillator, lucite pipes can be used.

1.9.8 Background

The background of an unshielded scintillator arises due to three major sources:

i. γ-radioactivity in the neighborhood, which can be reduced by shielding the counter.

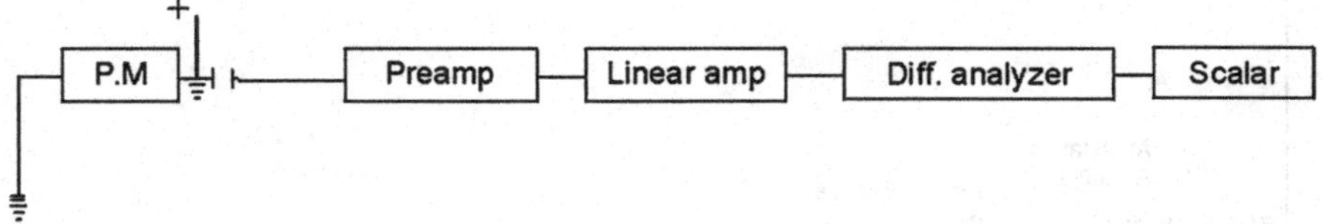

Fig. 1.19 Block diagram of electronic accessories connected to the PM

ii. Cosmic rays and other external sources such as reactor or accelerator, if any. This background is reduced by surrounding the scintillation counter with a set of GM tubes, the output of which is arranged in anti-coincidence with the output of the scintillation counter.
iii. The background noise normally is about 1 per sec for a pulse height equivalent to liberation of 8 photo electrons. This corresponds to a 5 to 10 keV electron incident on a NaI crystal. This can be reduced by cooling the counter. Alternatively, the phosphor is viewed by two photomultipliers in coincidence. In this case, the noise pulses which occur randomly alone contribute to chance coincidence.

1.9.9 Electronic Equipment

Figure 1.19 is a block diagram of the electronic equipment attached to the PM. Pulses from the anode of the PM which are small are passed through a pre-amplifier and then to a linear amplifier and through a "window" of the differential analyzer, and finally counted by the scalar. The high voltage can be varied to suite the given PM. The gain of linear amplifier can be varied so that the input fed to the differential analyzer is within the range of operation. The differential analyzer accepts pulses of height between V and $V + \Delta V$, where ΔV is the width of the window. The pulse height is proportional to E_γ.

1.9.10 Gamma Ray Spectroscopy with NaI (TL) Scintillator

At low γ-ray energy ($E_\gamma < 100$ keV) photoelectric absorption is the dominating process. As $\sigma_{ph} \propto Z^{4.5}$, most of the absorptions occur in Iodine, with the K-shell electron (ionization energy $E_k = 29$ keV). The vacancy caused by the ejection of electron is filled in by radiative transitions (mainly X-rays) from electrons belonging to upper levels. If the resulting X-rays get absorbed then full energy (E_γ) is available and this corresponds to photo peak in the pulse height distribution, Fig. 1.20. However, in few events the X-rays escape. Hence energy equal to $(E_\gamma - E_k)$ is available. This results in the "Iodine escape peak". The ratio of photons under escape peak to those under photo-peak depends on E_γ crystal size and experimental geometry.

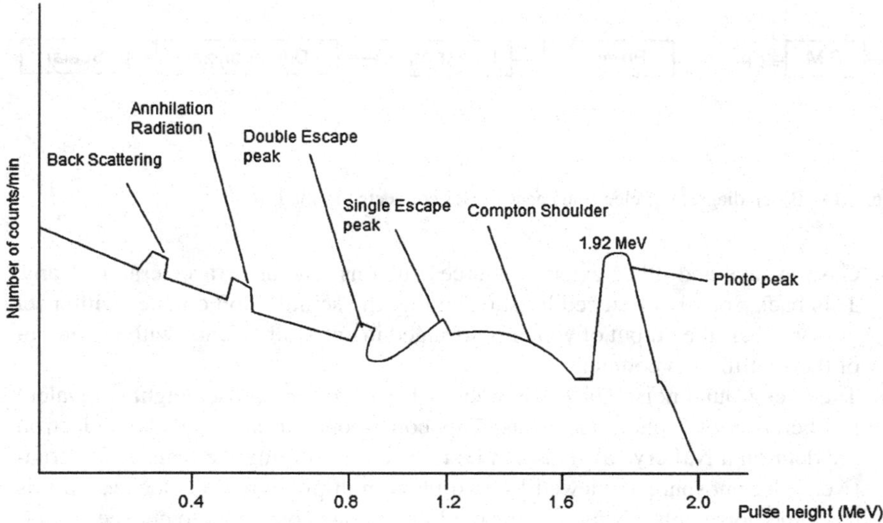

Fig. 1.20 Pulse height distribution of 3″ × 3″ NaI (TL) showing the positions of various peaks as discussed in the text

For $E_\gamma > 100$ keV, Compton scattering also becomes significant. The "escape peak" is not significant when the mean absorption length of the incident γ-rays becomes greater than that of iodine X-rays. In the case of single Compton scattering, the energy of escape radiation extends from E' to E_γ, where $E' = E_\gamma/(1+2\alpha)$ with $\alpha = E_\gamma/mc^2$ corresponds to the energy of the scattered photon at 180° (back scattering). The corresponding energy deposited ranges from $(E_\gamma - E')$ to 0. There will be a broad Compton distribution with the Compton edge occurring at energy $(E_\gamma - E')$. There can also be external Compton scattering from material outside such as the PM shielding. This gives rise to the "back scattering peak".

In the case of multiple Compton scattering the energy of escaping radiation extends from 0 to E_γ and the corresponding energy deposited is E_γ to 0. For $E_\gamma > 2mc^2$ (threshold for pair production) two other peaks are observed, a single escape peak at $(E_\gamma - mc^2)$ and a double escape peak at $(E_\gamma - 2mc^2)$.

Figures 1.21 and 1.22 show typical spectra obtained from γ-rays incident on NaI crystal from Cs-137 (661 keV) and Co-60 (1.17 MeV and 1.33 MeV), respectively.

The 661 keV photo peak, Compton shoulder, back scattering from material of the phosphor and noise are indicated. The decay scheme is also shown.

1.9.10.1 Energy Resolution

The photo peaks shown in Figs. 1.21 and 1.22, are not sharp. The energy resolution is defined by the quantity $\Delta E/E$, which is the ratio of full width at half maximum to the photon energy. It is important that the spread in photo peaks be as small as

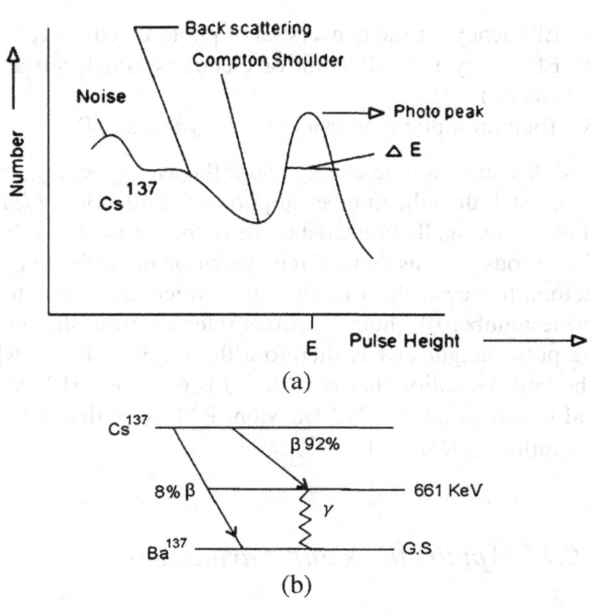

Fig. 1.21 (**a**) Pulse height spectrum of γ-rays from Cs-137 of energy 661 keV. (**b**) Decay scheme for Cs^{137}

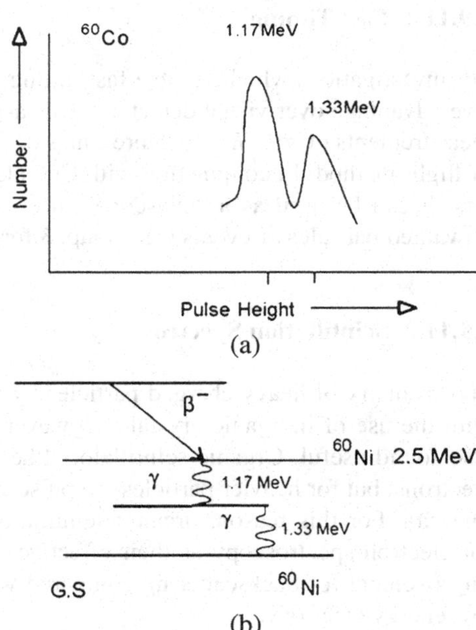

Fig. 1.22 (**a**) Pulse height spectrum of γ-rays of energy 1.17 MeV and 1.33 MeV. The origin of γ-rays is shown in the decay scheme. (**b**) Decay scheme for ^{60}Co

possible, otherwise γ-rays of neighboring energies cannot be resolved. Many factors contribute to the energy resolution of a scintillation counter. These factors are:

i. Fluorescent radiation conversion efficiency (f)
ii. Efficiency for the collection of light by the cathode (b)

iii. Efficiency for the conversion of photo-electrons (c)
iv. Efficiency for collection of electrons which are accelerated to the first dynode (p)
v. Total multiplication from all the dynodes (M)

If E is the particle energy and ε the average energy of the photons generated in the crystal, then the number of photons emitted is Ef/ε. It follows that the number of electrons finally collected at the output of the PM tube is equal to $(Ef/\varepsilon)bcpM$. For various reasons there will be variation in the factors f, b, c, p and M. Of various factors, however, the variation in c which arises due to the statistical fluctuations in the number of photo-electrons released from the photo cathode is decisive for the pulse height and is therefore the ultimate factor which limits the resolution. The best resolution that is achieved is 6 % for 661 keV γ-rays from Cs-137 using NaI phosphor and a 6292 Du Mont P.M. Note that $\Delta E \propto \sqrt{E_\gamma}$, so that the energy resolution, $\Delta E/E \propto 1/\sqrt{E_\gamma}$.

1.9.11 Applications and Advantages

1.9.11.1 Fast Timing

For investigations which involve fast timing, the scintillation counters have a decisive advantage over visual detectors. This aspect has been exploited in the lifetime measurements of π^+, K^+, capture times of μ^- and in the discovery of p^- by time of flight method. In conjunction with Cerenkov counter or other scintillation counters, it can be used as a "telescope" in coincidence or anti-coincidence to avoid unwanted particles or events (see Chap. 3 for the discovery of p^-).

1.9.11.2 Scintillation Spectroscopy

Spectrometry of heavy charged particles by scintillation technique is usually done with the use of inorganic crystals. However, a number of organic compounds are also found useful. Organic scintillators like anthracene have a linear response for electrons, but for heavier particles the pulse-height energy relationship exhibit nonlinearity. For this reason, organic scintillators are preferred to inorganic crystals for electron spectroscopy as their effective low atomic number causes substantial improvement for backscattering compared with inorganic crystals, except for very low energy electrons.

1.9.11.3 Gas Scintillation Counters

Gas scintillation counters have the merit of short decay times ($\sim 10^{-9}$ s), large light output per MeV independent of ionization density, and their availability in a wide range of Z and density. Fission fragments in the presence of heavy background of α's can be discriminated. Also, because of low stopping power and small pulse height for γ-rays of nuclear origin, relativistic charged particles can be separated from an intense γ radiation background.

Example 1.13 Calculate the energy limits of the Compton-scattered photons from annihilation radiation.

Solution The energy of annihilation radiation is $h\nu = m_0c^2 = 511$ keV.
The energy of the scattered photon is

$$h\nu' = \frac{h\nu}{1 + (h\nu/m_0c^2)(1 - \cos\theta)}$$

for $\theta = 0$,

$$h\nu' = h\nu = 511 \text{ keV}$$

for $\theta = \pi$,

$$h\nu' = \frac{h\nu}{3} = \frac{511}{3} = 170 \text{ keV}$$

Thus, the energy limits are 511 keV and 170 keV.

Example 1.14 A scintillation spectrometer consists of an anthracene crystal and a 10-stage photomultiplier tube. The crystal yields about 15 photons for each 1 keV of energy dissipated. The photo-cathode of the photomultiplier tube generates one photo-electron for every 10 photons striking it, and each dynode produces 3 secondary electrons. Estimate the pulse height observed at the output of the spectrometer if a 1 MeV electron deposits its energy in the crystal. The capacitance of the output circuit is 1×10^{-10} F.

Solution Number of photons emitted due to absorption of 1 MeV electron is

$$\frac{15 \times 10^6}{10^3} = 15000$$

Number of photo-electrons emitted $= \frac{15000}{10} = 1500$. Since the photomultiplier tube has 10 dynodes and each dynode produces 3 secondary electrons, the electron multiplication factor $M = 3^{10}$. The charge collected at the output is $q = 3^{10} \times 1500 \times 1.6 \times 10^{-19}$ Coulomb. The pulse height will be, $V = \frac{q}{C} = \frac{3^{10} \times 1500 \times 1.6 \times 10^{-19}}{1 \times 10^{-10}} = 0.14$ V.

Example 1.15 The peak response to the 662-keV gamma rays from ^{137}Cs shown in Fig. 1.23 occurs in energy channel 298, with half maximum points in channels 281 and 316. Calculate the standard deviation of the energy, assuming the pulse analyzer to be linear.

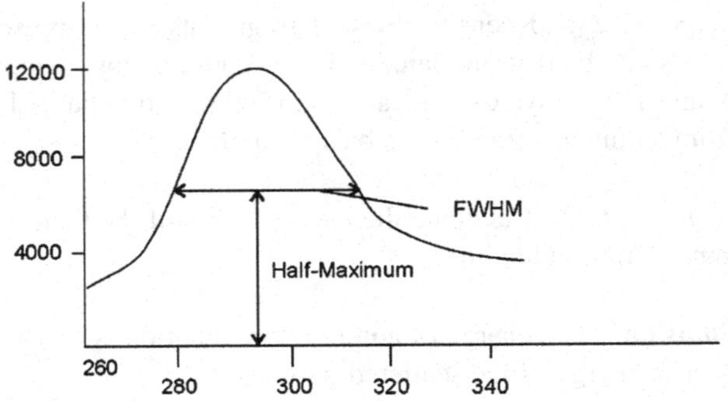

Fig. 1.23 Photo peak from 662 keV γ-rays of Cs-137

Solution The half width at half maximum, HWHM $= 1.177\sigma$. Full width at half maximum, FWHM $= 316 - 281 = 35$. HWHM $= \frac{1}{2}$FWHM $= \frac{35}{2} = 17.5$. The channel 298 corresponds to 662 keV. Hence 17.5 channels correspond to $\frac{17.5}{298} \times 662 = 38.87$ keV. Standard deviation $\sigma = \frac{\text{HWHM}}{1.177} = \frac{3887}{1.177} = 33$ keV.

1.10 Cerenkov Counters

Cerenkov counters based on the phenomenon of Cerenkov radiation work in two different modes.

1. The focusing mode which works as a differential mode
2. Non-focusing counter, also known as the threshold counter which detects the presence of particles whose velocities exceed some minimum value.

1.10.1 Focusing Type (Angular Selection Counters)

In the focusing type the photons in the Cerenkov cone in the angular range $\theta_1 < \theta < \theta_2$ are selected and directed on to a photomultiplier. It consists of three elements, radiator, focusing system and photomultiplier(s).

1.10.1.1 Radiator

This is the optical medium in which the Cerenkov radiation is produced. The Radiator generally consists of a cylinder of a solid optical medium such as lucite or glass. The cylindrical surfaces and the bases are optically polished. If a fast charged particle does not scatter or interact or appreciably slows down, the unique angle of Cerenkov photons with respect to the axis of the cone is maintained regardless of the number of reflections suffered from the cylindrical surface. The base at the entry is

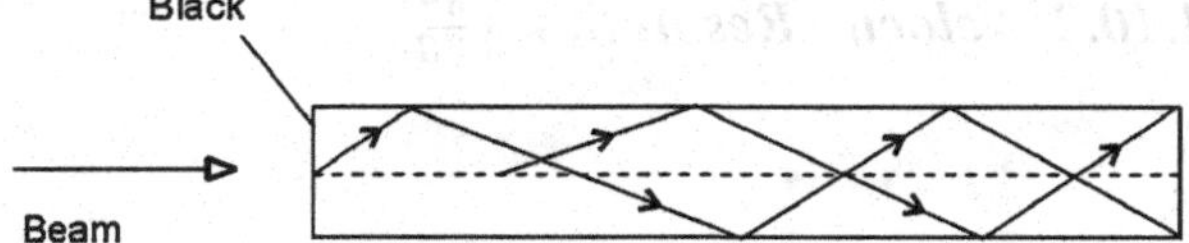

Fig. 1.24 Radiator or the generation of Cerenkov radiation

painted black in order to absorb Cerenkov radiation of particles going in the wrong direction, Fig. 1.24.

Liquids in thin cylindrical containers are also used, so also radiators in the pressurized containers. Gas radiators are specially suited for ultra relativistic particles ($\beta \to 1$) since the refractive index is close to unity. They enjoy a higher velocity resolution $d\theta/d\beta$. They also afford the variation of μ by simply changing the gas pressure.

1.10.1.2 Focusing System

It consists of a series of cylindrical or spherically symmetrical surfaces around the axis parallel to the direction of motion of particle, Fig. 1.25. In practice, two or more photomultipliers generally off the axis are used in coincidence to eliminate accidental coincidences due to tube noise, direct excitations of the photomultiplier by a charged particle, stray light coming from a part of baffle system and due to scattering or inelastic interaction of the incident particle or due to background particles.

1.10.1.3 Photomultipliers (PM)

The desirable characteristics of PM are:

i. Semi-transparent photo cathode of large cathode area be used with an end-window.
ii. PM must have high efficiency for converting photons into photo-electrons. A peak of conversion efficiency in the blue or ultraviolet is desirable.
iii. PM must provide high gain.
iv. It must provide a good signal-to-noise ratio. The dark current due to noise is reduced by cooling the cathode. Popular photomultipliers are 56 AVP (Philips), 6810 A and 7264 (RCA).

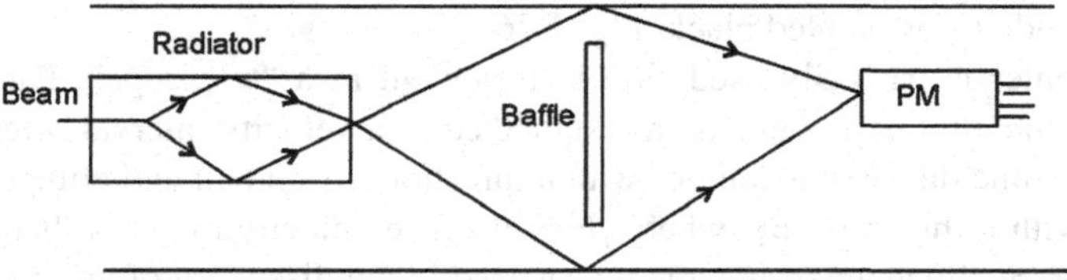

Fig. 1.25 Focusing system

1.10.2 Velocity Resolution $[\frac{\partial\theta}{\partial\beta}]$

$$\cos\theta = \frac{1}{\beta\mu} \tag{1.41}$$

$$\frac{\partial\theta}{\partial\beta} = \frac{1}{\mu\beta^2 \sin\theta} \tag{1.42}$$

For high resolutions, $\beta \to 1$ and $\mu \approx 1$. It follows that

$$\frac{\partial\theta}{\partial\beta} \cong \frac{1}{\sin\theta} \tag{1.43}$$

However, the intensity,

$$I \sim \text{Const} \cdot \sin^2\theta \tag{1.44}$$

Combining (1.43) and (1.44),

$$\frac{\partial\theta}{\partial\beta} \sim \frac{\text{Const}}{\sqrt{\text{Intensity}}} \tag{1.45}$$

We conclude from (1.43) that a good velocity resolution will be achieved by choosing small θ and therefore μ close to unity. However, from (1.44) it is obvious that θ must be sufficiently larger than zero to permit sufficiently large number of photons to be generated.

1.10.3 Non-focusing Counters

1.10.3.1 Velocity Selectors

A special class of non-focusing counter is used as a velocity interval selecting counter. Here the lower velocity limit is set by the threshold and the upper velocity limit by the internal reflection. Subsequent absorption of light occurs with a cone angle greater than θ_c, the critical angle for internal reflection. The back end of the cylindrical radiator is painted black, Fig. 1.26.

This counter is basically used as a last element in a "telescope". Particles of selected momentum in a magnetic are subjected to a velocity interval selection by requiring a coincidence in a counter with a threshold β_1 and an anti coincidence in a counter with a threshold β_2, where $\beta_2 > \beta_1$. The velocity range is then defined by $\beta_1 < \beta < \beta_2$. Such an arrangement was used in the discovery of antiproton (see Chap. 3).

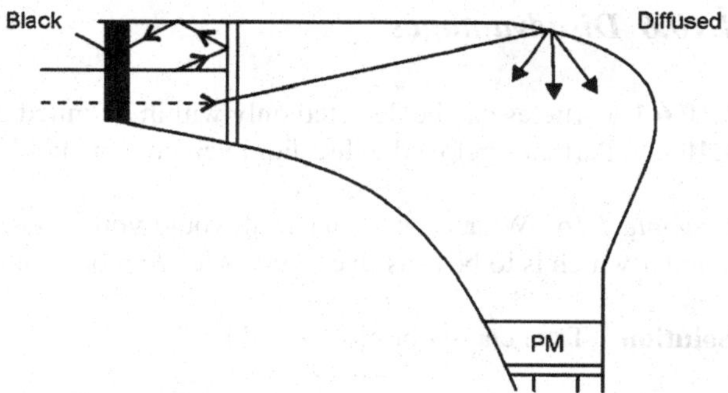

Fig. 1.26 Non-focusing counter

1.10.4 Total Shower Absorption Cerenkov Counter

These counters use lead loaded glass (density 4.49 g/cm^3, $\mu = 1.72$, Radiation length 2 cm) for the radiator element. As the electrons are of several MeV energy, mean number of Cerenkov photons emitted within the radiator is linearly proportional to incident energy. Hence, the mean number of photo-electrons generated at the photo-cathode will be nearly a linear function of the energy of the incident photon or electron.

1.10.5 Advantages

1.10.5.1 All focusing counters are highly directional. Also, non-focusing counters furnish front to back discrimination by painting the back end of the radiator black. Such a method has been used to measure Albedo for cosmic rays in the atmosphere.

1.10.5.2 As number of Cerenkov photons is proportional to z at a known velocity, particles of various charges can be separated by the pulse height.

1.10.5.3 In Cerenkov counters using gas radiator μ can be varied by varying pressure of the gas and consequently particles can be selected over a wide range of velocities.

1.10.5.4 Cerenkov counters can be used to detect short lived particles by placing them close to the source for velocity measurements to avoid intensity losses due to decays. In contrast, in the time of flight method extremely long time of flight paths are involved.

1.10.5.5 In Cerenkov counters, chance coincidences are enormously reduced compared to scintillation counters.

1.10.5.6 Cerenkov pulses are extremely fast. Hence jamming of pulses is avoided in contrast with the Scintillation counters in which the decay times for the scintillation process is of the order of 10^{-9} sec.

1.10.5.7 They eliminate low velocity particles.

1.10.6 Disadvantages

1.10.6.1 Particles can be detected only within a limited direction.
1.10.6.2 Particles below the threshold velocity can not be detected.

Example 1.16 What type of material would you choose for a threshold Cerenkov counter which is to be sensitive to 900 MeV/c pions but not to 900 MeV/c protons.

Solution Total energy of pion ($c = 1$)

$$E_\pi = \sqrt{p^2 + m^2} = \sqrt{900^2 + 140^2} = 910.8 \text{ MeV}$$

Pion velocity, $\beta_\pi = \frac{p_\pi}{E_\pi} = \frac{900}{910.8} = 0.988$.
For emission at threshold, refractive index

$$\mu_\pi = \frac{1}{\beta_\pi} = \frac{1}{0.988} = 1.012$$

Total energy of proton

$$E_p = \sqrt{p^2 + m^2} = \sqrt{900^2 + 938^2} = 1300 \text{ MeV}$$

Proton velocity,

$$\beta_p = \frac{P_p}{E_p} = \frac{900}{1300} = 0.692$$
$$\mu_p = \frac{1}{\beta_p} = \frac{1}{0.692} = 1.445$$

Hence the material must have a refractive index in the range $1.012 < \mu < 1.445$.

Example 1.17 Calculate the threshold energy for the production of Cerenkov radiation in Lucite for (a) electron (b) muon (c) proton (d) alpha. The refractive index of lucite is 1.5.

Solution

a.

$$\beta = \frac{1}{\mu} = \frac{1}{1.5} = 0.6667$$
$$\gamma = \frac{1}{\sqrt{1 - \beta^2}} = \frac{1}{\sqrt{1 - 0.6667^2}} = 1.3417$$
$$\text{KE(thres)} = (\gamma - 1)m = (1.3417 - 1) \times 0.511$$
$$= 0.1746 \text{ MeV or } 174.6 \text{ keV}$$

b. KE(thres) = $(1.3417 - 1) \times 106 = 36.2$ MeV
c. KE(thres) = $(1.3417 - 1) \times 938 = 320.5$ MeV
d. KE(thres) = $(1.3417 - 1) \times 3726 = 1273$ MeV

1.11 Photographic Emulsions

1.11.1 Composition

Photographic Emulsions or Nuclear Emulsions differ from ordinary optical emulsions by a higher silver-bromide content, smaller average crystal diameter (0.14 μm for L_4 emulsions, 0.27 μm for G_5 or Nikfi-R emulsions), and much greater thickness. The silver-halide (mainly silver bromide with 5 % silver-iodide) crystals are embedded in gelatin (HCNO). The gelatin is usually made from clippings of calf hide, ear and cheek or from pig skin and bone. The main function of gelatin is to keep the silver halide crystals well dispersed in the medium and to prevent clamping of the crystals. Atomwise, the AgBr group and HCNO groups compromise 25 % and 75 % respectively. But the interactions with medium and high energy particles take place with a frequency of 70 % in AgBr, 20 % in CNO and 5 % in H. The emulsion sheets called pellicles of standard size 400 μm or 600 μm are stacked with one on the top of the other before the exposure in order to increase the volume. A variety of emulsions of different crystal sizes have been manufactured which differ in sensitivity. The type G_5, L_5 (Ilford), NTB (Kodak), ET-7A (Fuji) and Nikfi-R with crystal size in the range 0.2 μm–0.28 μm are highly sensitised and are capable of recording relativistic particles ($\beta \sim 1$). K_2 and L_2 are less sensitised and record protons up to $\beta = 0.4$. K_1 is less sensitised and record less protons up to $\beta = 0.12$. K_0 is least sensitised and is used mainly for fission studies.

1.11.2 Latent Image

When a charged particle moves through emulsion energy is absorbed by the silver halide crystal, and under the action of reducing agent is converted into metallic silver. The physical condition which renders the crystal developable is called "latent image". The latent image will fade if too much time elapses between irradiation and development, similar to ordinary photography.

1.11.3 Processing

Stripped emulsions are first mounted on glass before processing. For uniform development, it is essential that the developer, for example amidol, permeates the thickness of emulsion. For this reason, the plates are bathed in the developer at low temperature (0–5 °C) so that the developer is permitted to penetrate but the development will not ensue. Now, if the temperature is raised to say 23 °C, the development ensues. This is called high temperature development. After the development stage, the plates are "fixed", washed and dried in alcohol.

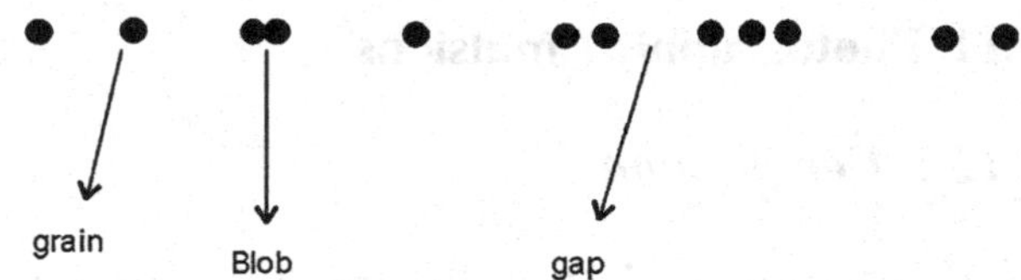

Fig. 1.27 A typical track in emulsion. Grain, Blob and Gap are indicated

1.11.4 Techniques

Events are analyzed with the aid of special type of microscopes with smooth movable stages and high power oil objectives and eyepieces with graticules capable of giving magnifications as high as 2700. After processing, normal emulsion shrinks by a factor of 2–2.5. The shrinkage factor is taken into account in the dip measurements of angles. For particles, which stop within the emulsion stack, Range-Energy-Relation of the type (1-101) is used.

Ionization measurements are made either by counting grains or blobs for relativistic particles or by counting blobs and gaps of length $> l$, for non-relativistic particles, and determining the exponent g from the relation

$$H = Be^{-gl} \tag{1.46}$$

where H and B are gap and blob density, respectively. Blobs are unresolved grains and gap is the space separating two successive grains or blobs, as in Fig. 1.27.

For energetic particles, the parameter $p\beta$ (momentum times velocity) can be found out from multiple scattering measurements by essentially measuring the y-coordinates of the track along the axis, at constant intervals called "cell's". The arithmetic average of second differences is given by,

$$\langle |D_2| \rangle = \langle |y_1 - 2y_{i+1} + y_{i+2}| \rangle \tag{1.47}$$

And $p\beta$ is given by the relation

$$p\beta = \frac{Kt^{\frac{3}{2}} \times 18.1}{|D_2|} \tag{1.48}$$

where $K = 28$ is the scattering constant, for $\beta > 1$, D_2 is in μm, $t =$ cell length in mm and p in GeV/c. The quantity $t^{\frac{3}{2}}$ arises due to the fact that the scattering angle $\theta = D/t$. The factor 18.1 arises due to conversion of degrees into radians. The choice of cell-length is such that the signal-to-noise ratio is greater than 2–3. Multiple scattering technique with constant cell method works provided the energy loss over the tracks is not significant. In order that the method be useful, it is important that the spurious scattering and distortion resulting from the processing of emulsions be small and that the stage noise, which arises due to the non-linear motion of the stage be negligible.

Charge of the particle can be determined from δ-ray counting (AAK, 1, 1) or from photometric measurements; for example, the fluxes of heavy primaries of Cosmic rays have been determined from emulsion exposures in balloons or rockets following this procedure.

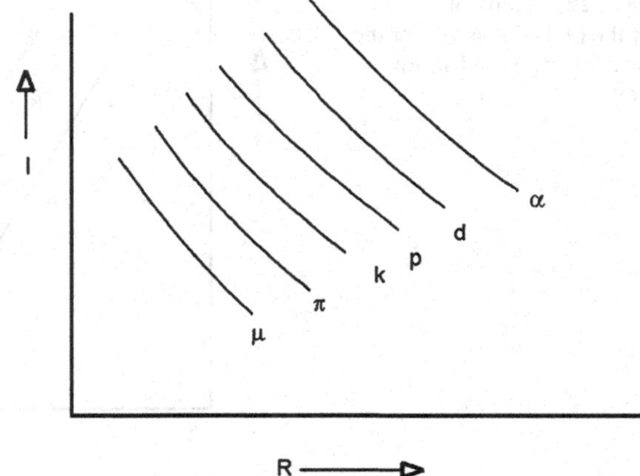

Fig. 1.28 Family of curves are obtained for $I-R$ plots for different particles

Particles are identified from their mass determinations. In this context we recall from (AAK, 1, 1),

i. Range measurement gives energy of the particle
ii. Ionization measurement gives the velocity
iii. Multiple Scattering measurements give $p\beta$.
iv. δ-ray density measurement gives z of the particle.

For singly charged particles, combination of any two parameters arising in (i), (ii) and (iii) uniquely fixes the mass of the particle since velocity must be eliminated. Thus, the plot of ionization (I) versus residual range (R) gives a family of curves for particles of different mass, Fig. 1.28. Notice that for the given I, the ranges are in the ratio of the masses. The method is very extensively used for particles which are brought to rest. Masses can be estimated with an accuracy of about 10 % from a single measurement.

The method can be extended for identifying particles which are not arrested in the emulsion stack, if an appreciable change in ionization over a known distance is determined.

At higher energies, combination of (ii) and (iii) in favorable cases permits the identification of particles, Fig. 1.29. At still higher energies, the curves cross each other and the identification becomes difficult or even impossible. On the other hand, energy measurements can seldom be made from multiple scattering method with an accuracy better than 10–15 % due to the presence of spurious scattering. At energies greater than few GeV, the measurements are rendered meaningless if the noise due to spurious scattering competes with the Coulomb's signal. Sometimes in favorable cases it has been possible to extend the energy measurements up to 15–20 GeV in cosmic ray jets by making relative scattering measurements—a method in which multiple scattering measurements are made with reference to a neighboring track due to an ultra relativistic particle so that spurious scattering and stage noise which affect both the tracks similarly are eliminated.

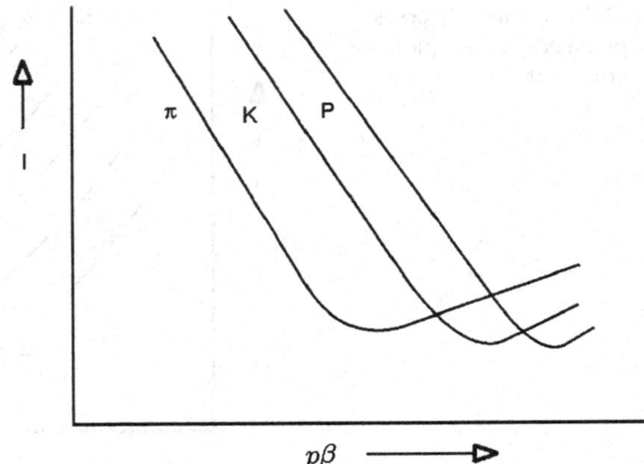

Fig. 1.29 Family of curves for the plot of I vs $p\beta$ for the particles, π, K and P are shown

1.11.5 Advantages

1.11.5.1 High Stopping Power and High Spatial Resolution

Owing to high stopping power, charged particles of moderate energies are easily arrested within the stack of emulsions. Accurate range measurements have permitted precise mass determination of several elementary particles immediately after their discovery. It is usually possible to determine the magnitude and direction of velocity (by blob counting), the rate of change of velocity (by observing variation of Ionization over measured distance), the product of velocity and momentum up to 1.5 GeV/c (from multiple scattering measurements), charge (from delta ray density), the moderation time (from counting number of decays from known traversals and velocity), and the interaction cross-section (from mean free path). Momentum can be measured by bending magnet external to the emulsion, or by employing pulsed magnetic field within the emulsion. Owing to high stopping power a large fraction of even such short-lived particles as hyper fragments and slow Σ^+ hyperons are brought to rest before their decay. Their ranges and moderation times are easily measured. In this way reliable measurements of life times of most of the unstable elementary particles have been made in emulsions and also some of the partial decay rates have been evaluated. Because of high spatial resolution time intervals as short as 10^{-16} sec have been measured, a good example being the mean lifetime determination of π^0. Reliable neutron spectroscopy in the energy range, 1–15 MeV is done with the emulsion technique.

1.11.5.2 High Angular Resolution

The angular resolution is unsurpassed. This aspect has been exploited in the determination of the magnetic moment of Λ^0.

1.11.5.3 Compactness

In situations where compactness of equipment is essential, emulsions can be conveniently used. For example, they can be sent in balloons or rockets to high altitudes and recovered conveniently after the required exposure. Further, they are economical.

1.11.5.4 Radiation Length

Because of high stopping power and short radiation length huge electromagnetic cascades can be contained in a large stack and the complete development and final degradation can be studied in detail.

1.11.5.5 Loading

It is possible to load emulsions with H_2O, D_2O, Li_2SO_4, $Th(NO_3)_4$, UO_2 etc. to study reactions with elements which are not contained in normal emulsions.

1.11.6 Limitations

1.11.6.1 Composition Invariability

The composition of nuclear emulsions can not be changed arbitrarily so that interaction studies are limited only to those nuclei which are present in normal emulsions, although loaded emulsions in limited concentration have been used with some difficulty. The presence of a large variety of nuclei in normal emulsions makes it difficult to identify the target nucleus, although in favorable cases it is possible to identify the groups (Ag, Br, I), (C, N, O), (H), the first one being heavy and the second one light. Events in hydrogen are invariably clean and can be identified confidently.

1.11.6.2 Minuteness of Volume

Because of minuteness of volume of emulsion under study in the microscope it is exceedingly difficult to find correlated events even 1 cm or so apart.

1.11.6.3 Continuous Sensitivity

Because of continuous sensitivity the background tracks are a source of nuisance. The best available emulsions from the stand point of sensitivity lack discrimination and all highly ionizing particle tracks are saturated.

1.11.6.4 Distortion and Spurious Scattering

Emulsion which has a gelatin base is subject to distortion in the processing regime. This can seriously affect the range and angle measurements. Spurious scattering can interfere with Coulomb's signal in multiple scattering measurements.

1.11.6.5 Scanning

It usually takes several months involving a large group of Physicists and scanners to scan and analyze events of statistical significance.

1.11.6.6 The Study of Elementary Interactions

Since only 5 % of the interactions take place with hydrogen and 95 % in complex nuclei of emulsion, the interactions in the latter are obscured by secondary effects. Although hydrogen density in emulsions is comparable with that in hydrogen bubble chamber, the latter is by far better suited in so far as the elementary interaction studies are concerned.

1.11.7 Discoveries Made with Photographic Emulsions

The years 1945–54 were the golden era of Nuclear Emulsions. At that time high energy accelerators were not yet available. The only source of high energy particles was in the form of cosmic rays. Major discoveries of fundamental importance included the particles π^+, π^-, π^0, K^+, K^- mesons, several decay modes of K^+ mesons (two-body and three-body decay modes), the hyperons, hyper fragment, double-hyper fragment, the composition of primary cosmic rays etc. Reliable mass measurements of various types of mesons and the $\Sigma^+$and Λ^0 hyperons, and their mean life times were first carried out in emulsions.

Example 1.18 The kinetic energy and momentum of a meson deduced from measurements on its track in a nuclear track emulsion are 200 MeV and 490 MeV/c, respectively. Determine the mass of the particle, in terms of the electron mass, and identify it.

Solution

$$\mathrm{KE} = (\gamma - 1)m = 200 \quad \text{(i)}$$

$$P = m\gamma\beta = 490 \quad \text{(ii)}$$

where we have put $c = 1$ and $\gamma = \frac{1}{\sqrt{1-\beta^2}}$

$$\beta = \frac{\sqrt{\gamma^2 - 1}}{\gamma} \tag{iii}$$

Using (iii) in (ii) and dividing the resultant equation by (i),

$$\frac{\sqrt{\gamma^2 - 1}}{\gamma - 1} = \frac{490}{200} = 2.45$$

or

$$\frac{\sqrt{\gamma + 1}}{\sqrt{\gamma - 1}} = 2.45 \tag{iv}$$

Solving (iv), $\gamma = 1.4$. Using the value of γ in (i), $m = 500\ \text{MeV}/c^2$.

Mass of meson in terms of electron mass, $m = \frac{500}{0.511} = 978.5m_e$.

The particle is a Kaon as its known mass is $966m_e$.

1.12 Expansion Cloud Chamber

1.12.1 Principle

We owe the invention of cloud chamber to C.T.R. Wilson (1912). It contains super-saturated vapor of a gas under pressure. When the chamber is allowed to expand adiabatically, dust particles and preferably ions form nucleation centers around which droplets are formed which grow in size and become visible under proper illumination. If a charged particle passes through the chamber simultaneously a trail of droplets in the form of a track is formed which can be photographed. The unwanted nucleation centres due to dust and the chemical compounds suspended in the gas are removed from the chamber either by electrostatic clearing field or by repeated expansions until dust particles settle on the bottom of the chamber where they stick to the wall.

1.12.2 The Stability of Charged Drops

The surface energy of the drop of radius 'a', surface tension S, charge q in an external medium of dielectric constant ε_1 is given by

$$E = 4\pi a^2 S + \frac{1}{2}\frac{q^2}{\varepsilon_1 a} \tag{1.49}$$

A change in radius a results in the change of energy δE and is given by

$$\delta E = \frac{d}{da}\left(4\pi a^2 S + \frac{1}{2}\frac{q^2}{a\varepsilon_1}\right)\delta a \tag{1.50}$$

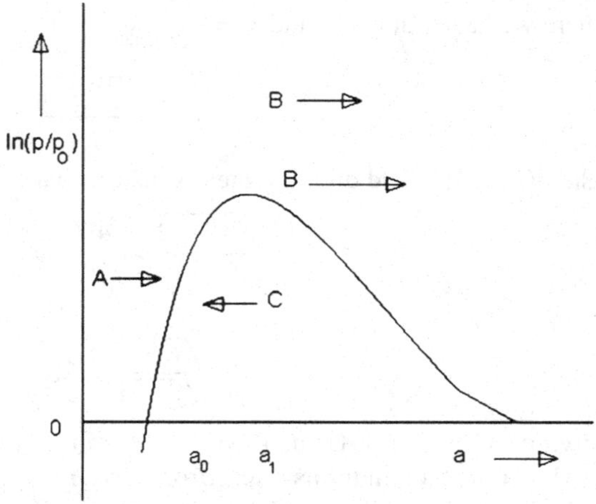

Fig. 1.30 Variation of $\ln(p/p_0)$ with radius of the drop a

This change of energy does not take place in corresponding condensation or evaporation at a plane surface. The change of energy may be equated to the work needed to bring the amount of vapor produced by evaporation at vapor pressure p in equilibrium with the drop, to the saturation pressure p_0 over a plane surface. We can then write:

$$\frac{d}{da}\left(4\pi a^2 S + \frac{1}{2}\frac{q^2}{a\varepsilon_1}\right)\delta a = 4\pi a^2 p \delta a \frac{R\theta}{M} \ln \frac{P}{P_0} \tag{1.51}$$

where ρ is the density of the drop, θ is the absolute temperature, M the molecular weight and R the gas constant. Equation (1.51) becomes

$$\frac{R\theta\rho}{M} \ln \frac{P}{P_0} = \frac{2S}{a} + \frac{ds}{da} - \frac{q^2}{8\pi\varepsilon_1 a^4} \tag{1.52}$$

If we assume that surface tension S is independent of the drop radius then we can set $dS/da = 0$ in (1.52). Taking into account the dielectric constant ε_2 of the condensed liquid, (1.52) is modified as

$$\frac{R\theta\rho}{M} \ln \frac{P}{P_0} = \frac{2S}{a} - \frac{q^2}{8\pi a^4}\left(\frac{1}{\varepsilon_1} - \frac{1}{\varepsilon_2}\right) \tag{1.53}$$

The variation of the equilibrium super-saturation, p/p_0 with drop radius a is shown in Fig. 1.30. a_1 is the equilibrium radius of the drop and a_0 is the value for which $p = p_0$. The values of a_0 and a_1 are given by the expressions

$$a_0^3 = \frac{q^2}{16\pi S}\left(\frac{1}{\varepsilon_1} - \frac{1}{\varepsilon_2}\right) \tag{1.54}$$

$$a_1^3 = \frac{q^2}{4\pi S}\left(\frac{1}{\varepsilon_1} - \frac{1}{\varepsilon_2}\right) \tag{1.55}$$

Each point in the diagram represents the condition of a drop of given radius in vapor at given pressure. The curve divides the space of the diagram into two parts.

Points above the curve represent the domain of drop growth (condensation) and those below correspond to evaporation. The curve itself is a stability curve. Consider a typical point A on the diagram. Condensation makes it move on the stability curve where further growth stops. The drop however, may be still invisible. On the other hand, a point like B does not hit the curve and moves from it indefinitely due to continuous growth. In a cloud chamber drops of class B do not exist initially but due to adiabatic expansion drops belonging to class A which are present pass over to class B if the super-saturation obtained is greater than the equilibrium value of radius a_1.

For water at 0 °C, for a drop carrying a single electronic charge, $a_1 \cong 6 \times 10^{-10}$ m, for which $(\frac{P}{P_0})a_1 = 4.2$. Super saturation is produced by a rapid, adiabatic expansion of the mixture of gas and vapor. Most of the cloud chambers work by volume expansion. If θ_1 is the initial absolute temperature, θ_2 the final temperature, V_1 the initial volume and V_2 the final volume, $\gamma = C_p/C_v$, the ratio of specific heats at constant pressure and constant volume of the gas mixture, then:

$$\frac{\theta_1}{\theta_2} = \left(\frac{V_1}{V_2}\right)^{\gamma-1} \quad (1.56)$$

The change in temperature obviously depends on the value of γ. Because of a larger value of γ monatomic gases are preferably used. Most of the cloud chambers work on volume expansion by a mechanical device. But pressure defined chambers have also been used, specially in chambers containing metal plates. The pressure expansion is accomplished with only a thin rubber diaphragm separating the pressure vessels.

1.12.3 Choice of Pressure

Cloud chambers have been operated at pressures ranging from 1 atmosphere to 50 atmospheres. At low pressure, the vapor occupies a substantial fraction of the total amount of gas present, causing a change in the value of γ and results in an appreciable amount of ionization. Operation of the cloud chamber in the range of pressures 0.1 to 2.0 atm does not present any serious difficulties. At higher pressures, it is more difficult to remove old droplets and necessary waiting time between expansions increases with the increasing pressure. Curvature measurements in a magnetic field are rendered less accurate due to Coulomb multiple scattering of a charged particle in the high density gas.

1.12.4 Choice of Gas

Generally, noble gases are preferred because of lower expansion ratio required. Argon or a mixture of 50 % Argon and 50 % Helium are commonly used as they give clear tracks and rapidly growing droplets.

1.12.5 Curvature Measurements

A magnetic field is normally employed for momentum determination by curvature measurement. A large field is used so that spurious curvatures due to scattering and molecular motion are relatively less important. The maximum detectable momentum is that for which the uncertainty in the curvature is equal to the true curvature. With a field of 10 kG, momentum as large as 50 GeV/c has been measured over a track of 40 cm.

1.12.6 Multiplate Chamber

Since the stopping power is so low in a gas and the probability of interaction so small that in certain specific experiments it becomes necessary to place a set of metal plates in parallel planes leaving space in between so that the tracks can be recorded. Such multiplate chambers have proved to be invaluable in the investigations of nuclear reactions, and the short lived unstable particles produced in the reactions. The range of particles can be measured if the particle stops in one of the plates. In a multi-plate chamber γ-rays are materialized with much greater probability. Also the details of nuclear reactions can be studied.

1.12.7 Counter Control

It is a great advantage to use counter control expansions if rare events are to be studied. The expansion is effected only when a charged particle passes through associated counters in coincidence or anti-coincidence. Since expansion speeds are of the order of 10 milliseconds, counter controlled tracks are broadened by diffusion to widths of the order of a mm.

1.12.8 Temperature Control

In order to make accurate momentum measurements the gas in the chamber should be stable. This is accomplished by maintaining a temperature gradient of about 0.01 °C from top to bottom.

1.12.9 Expansion Speed

The speed of expansion of cloud chamber is an important consideration only for counter control type in which the width of the track may become too wide for accurate measurements. Tracks about 1 mm wide are obtained with expansion times of 14 milliseconds in air at N.T.P.

1.12.10 Recycling Time

An expansion cloud chamber normally takes at least one minute to prepare for each expansion. This time is used up for slow clearing expansions in order that the motion of the gas may subside and the vapor to diffuse back through the gas. Such repetition rates can not match the frequency of pulsed accelerators and in this regard they are inferior to bubble chambers which are capable of recycling every few seconds.

1.12.11 Discoveries

Expansion cloud chambers have played important role in the fundamental discoveries in Cosmic rays and Particle Physics. The penetrating nature of cosmic radiation was first established in 1932. The positron was discovered in 1933. The muon was confirmed and the K^+ meson was discovered in 1944, and the two-pion decay mode in 1953. The so-called “V” events comprising kaons and hyperons were first identified in a cloud chamber as early as 1947 and the cascade hyperon in 1952.

1.13 Diffusion Cloud Chamber

1.13.1 Principle

A diffusion cloud chamber works on the principle of diffusion of condensable vapor from a warm region where it is not saturated to a cold region which is supersaturated and proper conditions are produced for the growth of droplets around ions similar to the expansion cloud chamber. The vapor density increases with temperature, while the gas density decreases with the increase in temperature. A given mixture of gas and vapor may thus have density that increases or decreases with the temperature depending on the composition. It is necessary to maintain density gradient in the chamber such that the density of the mixture is less at the top than at the bottom to avoid convection turbulence. Most of the diffusion chambers conveniently provide downward diffusion. The top is warm and it is here that vapor is introduced, and the bottom is cold. Figure 1.31 shows the essential details of a diffusion cloud chamber.

The body of the chamber containing the gas-vapor mixture is thermally connected to the base which is cooled to some low temperature T_0. A trough containing liquid which is vaporized by maintaining it at some temperature T_1 higher than T_0 is in contact with the glass top of the chamber. A temperature gradient is thus set up between top and bottom of the chamber. The gas at the top of the chamber is saturated with the vapor of the liquid contained in the trough. This vapor will diffuse towards the cooled base.

At the lower temperature the gas becomes supersaturated with the vapor. As the temperature decreases the value of super saturation increases. With a suitable gas-vapor ratio and the choice of T_1 and T_0, the super saturation (S) can exceed the

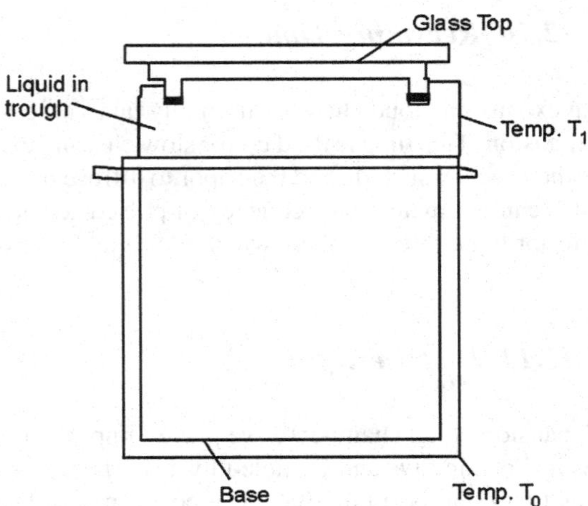

Fig. 1.31 Diffusion cloud chamber

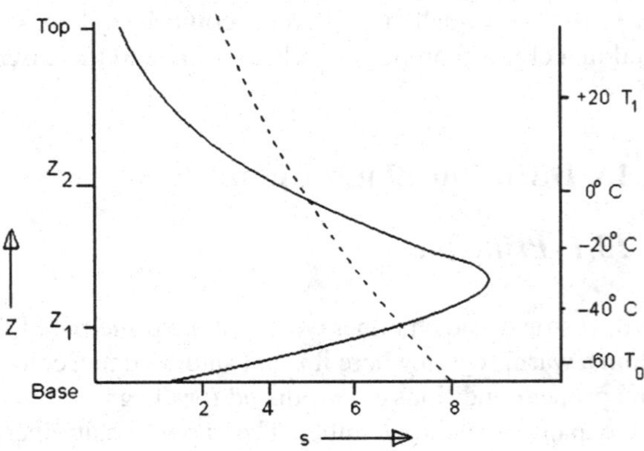

Fig. 1.32 The *solid line* is a plot of S vs. Z; the *dotted line* is the S vs. T curve

critical value necessary for condensation around ions. It can be shown (Thompson) that

$$\ln S = \frac{3M}{2RT\rho} \frac{(4\pi\sigma^4)^{\frac{1}{3}}}{(Kq^2)} \tag{1.57}$$

where M is the gram molecular weight, ρ the density, S the surface tension, K the dielectric constant of the liquid, T the absolute temperature, R the gas constant and q the charge carried by the ions. The saturation values resulting from certain temperature and vapor density distributions within the chamber are determined. For a given base temperature T, a typical curve for S versus Z (the chamber height) is shown in Fig. 1.32 by the solid line. The dotted curve (S vs T) calculated from formula (1.57) for a typical liquid is also shown. The values of Z where the two curves intersect correspond to maximum height (Z_2) and minimum height (Z_1) in the cloud chamber at which the dropwise condensation upon ions is expected to occur. The region between Z_2 and Z_1 is the sensitive depth of the chamber.

1.13.2 Choice of Liquid Vapor

The operation of the diffusion chamber depends on large changes of vapor pressure with temperature. The choice of liquids is dictated by this property, over a convenient range of temperatures. It is desirable that the liquid has a low latent heat of vaporization so that the thermal equilibrium in the chamber is not disturbed due to the heat released from drop formation. Further, the liquid must have a low value for the critical vapor pressure necessary to sustain drop growth on ions. Methyl alcohol is a satisfactory liquid over the range of temperatures +30 °C to −70 °C. Ethyl alcohol is only slightly inferior for this purpose, while water is avoided as it freezes at a temperature within the range normally employed.

1.13.3 Choice of Gas

Chambers containing heavy gases like air or argon are useful only up to two or three atmospheres. According to Shut if air is used at 3 atm, with base at dry ice temperature, the top must be at a temperature around 293 K. If the temperature at the top is higher than this value, the sensitive region tends to become unstable and turbulent and it leads to background fog. The stability of gas is important from the point of view of curvature measurements.

Chambers containing light gas like hydrogen, deuterium and Helium are unsuitable at atmospheric pressure but become useful at pressures of 20 to 30 atm.

1.13.4 Operating Temperatures

With few exceptions, diffusion chambers have been operated with dry ice for the base temperature (−70 °C) and top temperature up to about 30 °C.

1.13.5 Clearing Fields

Apart from dust particles, there will be background due to neutral condensation nuclei and droplets formed on ions which have diffused into the sensitive region. The application of electric field between a grid of wires located just above the sensitive region and the top plate of the chamber enables the ions formed in the non-sensitive region to be collected by either electrode depending on the polarity. This method reduces the number of ions diffusing in the region and therefore considerably improves the track-to-background contrast. Dust particles and reevaporation nuclei are removed by continuous production of super-saturation until the nuclei are carried to the bottom of the chamber where they stick to the wall.

1.13.6 Magnetic Field

Momentum of a particle is determined by employing a magnetic field. A uniform field along the vertical axis may be produced either by a pair of Helmholtz coils placed around the chamber or by placing the chamber between the poles of an iron cored magnet.

1.13.7 Photography

Tracks may be photographed with a cine-camera under continuous illumination. Filters are necessary to avoid heating of the gas in the chamber. Alternatively, single shot photography may be employed using standard flash camera. A time delay of 0.1 to 0.2 sec is necessary between the ion formation and photography as the drops must be allowed to grow to a size which will scatter adequate light, but the delay must be kept to a minimum so that the falling of drops under gravity does not lead to serious distortion of tracks and there by destroy the accuracy of measurements.

1.13.8 Advantages

1.13.8.1 Versatility

Diffusion cloud chambers, when operated under suitable conditions provide information on the charge, momentum, velocity, life time, flux, direction of motion and the details of nuclear interactions. They have been operated at low atmospheric and high pressures in conjunction with high energy accelerators, in mines, on mountains and in balloons 100,000 ft above the earth.

1.13.8.2 Use of Hydrogen

Diffusion chambers using hydrogen at high pressures can be conveniently used to investigate elementary interactions with protons in Particle Physics. In contrast it is difficult to operate expansion cloud chamber with hydrogen filling at high pressure.

1.13.8.3 Recycling Time

In the diffusion chamber the re-cycling time for heavy gases is only 5–10 s, in contrast with the expansion chamber where it is 1–2 minutes. In pressurized hydrogen chambers, the contrast is still greater. This then means that comparatively short time is spent in taking photographs of desired events, there by saving expensive running time of the particle accelerator.

1.13.8.4 Counter Control

Diffusion chambers may also be used with counter control. It is only necessary to trigger the camera to record the desired event.

1.13.8.5 Experiments with Accelerators

As the sensitive region lies in the horizontal plane, diffusion chambers are ideally suited for investigations in Nuclear Physics and Particle Physics at the accelerators as the beams from the accelerators are in the horizontal direction.

1.13.9 Limitations

1.13.9.1 Cosmic Rays Studies

The sensitive layer is not more than 3 inch thick and can not be made vertical for cosmic rays investigations.

1.13.9.2 Depletion of Vapor

The passage of high flux of ionizing particles causes depletion of vapor, resulting in gaps in tracks. This is a serious problem in most diffusion chambers.

1.13.9.3 Ionization Limitations

When the overall ionization within the chamber arising from various causes exceeds a certain level, the chamber will be rendered insensitive due to inadequate super saturation.

1.13.9.4 Foil Limitation

When foils are used in a diffusion chamber, similar to multi-plate expansion cloud chamber, their design is limited in that they should not interfere with the temperature gradient. Thin strips of material of poor thermal conductivity fulfill this condition. Difficulties of turbulence are not a serious problem in diffusion chamber in contrast with the expansion chambers.

1.13.10 Discoveries

Some of the important experiments made with diffusion chambers are concerned with $\pi^- p$ scattering, V-particle production in $\pi^- p$ collisions, confirmation of Associated production of Strange particles, first direct evidence on multiple meson production in n-p collisions and Internal pair production of mesic γ-rays produced in slow π^- capture in hydrogen.

1.14 Bubble Chamber

1.14.1 Principle

We owe the invention of bubble chamber to D.A. Glaser who conceived the idea that nucleation centers may be produced due to the passage of a charged particle in a superheated liquid so that a string of bubbles may be formed along the track. A superheated liquid is the one which is heated above the boiling point without actually boiling. The nucleation centre are formed due to the deposit of energy along the track of the ionizing particle. The superheated liquid is unstable and erupts into boiling shortly after formation of nucleation centers and the bubbles appear. The action of bubble chamber is the inverse of an expansion cloud chamber, with a gas bubble forming in a superheated liquid instead of a liquid drop forming in a superheated gas.

The liquid in a bubble chamber is superheated by a sudden expansion leading to a reduction of pressure. After the track is formed and photograph taken, the pressure is increased to the initial value by contracting the chamber. The bubbles collapse and the chamber is ready for another expansion. All this takes place in the matter of a few seconds such that the operation rate may advantageously match the frequency of particle beam from a pulsed accelerator in its duty cycle.

1.14.2 Bubble Chamber Liquids

A wide variety of pure liquids, liquid mixtures, and liquids containing dissolved gases have been used in bubble chamber. They range from hydrogen with a density of 0.0586 gm/cm^3 and radiation length of 1100 cm to xenon with density 2.3 gm/cm^3 and radiation length 3.7 cm.

1.14.2.1 Hydrogen

This is by far the most important liquid from the point of view of the elementary particle processes as the target contains pure protons. However, this advantage is offset by serious cryogenic problems as the operating temperature is around 28 K, and pressure 5–7 atm.

1.14.2.2 Deuterium

This is the lightest element containing neutron, and the liquid can be used in a chamber similar to hydrogen at operating temperature 32 K.

1.14.2.3 Helium

It is the lightest atom that has nuclear spin 0 and iso-spin 0. The operating temperature is 3–4 K and pressure $\frac{1}{4}$ to 1 atm. Its merit is that it is non-flammable. However, the cryogenic problems are even more serious than those for hydrogen.

1.14.2.4 Propane (C_3H_8)

This is the most commonly used organic liquid operating at 58 K and at a pressure of 21 atm. The propane chambers are easier to build and operated compared to cryogenic chambers. However, the fire hazard is comparable with hydrogen chambers. The radiation length is down by a factor of 10. Although the amount of hydrogen in a propane chamber is greater by a factor of 1.38 than in hydrogen bubble chamber, it is difficult to isolate events in hydrogen from the carbon target.

1.14.2.5 Heavy Liquids

Liquids heavier than propane have been used for the study of γ-rays from π^0 or Σ^0, through pair production. Suitable candidates are xenon (radiation length 3.7 cm), tungsten hexafluoride and freon, CF_3Br (radiation length $\sim$11 cm). The last one mentioned is inexpensive, non-flammable and non-corrosive.

1.14.3 Control of Temperature

The temperature of a bubble chamber must be controlled to about 0.1 °C to ensure reproducible conditions.

1.14.4 Magnetic Field

Magnetic field is an important accessory of a bubble chamber for momentum measurement. For liquid hydrogen, multiple coulomb scattering is unimportant compared to the deflection in a magnetic field at 10 kG. Gauss or higher. In organic liquids, multiple scattering is more important and consequently at least 20 kG. Gauss fields are employed for momentum measurements with an accuracy of 10 %. In heavy liquid chambers, multiple scattering is so severe that magnetic fields are rarely employed. Instead $\rho\beta$ may be estimated from multiple scattering measurements as in emulsions. The mean second difference is given by

$$|\overline{D}| = \left(\frac{l^3}{X_0}\right) \times \frac{12.4}{\rho\beta} \tag{1.58}$$

However, the accuracy in momentum measurements is poorer by a factor of 10 compared to emulsions. If the curvature of a particle track in a magnetic field of B kG. Gauss is used to determine the particle momentum p, multiple scattering introduces an error Δp in the measurement of the momentum given by

$$\frac{\Delta P}{P} = \frac{4.1}{B\beta \sin\varphi\sqrt{X_0 l}} \tag{1.59}$$

where l is the length of track in space, $\beta = v/c$, φ is the angle between the direction of the magnetic field and the momentum of the particle. In order to achieve an accuracy of 10 % in momentum measurement on a 10 cm track of a relativistic particle with a 15 kG. Gauss field, radiation length should not be less than 20 cm. For this reason magnetic fields are not used in heavy liquid bubble chambers.

1.14.5 Track Distortion

Liquid motion may displace the true track position following the bubble formation. In hydrogen bubble chamber a 1 GeV/c track can be distorted to the same extent as due to multiple Coulomb scattering.

1.14.6 Number of Tracks per Picture

In a single picture the higher density of bubble chamber liquids usually limits the number of tracks to 20 or 30 in order to obtain unambiguous events. This is in contrast with the cloud chamber in which 100 to 200 tracks can be used per picture. The bubble images are recorded by several stereoscopic cameras. Subsequent measurements of film are digitized, the tracks and vertices of event are reconstructed in three dimensions with the aid of a geometry program. Time resolution of the bubble chamber as determined from bubble size is measured in tens or hundreds of μS.

1.14.7 Identification of Particles

Measurement of curvature and residual range for particles which are arrested within the chamber permit the particles to be identified. This procedure is limited for protons of momentum less than 0.5 GeV/c. Ionization measurements are carried out by bubble counting and are most valuable for protons of momentum 0.5 to 1 GeV/c. δ-rays provide the best identification for protons between 1 to 3 GeV/c.

In heavy liquid bubble chambers curvature measurement is substituted by multiple Coulomb scattering measurements. They yield the value of $p\beta$ (momentum times velocity). The combination of bubble density and curvature or bubble density and multiple scattering furnish the identification of particles.

1.14.8 Advantages

1.14.8.1 Scanning

Photographs of events can be easily scanned because of the large field of view and depth of focus. A large amount of data can be collected and analyzed in a short time. For the observation of a rare event sometimes it is necessary to examine more than hundred thousand photographs as in the discovery of Ω^- and this has been possible in a bubble chamber of reasonably large volume. Further, bubble chambers are ideally suited to go with high energy accelerators.

1.14.8.2 Mass Measurements and Correlation Studies

The masses of particles can be estimated in most of the cases by the methods discussed in Sect. 1.11.4. Also the decays of neutral particles like K^0, Λ, Σ^0 etc. can be correlated. This is rarely possible in emulsions.

1.14.8.3 Fundamental Interactions with Protons

The most important use of hydrogen bubble chamber is that it allows the studies of fundamental interactions with single protons, notwithstanding serious cryogenic problems.

1.14.8.4 Special Target Nuclei

Unambiguous interactions with neutrons can be studied in the deuterium bubble chamber, and specific interactions like the production of the hyper fragments $^4\mathrm{He}_\Lambda$ and $^4\mathrm{H}_\Lambda$ can be studied in helium bubble chamber with K^- beams.

1.14.8.5 Background

The short life time of nucleation centers (of the order of 1 millisecond) has the consequence of reducing the background of unwanted "old tracks".

1.14.8.6 Gamma-Ray Detection

γ-rays can be detected through their materialization in heavy liquid chambers although their energy determination carries large uncertainty since radiation loss of electrons is both large and unpredictable. In xenon the average radiation energy loss amounts to 30 % per cm. Reliable momentum measurements require large X_0 but conversion efficiency requires small X_0. The problem is solved with the use of a large propane bubble chambers with intermediate radiation length. Since the estimates of momentum obtained from curvature measurements and multiple Coulomb scattering are independent, it is possible to combine both types of measurements.

1.14.8.7 Re-cycling Time

The re-cycling rates can be made to match the pulsed rate of accelerator beams.

1.14.9 Limitations

1.14.9.1 Tracks per Picture

Number of tracks per picture can not exceed 20 or 30 for clarity.

1.14.9.2 Counter Control

Since several milliseconds are required for the expansion, counter controlled expansions are not possible with the present techniques. For this reason bubble chambers have not been favored for cosmic ray studies where controlled expansion is often required. Also, bubble chambers can not be carried in balloons or rockets.

1.14.9.3 Spatial and Angular Resolution

The spatial and angular resolutions of tracks are inferior to those in emulsions.

1.14.9.4 Use with Colliding Beam Accelerators

As most of the modern accelerators are of colliding beam type, with a very low duty cycle of bubble chamber together with the geometry of the beams, they do not find a place for their utilization.

1.14.10 Discoveries

Numerous discoveries of fundamental nature in Particle Physics have been made with bubble chambers, particularly with hydrogen bubble chambers. These include the discoveries of the neutral cascade hyperon Ξ^0 (1959), Anti-hyperon $\overline{\Sigma}$ (1961), Ω^- (1962), the meson resonances ω, ρ, η (1961), K^* meson (1963), Y^* (1964), $\overline{\Omega^+}$ (1971), parity determination of K^- (1962) and K^0–$\overline{K^0}$ regeneration (1961).

1.15 Spark Chamber

1.15.1 Principle

To G. Charpak (1957) we owe the discovery of spark chamber in which electrical discharge occurs in narrow gaps of metal plates (maintained at large potential difference) following closely the trajectory of a charged particle.

1.15.2 Mechanism of the Spark Discharge

A free electron produced by ionization of the incident charge particle, in the presence of large field gradients, is accelerated until it strikes a gas molecule. With the repetition of this process number of free electrons increases exponentially. Simultaneously, space charge and diffusion result in a conical growth of the electron avalanche. This process continues until the space charge field at the head of the avalanche is of the order of external field. Positive-ion column left behind the head of avalanche produces photons resulting from the recombination. In turn more electrons are emitted from photo production. These electrons are rapidly accelerated into the space charge column as the space charge field and the external field act approximately in the same direction. This process results in further avalanches, and ultimately into self propagating streamers in the gap between the electrodes providing a highly conducting path through which a major discharge takes place. The initial electron avalanche lasts for 10^{-7} to 10^{-6} sec while the last part following the formation of streamers takes place in times of the order of 10^{-10} to 10^{-9} sec.

1.15.3 Detection Efficiency

1.15.3.1 Detection Efficiency for a Single Particle

For minimum ionizing particle at atmospheric pressure, about 8 ion pairs are formed per cm in Helium, 25 in Neon and 40 in Argon. According to Streamer theory, a

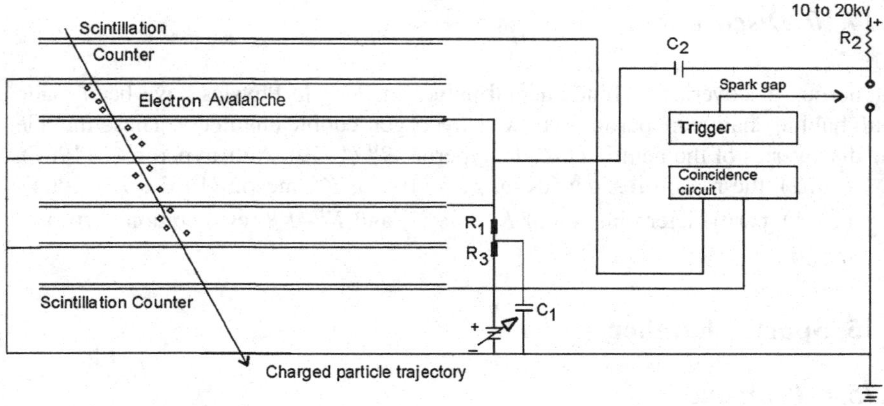

Fig. 1.33 Block diagram of Spark chamber

single free electron leads to a discharge. The detection efficiency under favorable conditions may be close to unity. However, for a number of reasons the efficiency is lowered. Factors which contribute to lowering of efficiency are:

i. Statistical fluctuations in ionization density specially at reduced pressure.
ii. Gas impurities.
iii. Delay in the application of high voltage pulse.
iv. Poor rise time of high voltage pulse.
v. Loss of energy from one gap by another gap not electrically decoupled.
vi. Loss of energy by one or more extra sparks in the same gap.

1.15.3.2 Multiple Track Efficiency

In many experiments conducted with spark chambers specially in high energy physics it becomes necessary to detect correlated particles. When two or more sparks are in the same gap, efficiency may decrease as the first spark discharges the supply capacitor, Fig. 1.33, before the other spark(s) develops. In noble gases the breakdown times are reduced enormously with the increase in gap gradient, uniformity of gap width and use of large values for C_2 capacitor (large discharge energies) generally improve the multi-track efficiency, at the expense of some definition of particle trajectory.

1.15.4 Time Resolution

Following the passage of a charged particle electrons and ions produced in the chamber gas recombine in times 10^{-5} to 10^{-4} sec. Electrons unrelated with the event under observation must be removed as rapidly as possible. This is accomplished by applying low D.C. voltage across each gap. At 1 atm of noble gases, electric fields of the order of 100 V/cm will clear 1 cm gap in about 10^{-6} sec. Alternatively, electronegative gases which capture free electrons can be introduced. Clearing times of the order of 10^{-6} sec with the introduction of 1 % SO_2 in neon can be achieved.

1.15.5 Spatial Resolution

Since ionization electrons of energy of several eV are produced, each electron diffuses in the time that elapses between the passage of the particle and the application of pulsed electric field. For typical delay times of 3×10^{-7} s the displacement is 0.8 mm, equal to the rms diffusion distance in neon. The apparent width of the spark is of the order of 1 mm.

1.15.6 Effect of Magnetic Fields

The use of large magnetic fields results in distinct changes in the performance of spark chamber:

i. The electron diffusion rate across the magnetic lines is reduced. Thus, for example, the diffusion length in neon is reduced by a factor of about 2 in a field of 14 kG.
ii. The clearing time for a given electric field is increased.
iii. A transverse displacement velocity perpendicular to both electric and magnetic field causes a displacement away from the true path. In a Multi-gap chamber the displacement reverses its sign in alternate gaps as the polarity of electric field changes. The displacement problem can be eliminated by the use of electronegative gases.

1.15.7 Recovery Time

In the wake of spark discharge, electrons, ions and excited atoms in the gas must be avoided before the chamber is ready for another operation. This recovery time is typically 10 msec. Recovery time is reduced with the use of electronegative gases and clearing fields.

1.15.8 Gas Composition

A small percentage of contamination like oxygen adversely affects the performance of spark chamber. The electronegative impurities are favored for reasons discussed earlier. Polyatomic gases like alcohol absorb ultraviolet arising from spark development. The inclusion of alcohol vapor to argon reduces spurious streamers, improves the quality of sparks and reduces the spark development time.

1.15.9 Chamber Construction

The chamber must be gas tight and electrically insulated.

1.15.10 Tracks Inclined to the Electric Field

For tracks inclined up to 40° to the electric field, sparks can follow the particle trajectory. With gaps up to 50 cm, and high voltage pulses 100 to 200 kV, sparks that following the particle trajectories to within 10 rad have been observed. Employing a magnetic field perpendicular to the electric field, particle momentum can be measured in a single gap.

1.15.11 Applications

1.15.11.1 Counter Controlled

The most important feature of spark chambers is that they can be counter-controlled with time resolutions less than 10^{-6} s. The limit is set not so much by clearing time as by the necessary delays in the photomultiplier transit time, in cables, in the triggering circuit and high voltage pulsing circuits.

1.15.11.2 Flexibility

As a track detector, spark chamber is quite flexible in regard to size, shape and construction material, it has been used as a large-volume track detector of low mass and has been successfully used for cosmic rays investigations in air planes and balloons. On the other hand, multi-ton narrow-gap chambers have been constructed. To enhance the interaction rates, chambers with graphite plates have been employed for nucleon polarization measurements. With the use of heavy metal plates such as brass, iron, lead and tantalum, they have been used as electron and γ-ray detectors. Hydrogen target is furnished with the use of covered polyethylene.

1.15.11.3 Momentum Analysis

Momentum of high energy beam particles or secondaries can be analyzed with great precision. The momentum limit of the system is about 2000 GeV/c—a value which is an order of magnitude higher than that available in the operation of bubble chamber.

1.15.11.4 Analysis of Rare Events

When used in conjunction with scintillation and Cerenkov counters in the telescopic arrangement, spark chambers are amenable for locating rare events against heavy background. In the counter telescopic arrangement excellent spatial and time resolution of high energy particle interactions can be achieved. The chamber can be arranged to suite the experimental requirements.

1.15.11.5 Stereographic Photography

The intensity of light from the sparks is so great that stereographic photography is feasible even for a complex assembly of chambers.

1.15.12 Disadvantages

When narrow-gap chambers are used for analysis of low momentum secondaries or for analysis of decay products at rest or large angle decays in flight, the trajectories are inaccurately measured specially for particles at grazing angles to the electrodes.

1.15.13 Discoveries

Spark chamber has been used for the electron momentum measurement in muon decay (1969). Distinction between muon neutrino (ν_μ) and electron neutrino (ν_e) was first established with the use of spark chambers by Schwartz, Lederman and Steinberger [2]. In 1964, Christenson et al. [3] observed CP violation ($\sim 2 \times 10^{-3}$) employing spark chamber in conjunction with Cerenkov counter.

In 1970, e^+e^- ring (Mark I) was successfully constructed by SLAC-LBL collaboration. The heart of the detector was a spark chamber, a multi-purpose large solid angle magnetic detector which worked in conjunction with time-of-flight counters for particle velocity measurements, shower counters for photon detection and electron identification, and proportional counters embedded in iron absorber slabs for muon identification. Experiments conducted in 1974 with Mark I confirmed the results of quark-parton model. Another important discovery was that of J/ψ resonance. Also, the decay mode $\psi' \rightarrow \psi\pi^+\pi^-$ was discovered by SLAC-LBL collaboration.

Further the discovery of Neutral Currents important for the electro-weak theory was made possible with the use of spark chamber. The neutrino detector at the Fermi Lab used 8 spark chambers and 16 liquid scintillator segments.

1.16 Calorimeters

1.16.1 Introduction

A device called calorimeter is a powerful and indispensable tool for the detection and energy measurement of very high-energy particles. In passing through the large mass of the detector, a particle is absorbed and produces secondaries which in turn produce tertiaries and so on. Thus, the particle energy is degraded through various processes and ultimately a substantial amount is lost to the calorimeter through ionization and excitation processes. The energy of the primary particle is determined by measuring the energy deposited in the ionization and excitation processes. Unlike other devices, like bubble chambers or ionization chambers, the particles are no longer available for further study. For this reason, calorimeters act as 'destructive' detectors. These devices permit the energy measurement of neutral hadrons (strongly interacting particles) as well as charged hadrons with great precision as the fractional energy resolution goes as $E^{-1/2}$, the precision at high energies (10–100 GeV) being comparable with what can be achieved from magnetic deflection. The position coordinates of the secondary particles may also be recorded under favourable conditions.

The characteristics of a calorimeter are determined by the nature of the dominant energy loss of the particles. For electrons and γ-rays, the energy loss is dominated by electromagnetic process—bremsstrahlung, pair production and Compton scattering. For hadrons, such as nucleons pions and kaons, nuclear interactions are primarily responsible for energy degradation. Thus Calorimeters fall into two categories—those which measure electron and γ-ray energies through electromagnetic processes and those which measure the energies of hadrons through nuclear interactions.

1.16.2 Electromagnetic Calorimeters

In the case of high energy electrons and photons a multiplicative process in the form of a cascade shower occurs, due to the combined effect of bremsstrahlung and pair production. A primary electron will radiate photons, which materialize to electron–positron pairs, which radiate and produce fresh pairs, the number of particles increasing exponentially with depth of the medium. The development and degradation of shower can be explained by a simplified model. Let the primary

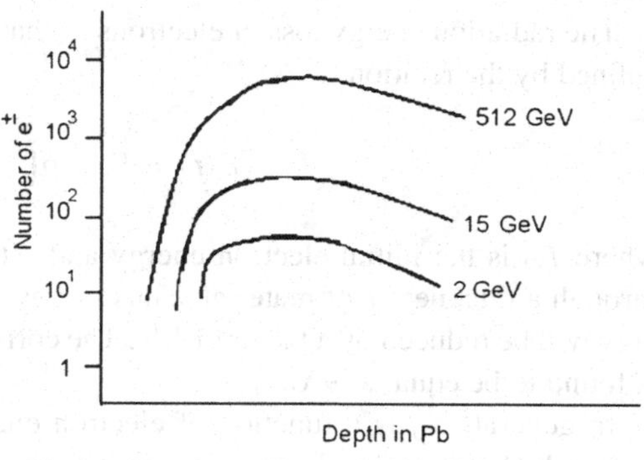

Fig. 1.34 The distribution of particles with depth in Pb for three primary energies

electron of energy E_0 after traversing one radiation length X_0, radiate half the energy, $E_0/2$ as one photon. Let after the next radiation length the photon convert to an electron–positron pair, each receiving half of the energy, i.e. $E_0/4$, and let the original electron further radiate a photon carrying half of energy $E_0/4$. Thus after two radiation length there will be a photon of energy $E_0/4$, two electrons and one positron, each of energy $E_0/4$. It follows that after t radiation lengths there will be $N = 2^t$ particles, with photons, electrons and positrons approximately equal in number. Neglecting ionization loss and the dependence of radiation and pair production cross-section on energy, the energy per particle at depth t will then be $(t) = E_0/2^t$. This process continues until $(t) = E_c$, the critical energy when ionization loss at once becomes important and no further radiation is possible. The shower will thus reach a maximum and cease abruptly. The maximum will occur at the depth

$$t = t_{\max} = \frac{\ln(E_0/E_c)}{\ln 2} \tag{1.60}$$

The number of particles at the maximum will be

$$N_{\max} = \exp[t_{\max} \ln 2] = \frac{E_0}{E_c} \tag{1.61}$$

The total integral tract length of charged particles (in radiation length) in the whole shower will be

$$L \simeq \frac{E_0}{E_c} \tag{1.62}$$

The shower particles continue to multiply until maximum number of particles is reached and the average particle energy is no longer high enough to continue the multiplication process. Beyond this point the shower decays as the particles lose energy by ionization and the photons by Compton scattering. The shower is characterized by an initial exponential rise, a broad maximum, and a gradual decline, see Fig. 1.34. Equations (1.60), (1.61) and (1.62) show

i. a maximum at a depth increasing logarithmically with primary energy E_0
ii. the number of shower particles at the maximum being proportional to E_0
iii. a total track-length integral being proportional to E_0.

The radiation energy loss of electrons is characterized by the radiation length X_0 defined by the relation.

$$E(t) = E_0 \exp\left(-\frac{t}{X_0}\right) \tag{1.63}$$

where E_0 is the initial electron energy and $E(t)$ the average energy after passing through a thickness t of material. Thus in traveling a thickness X_0 the electron energy will be reduced by a factor of $1/e$. The corresponding length for pair production is found to be equal to $9X_0/7$.

In general, X_0 is a function of electron energy and the nature of the absorber material. However for electron and photon energies above 1 GeV, X_0 is practically independent of energy.

The dependence of X_0 on the atomic number Z and mass number A is expressed by the formula

$$X_0\ \left(\mathrm{g\,cm^{-2}}\right) \simeq 716 \frac{A}{Z^2} \ln\left(183 Z^{-1/3}\right) \tag{1.64}$$

Since the logarithmic is a slow varying function, X_0 decreases as Z^{-2} so that calorimeters employ high-Z materials such as lead in order to minimize the overall size.

The multiple Coulomb scattering of electrons and positrons causes a radial spread of the shower from its axis (AAK, 1, 1). The processes of bremsstrahlung and pair production being predominantly peaked in the forward direction at high energies do not contribute significantly to the lateral (radial) spread. The radial spread is determined by R_M the Moliere unit which is defined by

$$R_M = (21\ \mathrm{MeV})\left(\frac{X_0}{E_c}\right) \tag{1.65}$$

where E_c is in MeV. About 90 % of the shower energy resides within R_M which is $\sim 3X_0$ for Pb.

In order to use compact electromagnetic detectors materials of high Z and small X_0 are used so that the shower may be contained within a small volume. Lead glass (55 % PbO, 45 % SiO_2) is an important material for electromagnetic shower detectors. Cerenkov light from relativistic electrons is used to measure the shower energy.

The energy resolution is mainly determined by the fluctuations of the energy deposited in the active material. The resolution is determined by the fluctuations $\sqrt{N}$ in the number N of particles at maximum, where $N \propto E$.

$$\frac{\Delta E}{E} = \frac{0.05}{\sqrt{E\ (\mathrm{GeV})}} \tag{1.66}$$

These calorimeters are usually built of layers of absorbing material sandwiched by layers of detectors such as scintillator on ionization detector.

1.16.3 Hadron-Shower Calorimeters

A hadron shower is generated where an incident hadron suffers an inelastic collision producing secondary hadrons mostly pions ($\pi^{\pm}, \pi^{0}$) which in turn lead through inelastic collisions tertiary generation of hadrons and so on.

Here the nuclear absorption length λ_0 is the corresponding parameter in the electromagnetic calorimeters and is given by

$$\lambda_0 = \frac{1}{N\sigma} = \frac{A}{N_{av}\rho\sigma_i} \tag{1.67}$$

where N is the number of nuclei/unit volume, N_{av} is the Avogadro's number, ρ is the density, A is the mass number, and σ_i is the inelastic cross-section. σ_i and hence λ_0 vary with energy near nucleon resonances but above 2 GeV is substantially constant. The scale of longitudinal development is determined by the nuclear absorption length λ_0 ranging from 80 g cm^{-2} for C through 130 g cm^{-2} for Fe to 210 g cm^{-2} for Pb. By comparison the scale is much larger than the radiation length for heavy elements, $X_0 = 1.76$ cm for Fe and 0.56 for Pb.

For this reason hadron calorimeters have larger dimensions than for electromagnetic calorimeters. As an example an iron-scintillator sandwich has typically dimensions of order of 2 m and 0.5 m respectively.

In an electromagnetic cascade the bulk of energy is ultimately transformed into ionization which shows up as 'visible' (observed) energy. In the case of hadrons, however 30 % of the energy is unobserved as it goes into nuclear fragments, production of slow neutrons, and pion neutrinos which escape from the calorimeter. A large fraction of the energy can be recovered by introducing layers of U^{238} into the calorimeter. Neutrons in the range of 1–10 MeV, typical for nuclear breakup cause fission in the U^{238} nuclei and their energy is thus converted into charged particles whose ionization is measured.

The energy resolution of a hadron calorimeter is usually inferior than the electromagnetic calorimeter. This is due to much greater fluctuations in the development of the shower. The production of π^0's in the early stage leads predominantly to electromagnetic cascade, of that of $\pi^{\pm}$'s predominantly to hadronic cascade, contributing substantially to the variation of energy that is deposited.

For hadron calorimeter

$$\frac{\Delta E}{E} \simeq \frac{0.5}{\sqrt{E}\ (\text{GeV})} \tag{1.68}$$

where E is the incident particle energy.

The lateral shower size is such that ~95 % of the energy is deposited within a cylinder of radius λ_0, the mean free path.

Hadron calorimeters, similar to electromagnetic calorimeters, are constructed from a stack of alternate layers of absorbing material, such as iron or lead, and detectors such as scintillators or proportional counters.

In high-energy physics, typical experiments involve simultaneous detection, measurement and identification of several particles, both charged and neutral from each interaction that occurs. It is usual to employ several types of detecting techniques, such as drift chambers for track position as well as ionization information, time of flight counters, muon detectors, electromagnetic and hadronic calorimeter etc. in a single assembly called a large hybrid detector, weighing 1000–10,000 tons.

1.17 Comparison of Various Experimental Techniques Used in High Energy Physics

Technique	Spatial Resolution	Time Resolution	Target	Interaction Rate	Triggable	Solid Angle
1. Scintillation Counter	$\geq$1 cm	$\approx 10^{-9}$ s	Separate target	10^6/sec	Yes	$\ll 4\pi$
2. Cerenkov Counter	small	$\approx 10^{-9}$ s	Separate target		Yes	$\ll 4\pi$
3. Photographic Emulsion	1 μm	None	Ag, Br, H, C, N, O	MFP 30 cm	No	4π
4. Cloud Chamber (Expansion)	200–500 μm	100 ms	Mixture of gases	One interaction μm per 100–1000 pictures	Yes	4π
5. Cloud Chamber (diffusion)	200–500 μm	Continuously sensitive	Mixture of gases	One interaction per 100–1000 pictures	Yes	$< 4\pi$
6. Spark Chamber	300 μm	10^{-6} s	Separate target	20 triggers/sec	Yes	$\ll 4\pi$
7. Bubble Chamber	30 μm	$\approx$1 ms	H_2, $D_2C_3H_8$, CF_3Br, Xe	1 interaction per picture	No	4π

1.18 Questions

1.1 Why ionization chambers with cylindrical geometry are to be preferred to plane electrode configuration?

1.2 All the G.M. tubes have plateau with a small upward slope. Explain.

1.3 G.M. tubes with halogen quenching have much longer life time than those with organic quenching vapors. Why?

1.4 What happens if the ionization potential of quenching gas is greater than that of the main gas? Explain.

1.5 What are 4π proportional counters?

1.6 What type of counter would you use to count α's against a background of β's and γ-ray ?

(a) pulsed ionization chamber
(b) proportional counter
(c) G.M. counter?

1.7 Show that the capacitance of a cylindrical ion chamber is, $C = \frac{2\pi\varepsilon}{\ln(b/a)}$.

1.8 G.M. counters were taken to high altitudes to study cosmic ray studies. When they passed through Van Allen Radiation belts, they had stopped counting. Explain.

1.9 In a surface barrier detector the energy resolution of particles or radiation is much better than in an ionized chamber. Explain.

1.10 Why are the surface barrier detectors thus called?

1.11 Why is germanium preferred to silicon for low energy γ-ray spectroscopy?

1.12 When P-31 is irradiated by slow neutrons in a nuclear reactor, it is converted into radioactive phosphorus P-32 ($T_{1/2} = 14.5D$). What should be the optimum time of irradiation so that a substantial amount of radioactive phosphorus is produced: (i) 1 week. (ii) 2 weeks. (iii) 10 weeks. (iv) 1 year. (v) 2 years?

1.13 The n-p scattering is known to be isotropic in the C.M. system at neutron energy of 3 MeV. What can you say about the energy distribution of recoil protons in Laboratory system?

1.14 Why must photomultiplier tubes be cooled to low temperatures in experiments with scintillation counters?

1.15 What is the difficulty in the use of inorganic phosphors with a large value of refractive index?

1.16 Name three sources of background radiation experienced by an unshielded scintillator.

1.17 If $E_\gamma > 2mc^2$ (threshold for e^+–e^- pair production) two new escape peaks are observed. Locate them.

1.18 What is a counter telescope arrangement?

1.19 What is time of flight method?

1.20 What type of scintillator is used for the detection of a few MeV neutrons: (i) Organic? (ii) Inorganic?

1.21 What is the dependence of velocity resolution $\frac{\partial\theta}{\partial\beta}$ on the intensity of light emitted in the Cerenkov counter?

1.22 The back-end of the cylindrical radiator of a Cerenkov counter is painted black. Explain.

1.23 How do the chance coincidences compare for scintillation counters and Cerenkov counters?

1.24 What is the function of gelatin in photographic emulsions?

1.25 In what respect photographic emulsions are superior than other detectors in respect to (i) compactness? (ii) sensitivity? (iii) scanning? (iv) spatial and angular resolution? (v) a study of elementary interactions?

1.26 Name three important discoveries that were made using photographic emulsion technique.

1.27 It is difficult to make curvature measurements in magnetic field in photographic emulsions. Explain.

1.28 Noble gases are generally used in expansion cloud chambers. Explain

1.29 What is the function of a multi-plate expansion cloud chamber?

1.30 In an expansion cloud chamber a small temperature gradient is maintained from top to bottom. Explain.

1.31 Name three properties for the choice of liquid vapor in the operation of a diffusion cloud chamber.

1.32 What kind of gases can be used at one or two atmospheres in a diffusion cloud chamber?

1.33 How do the re-cycling times compare in expansion cloud chamber and diffusion chamber?

1.34 Name two important discoveries made with diffusion cloud chamber.

1.35 In what way is the study of neutral unstable particles through their decays more convenient in bubble chamber than in emulsions?

1.36 Name two liquids which are used for the detection of γ-rays in a bubble chamber.

1.37 Name three important discoveries made with bubble chamber.

1.38 Many new accelerators are of colliding beam type. Are the bubble chambers useful in this area as well?

1.39 Which of the following devices are operated counter-controlled?: (i) Expansion cloud chamber. (ii) Bubble chamber. (iii) Spark chamber. (iv) Emulsions?

1.40 Name two important discoveries made with Spark chamber.

1.41 Why are hadron calorimeters inferior than electromagnetic calorimeters?

1.19 Problems

1.1 An ionization chamber is connected to an electrometer of capacitance 0.5 μF and voltage sensitivity of 4 divisions per volt. A beam of α-particles causes a deflection of 0.8 divisions. Calculate the number of ion pairs required and the energy of the source of the α-particles (1 ion pair requires energy of 35 eV, $e = 1.6 \times 10^{-19}$ Coulomb)
[Ans. $6.25 \times 10^5, 21.9$ MeV]

1.2 The dead time of a G.M. counter is 100 μs. Find the true counting rate if the measured rate is 10,000 counts per min.
[Ans. 10,169 per minute]

1.3 The dead time of a counter system is to be determined by taking measurements on two radioactive sources individually and collectively. If the pulse counts over a time interval t are, respectively, N_1, N_2 and N_{12}, what is the value of the dead time?

1.4 A certain G.M. counter has a dead time of 150 μs after each discharge, during which a further particle entering is not detected. If it's true counting rate is 32, 432 cpm, what is the expected observed counting rate?
[Ans. 30,000 cpm]

1.5 Suppose 200 mg of gold ($_{79}Au^{197}$) foil are exposed to a thermal neutron flux of 10^{12} neutrons cm^{-2} sec in a reactor. Calculate the activity and the number of atoms

of Au^{198} in the sample at equilibrium. (Thermal neutron activation cross section for Au^{198} is 98 barns and half-life for Au^{198} is 2.7 h.)
[Ans. 3.08×10^7 dis/sec: 2.24×10^9]

1.6 A beam of neutrons of kinetic energy 0.29 eV, traverses normally a foil of ${}_{92}U^{235}$ of thickness $0.05\ \text{kg m}^{-2}$. Calculate the attenuation of neutron beam by the foil ($\sigma_s = 2 \times 10^{-30}\ \text{m}^2$, $\sigma_a = 7 \times 10^{-27}\ \text{m}^2$, $\sigma_f = 2 \times 10^{-26}$).
[Ans. 91.4 %]

1.7 A S^{35} containing solution had a specific activity of 1 m. Curie per ml. A 20 ml sample of this solution mass was assayed.

(a) in a Geiger-Muller counter, when it registered 1500 cpm with a background count of 600 in 5 min; and
(b) in a liquid scintillation counter, where it registered 8200 cpm with a background count of 300 in 5 min.

Calculate the efficiency as applied to the measurement of radioactivity and discuss the factors responsible for the difference in efficiency of these two types of counter.
[Ans. 3.8×10^{-8}; 18.3×10^{-8}]

1.8 A counter arrangement is set up for the identification of particles which originate from bombardment of a target with 5.3 GeV protons. The negatively charged particles are subjected to momentum analysis in a magnetic field, which permits only those having momentum $p = 1.19$ GeV/c (±2 %) to pass through into a telescope system comprising two scintillation counters in coincidence with a separation of 12 m between the two detectors. Identify the particles, whose time of flight in the telescope arrangement was determined as $t = 51 \pm 1$ ns. What would have been the time of flight of π^- mesons?
[Ans. p^-, 941 MeV; 40 ns]

1.9 The peak resolution (full width at half maximum) of a germanium counter is 1.07 and 1.96 keV for γ-rays of energy 81 and 272 keV, respectively. Calculate the resolution at 662 keV.
[Ans. 3.06 keV]

1.10 The 661 keV -rays from a radioactive source ^{137}Cs are observed using a scintillation spectrometer and the pulse height spectrum shows a photo peak at 21.4 V. Determine the γ-ray energies from ^{60}Co if the photo peaks occur at 37.9 V and 43.0 V.
[Ans. 1.17 MeV, 1.328 MeV]

1.11 A sodium iodide crystal is used with a ten-stage photomultiplier to observe protons of energy 6 MeV. The phosphor gives one photon per 100 eV of energy loss. If the optical collection efficiency is 50 % and the conversion efficiency of the

photo-cathode is 5 %, calculate the average size and the standard deviation of the output voltage pulses when the mean gain per stage of the multiplier is 3 and the collector capacity is 10 pfs.
[Ans. 1.42 V]

1.12 A 400-channel pulse-height analyzer has a dead time $\tau = (17 + 0.5K)$ μs when it registers counts in channel K. How large may the pulse frequencies become if in channel 100 the dead time correction is not to exceed 10 %? Repeat the calculations for the channel 400.
[Ans. 1500/s; 460/s]

1.13 An anthracene crystal and a 12-stage photomultiplier tube are to be used as a scintillation spectrometer for β-rays. The phototube output circuit has a combined capacitance of 45 pf. If an 8-mV output pulse is desired whenever a 55-eV beta particle is incident on the crystal, calculate the electron multiplication required per stage. Assume perfect light collection and a photo-cathode efficiency of 5 %. (Assume 550 photons per beta particle.)
[Ans. 2.57]

1.14 In an experiment using a Cerenkov counter, one measures the kinetic energy of a given particle species as $E(\mathrm{kin}) = 420$ MeV and observes that the Cerenkov angle in flint glass (refractive index $\mu = 1.88$) is $\theta = \arccos(0.55)$. What particles are being detected (calculate their mass in m_e units).
[Ans. $280.5m_e$, Pion]

1.15 Calculate the number of Cerenkov photons produced by a particle travelling at $\beta = 0.95$ in water ($\mu = 1.33$) in the response range (3500–5500 A). Assuming 90 % light collection and 6 % photo-electrode conversion efficiency, what photomultiplier output signal would be produced?

1.16 A proton of kinetic energy 40 keV enters a region where there is uniform electric field of 4×10^4 V/m, acting perpendicularly to the velocity of the proton. What is the magnitude of a superimposed magnetic field that will result in no deflection of the proton?
[Ans. 144 gauss]

1.17 A singly charged particle of speed $0.3c$ gives a track of radius of curvature 100 cm under a uniform normal magnetic field of 1100 gauss. Calculate the rest mass of the particle, in terms of the electron mass m_e and identify it.
[Ans. $M = 205m_e$; It is a muon]

1.18 A fast (extremely relativistic) electron enters a capacitor at an angle α as shown in Fig. 1.35. V is the voltage across the capacitor and d is the distance between plates. Give an equation for the path of the electron in the capacitor.
[Ans. $y = x\tan\alpha - \frac{1}{2}Vx^2/mc^2d\cos^2\alpha$]

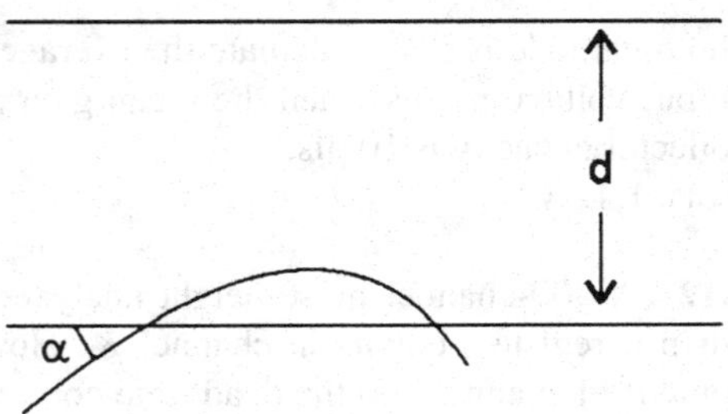

Fig. 1.35 Electron moving across a capacitor

1.19 A particle of known charge but unknown rest mass is accelerated from rest by an electric field E into a cloud chamber. After it has traveled a distance d, it leaves the electric field and enters a magnetic field B directed at right angles to E. From the radius of curvature of its path, show that the rest mass of the particle and the time it had taken to traverse the distance d, are given by $qB^2r^2/2Ed$; $t = Br/E$.

1.20 It is planned to use a mono-energetic k^- beam to perform reaction experiments with hydrogen target. Assuming that the beam transport system must have a minimum length of 30 m, calculate the minimum momentum of a kaon beam such that 5 % of the kaons accepted by the transport system arrive at the hydrogen target ($\tau = 1.2 \times 10^{-8}$ sec, $m_k = 494$ MeV/c^2).
[Ans. 1646 MeV/c]

References

1. F.S. Goulding, Y. Stone, Science **170**, 280 (1970)
2. Schwartz, Lederman, Steinberger (1962)
3. Christenson et al. (1964)

Chapter 2
High Energy Accelerators

2.1 Introduction

Accelerators are machines which accelerate charged stable particles like protons, deuterons, alpha particles, heavy ions and electrons to high energies. In 1920s the only way to study nuclear reactions was to use α-particles of a few MeV from radioactive substances as bombarding particles, as in the discovery of neutron in 1932.

2.2 Electrostatic Accelerators

The simplest machines, Van de Graff accelerator and Cockcroft-Walton generator used dc electric field for the acceleration. The acceleration takes place in one stroke due to drop of a huge electric potential. The highest energy achieved was 20 MeV protons. The first nuclear reaction, $^{+7}\text{Li} \to {}^4\text{He} + {}^4\text{He}$, was observed by Rutherford (1932) at the University of Cambridge, Cavendish Laboratory using Cockcroft-Walton generator which produced 300 keV protons. The Van de Graff accelerator gives current output of the order of μA, while the Cockcroft-Walton gives mA. The Van de Graff generator is free from the ac ripple. The terminal voltage is extremely stable, being constant within $\pm 0.1\ \%$, which is quite important for the study of nuclear reactions. The low-energy nuclear structure studies have been carried out with the aid of Van de Graff generator. Also this machine has been employed for the production of neutrons through the reactions caused by accelerating deuterons and bombarding them against target nuclei. Now a days this machine is mainly used as an auxiliary accelerator for ions to be further accelerated in larger machines.

2.3 High Energy Accelerators

The development of nuclear and particle physics largely depended on the availability of high energy accelerators until early 1950s. The only source of high energy

A. Kamal, *Particle Physics*, Graduate Texts in Physics,
DOI 10.1007/978-3-642-38661-9_2, © Springer-Verlag Berlin Heidelberg 2014

particles, mainly protons, was in the form of cosmic rays which lead to important discoveries such as positron, muon, pion, strange particles etc. However, it was not possible to study high energy nuclear interactions under controlled conditions.

With the introduction of high energy accelerators it became possible to produce beams of monoenergetic particles of specific energy to study their interactions in detail with the aid of a variety of radiation detectors.

High energies, say ~200 MeV corresponding to de Broglie wavelength of ~1 fm are needed for exploring the charge distribution in nuclei as in the electron scattering experiments. Energy in the range 1–6 GeV is needed for the production of many elementary particles and much greater energies are required for the observation of $W^{\pm}$ and Z^0 particles. To this end Lawrence and Livingstone built the 184 inch synchro-cyclotron at Berkley (1948), which produced pions, the first proton synchrotron called cosmotron was completed in 1952 designed to produce strange particles at 3 GeV, the Bevatron (1954) designed at Berkley to produce proton-antiproton pairs at 6.4 GeV, the SPS (super proton synchrotron) which produces protons of 400 GeV at CERN and the Tevatron at the Fermi National Accelerator Laboratory near Chicago. The last one is a proton synchrotron with 1 km radius and yields protons of 1000 GeV or 1 TeV.

2.4 Cyclic Accelerators

In the case of electrostatic accelerators particles gain energy in one stroke having dropped through huge potential difference. A vast improvement results if a high frequency ac be employed and a bunch of particles be permitted to receive succession of acceleration kicks. This is achieved by arranging the particles to pass through a cavity fed with radio frequency power and phased such that the particle is accelerated as it passes through. In practice, a single passage through a cavity can produce only a small acceleration. However, a particle on passing many times through one or more cavities would receive significant amount of energy. A guide magnetic field normal to the field constrains the ions to move in a circular path. Such machines are named cyclic accelerators.

2.5 Cyclotron Accelerator

This machine is used to accelerate particles to non-relativistic energies, say 25 MeV for protons. The ion source is located centrally between the two hollow metal "dees" Fig. 2.1. The dees form part of a radio frequency oscillator. The dees are thus called because of their shape. The frequency of the potential difference applied to the gap between the dees must be such that particles are accelerated each time they cross the gap. The dee assembly is in the vacuum and is placed between poles of a large electromagnet.

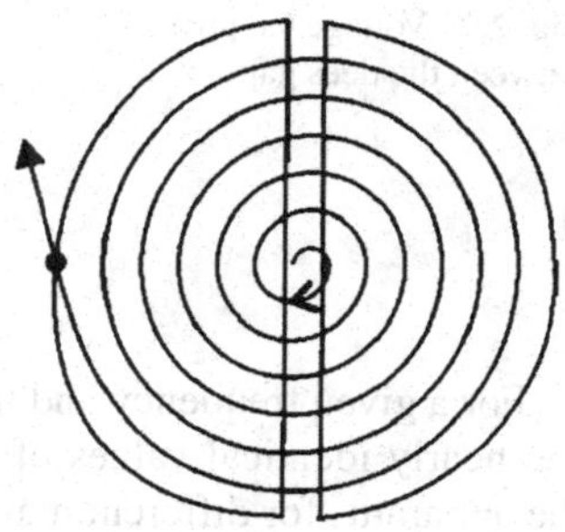

Fig. 2.1 Particle moving in dees assembly

Equating the centripetal force to the magnetic force

$$F = \frac{mv^2}{r} = qvB \tag{2.1}$$

where we have assumed that $\vec{B}$ is perpendicular to the plane of the orbit of instantaneous radius r.

The time necessary for semicircular orbit

$$t = \frac{\pi R}{v} = \frac{\pi m}{qB} \tag{2.2}$$

The frequency of the ac voltage ought to be

$$\nu = \nu_0 = \frac{1}{2t} = \frac{qB}{2\pi m} \tag{2.3}$$

which is called the cyclotron frequency or cyclotron resonance frequency for a particle of charge q and mass m moving in a uniform field B. Observe that the resonance frequency is independent of R, the radius of the orbit.

When the particles are inside the dees they are in a free-field region and so they feel no electric field and follow a circular path under the influence of magnetic field (Fig. 2.1). However, in the gap between the dees, the particles experience accelerating voltage and gain a small energy each cycle. The velocity occurs at the largest radius R.

$$v_{\max} = \frac{qBr}{m} \tag{2.4}$$

From (2.1), the momentum is given by

$$p = mv = qBr \tag{2.5}$$

or,

$$cp = cqBr$$

For Protons $cp = 3 \times 10^8 \times 1.6 \times 10^{-19} Br = 4.S \times 10^{-11} Br$ Joule or

$$p = 0.3Br \text{ GeV/c} \tag{2.6}$$

where B is in tesla and r is in metres.

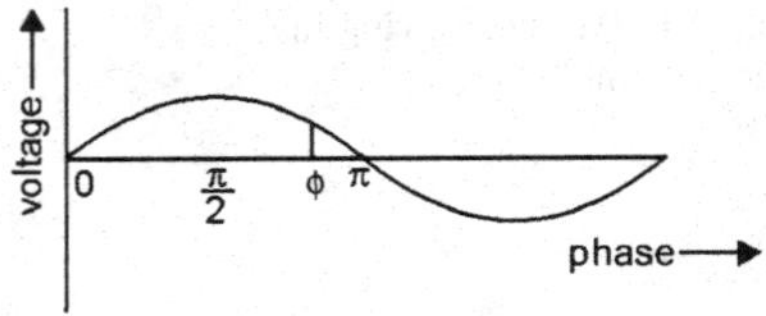

Fig. 2.2 Voltage vs. phase between the dees gap

For a given frequency and field a cyclotron can accelerate different particles having nearly identical values of q/m, such as deuterons and helium ions. By tuning the apparatus for different masses, it is possible to accelerate a number of ions up to carbon and nitrogen.

Equation (2.3) shows that υ_0 is independent of both the radius r and energy T. Thus the frequency of rotation is constant in a uniform, steady magnetic field provided the mass of the particle remains constant, that is at non-relativistic energies. However, energy of the particle T is a function of the radius r.

$$T = \frac{p^2}{2m} = \frac{q^2 r^2 B^2}{2m} \tag{2.7}$$

Each time the ion crosses the gap, it picks up an energy $qV \sin\phi$, where ϕ is the phase of the ac voltage which the ion experiences at each gap and V is the peak potential, Fig. 2.2. The energy gained per turn $= 2qV \sin\phi$, as normally there are two gaps in a cyclotron. In each turn the orbit expands but the orbital frequency υ_0 remains constant. If $t_{\max}$ is the time for the ion to reach the maximum radius from the centre, the number of orbits

$$N = \frac{t_{\max}}{t} = \frac{T}{2qV \sin\phi} \tag{2.8}$$

It is advantageous to build the cyclotron with larger fields and large radii. Notice that the amplitude of the ac voltage between the dees does not appear in any of these expressions. A larger voltage means that the particle receives a larger 'kick' in each orbit but it makes a smaller number of orbits and emerges with the same energy as it would with the smaller voltage.

2.5.1 Energy Limit

It was pointed out that the resonance frequency is independent of both orbital radius and the ion energy, so long as the mass of the particle is constant. When the energy gained is significant after few turns, the particle would start late at the gap compared to a synchronous particle. If this process is continued the particle may get out of the accelerating part of the cycle.

Let ω be the angular velocity of the ion and ω_o the oscillator's angular frequency. The time rate of change of ϕ can be written as

$$\frac{d\phi}{dt} = \omega - \omega_o \tag{2.9}$$

Also,

$$\omega = \frac{Bq}{m}$$

Let the total energy $E = T + mc^2$, where m is the rest mass.

Now,

$$\frac{dE}{dt} = \frac{\omega qV}{\pi}\sin\phi$$
$$\frac{d\phi}{dE} = \frac{d\phi/dt}{dE/dt} = \frac{\pi}{qV\sin\phi}\left(1 - \frac{\omega_o}{\omega}\right) \tag{2.10}$$

$$\therefore \quad \sin\phi d\phi = \frac{\pi}{qV}\left(1 - \frac{\omega_o m}{Bq}\right)dE = \frac{\pi}{qV}\left(1 - \frac{\omega_o E}{Bqc^2}\right)dE \tag{2.11}$$

Expression (2.11) can be integrated to find how ϕ varies with E. The condition for continued acceleration is that ϕ shall remain between the limits 0 and π. The integration is simple if B is constant. Actually, from the view point of focusing B must decrease with increase in r. For the purpose of calculation we suppose that B = constant, with the initial condition, $\phi = \phi_0$, $E = T + mc^2$, where mc^2 is the rest mass energy. Integrating (2.11)

$$\cos\phi_0 - \cos\phi = \frac{\pi}{qV}\left[E - \frac{\omega_o E^2}{2Bqc^2}\right]_{mc^2}^{\tau+mc^2}$$
$$= \frac{\pi}{qV}\left[\left(1 - \frac{\omega_o m}{Bq}\right)T - \frac{1}{2}\frac{\omega_o T^2}{qBc^2}\right] \tag{2.12}$$

The right side of (2.12) when plotted as a function of T is a parabola passing through the origin whose shape can be changed by varying the parameter ω_o/B. This value is adjusted to give maximum output so that it is set at the optimum. The best condition is achieved when the parabola rises and then falls, so that ϕ first increases and then decreases rather than changing always in one direction. Obviously, the widest range is allowed if ϕ starts at zero, goes to its upper limit of π and then returns to zero. The parabola must then cross the T-axis at final energy given by the equation

$$\left(1 - \frac{\omega_o m}{Bq}\right)T_f - \frac{1}{2}\frac{\omega_o T_f^2}{qBc^2} = 0$$

It easily follows that

$$\frac{qB}{m\omega_o} = 1 + \frac{1}{2}\frac{T_f}{mc^2} \tag{2.13}$$

Setting $\phi_0 = 0$ and $\phi = \pi$ in (2.12)

$$\frac{2qV}{\pi} = \left(1 - \frac{\omega_o m}{Bq}\right)T_M - \frac{1}{2}\frac{\omega_o T_M^2}{qBc^2} \tag{2.14}$$

The maximum of the parabola is obtained by maximizing (2.12)

$$\left(1 - \frac{\omega_o m}{Bq}\right) - \frac{\omega_o T_M}{qBc^2} = 0$$

Or,

$$\frac{qB}{\omega_o m} = 1 + \frac{T_M}{mc^2} \tag{2.15}$$

Comparing (2.13) and (2.15) we find

$$T_M = \frac{T_f}{2} \tag{2.16}$$

Substituting (2.16) in (2.14) and solving for T_f, we find

$$T_f = \frac{4qV}{\pi} \pm \sqrt{\frac{16q^2V^2}{\pi^2} + \frac{16qV}{\pi}mc^2}$$

qV will be of the order of 50 keV and $m_0c^2 = 940$ MeV for protons. Hence $mc^2 \gg qV$

$$\therefore \quad T_f \simeq 4\sqrt{qVmc^2\pi}$$

Thus the condition for relativistic limit on cyclotron energies is

$$T \leq 4\sqrt{qVmc^2\pi} \tag{2.17}$$

For protons, $mc^2 = 940$ MeV and for a typical value $qV = 50$ keV, the condition that a conventional cyclotron would work is $T \leq 16$ MeV. The ideal limit set by (2.17) is not realized in practice for several reasons.

i. Because of forced harmonic oscillations, ϕ becomes $\pi/2$ in two or three turns. Hence ϕ_0 becomes $\pi/2$ in spite of the initial conditions.
ii. Focusing (defocusing) effect of electric accelerating field has been ignored.

This turns out to be defocusing over a considerable part of the path. This is usually compensated for by providing positive magnetic focusing by the use of a magnetic gradient extending to the central region. The additional resonance errors are then taken care of by providing a higher dee voltage. Consequently, the final energy limit is usually twice or thrice that given by (2.17)

In these accelerators currents are typically of the order of tens of microamperes which are adequate for detailed studies of nuclear reactions. In fact from 1930s until 1960s when large Van de Graff accelerators became available, the fixed frequency fixed magnetic field cyclotrons played a major role for nuclear structures studies with nuclear reactions.

2.5.2 Fringing Field

As the beam in a cyclotron travels outward toward the edge of the machine the magnetic lines are curved rather than straight vertical lines as in Fig. 2.3. This is called fringing field. It has two-fold effect. One is favorable in that the curvature of

Fig. 2.3 Fringing field

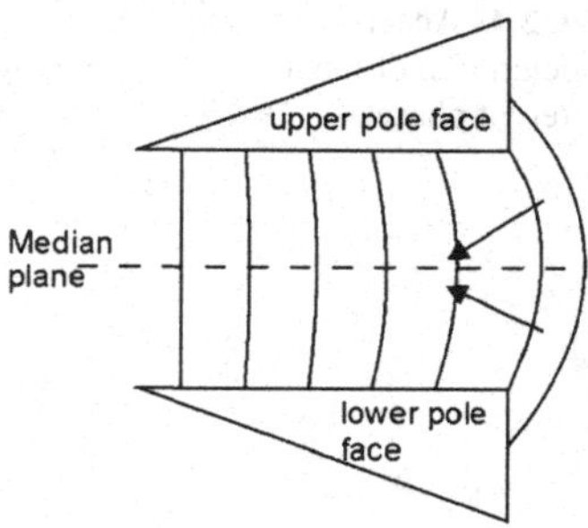

the field gives a net force component toward the median plane to provide focusing and to oppose the tendency of the beam to diverge. The other one is adverse in that the field loses its uniformity and the resonance condition (2.3) can no longer be maintained if the frequency is held constant.

A more serious difficulty arises due to the relativistic behaviour of the accelerated particles. This sets an upper limit to the size of fixed-field, fixed-frequency cyclotron. Thus the maximum energy that can be obtained is about 40 MeV for protons corresponding to $\gamma = \frac{1}{\sqrt{1-\beta^2}} = 1.04$.

Heavy ions like ^{4}He, ^{7}Li, ^{12}C etc. can also be accelerated by this machine. However electrons are not at all suitable as they become relativistic even at nominal energies.

2.6 Synchrocyclotron

It was pointed out that in a cyclotron frequency ν will no longer be a constant of motion if the mass varies as it would at relativistic energies. Consequently, the ions will not be able to cross the gaps at the right time. This difficulty is overcome by the use of a frequency-modulated cyclotron called synchrocyclotron. This consists of modulating the applied frequency, that is imposing a periodic time variation of the oscillator frequency such that the applied frequency matches with the ion frequency through out so that a particle will always be in step with the accelerating field. This is accomplished by using the principle of phase stability.

2.6.1 Phase Stability in Circular Motion

Resonance accelerators operate on the principle of providing successive small accelerations synchronized with the motion of the particles. The synchronization must be maintained for a very large number of accelerations. When the applied frequency is modulated cyclically a short bunch of ions will be accelerated to high energies in each frequency sweep resulting in a sequence of such bursts occurring at the modulation frequency. This is in contrast with the behavior of a conventional cyclotron

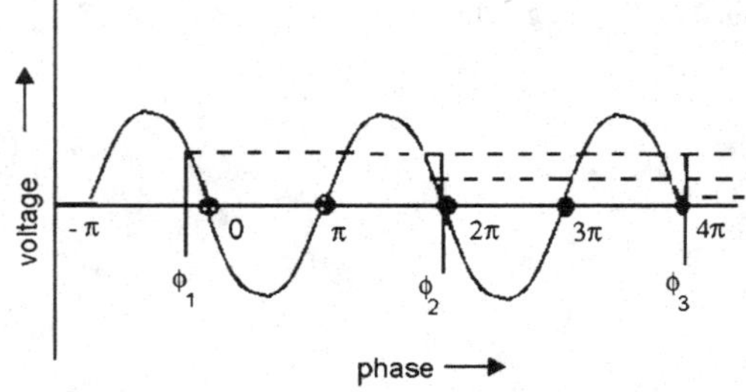

Fig. 2.4 Acceleration and deceleration of ion at different phases

in which ions are accelerated continuously. The reduced effective duty cycle may yield a low average ion output ($\sim$1 %) than in the conventional cyclotron but it has the merit of avoiding the resonance limit due to relativistic effect and permits the ions to be accelerated to much higher energies ($\sim$500 MeV protons). The particles are kept in resonance with the accelerating voltage by means of a stable oscillation in phase about a mean value. This is the principle of phase stability and is exploited in synchro-cyclotron as well as synchrotron. In the synchrocyclotron the magnetic field is kept constant but the frequency of the accelerating electric field is slowly decreased. The radius of the orbit increases, so also the particle energy. The mechanism of phase stability can be explained in the following way. Considering a high frequency ac impressed across the gap of the dees, the frequency being identical with the natural rotational frequency of the ions given by (2.3). A particle which crosses the gap at zero phase ϕ_0 when the electric field is crossing zero in the direction of changing from accelerating to decelerating is indicated by the points $0, 2\pi, 4\pi, \ldots,$ Fig. 2.4. This particle will neither gain nor lose energy. It will be in precise resonance and will continue to rotate in the same orbit with constant frequency. To show that this orbit is stable consider a particle crossing the gap at an early phase such as ϕ_1. In the first few turns this particle will gain energy, the frequency will decrease since

$$\nu = \frac{qB}{2\pi m\gamma} = \frac{qB}{2\pi m(1 + T/mc^2)} = \frac{\nu_0}{1 + T/mc^2} \tag{2.18}$$

where ν_0 is the non-relativistic cyclotron frequency which can be considered as a constant of motion. The reduced frequency will cause the particle to be delayed slightly in subsequent traversals of the gap, indicated by the points $\phi_2, \phi_3, \ldots$. It will continue to migrate in phase until it crosses the gap at zero phase. However, it would have accumulated excess of energy so it will continue its migration in phase in the decelerating part of the cycle, Fig. 2.5. Here the situation is reversed. It starts losing energy; its frequency increases and returns to the zero phase position. This motion represents an oscillation in phase about the phase of the stable orbit which satisfies the resonance condition (2.18). It also represents an oscillation in energy about the equilibrium value. Since the orbit radius changes with energy, it results in radial oscillation about the equilibrium radius. When the particle energy exceeds the equilibrium energy the orbit expands. Conversely at smaller energy the orbit would shrink. The principle of phase stability is based on the existence of phase oscillations centered around the equilibrium values of phase, energy and radius. For

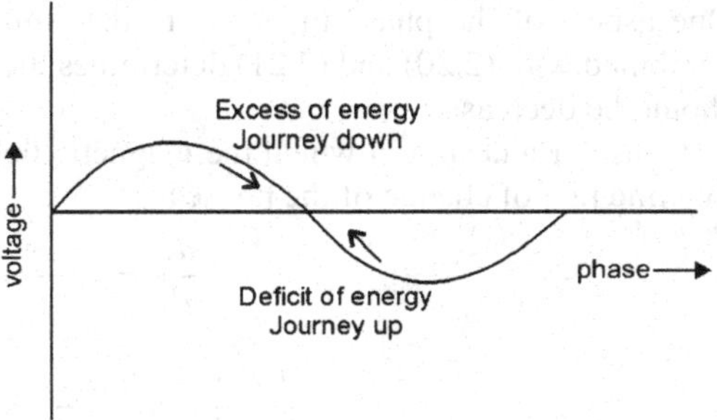

Fig. 2.5 Accelerating and decelerating ion through acquisition and loss of energy

a given magnetic field the particles will bunch and spread around the entire circuit over a phase bracketed by π and $-\pi$.

Each particle migrates slowly forward or backward along the length of the bunch. By a similar argument it can be shown that an ion lying on the accelerating part of the cycle will be completely unstable and would be lost.

A bunch of ions requires many hundreds or thousands of revolutions to complete a phase cycle. The stable orbit with fixed energy may be called a stationary orbit. Consider the effect of a small decrease in frequency of the applied electric field, with the magnetic field kept constant. The lengthening of time cycle, with the particle frequency remaining constant, causes the particles to cross the gap at earlier phases and to acquire energy until it reaches the energy which resonates with the new frequency given by (2.18). An increased energy in the nearly uniform magnetic field implies that the orbit expands. Now if the frequency is decreased slowly and continuously, the particles will follow this change, gain in energy with increase in orbit radius at a steady rate determined by the rate of frequency modulation. In executing several hundred revolutions to complete the phase cycle the particle will gain on an average energy equal to $V_e e$ corresponding to the potential difference across the gap at equilibrium phase.

Let the angular velocity of the ion be ω.

Then

$$\omega = \frac{Bq}{m} = \frac{Bqc^2}{E} \tag{2.19}$$

where $E = T + mc^2$ is the total energy.

Let ω_s be the frequency of synchronous ions

$$\omega_s = \frac{Bqc^2}{E_s} = \frac{Bqc^2}{T_s + mc^2} \tag{2.20}$$

The rate at which ω_s should change with time is determined by the rate at which the energy increases with time due to electric acceleration across the dee gap. This acceleration gives an energy increase $2qV \sin\phi_s$ per turn, where V is the maximum electric voltage across the dee gap and ϕ_s is the phase of synchronous ions. We can then write

$$\frac{2\pi}{\omega_s}\frac{dE_s}{dt} = 2qV \sin\phi_s \tag{2.21}$$

One aspect of the phase theory is to determine the optimum value of ϕ_s which combined with (2.20) and (2.21) determines the rate at which the applied frequency should be decreased

Consider a design in which the magnetic field does not vary. We can then write the time rate of change of the phase as

$$\frac{d\phi}{dt} = \omega - \omega_s \tag{2.22}$$

or

$$\frac{1}{\omega}\frac{d\phi}{dt} = 1 - \frac{\omega_s}{\omega} = 1 - \frac{E}{E_s} \tag{2.23}$$

where we have used (2.19) and (2.20).

Let us have a change of variable

$$\frac{d}{d\alpha} = \frac{1}{\omega}\frac{d}{dt} \tag{2.24}$$

Equation (2.23) then becomes

$$\frac{d\phi}{d\alpha} = 1 - \frac{\omega_s}{\omega} = 1 - \frac{E}{E_s} \tag{2.25}$$

Combining (2.24) and (2.25)

$$\frac{d}{d\alpha} = \frac{1}{\omega_s}\left(1 - \frac{d\phi}{d\alpha}\right)\frac{d}{dt} \tag{2.26}$$

multiplying both sides of (2.25) by E_S^2 and differentiating with respect to α,

$$\frac{d}{d\alpha}\left(E_S^2\frac{d\phi}{d\alpha}\right) = \frac{d}{d\alpha}(E_S^2 - EE_S) \tag{2.27}$$

Combining (2.26) and (2.27) and simplifying using (2.25)

$$\begin{aligned}\frac{d}{d\alpha}\left(E_s^2\frac{d\phi}{d\alpha}\right) &= \frac{1}{\omega_s}\left(1 - \frac{d\phi}{d\alpha}\right)(2E_s - E)\frac{dE_s}{dt} - E_s\frac{dE}{d\alpha} \\ &= \frac{dE_s}{dt}\frac{1}{\omega_s}\left(1 - \frac{d\phi}{d\alpha}\right)E_s\left(1 + \frac{d\phi}{d\alpha}\right) - E_s\frac{dE}{d\alpha}\end{aligned}$$

Assume that the phase and energy deviations from the synchronous values are small. Equation (2.25) shows that $d\phi/d\alpha$ is small so that its square may be neglected.

$$\frac{d}{d\alpha}\left(E_s^2\frac{d\phi}{d\alpha}\right) = \frac{E_s}{\omega_s}\frac{dE_s}{dt} - E_s\frac{dE}{d\alpha}$$

But

$$\frac{1}{\omega_s}\frac{dE_s}{dt} = \frac{qV\sin\phi_s}{\pi}$$

Also,

$$\frac{dE}{d\alpha} = \frac{1}{\pi}qV\sin\phi$$

$$\frac{d}{d\alpha}\left(E_S^2\frac{d\phi}{d\alpha}\right) = E_S\frac{qV}{\pi}(\sin\phi_s - \sin\phi) \tag{2.28}$$

This is the equation of phase oscillations. If α is replaced by t, it is identical with the equation of motion of a pendulum, acted upon by a constant torque. The equilibrium position is at $\phi = \phi_s$ since (2.28) shows that no force acts at that point.

The solution of (2.28) is simplified considerably if small oscillations are considered and define a new variable

$$\theta = \phi - \phi_s \tag{2.29}$$

Since θ is assumed to be small, $\sin\theta \to \theta$ and $\cos\theta \to 1$. Using these approximations in (2.28)

$$\begin{aligned}\frac{d}{d\alpha}\left(E_s^2 \frac{d\theta}{d\alpha}\right) &= E_s \frac{qV}{\pi}.2\cos\left(\frac{\phi_s + \phi}{2}\right)\sin\left(\frac{\phi_s - \phi}{2}\right)\\ &\simeq -E_S \frac{qV}{\pi}\theta\cos\phi_s\end{aligned} \tag{2.30}$$

Using (2.25) and (2.26)

$$\frac{1}{\omega_s}\left(1 - \frac{d\theta}{d\alpha}\right)\frac{d}{dt}\left[E_S^2 \frac{1}{\omega_s}\left(1 - \frac{d\theta}{d\alpha}\right)\frac{d\theta}{dt}\right] = -E_S \frac{qV}{\pi}\theta\cos\phi_s \tag{2.31}$$

$$\frac{d}{dt}\left(E_S^2 \frac{d\theta}{dt}\right) = -\theta \frac{\omega_s^2 E_s q V}{\pi}\cos\phi_s \tag{2.32}$$

Neglecting terms which are squares in $\frac{d\theta}{dt}$ in (2.31)

$$\frac{d^2\theta}{dt^2} = -\theta \cdot \frac{\omega_s^2}{E_s}\frac{qV}{\pi}\cos\phi_s \tag{2.33}$$

Equation (2.33) represents simple harmonic motion.

$$\theta = A_1 \sin\left[\sqrt{\frac{\omega_s^2}{E_s}\frac{qV}{\pi}\cos\phi_s}\,(t - t_1)\right] \tag{2.34}$$

where A_1 and t_1 are the constants of integration. The frequency of phase oscillations is therefore given by

$$\omega_\phi = \omega_s \sqrt{qV\cos\phi_s \pi E_s} \tag{2.35}$$

It can be shown that

$$E_s^2 A_1^2 \omega_\phi = \text{const.}$$

$$\therefore \quad A_1 \propto \frac{1}{E_s\sqrt{\omega_\phi}}$$

$$A_1 \propto \frac{1}{E_s^{3/4}(qV\cos\phi_s)^{1/4}} \tag{2.36}$$

Thus the amplitude of the phase oscillations decreases as the energy increases so that the oscillations are damped. This then means that if the ion survives the first phase oscillations, it will survive all subsequent phase oscillations and be accelerated to high energy.

For large oscillations the small angle approximation (2.29) is not valid. Equation (2.28) shows that the quantity $d^2\phi/d\alpha^2$ changes sign when $\phi = \pi - \phi_s$. For ϕ values larger than this the phase acceleration is no longer restoring so that this is the limit of stability of phase oscillation.

The equation of motion for the largest stable oscillation is obtained by integrating (2.28) and evaluating the constant of integration from the condition

$$\frac{d\phi}{d\alpha} = 0 \quad \text{when } \phi = \pi - \phi_s \tag{2.37}$$

Neglecting the variation of E_s with time and using the identity

$$\frac{d^2\phi}{d\alpha^2} = \frac{d\phi}{d\alpha} \cdot \frac{d}{d\phi}\left(\frac{d\phi}{d\alpha}\right)$$

Equation (2.28) is then transformed to

$$\frac{d\phi}{d\alpha} d\left(\frac{d\phi}{d\alpha}\right) = \frac{qV}{\pi E_s}(\sin\phi_s - \sin\phi)d\phi$$

Integrating both sides:

$$\left(\frac{d\phi}{d\alpha}\right)^2 = \frac{2qV}{\pi E_s}[\phi \sin\phi_s + \cos\phi + c_1]$$

where c_1 is the constant of integration. Using (2.37)

$$c_1 = \cos\phi_s - (\pi - \phi_s)\sin\phi_s$$

$$\left(\frac{d\phi}{d\alpha}\right)^2 = \frac{2qV}{\pi E_s}\left[\phi \sin\phi_s + \cos\phi + \cos\phi_s - (\pi - \phi_s)\sin\phi_s\right]$$

Using (2.25)

$$(E_s - E)^2 = \frac{2qVE_s}{\pi}\left[\phi \sin\phi_s + \cos\phi + \cos\phi_s - (\pi - \phi_s)\sin\phi_s\right] \tag{2.38}$$

Equation (2.38) gives the relation of the energy and phase for the largest stable phase oscillation. If a given deviation of energy from the synchronous energy causes the phase to be larger than the value given by (2.38), the ion will be lost. Conversely, the ion will also be lost if for a given phase the energy deviation is larger than the value given by (2.38). Equation (2.38) permits the determination of the condition for acceptance of ions emitted from the ion source into a stable phase. Let the initial phase by $\phi = \pi/2$ at $T = 0$, inserting this in (2.38)

$$T_s = \pm\sqrt{2qVmc^2\pi}\sqrt{\left(\phi_s - \frac{\pi}{2}\right)\sin\phi_s + \cos\phi_s} \tag{2.39}$$

All ions emitted from the ion source are accepted whose synchronous energy is within the limits given by (2.39). The corresponding electric frequency for which the ions are accepted is obtained by substituting (2.39) in (2.20), i.e.

$$\omega_o = \frac{Bqc^2}{T_s + mc^2}$$

All other ions are lost.

2.6.2 *Number of Orbits*

In the synchrocyclotron as the frequency is slowly decreased, the radius of the synchronous orbit will increase and so also its energy. After each crossing of the gap the decreasing frequency causes particles to appear earlier than that for synchronous orbit. These particles are both accelerated and bunched by the phase-stability effect.

As the accelerations in each crossing of the gap are small many more orbits are required in a synchrocyclotron than in a conventional cyclotron. An approximate estimate of the number of orbits involved is given by the ratio of the time taken by the particles to travel from the centre to the edge of the field to the time taken to execute a single orbit.

$$\text{Number of orbits} = \frac{\text{total time}}{\text{time/orbit}} \sim \frac{\text{orbital frequency}}{\text{modulation frequency}}$$

The time taken for the particles to travel from the centre to the edge is given by the reciprocal of the modulation frequency with which the cyclotron frequency is altered from the maximum to minimum value. As an example for the 184 inch synchrocyclotron at Berkley, California the cyclotron frequency works in the range 36 to 18 MHz, the modulation frequency being 64 Hz. Thus the number of orbits is of the order of 10^5. At the largest radius the frequency is 20 MHz. The protons are extracted with 740 MeV for which the proton mass is 1.8 times the rest mass, the field strength being 2.3 T.

The Berkley synchrocyclotron which was first commissioned in 1946, is the highest energy synchrocyclotron and produces protons with a mean current of the order of 0.1 μA. Of course the current is pulsed and not continuous. Other comparable synchrocyclotrons are located at Dubna in Russia and CERN in Geneva.

The main differences between a conventional cyclotron and synchrocyclotron are:

1. The former gives continuous ion current of the order of tens of μA, while the latter gives pulse current of the order of 0.1 μA. The ions come off as spurts lasting about 50 μsec with a repetition rate of 100 sec^{-1}.
2. The former operates with fixed ac frequency while the latter is frequency modulated
3. In the former the particles may describe orbits of the order of 10^2 before acquiring maximum energy instead of $\sim 10^5$ in the latter. Consequently, in the synchrocyclotron the potential needed across the dees is 10^3 less.
4. In a conventional cyclotron the maximum energy is limited to 40 MeV protons because of relativistic effect while for synchrocyclotron the energy is pushed to about 700 MeV.

2.7 AVF (Azimuthally Varying Field) Cyclotron

An alternative way to boost cyclotron energy is to increase the magnetic field with increasing orbital radius to compensate for the increasing relativistic mass of the

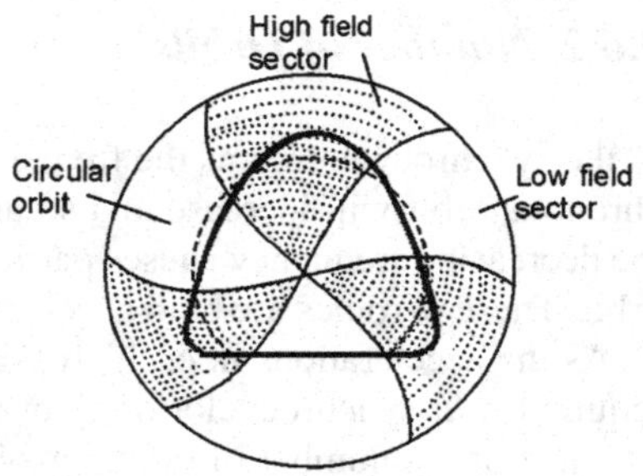

Fig. 2.6 AVF (Azimuthally Varying Field) cyclotron

orbiting particles. However, this has the undesirable effect of defocusing the beam due to the curvature of field lines. This setback is avoided by dividing the magnetic field into sectors of high and low field, Fig. 2.6. This type of cyclotron is called a sector-focusing or AVF (azimuthally varying field) cyclotron. At the boundaries between the high-field and low field sectors, there is an azimuthal component to the field that provides a focus of the beam in the midplane. The focusing action must be stronger than defocusing due to radially increasing field.

The major advantage of AVF cyclotron over synchrocyclotron is the continuous beam capable of producing large beam currents (of the order of 100 μA). These accelerators are mainly used to accelerate protons up to about 500 MeV which upon hitting suitable targets produce large intensities of pion beams. They also find useful applications for studying heavy ion reactions.

Example 2.1 A synchrocyclotron accelerates protons to 400 MeV. Use the data: $B = 20$ kG, $V = 10$ kV and $\phi_s = 30°$ to find

1. The radius of the orbit at extraction
2. Energy of ions for acceptance
3. The initial electric frequency limits
4. The range of frequency modulation

Solution

1.

$$p(\mathrm{GeV}/c) = 0.3BR$$

$$R = \frac{\sqrt{T^2 + 2Tmc^2}}{0.3B}$$

Putting $T = 0.4$ GeV, $mc^2 = 0.938$ GeV, $B = 2.0$ T, we find $R = 1.59$ m.

2. $T_s = \pm\sqrt{2\, eVmc^2\pi}\sqrt{(\phi_s - \frac{\pi}{2})\sin\phi_s + \cos\phi_s}$

Putting

$$eV = 10\ \mathrm{keV} = 0.01\ \mathrm{MeV}$$

$$mc^2 = 938 \text{ MeV}$$
$$\phi_s = 30°$$

we find $T_s = \pm 1.45$ MeV.

3. The initial electric frequency

$$f = \frac{Bqc^2}{2\pi(mc^2 + T_s)}$$

Putting

$$B = 2.0 \text{ T}, \qquad q = 1.6 \times 10^{-19} \text{ Coulomb}, \qquad c = 3 \times 10^8 \text{ m s}^{-1}$$
$$mc^2 = 938 \text{ MeV} = 938 \times 1.6 \times 10^{-13} \text{ J},$$
$$T_s = \pm 1.45 \text{ MeV} = \pm 1.45 \times 1.6 \times 10^{-13} \text{ J}$$

we find $f_1 = 30.509Mc$, $f_2 = 30.604Mc$.

4. The electric frequency for 400 MeV protons ought to be

$$f = \frac{B}{2\pi} \frac{qc^2}{(mc^2 + T)}$$

Putting, $T = 400$ MeV, we find $f = 21.42Mc$.

Thus the range of modulation frequency is

$$30.6Mc\text{–}21.4Mc$$

2.8 Synchrotron

2.8.1 Proton Synchrotron

If a greater energy is to be extracted from a synchrocyclotron or AVF cyclotron then magnets of larger diameter must be employed. Now the magnet is the main component in the construction of these machines. It turns out that the cost of the magnet goes roughly as the cube of energy. This then means that the cost of these machines even for a few GeV energy range would be exorbitantly high. The breakthrough was made with the invention of synchrotron accelerator which consists of an evacuated doughnut (torus) in which the protons are accelerated by RF cavities and constrained to move in circular or nearly circular path by the application of magnetic field normal to the orbit. Both the field B and the RF must increase and be synchronized with the particle velocity as it increases. Thus the name synchrotron. In the synchrotron the particles are maintained in the doughnut immersed in a magnetic field.

The magnet is thus in the form of a ring and need not cover the whole circular area as in the cyclotron. As the magnetic field is applied only at the circumference of the machine the cost of the magnet is enormously reduced.

In a magnetic field of strength B, a particle of charge e moves in a circular arc of radius R with momentum $p = eRB$. The total relativistic energy of the particle is

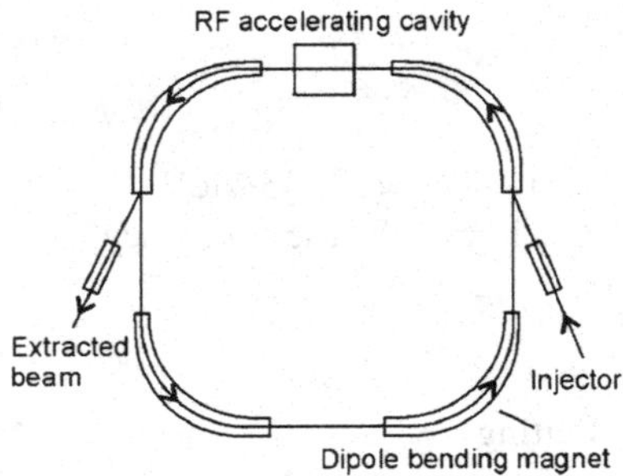

Fig. 2.7 Basic structure of the proton synchrotron

$$\begin{aligned} E &= \sqrt{p^2c^2 + m^2c^4} \\ &= \sqrt{e^2R^2B^2c^2 + m^2c^4} \end{aligned} \tag{2.40}$$

The cyclotron resonance condition (2.3) can be written as

$$\nu = \frac{eB}{2\pi m} = \frac{eBc^2}{2\pi E} = \frac{eBc^2}{2\pi\sqrt{e^2R^2B^2c^2 + m^2c^4}} \tag{2.41}$$

For a given R, (2.41) gives the relationship between B and υ required for synchronization.

For protons which are not ultrarelativistic, it is necessary to increase the frequency as the magnetic field increases. In large machines the problem of employing extremely weak magnetic field at injection is avoided by the use of a linear accelerator which accelerates protons of few MeV energy. The protons are then accelerated to the maximum value over a cycle of a few seconds. Then a new cycle begins. Thus the beam is produced in discrete pulses. The protons while orbiting for injection, acceleration and extraction.

The first proton synchrotron called "Cosmotron" designed to produce protons of energy 3 GeV was completed in 1952 at the Brookhaven National Laboratory, Fig. 2.7. It had the mean radius of 10 m. The injection energy was 3.5 MeV and protons were accelerated with RF which varied from 0.37 to 4.0 MHz under the maximum magnetic field of 1.4 T.

During the acceleration 5×10^{10} protons were produced after completing about 3×10^6 orbits.

Figure 2.7 shows the basic structure of the proton synchrotron, with four dipole sections to bend the beam and one RF cavity to produce the acceleration.

In 1954 a slightly larger proton synchrotron was completed at the Lawrence Radiation Laboratory at Berkley, California. It had a radius of 18 m and maximum field strength of 1.6 T. It was named Bevatron (1 BeV $= 10^9$ eV, which is now called 1 GeV) and was designed to produce an energy of 6.4 GeV and was targeted at the production of proton-antiproton pairs for which the threshold energy is 5.64 GeV in proton-hydrogen collisions.

The highest energy (500 GeV) is produced by the proton synchrotron at the Fermi National Accelerator Laboratory (FNAL) near Chicago as well as by 400 GeV super-proton synchrotron (SPS) at CERN (Centre for European Research Nuclear).

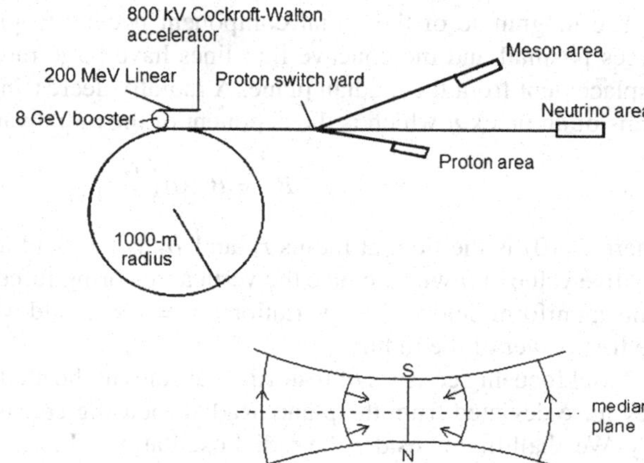

Fig. 2.8 Layout of Fermilab proton synchrotron

Fig. 2.9 Fringing field

The layout of Fermilab is shown in Fig. 2.8. It has an orbital radius of 1000 m. Three auxiliary accelerators used for pre-acceleration, before injection, are sequentially arranged. The order is, 0.8 MeV Cockcroft-Walton, followed by a 200-MeV drift tube linear accelerator and then an 8-GeV "booster" synchrotron. The bending magnets use a peak field of 1.4 T and the RF cavities use a frequency of 53 MHz. The pulses obtained are of roughly 1-sec width every 12 sec and the power required for the magnet is 36 MW.

Introduction of super-conductivity magnets permits the doubling of beam energy that is 1000 GeV or 1 TeV. Hence this machine is also called Tevatron (1 TeV = 10^{12} eV).

2.8.1.1 Betatron Oscillations

The particles orbiting the synchrotron do not travel in ideal circular orbits but make excursions across the circular path in both horizontal and vertical planes. These are called betatron oscillations. They may originate from the natural divergence of the originally injected beam from small misalignments of magnetic field, scattering of particles in the residual gas of doughnut etc. We shall now derive an expression for the frequency of radial (horizontal) oscillation. It is essential that restoring forces must be provided for particles which deviate from the central orbit. In an absolutely uniform magnetic field if a particle makes a small angle with the plane of the orbit, it would describe a helix and the axial displacement would cause the particle to be lost against the side of the called fringing field. In other regions the pole faces can be properly shaped so that the gap length increases and the flux density decreases with increasing radius. Figure 2.9 shows the magnetic field between the pole faces of a magnet and the direction of the restoring forces exerted on charged particles circulating above and below the median plane. On the median plane the radial component of the field is zero. But for other locations both the components B_z and B_r exist.

The magnitude of the radial component needed to provide adequate restoring forces is small and the concave flux lines have large radii of curvature relative to displacement from the median plane. A radially decreasing field can be specified in terms of an index n which is the exponent of the radial variation.

$$B_z = B_z(0)\left(\frac{r_o}{r}\right)^n \tag{2.42}$$

where $B_z(0)$ is the field at radius r_o and B_z is the field at a near by radius r. Any positive value of n will produce the vertical restoring forces. A value of $n = 0$ represents a uniform field with no variations; $n = 1$ is a field which varies inversely with the first power of the radius.

Axial focusing consists of restoring particles to the median horizontal plane when they have deviated from the plane. Radial focusing consists of restoring them radially. We shall first consider the radial oscillations. In the equilibrium orbit ($r = r_o$) the centripetal force is supplied by the magnetic force and

$$\frac{mv^2}{r_o} - evB_z(0) = 0 \tag{2.43}$$

Suppose now the particle is displaced radially in the median plane so that its instantaneous radius is $r = r_o + x$, where $x \ll r_o$. The net outward force acting on the particle is

$$m\ddot{x} = \frac{mv^2}{r_o + x} - B_z ev$$

or

$$\begin{aligned}\ddot{x} &= \frac{v^2}{r_o}\left(1 + \frac{x}{r_o}\right)^{-1} - \frac{e}{m}vB_z(0)\left(\frac{r_o}{r_o + x}\right)^n \\ &= \frac{v^2}{r_o}\left(1 - \frac{x}{r_o}\right) - \frac{v^2}{r_o}\left(1 - \frac{nx}{r_o}\right) \\ \ddot{x} &= -x\omega_o^2(1 - n)\end{aligned} \tag{2.44}$$

where we have used (2.43), and ω_o is the angular frequency for the stable orbit. Equation (2.44) represents simple harmonic motion whose angular frequency is

$$\omega_r = \omega_o\sqrt{1 - n} \tag{2.45}$$

Similarly, it can be shown that for the vertical oscillations the angular frequency is given by

$$\omega_z = \omega_o\sqrt{n} \tag{2.46}$$

We, therefore, conclude that to achieve both radial and vertical focusing the parameter n which characterizes the field variation at the orbit must satisfy the condition $1 > n > 0$. It is seen that both the frequencies are less than ω_o. In the cyclotron radial oscillations are relatively unimportant. In the betatron and synchrotron in which the beam passes through the doughnut the amplitude of both radial and vertical oscillations must be small. The farther n is from zero, the stronger is the vertical restoring force and weaker the radial restoring force. For the special case $n = 0.5$ they are equal and are given by $\omega_r = \omega_z = 0.7\omega_o$.

Coupling can occur between the two transverse modes of betatron oscillations, the radial and the vertical. Coupled resonances occur when $\omega_r/\omega_z = n$ or $\frac{1}{n}$; where $n =$ integer or reciprocal of integer. The values of $n = 0$, 0.1,0.2, 0.5, 0.8, 0.9, 1.0 should be avoided. A typical value of n for synchrotrons is 0.6.

2.8.1.2 Strong Focusing

The amplitude of these oscillations must be kept at minimum not only to avoid losses of particles but also from the point of view of supplying power to the magnet. Taking $z = 0$ at the median plane, it is obvious that $z_{\max} = z(0)/\omega_z$. Since $\omega_z = \omega_o(n)^{1/2}$, a large value of n is required to keep the value of $z_{\max}$ low. However we know that radial stability requires $n < 1$.

This difficulty was overcome by the suggestion of Christifilos [1] and independently by Courant, Livingston and Snyder [2] that the magnet be built of successive segments with n large and positive and n large in magnitude but negative.

The first segment focuses vertically but defocuses horizontally. The second one does just the opposite.

2.8.1.3 Focusing and Beam Stability

In all the cyclic accelerators, protons make typically 10^5 to 10^6 rotations, receiving an RF "kick" of the order of 0.1 MeV per turn before attaining maximum energy. In the acceleration regime they may cover a total path length, say 10^6 km. It is therefore very important that the proton bunch be stable and focused. Otherwise, particles may diverge and hit the wall of the doughnut and get lost. In the recent machines two types of magnets are in use, bending magnets which produce a uniform vertical dipole field over the width of the doughnut and constrain the protons to move in a circular path, Fig. 2.10(a) and focusing magnets produce a quadrupole field with four poles, Fig. 2.10(b). In the figure shown, the field is zero in the centre and increases rapidly as one moves outward. A proton moving downward (into the paper) will experience magnetic forces indicated by the arrows. For the arrangement shown the magnet is vertically focusing (force toward the centre) and horizontally defocusing (force away from the centre). If quadrupoles be arranged with the poles reversed, then the focusing and defocusing effects would be interchanged. Just as a light beam on passing through a succession of diverging and converging lenses of the same focal power will produce a net effect of focusing in both planes, same result will be produced with the quadrupole magnets. This is the principle of AGF (alternate gradient focusing). The field at the centre of the beam tube has identical values in all sectors, but in one set it decreases with r but in the alternate it increases. The variation of the field with r is quite dramatic: $B_z \propto r^{+n}$ and $B_z \propto r^{-n}$ in the alternate sectors, with $n \sim 300$. Figure 2.11(a) shows a section of alternate focusing and defocusing magnets. The whole ring is made by the periodic repetition of the elementary sections. Figure 2.11(b) shows the displacement of the beam along the

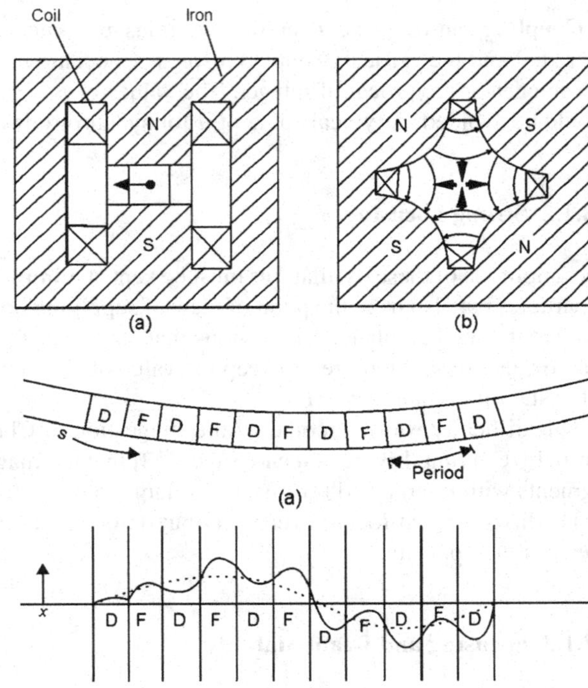

Fig. 2.10 (**a**) Bending magnets. (**b**) Focusing magnets

Fig. 2.11 (**a**) Section of alternate focusing and defocusing magnets. (**b**) Displacement of beam

circumference s under the focusing and defocusing action. Phase stability is not independent from orbit stability in the case of synchrotron because a particle receiving slightly greater "kick" in the gap will move to larger r.

2.8.1.4 Synchrotron Oscillations

In addition to the transverse betatron oscillations, protons also undergo longitudinal oscillations, called synchrotron oscillations as the particles get out of step with the ideal synchronous phase for which the increase in energy per turn exactly matches with the increase in magnetic field. Thus a particle at A which arrives late compared to the synchronous particle at C, would receive a smaller RF kick and get into a smaller orbit and next time earlier (Fig. 2.12). On the other hand, a particle B which arrives early will receive a larger kick, move into a larger orbit and eventually acceleration is contracted by the focusing to 1 mm or so and is extracted by a fast kicker magnet.

The extracted high energy proton beam is made to bombard a suitable target which produces all kinds of particles. Among positive charge we can expect p, π^+, K^+, Σ^+ and their decay products μ^+ and e^+, among negative charge, π^-, K^-, p^-, Σ^-, Ξ^-, Ω^-, μ^-, and e^- while neutrals consist of π^0, K^0, Λ, Ξ^0, ν, γ.

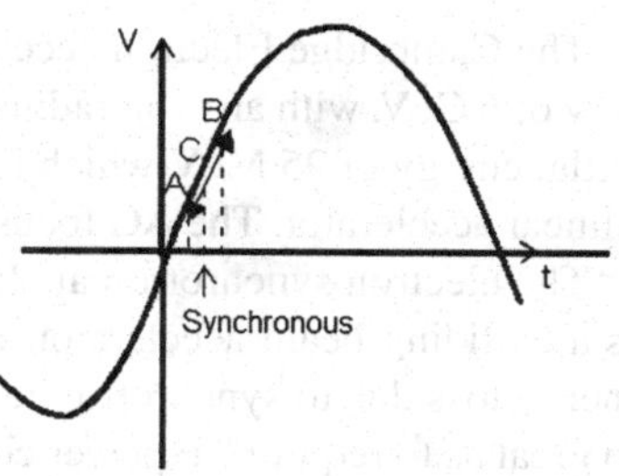

Fig. 2.12 Synchrotron oscillations of protons

The charged particles are separated by employing dipole magnets which act like prisms in optics. The focusing is done by the use of quadrupole fields which act like lenses. An example of a typical layout for the transportation of beams is shown in the experiment for the discovery of antiproton (Chap. 3).

2.8.2 Electron Synchrotron

Electrons can also be accelerated by a synchrotron. The electron synchrotron works on the same principles as proton synchrotron. The main difference is that relativistic energies are quickly reached for $cp \gg mc^2$ so that the orbital frequency of electrons is almost constant. Consequently, magnetic field need not be increased once the electrons have acquired energy like 25 MeV or so initially injected by a linear accelerator. The second difference is that the electron energy from a synchrotron is limited due to production of synchrotron radiation which is absent in a proton synchrotron. Under the circular acceleration, an electron emits synchrotron radiation, the energy radiated per particle per turn being

$$\begin{aligned}\Delta E &= \frac{e^2\beta^3}{3\varepsilon_o R}\left(\frac{E}{mc^2}\right)^4 \\ &= \frac{e^2\beta^3\gamma^4}{3\varepsilon_o R}\end{aligned} \tag{2.47}$$

where $\gamma = E/mc^2 = (1-\beta^2)^{-1/2}$ and $\beta = v/c$.

Putting the numerical values for ε_0 and e, and setting $\beta = 1$

$$\Delta E = 0.08856\frac{E^4}{R} \tag{2.48}$$

where ΔE is in MeV, E is in GeV and R is in meters.

Thus, for relativistic protons and electrons of the same energy, the energy loss is $(m/M)^4 \simeq 10^{13}$ times smaller for protons than electrons. For an electron of energy 10 GeV circulating in a ring of radius 1 km, the energy loss per turn is of the order of 1 MeV rising to 14 MeV per turn at 20 GeV. Thus, even with very large rings and low guide fields, synchrotron radiation and the need to compensate this loss with large amount of RF power becomes a serious problem.

The Cambridge Electron accelerator operated in the period 1962–1968 at an energy of 6 GeV, with an orbit radius of 36 m and maximum magnetic field 0.76 T. The initial energy of 25 MeV which is already relativistic for electrons was provided by a linear accelerator. The AG focusing was provided by the use of 48 magnet sectors.

The electron synchrotron at Hamburg, Germany, called DESY works at 35 GeV as a colliding beam accelerator. For energies much higher then 150–200 GeV the energy loss due to synchrotron radiation in a circular electron accelerator would be so great as to require RF power compensation quite unacceptable. This was actually the motivation for the construction of the two mile 30 GeV linear accelerator.

The Electron synchrotron has to a large extent replaced the betatron as a source of very high-energy electrons. It has two important advantages, first, the magnet structure is much lighter, consisting essentially of a ring magnet to provide a guide field across the doughnut-shaped vacuum chamber. Second, the large and heavy laminated core needed in the betatron for induction to produce the accelerating electric field is replaced by a compact cavity resonator which supplies the RF field in the synchrotron. Also the correction for radiation losses is applied automatically in the synchrotron by phase shifts which supply additional accelerating potential. The complicated compensating devices necessary in the betatron are not required.

2.9 Betatron

As the name suggests, this machine is used to accelerate electrons in a doughnut under varying magnetic field. If a magnetic flux ϕ changes in an electromagnet an emf is produced which accelerates the electrons and at the same time holds them is an orbit with a fixed radius, as in Fig. 2.13. Using Faraday's Law of induction,

$$\begin{aligned}\text{The work done} &= \text{(induced voltage)(charge)}\\ &= \text{(average force)(distance)}\end{aligned}$$

$$\therefore \quad e\frac{d\phi}{dt} = 2\pi R\frac{dp}{dt}$$

$$e\frac{d\phi}{dt}\Big/2\pi R = \frac{dp}{dt} = \frac{d}{dt}(BeR)$$

where B is the magnetic field and R is the radius.

Integrating and assuming that the radius of orbit is constant and the flux is zero initially

$$\phi = 2\pi R^2 B \tag{2.49}$$

we may write

$$\Delta\phi = 2\pi R^2 \Delta B \tag{2.50}$$

Fig. 2.13 The cross-section of the electromagnet for betatron

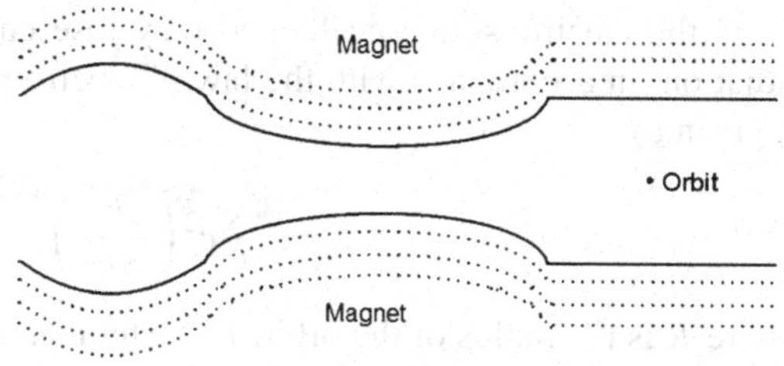

This tells us what must be the change in flux through the orbit in terms of the change in magnetic field strength at the orbit. The energy of the particle does not enter the calculation. We can write

$$\begin{aligned}\Delta T &= e\frac{\Delta\phi}{\Delta t} = \frac{e\Delta\phi}{2\pi R/v}\\ \Delta T &= \frac{ec}{2\pi R}\Delta\phi \quad (\because v \sim c)\end{aligned} \tag{2.51}$$

Therefore, in one quarter cycle of the low frequency magnet current during which time the flux changes from ϕ_i to ϕ_f, the electrons acquire the energy.

$$T = \frac{ec}{2\pi R}(\phi_f - \phi_i) = (B_f - B_i)Rce \tag{2.51a}$$

Thus, electrons can be accelerated in a circular magnetic accelerator either by betatron induction or by RF electric field applied to suitable electrodes. At high energy electrons lose the energy by radiation. Otherwise the only significant energy loss is caused by particles striking the residual gas molecules in the chamber or by hitting the wall. In a betatron the electron gains energy as it travels in its orbit. In most of betatrons the energy gained per turn is of the order of a few hundred electron volts. Hence a large number of turns are required, and the beam must be focused. Otherwise, it would spread out and dissipate by hitting the walls of the doughnut. While in the cyclotron radial oscillations are relatively unimportant in the betatron the magnitude of both the radial and vertical oscillation must be small, because if the oscillations exceed the dimensions of the cavity the beam would hit the wall and be lost. The betatron oscillations, however, are damped as the field grows. In a betatron the energy necessary to energize the magnet flows between the magnet and a large bank of capacitors at a frequency of about 60 cps. Hence the magnet must be laminated in order to avoid large eddy currents.

Electrons are injected with a special gun at energies of the order of 50 keV After acceleration the orbit is either expanded or displaced and the electrons strike a target, producing a beam of bremsstrahlung. If the target is thin, this beam has a small angular aperture of the order of mc^2/E and can be very intense. The practical limit for electron energy is about 300 MeV. Thus, betatron is a convenient source of energetic γ-rays.

As the electrons are accelerated they lose energy by emitting electromagnetic radiation, in accordance with the law of electrodynamics. The energy loss per turn in given by

$$\Delta E = \frac{4\pi}{3} e^2 \left(\frac{E}{mc^2}\right)^4 \times \frac{1}{4\pi \varepsilon_o R} \tag{2.52}$$

where R is the radius of the orbit, E the total energy, and ε_0 is the permittivity. The radiation loss forces the electrons to spiral inward toward the centre of the betatron where they can be caught by a target that scraps the inner side of the beam. For a 100 MeV betatron having $R = 1.0$ m, the energy loss per turn will be 8.8 eV per turn.

The energy is radiated in a continuous spectrum that has a maximum for frequencies near

$$\nu_{\max} = \frac{3}{2}\left(\frac{E}{mc^2}\right)^3 \frac{\omega_o}{2\pi} \tag{2.53}$$

For $R = 1.0$ m, $E = 100$ MeV; $\lambda_{\max} = 555$ nm which is in the visible region. The light is indeed observed.

Example 2.2 In a betatron of radius 82.5 cm the maximum magnetic field used is 4000 G. The frequency used was 66 cps.

Find

a. the electron energy
b. number of revolutions
c. total path length
d. radiation loss/turn

Solution

(a)

$$\begin{aligned} T = BRec &= (0.4)(0.825)\left(1.6 \times 10^{-19}\right)\left(3 \times 10^8\right) = 1.584 \times 10^{-11}\ \text{J} \\ &= \frac{1.584 \times 10^{-11}}{1.6 \times 10^{-13}}\ \text{MeV} = 99\ \text{MeV} \end{aligned}$$

(b) $(2\pi RN)(4f) = c$

$$N = \frac{c}{8\pi Rf} = \frac{3 \times 10^8}{(8\pi)(0.825)(66)} = 2.19 \times 10^5$$

The factor 4 arises due to the fact that the duty cycle is over a quarter of a period.

(c)

$$\begin{aligned} \text{Total pathlength} &= 2\pi RN = (2\pi)(0.825)\left(2.19 \times 10^5\right) \\ &= 11.36 \times 10^5\ \text{m} = 1136\ \text{km} \end{aligned}$$

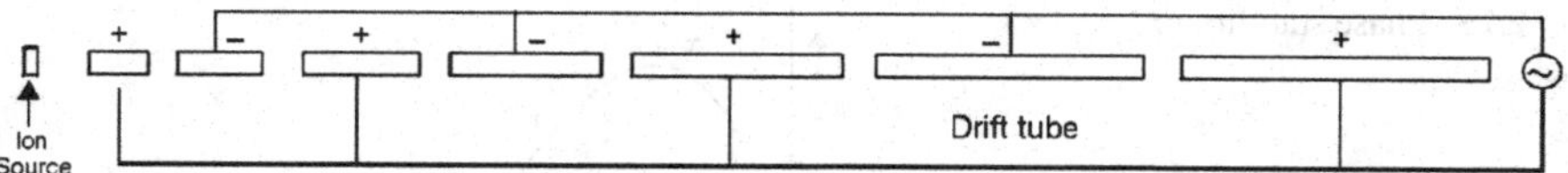

Fig. 2.14 A simple sketch of Linac

(d) By (2.48)

$$\Delta E = 0.08856\frac{E^4}{R}$$
$$= \frac{0.08856 \times (0.0995)^4}{0.825} = 10.5 \times 10^{-6}\ \text{MeV} = 10.5\ \text{eV}$$

2.10 Linear Accelerators

2.10.1 Principle

The linear accelerator (linac) also works on the principle of repeated accelerations. The linac consists of an evacuated pipe containing a set of drift tubes The particles are accelerated in a straight line rather than in circular orbits. This avoids the use of magnetic field. Oscillating electric field is applied to a periodic array of electrodes. The electrodes are connected alternately to the RF oscillator, Fig. 2.14 and the electrode spacing is arranged in such a way that the particles spend same time inside the drift tube and arrive at each gap in the same phase and so experience an acceleration at each gap.

The separation L between accelerating gaps is the distance traversed by the particles during one half cycle of the applied electric field. Hence the length is given by

$$L = \frac{1}{2}\frac{v_n}{f} \tag{2.54}$$

where v_n is the particle velocity in the nth gap. Each time a particle crosses a gap it finds an accelerating field, so the energy increases by an amount $V_e e$ where V_e is the voltage across the gap at the instant the particle crosses. In the nth gap, the energy gained would be eV_n and the velocity would be $\sqrt{2eV_e n/m}$. In order the particle will be in exact resonance and obtain the same energy increment $V_e e$ on each traversal the length of the nth drift tube is given by,

$$L_n = \frac{1}{2f}\sqrt{\frac{2eV_e}{m}}\sqrt{n} \tag{2.55}$$

At low energies the successive gap separations must increase in a sequence proportional to the square root of integers. At relativistic energies the particle velocity $v \to c$ and the length of the drift tubes becomes constant

$$L_n = \frac{c}{2f} = \frac{\lambda}{2} \tag{2.56}$$

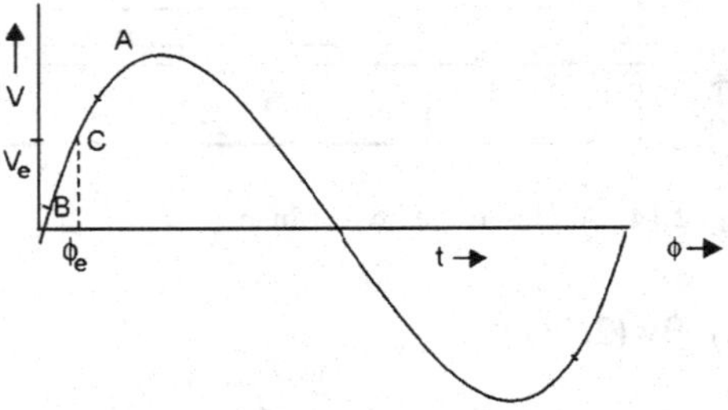

Fig. 2.15 Phase stability in line

where λ is the wavelength of applied RF to the electrodes. The electrodes would be spaced at half wavelength intervals. The particles are accelerated only in the gaps,and not within the electrodes which are in the form of drift tubes. The name connotes the meaning that the particle drifts in the tube without gaining energy.

2.10.2 Total Length

The total accelerating length is given by

$$L = \sum_{n=1} L_n \simeq \int L_n dn \tag{2.57}$$

Using (2.55) in (2.57)

$$L = \frac{2}{3f}\sqrt{\frac{eV_e}{2m}}n^{3/2} \tag{2.58}$$

Now final energy E is neV_e. Eliminating n from (2.58)

$$E = \left(\frac{m}{2}\right)^{1/3} (3fLeV_e)^{2/3} \tag{2.59}$$

Thus to obtain large energies one must choose heavy ions of high charge, use high frequencies and large accelerating potentials and make the machine very long. However, with heavy ions less energy is available in the centre of mass.

2.10.3 Phase Stability

Consider a particle crossing a gap at a phase denoted by the point A on the rising part of the voltage-time curve (Fig. 2.15), when the voltage is higher than V_e. This particle will gain more than the correct amount of energy, will have higher velocity and therefore a shorter time to arrive at the next gap. The resultant phase-shifts at the second, third ... gaps will reduce the energy per acceleration until it reaches the equilibrium value. The accumulated excess energy will continue the phase-shift so that the particle receives less than the reverse of that required in synchrocyclotron.

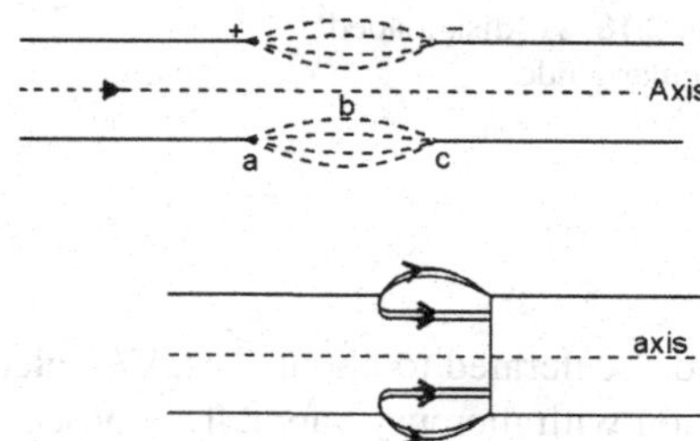

Fig. 2.16 The electric field in the gap between two drift tubes

Fig. 2.17 Field lines with grid

2.10.4 Radial Focusing

Consider an ion which at the time it the gap is away from the tube axis i.e. along *abc* in Fig. 2.16; the ion is accelerated the region *ac*. The electric lines of force indicate that the ion is accelerated along the axis as well as radially directed toward the axis. As the ion is under acceleration it would spend more time in the region *ab*, the first half of the gap than in region *bc*, the second half. The result is that the two radial displacements do not cancel and therefore there is slightly more focusing effect than defocusing effect, that is more deviation toward the axis than away from the axis.

Radial focusing is possible only if the passage is made during the quarter cycle in which the accelerating potential is decreasing toward zero. But this is incompatible with the requirement of phase stability. To circumvent this difficulty a metallic grid is placed across the entrance end of each drift tube. This results in lines of force such as those shown in Fig. 2.17. We see that radial focusing now occurs for particles crossing at any time during the accelerating half cycle, phase stability conditions remaining unchanged.

The accelerators generally use wave guides field to establish oscillating electric field. The electromagnetic field can form a standing wave in the cavity which acts as a resonator. The ion source is continuous but only those ions within a certain time bunch are accelerated. Such ions traverse the gap between successive tubes when the field is from left to right. They are inside the tube which is a field free region. When the voltage changes sign, the tubes form resonant cavities fed by a series of klystron oscillators synchronized in time to provide continuous acceleration. The oscillating frequencies used are in the microwave region. Linacs which accelerate protons to 50 MeV or so are used as auxiliary machines to inject protons in high energy synchrotrons. The linacs are used to accelerate not only protons to high energies but also electrons and heavy ions. In circular electron accelerators radiation energy losses become serious at high energies. In this respect linear accelerators have the obvious advantage. For heavy ions q/m is often low because the ions are only partially ionized. Usual magnetic fields would bend the trajectories of these ions only slightly and it would be more costly to use a circular machine than a linear accelerator. Energies of the order of 10 MeV/nucleon have been achieved for heavy ions (C, N, Ne etc.) with linear accelerators. Positive ions of small charge are first accelerated in a Cockcroft-Walton type of accelerator having a potential drop of about 400 kV. The ions are then "bunched" and fed into a linear accelerator,

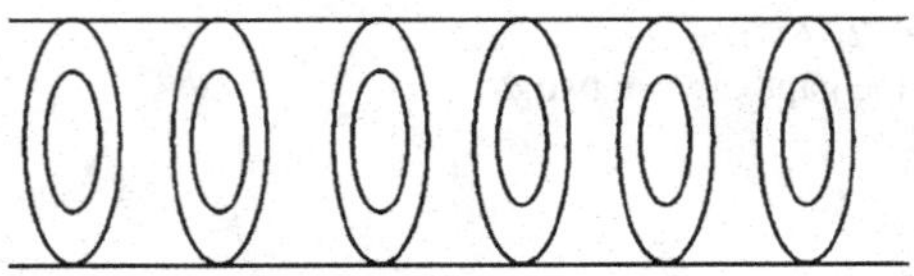

Fig. 2.18 A 'disk loaded' circular guide

and accelerated to about 1 MeV/nucleon. The ions are further ionized through collision with mercury vapor. This process takes place in the "stripper". They are then accelerated to about 10 MeV/nucleon in the linear way.

2.11 Electron Linear Accelerators

In the case of electron linear accelerator electrons of a few MeV energy travel essentially with the velocity of light, so that after the first metre or so the tubes have to be of uniform length. At high energies the electrons may be imagined to travel riding the crest of a traveling wave down the accelerator. Because the resistive losses are high power is fed into the length of the accelerator at regular intervals to maintain the traveling wave. This is the reason for operating linacs in the pulse mode rather then in the continuous fashion. In the pulsed mode the power need to be supplied only for a few percent of the time. The cavity is designed in such a way that the phase velocity of the traveling wave exactly matches the velocity of a particle as it is accelerated along the cavity. This is accomplished by using the "disk-loaded" configuration as in Fig. 2.18. The phase velocity of the wave is determined by the dimensions of the disks.

The largest linac is at Stanford. It is 3 km long and accelerates electrons to 25 GeV using 240 klystrons which give 2 μS short bursts of intense power with a current of 30 μA.

The chief advantage of the linear accelerator lies in the excellent collimation of the emergent beam in contrast with the spreading beams from circular accelerators. At high energies linear accelerators become physically long, and a large number of rf power sources is required to supply the extended electrode system.

Some differences in proton and electron linear accelerators may be mentioned. Because of the low proton velocities during most of the acceleration, proton accelerators use relatively low frequencies (50 to $200Mc$). Electrons attain the velocity at modest energies and can be accelerated in small structures which are resonant at microwave frequencies ($3000Mc$). These structures act as waveguides, rather than arrays of electrodes, and particle acceleration can be described in terms of the wave properties of the field. In this kind of waveguide the electrodes are replaced by diaphragms which load the waveguide so that the phase velocity equals the electron velocity. The applied RF electric fields set up standing or traveling waves in the loaded waveguide. A bunch of electrons riding this wave will pick up energy continuously, similar to the motion of a surfboard down the advancing front of a water wave.

The technical problems of producing a high-intensity electron beam from hot cathodes are by far simpler than those for positive ions sources. Furthermore, proton beam tends to get diverged due to the repulsive space-charge forces on each other, while these forces are absent in the case of relativistic electrons.

The cost of the linear accelerator is approximately linear with the energy as the length is extended. This is to be compared with the cost of a magnet for synchrotron which increases roughly as T^3 for relativistic particles. However, the development of circular accelerators, from the cyclotron with the solid core magnet to the synchrotron with its lighter ring magnet, to the alternating gradient synchrotron with its still further decrease in magnet's dimensions and cost have given an edge over the linear one.

2.12 Colliding Beam Accelerators

The accelerators described so far provide high energy beams of protons, electrons or heavy ions. On extracting these beams of particles and allowing them to hit external targets, secondary beams of hadrons (π, K, p, p^-) or leptons (μ, ν) may be produced. Such experiments are called fixed-target experiments. Now, the energy W available in the centre of mass system (CMS) is the one which is of consequence for particle production. It is obvious that the more massive are the newly produced particles, the higher will be the threshold energy, which requires a larger value of W.

Let a proton of mass M moving with kinetic energy T strike another proton at rest and new particle(s) of mass m be produced. Denoting the quantities in the CMS by asterisk (*), and using the invariance of

$$
\begin{aligned}
&E^2 - \left|\sum \overline{p}\right|^2 (c = 1),\\
&E^2 - (p_1 + p_2)^2 = E^{*2} - \left(p_1^* + p_2^*\right)^2 = E^{*2}\\
&\quad (\because p_2^* = -p_1^*)\\
&E^{*2} = S = E^2 - p_1^2 = (E_1 + M)^2 - \left(E_1^2 - M^2\right) = 2E_1 M + 2M^2\\
&\quad (\because p_2 = 0,\ E_1 = T + M,\ E = E_1 + E_2,\ E_2 = M)
\end{aligned}
\tag{2.60}
$$

Thus at high energies where $E_1 \gg M$, the energy available in the centre of mass increases only as the square root of the particle energy E_1, much of the remaining energy being associated with the centre of mass motion itself and is not available for particle production. A significant fraction of energy goes into the motion of the particles after the interaction, otherwise linear momentum is not conserved. It shows up in the production of high energy secondary beams.

We can easily calculate the threshold energy T_{th} for the production of new particles of mass m by requiring that at threshold all the particles in the CMS must be at rest. Putting $E^* = 2M + m$ and $E_1 = T + M$, in (2.60) and simplifying

$$T_{th} = 2m\left(1 + \frac{m}{4M}\right) \tag{2.61}$$

For pion production, $m = 140$ MeV, $M = 938$ MeV. In the reaction $p + p \rightarrow p + n + \pi^+$, $T_{th} = 290$ MeV. The energy utilized is only $m/T_{th} = 140/290$ or 48 %.

For proton antiproton pair production, in the reaction

$$p + p \rightarrow p + p + p + p^-, \qquad m = 2M, \qquad T_{th} = 6M = 5.63 \text{ GeV}$$

the energy utilized in only $2M/T_{th} = 33$ %.

For the production of massive particles like W or Z (carriers of weak interaction), $m \sim 90$ GeV, $T_{th} = 4300$ GeV or 4.3 TeV, the energy utilized is only 2 %. Now the energy available from the largest proton synchrotron is only 1 TeV (with superconductor magnets). It was therefore not possible to produce such large energies with conventional synchrotrons and carry out the experiments with fixed targets to produce very heavy particles. Such considerations have motivated the introduction of a new class of machines called colliding beam accelerators. In these accelerators, two relativistic particles, such as protons of energy E_1 and E_2 and momenta p_1 and p_2 circulate in opposite directions in a storage ring, then the value of W is given by

$$\begin{aligned} S = W^2 = E^2 - \left|\sum \overline{p}\right|^2 &= (E_1 + E_2)^2 - (p_1 - p_2)^2 \\ &= E_1^2 - p_1^2 + E_2^2 - p_2^2 + 2(E_1E_2 + p_1p_2) \\ &= 2(E_1E_2 + p_1p_2) + 2M^2 \quad \text{or} \\ S \simeq 4E_1E_2 \quad &(\because E_1 \text{ or } E_2 \gg M) \end{aligned} \tag{2.62}$$

If $E_1 = E_2$, the CMS of the collision is at rest in the laboratory. In other words, the LS system is reduced to the CMS and the entire energy is now available for the production of new particles, and $W \propto E$ instead of $W \propto E^{1/2}$ as in the case of fixed target. Comparing (2.60) with (2.62), the value of W is identical with that for fixed-target machine of energy $E = 2E_1E_2/M$. Thus for a colliding beam machine of energy $E_1 = E_2 = 50$ GeV is equivalent to $E = 5300$ GeV for the fixed-target machine.

Colliding-beam machines have some disadvantages as well. The colliding beam particles must be stable. The collisions are limited to pp, pp^-, e^+e^-, ep types. In a pp or ep collider, two separate beam pipes and two sets of magnets are required, while e^+e^- and pp^- colliders make use of the same set of magnets and RF cavities to circulate them. If electrons circulate say clockwise then positrons would circulate in counter-clockwise direction in the same pipe and magnet ring.

The other disadvantage is that the collision rate R in the intersection is low. The reaction rate is given by

$$R = \sigma L \tag{2.63}$$

where σ is the microscopic cross-section for the interaction and L is the luminosity (in units of cm^{-2} sec^{-1}). For two oppositely directed beams L is given by

$$L = \nu n \frac{N_1 N_2}{A} \tag{2.64}$$

where N_1 and N_2 are the number of particles in each bunch, n is the number of bunches in either beam, ν is the revolution frequency and A is the cross-sectional

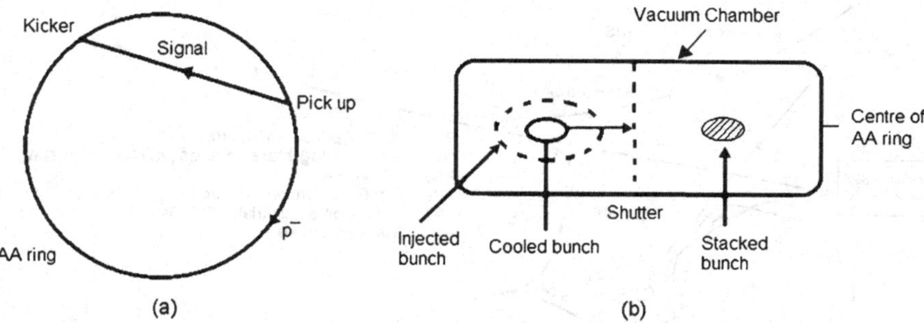

Fig. 2.19 (**a**) Deflection of deviated particles towards the ideal orbit. (**b**) Cooling and extraction of magnetically deflected antiprotons

area of the beams, assuming a complete overlap. Clearly L is largest if A is least. Typical L-values are $\sim 10^{31}$ cm^{-2} sec^{-1} for e^+e^- colliders and $\sim 10^{30}$ cm^{-2} sec^{-1} for p^-p accelerators.

These values may be compared with that for a fixed-target machine in a typical experiment in which a beam of 10^{12} protons sec^{-1} from a proton synchrotron traversing a liquid-hydrogen target 1 m long; yields a luminosity of $L \sim 10^{37}$ cm^{-2} sec^{-1}.

2.12.1 Cooling in p^-p Colliders

In the case of e^+e^- colliders it is not difficult to obtain intense source of positrons to circulate. However, the production of intense beam of antiprotons is a lot more difficult. First, the antiprotons are created in pairs with protons in high energy collisions of protons with complex nuclei with low yield at various momenta and emission angles. The random and chaotic behavior of antiprotons must some how be transformed into an orderly way so that a good number of antiprotons may be stored and be ready for acceleration. Toward this end, the beam must be "cooled" in order to reduce its divergence and longitudinal momentum spread. This technique which was crucial in the discovery of $W^\pm$ and Z^0 particles and was developed at CERN by Van der Meer uses a statistical method called stochastic cooling (1972). A bunch of 10^7 antiprotons are produced from a pulse of 10^{13} protons at 26 GeV bombarding a Cu target and are injected into the outer half of a wide-aperture toroidal vacuum chamber divided by a mechanical shutter and placed inside a special accumulator-magnet ring, Fig. 2.19(a).

A pickup coil in one section of the ring senses the average deviation of particles from the ideal orbit, and a correction signal is sent across a chord to a kicker in time to deflect them as they come round, toward the ideal orbit Fig. 2.19. After $2s$ of circulation, lateral and longitudinal spreads are reduced by an order of magnitude. The shutter is opened and the "cooled" bunch of antiprotons is magnetically deflected and stacked in the inside half of the vacuum chamber where they are further cooled

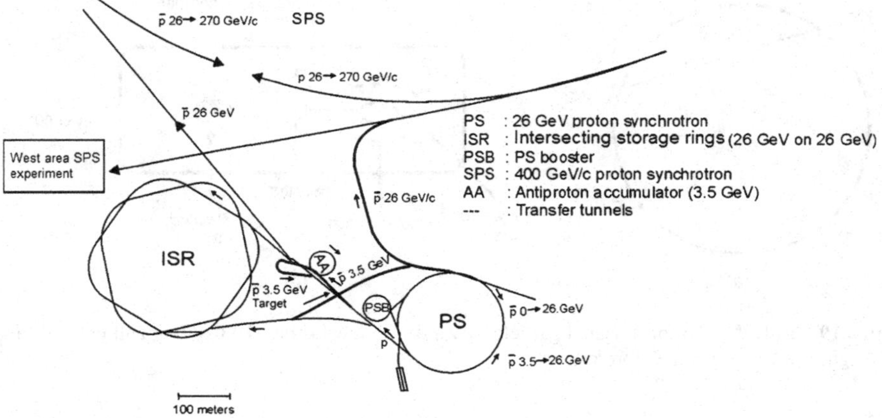

Fig. 2.20 Layout of the accelerator complex at CERN (CERN Service)

Fig. 2.19(b). The process is repeated until, after a day or so, 10^{12} antiprotons have been stacked and are then extracted for acceleration in the main SPS (collider) ring.

Figure 2.20 shows the layout of the accelerator complex at CERN, near Geneva. The 26 GeV PS accelerator has been used to fill the ISR with oppositely circulating proton beams, as an injector to the 450-GeV SPS and also to make secondary antiprotons, which are transferred, stacked, and "cooled" in a 3.5-GeV antiproton accumulator before being accelerated in the PS and injected into the SPS, used as a 310-GeV + 310-GeV pp^- collider.

We summarize below the advantages and disadvantages of fixed target accelerators and the colliders.

2.13 Fixed Target Accelerators

2.13.1 Advantages

1. Easy to produce many interactions as the beam particles have a high probability of hitting the target nuclei.
2. With the aid of a bubble chamber or vertex detector the point of interaction can be located.
3. particles like π, K, p^-, μ can be produced in the form of beams.
4. Short lived particles even with lifetimes of the order of 10^{-10} sec as for Σ^- hyperons, have been produced because of time dilation.

2.13.2 Disadvantages

1. Not all the energy available in the collision can be used to create new particles. Most of the energy is wasted.

2.14 Colliders

2.14.1 Advantages

1. Entire kinetic energy of accelerated particles can be used to produce new particles. One essentially works in the centre of mass system itself.

2.14.2 Disadvantages

1. Difficult to locate the actual point at which the interaction took place.
2. Short lived particles will not be able to travel long distances (using time dilation) before decaying.
3. High energy beams of secondary particles can not be produced.
4. The reaction rate is lowered considerably. This is measured by the luminosity which is the number of reactions per second per cm^2 of the beam area. The highest luminosity achieved so far has been $1.7 \times 10^{31}\ \mathrm{sec}^{-1}\ \mathrm{cm}^{-2}$ at the Fermi Lab.
5. Colliding particles must be stable. All Colliders are of the type pp, e^+e^- or pp^-, ep.

Example 2.3 Calculate the orbit radius for a synchrotron designed to accelerate protons to 5 GeV assuming a guide field of 1.5 kG.

Solution

$$p = 0.3Br = 0.3 \times 1.5r = 0.45r$$
$$p = \sqrt{T^2 + 2Tm} = \sqrt{5^2 + 2 \times 5 \times 0.94} = 5.865\ \mathrm{GeV}/c$$
$$\therefore \quad r = \frac{5.865}{0.45} = 13\ \text{metre}$$

Example 2.4 What percentage depth of modulation must be applied to the dee voltage of a synchro-cyclotron in order to accelerate protons to 300 MeV assuming that the magnetic field has 4 % radial decrease in magnitude.

Solution

$$\omega_0 = \frac{B_0 e}{m} \quad \omega = \frac{0.96 B_0 e}{m+T}$$
$$\frac{\omega}{\omega_0} = 0.96 \frac{m}{m+T}$$
$$\therefore \quad \frac{\omega_0 - \omega}{\omega_0} = \frac{T + 0.06m}{m+T} = \frac{300 + 0.06 \times 940}{940 + 300} = 0.287$$

Therefore, percentage depth of modulation is 28.7 %.

Example 2.5 It is required to collide with an energy in their CMS, of $2Mc^2$ in excess of their rest energy $2Mc^2$. This can be achieved by firing protons at one another with two accelerators each of which imparts a kinetic energy of Mc^2. Alternatively protons can be fired from an accelerator at protons at rest. How much energy must this single accelerator be capable of imparting to a proton? What is the significance of this result for experiments in high energy nuclear physics?

Solution Using the invariant quantity

$$E^2 - p^2 = E^{*2} - p^{*2} = E^{*2} \quad (\because p^* = 0)$$
$$(T + 2M)^2 - \left(T^2 + 2TM\right) = (2M + 2M)^2 = 16M^2$$
$$\therefore \quad T = 6Mc^2$$

The result is significant for the design of colliding beam experiments in which it is possible to get the same CMS energy with relatively two small accelerators rather than a single gigantic accelerator.

Example 2.6 A collimated beam of π-mesons of 140 MeV energy passes through a liquid hydrogen bubble chamber. The intensity of the beam in found to decrease with distance s along its path as $\exp(-as)$ with $a = 0.074\ \text{m}^{-1}$. Hence calculate the life time of the π-meson. (Rest energy of π-meson 140 MeV.)

Solution

$$I = I_o e^{-s/v\tau} = I_o e^{-as}$$
$$\tau = \frac{1}{av} = \frac{1}{a\beta c}$$

Now

$$\gamma = 1 + \frac{T}{M} = 1 + \frac{140}{140} = 2$$
$$\beta = \frac{\sqrt{\gamma^2 - 1}}{\gamma} = \frac{\sqrt{(2)^2 - 1}}{2} = 0.866$$
$$\tau = \frac{1}{0.074 \times 0.866 \times 3 \times 10^8} = 5.2 \times 10^{-8}\ \text{sec}$$
$$\tau_o = \frac{\tau}{\gamma} = \frac{5.2 \times 10^{-8}}{2} = 2.6 \times 10^{-8}\ \text{sec}$$

Example 2.7 A beam of kaons, $m = 494\ \text{MeV}/c^2$, of energy 100 MeV have a mean track length of 2.39 m up to the point of decay. Estimate their mean life time.

Solution

$$\tau_o = \frac{\tau}{\gamma} = \frac{R}{\beta c \gamma} = \frac{R}{c\sqrt{\gamma^2 - 1}}$$

Now,

$$\gamma = 1 + \frac{T}{m} = 1 + \frac{100}{494} = 1.2$$

$$\tau_o = \frac{2.39}{3 \times 10^8 \times \sqrt{(1.2)^2 - 1}} = 1.2 \times 10^{-8}\ \text{sec}$$

Example 2.8 A singly charged particle of speed $0.3c$ gives a track of radius of curvature 1 m under uniform normal magnetic field of 1100 gauss. Calculate the rest mass of the particle, in terms of the electron rest mass m_e.

Solution

$$p = 0.3BR = (0.3)(0.11)(1.0) = 0.033\ \text{GeV}/c = 33\ \text{MeV}/c$$

$$m = \frac{p}{\gamma\beta}$$

$$\gamma = \frac{1}{\sqrt{1-\beta^2}} = \frac{1}{\sqrt{1-(03)^2}} = 1.048$$

$$m = \frac{33}{1.048 \times 0.3} = 104.96\ \text{MeV}/c^2 = \frac{10.4.96}{0511} m_e = 205.4 m_e$$

It is a muon.

Example 2.9 An electron synchrotron with a radius of 1 m accelerates electrons to 300 MeV. Calculate the energy lost by a single electron per revolution when it has reached maximum energy.

Solution Radiation loss per turn

$$\Delta E = \frac{4}{3}\pi \frac{e^2}{R}\left(\frac{E}{mc^2}\right)^4 \frac{1}{4\pi\varepsilon_0}$$

$$\Delta E = \frac{4}{3}\pi \times \frac{(1.6 \times 10^{-19})^2}{1.0}\left(\frac{300 + 0.511}{0.511}\right)^4 \times 9 \times 10^9\ \text{J} = 721\ \text{eV}$$

Example 2.10 The orbit of electrons in a betatron has a radius of 1 m. If the magnetic field in which the electrons move is changing at the rate of 100 Weber m^{-2} sec^{-1}, calculate the energy acquired by an electron in one rotation. Express your answer in electron volts.

Solution

$$\Delta T = e\frac{\Delta\phi}{\Delta t} = 1.6 \times 10^{-19} \times 100 \text{ Joule}$$
$$= 100 \text{ eV}$$

Example 2.11 If the acceleration potential (peak voltage 50,000) has a frequency of 1.5×10^7 *c*/sec, find the value of the field strength for cyclotron resonance when deuterons and alpha particles respectively are accelerated. For how long are the particles accelerated, if the radius of orbit at ejection is 30.0 cm? Deuteron rest mass = 2.01 a.m.u., a particle rest mass = 4.00 a.m.u.

Solution

$$B = \frac{2\pi m f}{q}$$

For deuterons,

$$B = \frac{(2\pi)(2.01 \times 1.66 \times 10^{-27})(1.5 \times 10^7)}{1.6 \times 10^{-19}}$$
$$= 1.9644 \text{ Tesla}$$

For *a* particles,

$$B = \frac{(2\pi)(4.0 \times 1.66 \times 10^{-27})(1.5 \times 10^7)}{2 \times 1.6 \times 10^{-19}}$$
$$= 1.9546 \text{ Tesla}$$

Time period

$$t = \frac{1}{f} = \frac{1}{1.5 \times 10^7}$$

$$\text{Deuteron energy} = \frac{1}{2}\frac{q^2 B^2 r^2}{m} = \frac{1}{2} \times \frac{(1.6 \times 10^{-19} \times 1.9644 \times 0.3)^2}{2.01 \times 1.66 \times 10^{-27}}$$
$$= 0.1332 \times 10^{-11} \text{ J}$$
$$= 13.32 \text{ MeV}$$

In each revolution, the energy picked up by deuteron

$$= 2 \times 50 \times 10^3 \text{ eV} = 10^5 \text{ eV}$$

Hence No. of orbits $N = \frac{13.32 \times 10^6}{10^5} = 133.2$.

Total time spent by deuterons

$$= t.N = \frac{1}{1.5 \times 10^7} \times 133.2 = 8.88 \times 10^{-6} \text{ sec}$$
$$= 8.88 \text{ μsec}$$

$$\alpha \text{ energy} = \frac{1}{2} \times \frac{(3.2 \times 10^{-19} \times 1.9546 \times 0.3)^2}{4.0 \times 1.66 \times 10^{-27}} = 0.2651 \times 10^{-11} \text{ J}$$
$$= 16.57 \text{ MeV}$$

In each revolution, the energy picked up by $\alpha = 2 \times (2 \times 50 \times 10^3) = 2 \times 10^5$.

Hence No. of orbits $N = \frac{16.57\times10^6}{2\times10^5} = 82.85$.

Total time spent by α's $= tN = 82.85 \times \frac{1}{1.5\times10^7} = 5.52 \times 10^{-6}$ sec $= 5.52$ μsec.

Example 2.12 Show that the radius R of the final orbit of a particle of charge q and rest mass m moving perpendicularly to a uniform field of magnetic induction B with a kinetic energy n times its own rest mass energy is given by

$$R = \frac{mc\sqrt{n^2+2n}}{Bq}$$

Solution

$$\begin{aligned}
p &= qBR \\
qBR &= m\gamma\beta c = mc\sqrt{\gamma^2 - 1} \\
T &= (\gamma - 1)mc^2 \\
n &= \frac{T}{mc^2} = \gamma - 1 \\
\gamma^2 - 1 &= (n+1)^2 - 1 = n^2 + 2n \\
R &= \frac{mc}{Bq}\sqrt{n^2 + 2n}
\end{aligned}$$

Example 2.13 Protons of kinetic energy 40 MeV are injected into a synchrotron when the magnetic field is 150 G. They are accelerated by an alternating electric field as the magnetic field rises. Calculate the energy at the moment when the magnetic field reaches 10 kG (rest energy of proton = 938 MeV).

Solution Following the results of Examples 2.10–2.12,

$$R = \frac{mc}{Bq}\sqrt{n^2 + 2n} = \frac{mc}{150q}\sqrt{\left(\frac{40}{938}\right)^2 + \frac{2 \times 40}{938}} = 1.9676 \times 10^{-3}\frac{mc}{q}$$

$$R = \frac{mc}{10^4 q}\sqrt{n^2 + n}$$

$$\therefore \quad \sqrt{n^2 + n} = 19.676$$

Solving for n,

$$\begin{aligned}
n &= 19.18 \\
T &= 19.18 \times 938 = 17993 \text{ MeV} \\
&= 18 \text{ GeV}
\end{aligned}$$

Example 2.14 A synchrotron accelerates protons to a kinetic energy of 1 GeV. What kinetic energy could be reached by deuterons or ^{3}He were accelerated in this machine? Take the proton mass to be equivalent to 1 GeV.

Solution Using the results of Example 2.12,

$$\frac{q}{m} = \frac{c}{B_o R}\sqrt{n^2 + 2n}$$

For protons, put $q = 1, m = 1, n = 1$

$$\frac{c}{B_o R} = \frac{1}{\sqrt{3}}$$

For deuterons, $\frac{q}{m} = \frac{1}{2} = \frac{1}{\sqrt{3}}\sqrt{n^2 + 2n}$.

Solving for n, $n = 0.3228$, KE reached $= 1876 \times 0.3228 = 605.6$ MeV.

For ^{3}He,

$$\frac{q}{m} = \frac{2}{3} = \frac{1}{\sqrt{3}}\sqrt{n^2 + 2n}$$

$$n = 0.527$$

KE reached $= 2814 \times 0.527 = 1483$ MeV.

Example 2.15 A cyclotron of diameter is 1.5 m operates with a frequency of 6Mc/sec. Find the magnetic field which satisfies the resonance conditions for (a) protons (b) deuterons (c) alpha-particles. Also what energies will these particles attain?

Solution

(a)

$$B_p = \frac{2\pi f m_p}{q} = \frac{(2\pi)(6 \times 10^6)(1.66 \times 10^{-27})}{1.6 \times 10^{-19}} = 0.39 \text{ Wb/m}^2$$

$$T^p_{\max} = \frac{1}{2}\frac{(Bqr)^2}{m} = \frac{1}{2}\frac{[(0.39)(1.6 \times 10^{-19})(0.75)]^2}{1.66 \times 10^{-27}}$$

$$= 0.06597 \times 10^{-11} \text{ J} = 4.12 \text{ MeV}$$

(b)

$$B_d = \frac{m_d}{m_p}B_p = \frac{3.34 \times 10^{-27}}{1.66 \times 10^{-27}} \times 0.39 = 0.78 \text{ Wb/m}^2$$

$$T^d_{\max} = \left(\frac{B_d}{B_p}\right)^2\left(\frac{m_p}{m_d}\right)T^p_{\max} = \frac{m_d T^p_{\max}}{m_p} = 2 \times 4.12 = 8.24 \text{ MeV}$$

(c)

$$B_\alpha = \frac{m_\alpha}{q_\alpha} \cdot \frac{q_p}{m_p} \cdot B_p = 0.78 \text{ Wb/m}^2$$

$$T^\alpha_{\max} = \left(\frac{B_\alpha q_\alpha}{B_p q_p}\right)^2 \frac{m_p}{m_\alpha} T^p_{\max} = (2 \times 2)^2\left(\frac{1}{4}\right)4.12 = 16.48 \text{ MeV}$$

Example 2.16 The following conditions are obtained in a betatron: Maximum magnetic field at orbit = 5,000 G, operating frequency = 50*c*/sec, stable orbit diameter = 1.5 m.

Determine the energy gained by an electron in one revolution, and the final energy of the electron.

Solution

$$\text{Energy gained per orbit} = e\frac{d\varphi}{dt} = \frac{2e\omega\varphi_0}{\pi}$$
$$= 4ef\varphi_0 = 4 \times 50 \times 0.5 = 100\ \text{eV}$$

Final energy of electron

$$T = \frac{ec\varphi_0}{2\pi r_0} = \frac{3 \times 10^8 \times 0.4}{2\pi \times 0.75} = 25.478 \times 10^6\ \text{eV} = 25.48\ \text{MeV}$$

Example 2.17 Calculate the magnetic field, *B* and the Dee radius of a cyclotron which will accelerate protons to a maximum energy of 6 MeV if a radio frequency of 10 MHz is available.

Solution

$$B = \frac{\omega m}{q} = \frac{(2\pi)(10 \times 10^6)(1.67 \times 10^{-27})}{1.6 \times 10^{-19}} = 0.655\ \text{Wb/m}^2$$

$$r = \sqrt{\frac{2T}{m\omega^2}} = \sqrt{\frac{2 \times 6 \times 1.6 \times 10^{-13}}{1.67 \times 10^{-27} \times (2\pi \times 10 \times 10^6)^2}} = 0.54\ \text{m}$$

Example 2.18 If the frequency of the dee voltage at the beginning of an accelerating sequence is 15*Mc*/sec, what must be the final frequency if the protons in the pulse have an energy of 500 MeV?

Solution

$$\omega_0 = \frac{qB}{m} \quad \text{(in the beginning)}$$
$$\omega = \frac{qB}{m\gamma} = \frac{\omega_0}{\gamma} \quad \text{(in the end)}$$
$$\gamma = 1 + \frac{T}{m} = 1 + \frac{500}{938} = 1.533$$
$$f = \frac{f_0}{\gamma} = \frac{15}{1.533} = 9.78 Mc/\text{sec}$$

Example 2.19 In a synchrotron, the magnetic flux density decreases from 15 kG at the centre of the magnet to 14 kG at the limiting radius of 2.0 m. Find (a) the range of frequency modulation required for deuteron acceleration (b) the maximum kinetic energy of the deuterons. (The rest energy of deuteron is 3.34×10^{-27} kG.)

Solution

(a)

$$f_0 = \frac{Bq}{2\pi m} = \frac{1.5 \times 1.6 \times 10^{-19}}{2\pi \times 3.34 \times 10^{-27}} = 11.44 \times 10^6 c/\text{sec} = 11.44 Mc/\text{sec}$$

$$f = \frac{1.4 \times 1.6 \times 10^{-19}}{2\pi \times 3.34 \times 10^{-27}} = 10.68 \times 10^6 c/\text{sec} = 10.68 Mc/\text{sec}$$

Range of frequency modulation is 11.44–10.68Mc/sec

(b)

$$T_{\max} = \frac{1}{2}\frac{q^2B^2r^2}{m} = \frac{1}{2} \times \frac{(1.6 \times 10^{-19})^2(1.4)^2(2.0)^2}{3.34 \times 10^{-27}} = 3 \times 10^{-11} \text{ J}$$

$$= \frac{3 \times 10^{-11}}{1.6 \times 10^{-13}} \text{ MeV} = 187.5 \text{ MeV}$$

Example 2.20 In a collider experiment two accelerators are used, each providing 10 GeV protons. In the fixed target experiment what would the energy of the protons need to be so as to produce the same collision energy as before? The rest energy of the proton is 940 MeV.

Solution Using the invariance, $E^{*2} - |\vec{p}_1^* + \vec{p}_2^*|^2 = E^2 - |\vec{p}_1 + \vec{p}_2|^2$

$$(2 \times 10 + 2M)^2 - 0 = (T + 2M)^2 - \left(T^2 + 2TM\right)$$

Solving,

$$T = 40 + \frac{400}{2M} = 40 + \frac{400}{2 \times 0.94} = 253 \text{ GeV}$$

Example 2.21 A cyclotron is powered by a 50,000 volts, 6Mc/sec radio frequency source. If its diameter is 1.5 m what magnetic field satisfies the resonance conditions for (a) protons (b) deuterons (c) alpha-particles? Also what energies will these particles attain?

Solution

(a)

$$B = \frac{2\pi f m}{q} = \frac{(2\pi)(6 \times 10^6)(1.67 \times 10^{-27})}{1.6 \times 10^{-19}}$$

$$= 0.3933 \text{ Tesla}$$

$$T_{\max} = \frac{1}{2}\frac{(Bqr)^2}{m} = \frac{(0.3933)^2(1.6. \times 10^{-19})^2(0.75)^2}{2 \times 167 \times 10^{-27}}$$

$$= 6.67 \times 10^{-13} \text{ J} = 4.17 \text{ MeV}$$

(b)

$$B = \frac{(2\pi)(6\times10^6)(3.34\times10^{-27})}{1.6\times10^{-19}} = 0.786 \text{ Tesla}$$

$$T_{max} = \frac{1}{2}\times\frac{(0.786)^2(1.6\times10^{-19})^2(0.75)^2}{3.34\times10^{-27}}$$
$$= 13.32\times10^{-13}\text{ J} = 8.32 \text{ MeV}$$

(c)

$$B = \frac{(2\pi)(6\times10^6)(6.64\times10^{-27})}{2\times1.6\times10^{-19}} = 0.782 \text{ Tesla}$$

$$T_{max} = \frac{1}{2}\frac{(0.782)^2(3.2\times10^{-19})^2(0.75)^2}{6.64\times10^{-27}}$$
$$= 26.52\times10^{-13}\text{ J} = 16.66 \text{ MeV}$$

Example 2.22 At what radius do 50 GeV protons circulate in a synchrotron if the guide field is 1 weber m^{-2}?

Solution

$$p = \sqrt{T^2 + 2Tm} = 0.3BR \text{ GeV}/c$$

$$R = \frac{\sqrt{50^2 + 2\times50\times0.94}}{0.3\times1} = 169.8 \text{ m}$$

Example 2.23 A beam of electrons is deflected through a semicircular path of radius 10 cm by means of a magnetic field of 5000 gauss. Determine the energy of the electrons.

Solution

$$p = 0.3Br = 0.3\times0.5\times0.1 = 15\times10^{-3}\text{ GeV}/c$$
$$= 15 \text{ MeV}/c$$
$$E = \sqrt{p^2 + m^2} = \sqrt{15^2 + 0.51^2} = 15.008 \text{ MeV}$$
$$\simeq 15 \text{ MeV}$$
$$T = 15 - 0.51 = 14.5 \text{ MeV}$$

Example 2.24 A 5 MeV electron is moving perpendicular to a magnetic field of 5000 gauss. Determine the radius of curvature of its path.

Solution

$$p = \sqrt{T^2 + 2Tm} = \sqrt{5^2 + 2\times5\times0.51}$$
$$= 5.486 \text{ MeV}/c$$
$$p = 300Br = 300\times0.5r = 150r = 5.486$$
$$r = 0.0365 \text{ m} = 3.65 \text{ cm}$$

Example 2.25 Electrons are accelerated to an energy of 5 MeV in a linear accelerator, and then injected into a synchrotron of radius 12.5 m, from which they are accelerated with an energy of 5 GeV. The energy gain per revolution is 1 keV

a. Calculate the initial frequency of the RF source. Will it be necessary to change this frequency?
b. How many turns will the electron make?
c. Calculate the time between injection and extraction of the electrons.
d. What distance do the electrons travel within the synchrotron.

Solution

(a) At injection

$$\gamma = 1 + \frac{5}{0.511} = 10.78$$
$$\beta = \frac{\sqrt{\gamma^2 - 1}}{\gamma} = \frac{\sqrt{10.78^2 - 1}}{10.78} = 0.996$$

At extraction,

$$\gamma = 1 + \frac{5000}{0.511} \simeq 10^4$$
$$\beta \simeq 1.0$$

Initial frequency, $f = \frac{\beta c}{2\pi r} = \frac{0.996 \times 3 \times 10^8}{2\pi \times 12.5} = 3.806 \times 10^6 c/\text{sec}$.

Final frequency would be $3.822 \times 10^6 c/\text{sec}$.

Since the initial and final frequencies are nearly equal, there is hardly any need to change the RF frequency.

(b) Total energy increase of the electrons is

$$E_f - E_i = 5000 - 5 = 4995 \text{ MeV}$$

Since the energy gain per revolution is 1 keV, the number of turns is:

$$n = \frac{4995}{1 \times 10^{-3}} = 4.995 \times 10^6 \text{ turns}$$

(c) The period of rotation

$$T' = \frac{1}{f} = \frac{1}{3.8 \times 10^6} = 0.263 \times 10^{-6} \text{ sec}$$

Time between injection and extraction is

$$t = nT' = (4.995 \times 10^6)(0.263 \times 10^{-6}) = 1.31 \text{ sec}$$

(d) The total distance travelled

$$d = 2\pi r n = 2\pi \times 12.5 \times 4.995 \times 10^6$$
$$= 392.1 \times 10^6 \text{ metres}$$
$$= 3.92 \times 10^5 \text{ km}$$

Example 2.26 The Stanford linear accelerator produces 50 pulses per second of about 10^{11} electrons with a final energy of 1 GeV. Calculate (a) the average beam current (b) the power output.

Solution

(a) Beam current

$$i = \frac{q}{t} = 50 \times 10^{11} \times 1.6 \times 10^{-19}$$
$$= 8 \times 10^{-7} \text{ amp}$$
$$= 0.8\ \mu\text{A}$$

(b) power output

$$W = iV$$
$$= (8 \times 10^{-7})(10^9)$$
$$= 800 \text{ watts}$$

Example 2.27 Protons of 1 MeV energy enter a linear accelerator which has 99 drift tubes connected alternately to a 200 MHz oscillator. The final energy of the protons is 50 MeV. (a) What are the lengths of the second cylinder and the last cylinder. (b) How many additional tubes would be needed to produce 70 MeV protons in this accelerator?

Solution

(a) The accelerator has 98 gaps. The energy gain per gap is

$$\Delta E = \frac{50 - 1}{98} = 0.5 \text{ MeV/gap}.$$

After crossing the first gap, the protons are still non-relativistic and

$$v_2 = \sqrt{\frac{2K}{m}} = \sqrt{\frac{2 \times (1 + 0.5) \times 1.6 \times 10^{-13}}{1.6725 \times 10^{-27}}}$$
$$= 1.694 \times 10^7 \text{ m/sec}$$

The time within each drift tube is the same

$$t = \frac{1}{2f} = \frac{1}{2 \times 2 \times 10^8} = 0.25 \times 10^{-8} \text{ sec}$$

Therefore, the length of the second tube is

$$l_2 = v_2 t = 1.694 \times 10^7 \times 0.25 \times 10^{-8}$$
$$= 0.0424 \text{ m} = 4.24 \text{ cm}$$

For the final tube, the protons have already the final energy:

$$\ell_f = v_f t$$

v_f can be calculated from the relativistic formula

$$\gamma = \frac{E_o + K}{E_o} = \frac{938.25. + 50}{93825} = 1.053$$

$$\beta = \frac{\sqrt{\gamma^2 - 1}}{\gamma} = \frac{\sqrt{1053 - 1}}{1.053} = 0.316$$

$$\begin{aligned} \ell_f &= 0.316 \times 3 \times 10^8 \times 0.25 \times 10^{-8} \\ &= 0.2371 \text{ metres} \\ &= 23.71 \text{ cm} \end{aligned}$$

(b) To produce 70 MeV protons, number of additional tubes required is $\frac{70-50}{0.5} = 40$.

Example 2.28 In an electron-positron collider the particles circulate in short cylindrical bunches of radius 1 mm transverse to the direction of motion. The number of particles per bunch is 8×10^{11} and the bunches collide at a frequency of 1.4 MHz. The cross-section for $\mu^+\mu^-$ production at 10 GeV total energy is 1.5×10^{-33} cm^2; how many $\mu^+\mu^-$ pairs are produced per second?

Solution Number of beam electrons/cm^2/s,

$$\begin{aligned} N_1 &= nf/\pi r^2 \\ &= \frac{8 \times 10^{11} \times 1.4 \times 10^6}{\pi (0.1)^2} = 3.567 \times 10^{19} \end{aligned}$$

Number of beam positrons/bunch, $N_2 = 8 \times 10^{11}$.

Expected production rate of $\mu^+\mu^-$ pairs/second

$$\begin{aligned} & N_1 N_2 \sigma \left(e^+ e^- \to \mu^+ \mu^-\right) \\ &= \left(3.567 \times 10^{19}\right)\left(8 \times 10^{11}\right)\left(1.5 \times 10^{-33}\right) \\ &= 0.043 \end{aligned}$$

2.15 Questions

2.1 Why electrons cannot be accelerated in a cyclotron?

2.2 Protons are accelerated in a cyclotron using a hydrogen source. If now the hydrogen source is replaced by a helium source leaving the magnetic field and applied *ac* unaltered would the alpha particles resonate. It not, why not?

2.3 In what way electron synchrotron is different from proton synchrotron?

2.4 What are the advantages of a synchrotron over a synchro-cyclotron?

2.5 What are betatron oscillations?

2.6 Are betatron oscillations important in a cyclotron?

2.7 Coupled resonances between the two transverse modes of betatron oscillations, the radial and the vertical can occur with the ratio $\frac{w_r}{w_z} = \sqrt{\frac{1-n}{n}}$. What values of n should be avoided, and what value accepted?

2.8 How do the conditions for phase stability differ in synchrotron and linear accelerator?

2.9 Give the amplification of 'AGF' and 'AVF'.

2.10 What is the major advantage of AVF cyclotron over synchrocyclotron.

2.11 Distinguish between betatron oscillations and synchrotron oscillations.

2.12 What is the main application of betatron?

2.13 What is the synchrotron radiation?

2.14 What is the advantage of electron linear accelerator over electron synchrotron?

2.15 In the linear accelerator, are the particles accelerated in the gaps or in the drift tubes?

2.16 List the advantages and disadvantages of fixed target accelerators.

2.17 List the advantages and disadvantages of colliders.

2.18 What are storage rings?

2.19 What do you understand by 'cooling' in $p^{-}-p$ colliders?

2.20 Describe various stages in the working of a super-proton accelerator, starting from very low energy up to the maximum possible energy.

2.16 Problems

2.1 In a betatron of diameter 2 m operating at 60c/sec the maximum magnetic field is 1.0 weber/m^2. Calculate

(a) the average energy gained per revolution
(b) the number of revolutions made each quarter cycle
(c) the maximum energy of the electrons

[Ans. (a) 1.5 keV (b) 1.99×10^5 (c) 300 MeV]

2.2 If m is the rest mass of a particle, p its momentum and T_0 its proper life time then show that the distance traveled in one life time is $d = \frac{pT_0}{m}$.

2.3 Deuterons are accelerated in a conventional cyclotron. Following data are given:

$$\text{The magnetic field strength} = 15000 \text{ gauss}$$
$$\text{Radius of the Dee} = 30''$$

Calculate the resonance frequency for deuterons and also maximum energy of the deuterons in MeV obtainable from the cyclotron.
[Ans. 11.44Mc/sec, 31.3 MeV]

2.4 Protons are accelerated in a conventional cyclotron whose Dees have a radius of 32 cm. The magnetic field strength is 6500 gauss. Calculate the resonant frequency for protons and the maximum energy in MeV of the protons issuing from the cyclotron.
[Ans. 9.916Mc/sec, 2.07 MeV]

2.5 At what velocity do the classical and relativistic kinetic energies of a particle differ by 5 %?
[Ans. $\beta = 0.2575$]

2.6 A synchro-cyclotron has a pole diameter of 5 m and a magnetic field of 1.4 Wb/m^2. What is the maximum energy carried by electrons which are struck by the protons extracted from this accelerator?
[Ans: 1.167 MeV]

2.7 What radius is needed in a proton-synchrotron to obtain particle energies of 25 GeV, assuming that a guide field of 1.2 webers per square metre is available? What is the radius for 500 GeV?
[Ans. 72 m, 1.389 km]

2.8 A cyclotron of diameter 2.0 m has a magnetic guide field of 0.5 Wb/m^2. Calculate the energy to which (a) protons (b) deuterons can be accelerated.
[Ans: 12 MeV, 24 MeV]

2.9 In a synchrocyclotron, the magnetic flux density decreases from 15 kG at the centre of the magnet to 14.5 kG at the limiting radius of 2.0 m. Calculate the range

of frequency of modulation required for deuteron acceleration and the maximum kinetic energy of the deuterons (the rest mass of the deuteron is 3.34×10^{-27} kg)
[Ans: 11.44 to 11.06Mc, 201.4 MeV]

2.10 A cyclotron has a magnetic field of 16 kG. The extraction radius is 60 cm. Calculate (a) the frequency of the rf necessary for accelerating deuterons and (b) the energy of the extracted beam.
[Ans: (a) 12.2Mc (b) 22 MeV]

2.11 Protons are accelerated in a synchrotron in the orbit of 5 m. At one moment in the cycle of acceleration, protons are making one revolution per microsecond. Calculate the value at this moment of the kinetic energy of each proton in MeV.
[Ans. 5.21 MeV]

2.12 A cyclotron is designed to accelerate protons to an energy of 5 MeV. The magnetic field used is 15 kG. Find the extraction radius.
[Ans: 21.47 cm]

2.13 In the fixed frequency cyclotron the energy to which charged particles can be accelerated is limited because of relativistic increase of mass. Determine the percentage increase of the magnetic field at the extraction radius to restore the resonance condition for protons of 25 MeV.
[Ans: 2.665 %]

2.14

(a) Calculate the threshold energy for the reaction $e^+ + e^- \rightarrow p + p^-$, in the fixed target experiment.
(b) Find the minimum energy requirement in the collider experiment, if e^+ and e^- are moving with equal energy ($M_p c^2 = 938$ MeV, $M_e c^2 = 0.51$ MeV).

[Ans: (a) 3444 MeV (b) 937.5 MeV for e^+ or e^-]

2.15 Calculate the threshold energy for the creation of an electron-positron pair in the photon-electron collision.
[Ans: $4m_e c^2$]

2.16 Show that a beam of 1 GeV/c K^- mesons when transported over a distance of 10 m will provide useful intensity while a beam of Σ^- hyperons will not. Take masses of K^- mesons and Σ^- hyperons to be 0.5 and 1 GeV/c^2, respectively and their lifetimes 1.2×10^{-8} and 1.6×10^{-10} sec.

2.17 A betatron is designed for an orbit radius of 40 cm. Electrons are injected with negligible energy where the magnetic flux density is zero and increasing, and

extracted when $B = 1990$ gauss. Find the kinetic energy of the electrons at extraction.
[Ans. 23.34 MeV]

References

1. Christifilos (1950)
2. Courant, Livingston, Snyder (1952)

Chapter 3
Elementary Particles

3.1 Concepts and Nomenclature

A nucleus consists of protons (p) (mass 938 MeV/c^2) and neutrons (n) (mass 939 MeV/c^2). If they are not to be distinguished, they are jointly called *nucleons*. They participate in strong (nuclear) interactions.

Hyperons ($\Lambda, \Sigma, \Xi, \Omega$) are particles heavier than nucleons and like nucleons they participate in strong interactions, but are unstable against decay with mean lifetimes of the order of 10^{-10} sec.

If nucleons and hyperons are not to be distinguished they are jointly called *baryons*. In Greek Baros is for 'heavy'. Each baryon is assigned the baryon number B (similar to the mass number A) which is conserved in all types of interactions, strong, electromagnetic and weak.

The family of pions (π meson) of mass ~140 MeV/c^2 and kaons (K-mesons) of mass ~495 MeV/c^2 constitute *mesons*. In Greek meson is for 'intermediate'. They are thus called because their masses are in between those of electron and proton. They mediate nuclear forces which bind the nucleons in the nucleus. Kaons and hyperons constitute 'strange particles'.

Baryons and mesons are collectively called *hadrons*. In Greek 'hadron' is for large. All of them are subject to nuclear interactions. The family of *Leptons* consists of the weakly interacting particles, neutrinos (ν); (mass zero), electrons (e) (mass 0.511 MeV/c^2), muons (μ) (mass 106 MeV) and τ leptons (mass 1784 MeV/c^2). Historically "lepton" was intended for light particles like ν, e and μ but now a massive particle τ and its associated neutrino are included in this category. These particles are subject to electromagnetic interaction if charged but are not affected by nuclear force. It is found that the neutrinos accompanying electron, muon and tau-lepton (1974) are different. Each is labeled accordingly, by attaching a subscript ν_e, ν_μ and ν_τ. Thus, to date there are three different types of neutrinos.

Each particle has its antiparticle. An antiparticle has the same mass and spin and lifetime but opposite charge and magnetic moment. Thus positron (e^+) (mass 0.511 MeV/c^2) is the antiparticle of electron, antiproton (p^-) is the antiparticle of

A. Kamal, *Particle Physics*, Graduate Texts in Physics,
DOI 10.1007/978-3-642-38661-9_3, © Springer-Verlag Berlin Heidelberg 2014

proton (p) and so on. Each of the three neutrinos ν_e, ν_μ, $\nu_\mathcal{T}$ have the antiparticles $\overline{\nu_c}, \overline{\nu_\mu}, \overline{\nu_\mathcal{T}}$. In the case of photon (γ) and neutral pion π^0, the particle and the antiparticle are the same. They are said to be self-conjugate.

3.1.1 Iso-spin

The concept of isospin was introduced in AAK1,4. It is thus named because of its resemblance to ordinary spin. Although there is no connection between the two, the algebra underlying both is the same. Hadrons with similar properties (mass and spin) but in different charge states are grouped together in the form of isospin multiplets and each member of the given isospin multiplet is assigned the isospin T, such that $2T + 1 = n =$ number of charge states. For nucleon $T = 1/2$, corresponding to two charge states proton and neutron, for pion $T = 1$ since $n = 3$, corresponding to π^+, π^0, π^- for Λ, $T = 0$ since $n = 1$. π^+ and π^- which form a particle-antiparticle are said to be charge conjugate while π^0 is an antiparticle of itself is said to be self conjugate. For a given multiplet, various charge states are distinguished by their projection along the third axis T_3 of isospin space. Thus, for nucleon doublet proton has $T_3 = +1/2$ and neutron $T_3 = -1/2$. The isospins T_1 and T_2 combine to give $T_1 + T_2, T_1 + T_2 - 1, \ldots, |T_1 - T_2 + 1|, |T_1 - T_2|$. Thus π and N have combined isospin $3/2$ or $1/2$ and the third component $I_3 = +3/2, +1/2, -1/2, -3/2$. Both I and I_3 are conserved in strong interactions. The electromagnetic interactions allow conservation of I_3 but violate I. Weak interactions violate both I and I_3.

Baryons and Leptons are fermions (particles with half integral spins, in units of $\hbar = h/2\pi$), while mesons are bosons (integral spin).

According to Pauli's principle, no quantum state could be occupied by more than one electron. We can show that the Pauli principle implies that only antisymmetric overall wave functions are possible for the electron or other fermions.

Let the combined wave functions of two particles be written as $\psi(\boldsymbol{x}_1, \boldsymbol{x}_2)$ where the quantum numbers for each are grouped as a vector $\boldsymbol{x}$. We now write $\psi(\boldsymbol{x}_1, \boldsymbol{x}_2)$ in terms of linear superposition of product wave functions $\psi_1(\boldsymbol{x}_1)\psi_2(\boldsymbol{x}_2)$. We can write

$$\psi(\boldsymbol{x}_1, \boldsymbol{x}_2) = \frac{\psi_1\psi_2 + \psi_2\psi_1}{\sqrt{2}} \tag{3.1}$$

and

$$\psi(\boldsymbol{x}_1, \boldsymbol{x}_2) = \frac{\psi_1\psi_2 - \psi_2\psi_1}{\sqrt{2}} \tag{3.2}$$

where $\sqrt{2}$ is the normalization factor.

Since the physical situation described by the wave function must be the same when the identical particles are interchanged,

$$|\psi(x_1, x_2)|^2 = |\psi(x_2, x_1)|^2 \tag{3.3}$$

so that

$$\psi(x_1, x_2) = \pm e^{i\phi}\psi(x_2, x_1) \tag{3.4}$$

where $e^{i\phi}$ is the unobservable phase factor which may be set equal to unity, since if we repeat the interchange of particles, we come back to the original state and the phase factor becomes

$$e^{2i\phi} = 1 \quad \text{or} \quad e^{i\phi} = \pm 1$$

Thus

$$\psi(x_1, x_2) = \pm\psi(x_2, x_1) \tag{3.5}$$

Now the wave function given by (3.5) is satisfied by (3.1) which is symmetric as well as (3.2) which is antisymmetric. However, if $\psi_1 \equiv \psi_2$ than expression (3.2) vanishes while (3.1) does not. Thus, Pauli's principle requires the wave function for Fermions to be antisymmetric. On the other hand, for bosons the total wave function is symmetric and Pauli's principle is no longer valid.

3.2 Historical Development

In the early part of last century, particle physics was much simpler. At the time of Ernest Rutherford (1910) and Neil Bohr (1913) only a few particles were known. Electron was already discovered by J.J. Thompson. It was assumed that the orbital electrons and the central positive core called the nucleus provided the constituents of the neutral atom. In 1920 already doubts were raised by Rutherford about the correctness of the model of nucleus.

The old model for the nucleus (protons + electrons) gave wrong results for the statistics of nuclei. Also it predicted very high values for electron energy which excluded their existence within the nucleus. These difficulties were resolved with the discovery of neutron by Chadwick [2]. For the neutron the required momentum ($\sim$100 MeV/c) as given by uncertainty principle will be identical with that for electron, but the corresponding energy will be reasonably small (of the order of 10 MeV). In the old atomic model the nitrogen nucleus $^{14}_{7}$N had 21 Fermions ($14p + 7e^-$) each of spin 1/2 so that the spin for $^{14}_{7}$N could be $1/2, 3/2$ etc. However the studies of alternate intensities of the rotational band spectrum had established its spin as 1. Now, with the proton + neutron model the number of particles would be 14. The combination of spins from 14 (even) particles can result in the total angular momentum $J = 0, 1, 2$, etc. and hence consistent with the experimental observation. Thus, the picture that emerged was that atom consists of a relatively heavy core called the nucleus consisting of protons and neutrons and that the electrons revolved in various orbits around the nucleus, the number of orbital electrons being equal to the number of protons to make the atoms neutral. With the advent of quantum mechanics well defined electron orbits were abandoned and were replaced by electron clouds.

The discovery of neutron provided necessary clues regarding the nuclear β-decay. The theory was provided by Enrico Fermi. The observations on the apparently missing energy in the beta decay led W. Pauli [3] to postulate the existence of neutrino (ν), a massless neutral freely interacting particle.

Photons were assumed to mediate the electromagnetic field between charges and radiation. Phenomena involving the charges and radiation could be explained within the preview of quantum electrodynamics to a high degree of precision. Photons are massless particles of spin 1 and travel with the velocity c (3×10^8 m s^{-1}).

According to Dirac's relativistic theory of electron there will be a sea of negative energy states which are normally filled up. However if a photon of energy at least equal to $2mc^2$ or 1.02 MeV (threshold energy) interacts then the negative energy level is accessible, resulting in the emission of a e^+–e^- pair. Anderson and Blacket independently discovered positron (e^+) in a cloud chamber in 1932.

Thompson's electron (e^-), Anderson's Positron (e^+) and Einstein's Photon (γ) required roles in Dirac's relativistic theory. Computation of interaction cross-sections for Compton scattering ($\gamma + e^- \rightarrow \gamma + e^-$), Moller scattering ($e^- + e^- \rightarrow e^- + e^-$) or Bhabha scattering ($e^+ + e^- \rightarrow e^+ + e^-$) were done as extension of theory.

After the discovery of neutron and its natural place in the nucleus the question arose, what kind of force holds the nucleons together? This question was answered by Yukawa, a Japanese Physicist, who predicted that strong short range nuclear forces resulted from the constant exchange of heavy quanta called mesons between various nucleons. The strong (nuclear) force is one of the four fundamental forces, the others being electromagnetic force, weak force and gravitational force. The strong force manifests itself in meson production in high energy collisions. The electromagnetic force is revealed in interactions between electronic charges and radiation field, for example in photoelectric effect and Compton scattering. The weak force is exemplified in beta decay and gravitational force arises by virtue of the mass of bodies. From the point of view of strength at short distances, strong force is by far the strongest, next comes the electromagnetic force which is down by a factor of 100, and then the weak force which is weaker by a factor 10^{12} and finally gravitational force which is weaker by a factor of 10^{36}.

The predicted mass of meson for a range of 10^{-15} m turns out to be $200m_e$. In, 1936 μ mesons (muons) of approximate mass of $200m_e$ were discovered by Anderson and Neddermeyer [1] in cosmic ray studies. Apparently they were suitable candidates for Yukawa's particle. But subsequent studies revealed that they had weak interactions with matter and could not be identified as Yukawa particle.

In 1947, Powell, Lattes and Occhialini discovered π meson (pion) of mass about $300m_e$ in photographic emulsions exposed to cosmic rays at high mountains. They interacted readily and were produced artificially at the Lawrence Lab (Berkley California) with the use of synchro-cyclotron in 1948. π meson was therefore identified as Yukawa's particle. Both pion and muon are short lived with mean lifetimes of the order of 10^{-8} sec and 10^{-6} sec respectively.

By 1947–48, most of the expected particles were discovered, the quantum of nuclear field was detected and the prospects of a comprehensive understanding of

Nuclear physics appeared in sight. But in the same year π meson was discovered, a still heavier unstable particle was discovered by G.D. Rochester and C.C. Butler in a cloud chamber. The event consisted of a neutral particle of mass $\sim 800 m_e$ that was observed to decay into two singly charged particles. Such events were called V events as they leave V shaped tracks in a cloud chamber. Further studies revealed that these heavier short-lived particles were produced copiously, with a frequency of about several percent relative to pions. Such Copious production suggested that these new unstable particles also had strong interactions and should also play a significant role in our understanding of Nuclear physics. By 1954 two types of neutral V particles were established, one called V_1^0. In the modern terminology this is Λ hyperon of mass of about $2200 m_e$ which decays into p and $\pi (\Lambda \rightarrow p + \pi^- + 40$ MeV), and the other one V_2^0 called θ^0 a particle (now called K^0 of mass of 498 MeV/c^2) with the decay mode $\theta^0 \rightarrow \pi^+ + \pi^-$. The mean life times of these particles were of the order of 10^{-10} sec.

In the period 1948–53, a number of heavy mesons (now called K-mesons) decaying into a bewildering variety of modes were discovered. In 1949, the Bristol group (Brown et al.) observed the decay of a heavy meson named τ-meson (now called $K\pi 3$) which came to rest in photographic emulsion and decayed into three charged pions as evidenced by the coplanarity of the three pion tracks and from the zero vector sum of their momenta within the errors of measurements. The Q-value for the decay scheme $\tau^+ \rightarrow \pi^+ + \pi^+ + \pi^- + 75$ MeV, corresponds to a mass of $966 m_e$ for the τ^+ meson—a value which is identical with that of θ^0 meson.

Other decay modes for K-mesons which involve two, three or four particles were also discovered. O'ceallaigh at Bristol found evidence for the decay of a K^+-meson into μ^+ and one or more neutrals.

In the mean time, the charged and neutral sigma hyperons of mass $\sim$1190 MeV/c^2 were discovered. The charged hyperons were found to decay with a mean life times of the order of 10^{-10} sec and the neutral ones with $\sim 10^{-20}$ sec.

$$\begin{aligned} \Sigma^+ &\rightarrow p + \pi^0 \\ &\rightarrow n + \pi^+ \\ \Sigma^- &\rightarrow n + \pi^- \\ \Sigma^0 &\rightarrow \Lambda + \gamma \end{aligned}$$

Evidence for Σ^+ was given by Bonetti et al. in photographic emulsion and York et al. in a cloud chamber. W.D. Walker discovered the Σ^0.

Further, a still heavier hyperon $\Xi^- (\sim 1320$ MeV/c^2) called cascade hyperon which decays in two steps with a lifetime of the order of 10^{-10} sec was established by Caltech.

$$\begin{aligned} &\Xi^- \rightarrow \Lambda + \pi^- \\ &\qquad\quad \hookrightarrow p + \pi^- \end{aligned}$$

Subsequently, its neutral counterpart Ξ^0 was also discovered.

The proliferation of new particles, many with several patterns of decay, produced great confusion. The primary source of confusion was whether each decay mode represented a new particle or was simply an alternative decay of a previously known particle

While the above mentioned heavy unstable particles which were not predicted were discovered in cosmic ray work or at the accelerators using cloud chambers or photographic emulsions, one other heavy hyperon called omega minus (Ω^-) which was predicted was discovered in 1964 in a large bubble chamber.

It was pointed out that the heavy unstable particles are produced in high energy collisions of nucleons and pions with appreciable cross-sections (several percent of geometrical cross-section). On the other hand they decay to systems of pions and nucleons with lifetimes of the order of 10^{-10} sec or longer-extremely long on the nuclear scale ($\sim 10^{-23}$ sec). For example the production of Λ through the interaction $\pi^- + p \rightarrow \Lambda$, and its decay $\Lambda \rightarrow p + \pi^-$ which is like a reversed process take place in great contrast. This contrast between the strength of the production process and the decay interaction was a puzzle for quite some time. This led the theoreticians to caricature them as "strange particles".

In the early history of these heavy unstable particles it was suggested that they possessed very large angular momentum and the decay was retarded by the angular momentum barrier. But this suggestion proved to be untenable. Pais [4] was led to the hypothesis of associated production of strange particles. According to this hypothesis, the strong production interactions are always such that at least two strange particles are involved in the process, while the interactions concerned with only one strange particle, as in the decay process, are extremely weak. Thus Λ in the π–N collisions is not produced singly but in association with a K-meson.

$$\pi^- + p \rightarrow \Lambda + K^0$$

There have been no exceptions to the rule of associated production. By virtue of associated production hypothesis one strange particle may transform into another as in

$$K^- + p \rightarrow \Sigma^+ + \pi^-$$

The associated production interactions do not come in conflict with slow decay of strange particles since they can only transform one strange particle into a system involving another strange particle. The strong interaction $\Lambda \rightleftarrows N + \overline{K^0}$ (where $\overline{K^0}$ denotes the antiparticle to K^0 particle) implied by $\pi^- + p \rightarrow \Lambda + K^0$, can effect transition for Λ particle to a virtual $N + \overline{K^0}$ state but it does not lead to a rapid decay process for the Λ particle, as Λ mass is not sufficiently heavy to allow such a process to be a real process.

After the discovery of e^+, it was assumed that every particle will have its antiparticle. To this end the bevatron (a proton synchrotron) at Berkley, California was to produce high energy protons which upon collision with a suitable target would produce antiprotons (p^-) among other negatively charged particles. (The threshold for p^- production is $6Mc^2$ or 5.6 GeV for free hydrogen target.) The main difficulty was in regard to the small proportion of p^- against a heavy background of π^-, μ^-

and K^-. The search for p^- was not any better than that of a needle in a haystack. With the aid of sound electronic techniques negatively charged particles of the mass of protons (within 10 %) were detected by Amaldi, Chamberlain and Segre in 1955. In the next few years the antiparticles of the known baryons were discovered, $\overline{n}$ by W. Powell and E. Segre et al. at Berkeley, $\overline{\Lambda}$ by D. Prowse and M. Baldo-Ceolin, $\overline{\Sigma^0}$ by J. Button, $\overline{\Xi}^+$ by H.N. Brown and $\overline{\Omega^+}$ by Firestone.

In early sixties, a new class of particles was observed, for example ω^0 (783 MeV/c^2), ρ(770 MeV/c^2), $\Delta 1232$, K^*(892 MeV/c^2) which are so short lived ($\sim 10^{-23}$ sec) that their masses can be determined only indirectly. The mass of these particles shows enormous spread ($\sim$10 %) whose width gives the mean lifetime by the use of the uncertainty relation $\Delta E \cdot \Delta t = h$. For $\Delta E \sim 100$ MeV, $\Delta t \sim 10^{-23}$ sec. Such short lived particles may be regarded as the excited states of baryons and mesons and are named resonant sates or simply *resonances*. They may be regarded as the analogues of myriad of excited states of atoms. The nuclear resonances decay to the system of mesons *and/or* baryons.

The first resonance in particle physics was discovered by H. Anderson, E. Fermi, E.A. Long and D.E. Nagle, at the Chicago cyclotron in 1952, working on π–p scattering. For both $\pi^+ p$ and $\pi^- p$ scattering the cross-section showed a peak at incident pion kinetic energy of 180 MeV.

This resonance at 180 MeV corresponds to an energy of 1225 MeV in the CMS. This was now interpreted as the formation of the $\Delta 1232$ resonance. These resonances which are excited states of hadrons (mesons and baryons) decay through a strong interaction to familiar hadrons in a time of the order of 10^{-23}–10^{-24} sec. The number of particles and resonances had now reached over a few hundred. Efforts were now made to discover families into which various baryons and mesons and their resonances could be classified so that the basic number is reduced and the Particle physics may be better understood. This work was similar to the work done by Balmer and Rydberg by arranging various spectral lines in the form of series before the advent of Bohr's theory of hydrogen atom. In 1964, Gell-Mann and Zweig independently discovered a broken symmetry called $SU(3)$ symmetry (AAK1,5) within whose framework a large number of hadrons could be arranged in supermultiplets with common spin and parity. The discovery of the hyperon omega minus (Ω^-) in a bubble chamber with the predicted properties was a triumph of SU(3) symmetry.

In 1974 a new class of resonances was discovered. They were characterized by very narrow width and therefore decayed by weak interaction. At the Brookhaven National Laboratory, the collision of 28 GeV protons in Be target, to e^+e^- pairs revealed a narrow peak at a total mass of 3.1 GeV/c^2. This was named J resonance. Independently in the SLAC experiment in the e^+e^- collisions in the storage ring a narrow peak was discovered in the partial cross-section for the process

$$\begin{aligned} e^+e^- &\rightarrow \psi \rightarrow \text{hadrons} \\ &\rightarrow e^+e^-, \mu^+\mu^- \end{aligned}$$

at $E_{cm} = 3.1$ GeV.

This was named as J/ψ resonance the width of the resonance being $\Gamma \sim 10$ keV corresponding to $\tau < 10^{-10}$ sec. Γ is 10^4 smaller than the width expected for conventional resonance of mass 3.1 GeV/c^2.

Many more resonances of this type were subsequently discovered. These are called *Charmed Particles* (AAK1,5).

The building blocks of hadrons are believed to be *quarks*. In the SU(3) model the baryons super multiplets are assumed to be built from 3 quarks. The quarks are assigned fractional baryon number and fractional charge. They are believed to have three different flavors, *u* (up), *d* (down) and *s* (strange). The baryons are assumed to be built from three quarks and mesons from a quark and an antiquark combination. Although free quarks have not been discovered, their existence is almost certain. When the theory was proposed (1962) in one of the baryon super multiplets called decuplet, nine members were already known but the tenth one was missing. Gell-Mann and Ne'eman predicted not only the existence of the missing particle but also gave its mass, charge, spin and strangeness. These predictions were soon verified with the discovery of Ω^-.

A new lepton, named the τ particle, was discovered in the study of e^+–e^- interactions in the SPEAR storage ring at the Stanford Linear Accelerator center (SLAC) in 1975 [5]. The threshold for τ^+, τ^- pair production was found to be 3.56 GeV, leading to a mass of 1.78 GeV/c^2 for the τ particle. The production and decay of τ particles are

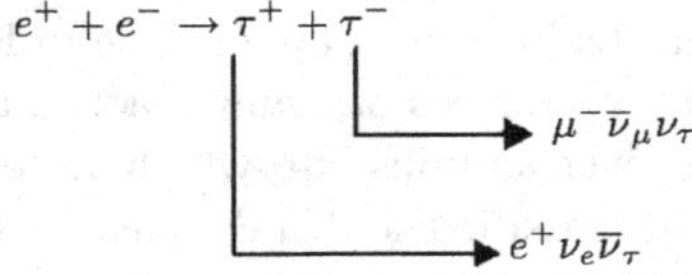

The neutrino ν_τ and its antineutrino $\overline{\nu}_\tau$ are found to be different from ν_e and ν_μ accompanying electron and muon and their antiparticles, which are labeled accordingly. Thus, in all there are three different types of neutrinos, ν_e, ν_μ and ν_τ and their three antineutrinos.

The efforts to classify hadrons into various families have been paralleled by unifying the fundamental forces. One such effort has been successfully accomplished for electromagnetic and weak forces by Weinberg, Salam and Glashow. The two forces are now jointly called electro-weak force just as electricity and magnetism are regarded as two different aspects of the same discipline, that is electromagnetism. The mediating particles of electroweak force are the massive bosons, $W^\pm$ ($\sim$80 GeV/c^2). Their discovery (1983) was crucial to the theory.

Efforts are also being made to unify strong and electro-weak interaction. These are called grand-unification-theories (GUTS). These theories have an important role to play in our understanding of the early universe immediately following the big-bang to which we shall return later (Chap. 6).

To sum up, compared to the simple ideas prevalent in the times of Rutherford when the number of particles known was small, we have moved away to a complex

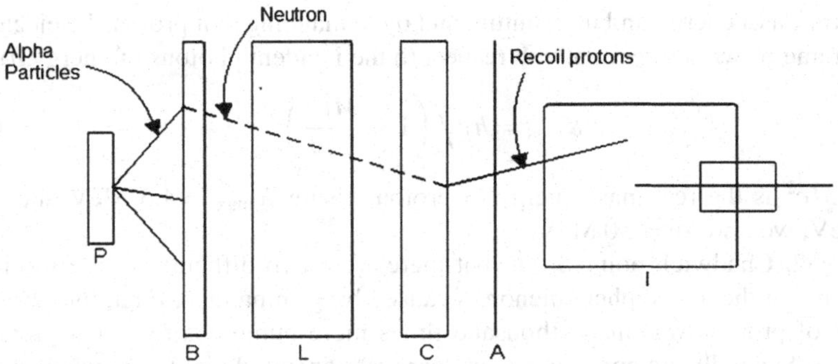

Fig. 3.1 Detection of a radiation by the ionization chamber I when a piece of Beryllium B is bombarded with alpha particles from polonium source P

world of particles. According to the modern theories there are only two basic particles quarks and leptons, in terms of which the rest of particles are to be understood.

3.3 Discovery and Properties of Elementary Particles

3.3.1 Neutron

3.3.1.1 Discovery

The existence of neutron, a neutral particle of almost the mass of proton was speculated by Rutherford in 1920. But not until 1932 was it actually discovered by Chadwick. Earlier in 1930, Bothe in Germany reported that beryllium bombarded with alpha particles from polonium produced a highly penetrating radiation which was assumed to be gamma rays. In 1932, Curie Joliot and F. Joliot bombarded a piece of Beryllium B with alpha particles from polonium source P as in Fig. 3.1. The resultant radiation was detected by the ionization chamber I. A lead absorber L had the effect of attenuating the gamma rays from polonium source. Insertion of absorbers of Al, Cu, Ag, Pb in front of the ionization chamber at C had the effect of reducing the ionization current, but with paraffin, water and cellophane, an increase in the ionization chamber current was recorded. In particular, for paraffin the intensity was double. Assuming that the increased intensity was caused by recoil protons ejected from hydrogenous substances, Curie and Joliot intercepted them with thin foils (0.2 mm) of aluminium at A and showed that this was enough to destroy the effect of enhanced intensity. From the Range-Energy Relation for protons in aluminium, they concluded that the maximum energy of recoil protons $K_{\text{max}} = 4.5$ MeV. Application of magnetic field revealed that the phenomenon could not be attributed to slow electrons, which if present could be easily deflected by a magnetic field. Assuming that energetic gamma rays were responsible for the ejection of protons from hydrogenous substance, maximum energy of protons K_{max} could be found from the conservation of energy and momentum and by demanding that protons be ejected in the extreme forward direction with respect to the incident photons of energy $h\nu$.

$$K_{\max} = h\nu \Big/ \left(1 + \frac{Mc^2}{2h\nu}\right) \tag{3.6}$$

where Mc^2 is the rest mass energy of proton. Using $K_{\max} = 4.5$ MeV and $c^2 =$ 940 MeV, we find $h\nu = 50$ MeV.

In 1932, Chadwick pointed out that there were two difficulties with the interpretation that the above phenomenon is caused by gamma rays. First, the observed number of protons was many thousand times more numerous than that predicted by theory. Secondly, no conceivable nuclear reaction could yield as much energy as 50 MeV for the gamma rays. Chadwick assumed that the phenomenon was caused by massive particles which were uncharged and could, therefore, pass freely through matter.

He set forth to determine the mass of neutron. Let the mass of neutron be m times the mass of proton which is taken as unity. Let neutron of velocity v_0 make a head-on elastic collision with the hydrogen nucleus. This ensures that the proton would be ejected with maximum energy. After collision, proton will move with velocity v_H and neutron with velocity v. Conservation of energy yields

$$\frac{1}{2}mv_0^2 = \frac{1}{2}v_H^2 + \frac{1}{2}mv^2 \tag{3.7}$$

In the event of head-on collision, the proton will necessarily move along the direction of incidence. Neutron will also continue to move in the direction of incidence if heavier than proton and in the opposite direction if lighter than proton. Taking both the possibilities into account, momentum conservation gives,

$$mv_0 = v_H \pm mv \tag{3.8}$$

Eliminating v between (3.7) and (3.8),

$$v_H = \frac{2v_0 m}{m+1} \tag{3.9}$$

Similarly, when nitrogen target is used,

$$v_N = \frac{2v_0 m}{m+14} \tag{3.10}$$

Dividing (3.9) by (3.10)

$$\frac{v_H}{v_N} = \frac{m+14}{m+1} \tag{3.11}$$

Chadwick used $v_H = 3.3 \times 10^9$ cm/sec obtained from his own experiment, and $v_N = 4.7 \times 10^8$ cm/sec as obtained by Feather from cloud chamber measurements. Inserting these values in (3.11) gives $m = 1.15$ units, with an estimated error of 10 %. Thus, neutron mass was shown to be close to that of proton within experimental error. Further, the initial velocity v_0 obtained from (3.9) or (3.10) is 3.2×10^9 cm/sec, corresponding to the kinetic energy or about 6 MeV.

Chadwick assumed that neutron was produced in the interaction of 5.3 MeV alpha particles with Beryllium-9 nucleus.

$$^{9}_{4}\text{Be} + ^{4}_{2}\text{He} \rightarrow ^{12}_{6}\text{C} + ^{1}_{0}n \tag{3.12}$$

From the known masses of the nuclei involved in the above reaction, the Q value could be found out and it was concluded that when alpha particles of energy 5.3 MeV were used, a maximum of 8 MeV was available for the neutrons that are produced, a value which is not too much off from the quoted value of 6 MeV.

He repeated the experiment with boron target and assumed neutrons to be produced by the reaction:

$$^{11}\text{B} + ^{4}\text{He} \rightarrow ^{14}\text{N} + ^{1}n \tag{3.13}$$

Here a much better value of the neutron mass could be determined. The accepted value for the neutron mass in $1838.86m_e$ which is only 0.14 % larger than proton.

3.3.1.2 Mass

A free neutron is unstable against decay. The mass of neutron is determined from the measurement of the end-point energy in free neutron decay.

$$n \rightarrow p + \beta^- + \overline{\nu}$$

$$\begin{aligned} m_n c^2 &= m_p c^2 + m_e c^2 + T_{\max} \\ &= 938.280 + 0.511 + 0.782 \\ &= 939.573 \pm 0.003\ \text{MeV} \end{aligned}$$

3.3.1.3 Mean Life Time

The mean life time is found to the 898 sec.

3.3.1.4 Neutron Sources

Po-Be Source

If ^{210}Po be mixed with ^{9}Be then alphas from ^{210}Po upon hitting the nuclei of ^{9}Be would produce neutrons through the reaction.

$$^{9}_{4}\text{Be} + ^{4}_{2}\text{He} \rightarrow ^{13}_{6}\text{C}^* \rightarrow ^{12}_{6}\text{C} + ^{1}_{0}n + 5.7\ \text{MeV}$$

The neutrons have energy spectrum, ranging from 6.69 MeV to 10.85 MeV. The α's at 5.3 MeV from ^{210}Po are elegantly suitable because at this energy the neutron yield from the above reaction is relatively high compared to other energies.

Alternatively Ra-Be source can be used but has distinct disadvantages in that radon as well as γ-rays are emitted.

Photo-Neutron Source

When energetic photons above 2.18 MeV are fired against deuterium target, monoenergetic neutrons are obtained through the reaction

$$\gamma + {}^2\mathrm{H} \rightarrow p + n$$

Also, energetic neutrons are obtained from photo-nuclear reactions with beryllium target.

$$\gamma + {}^9\mathrm{Be} \rightarrow n + \alpha + \alpha$$

Nuclear Reactions of the Type (p, n)

When energetic protons from accelerator are fired against a suitable target like ${}^7\mathrm{Li}$ or ${}^3\mathrm{H}$ neutrons are produced.

$$ {}^7\mathrm{Li} + p \rightarrow {}^7\mathrm{Be} + n - 1.63\ \mathrm{MeV}$$

$$ {}^3\mathrm{H} + p \rightarrow {}^3\mathrm{He} + n - 0.73\ \mathrm{MeV}$$

If neutrons are accepted at a certain angle then they are monoenergetic. Also at high energies protons on bombarding various targets can produce knock-on neutrons.

Deuterons Stripping Reactions of the Type (d, n)

Special interest is centred around deuteron induced reactions (stripping). Emerging neutrons are strongly collimated in the forward direction in a narrow cone. Neutron energy is centered around half of the deuteron energy. This is to be distinguished from the direct reactions in which neutrons have a much larger spread in energy and angle.

The Nuclear Reactor as an Intense Neutron Source

In a nuclear reactor neutron fluxes as large as $10^{14}\ \mathrm{m}^{-2}\,\mathrm{s}^{-1}$ may be obtained. The neutron energies range from thermal (fraction of eV) to a few MeV.

3.4 Positron

3.4.1 Discovery

Dirac's relativistic theory of electron predicted the antiparticle of electron, i.e. a particle of exactly the same mass as electron, same spin but opposite magnetic moment and positively charged. Such a particle was independently observed by Carl

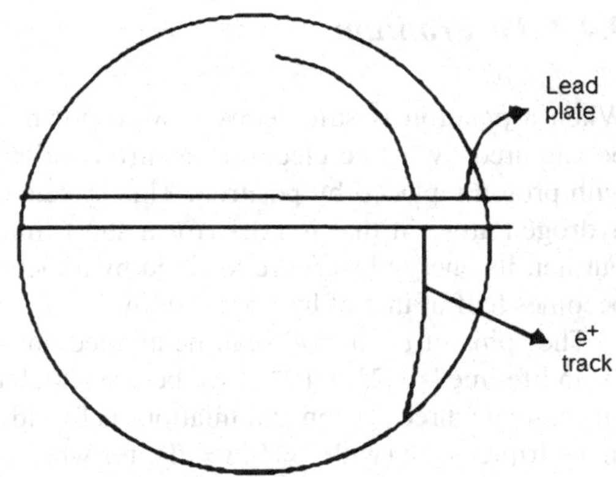

Fig. 3.2 Track of a particle in a strong magnetic field

Anderson and R.A. Millikan in a cloud chamber in 1930. A track shown in Fig. 3.2 represents the photograph of a track of a particle of the same mass similar to electron in a strong magnetic field. The nature of curvature is such that either electron travels in the clockwise sense or a positively charged particle in the counter clockwise sense. As a lead plate of a few mm thickness intercepts the path energy loss in the plate leads to a smaller radius of curvature in the upper half compared to that in the lower half, the sense of direction is counter clockwise, thereby establishing the positive charge of the particle. The interpretation of the particle being proton is ruled out as it would have stopped within the plate.

By 1932 Anderson, out of a group of 1300 photographs of cosmic-ray tracks in a vertical Wilson chamber registered 15 tracks due to positive particles which could not have a mass as great as that of proton. From the observed energy loss and ionization produced it was concluded that charge is less than twice and was probably exactly equal to that of proton The ionization-energy measurements showed that the tracks are caused by particles of electron mass. While the paper was in preparation [6] press reports had announced that P.M.S. Blackett and G. Occhialini in an extensive study of cosmic ray tracks ($\gamma \rightarrow e^+ + e^-$) had also obtained evidence for the existence of light positively charged particles confirming this discovery.

3.4.2 Sources

(a) Positrons are produced from the materialization of γ-rays into e^+–e^- pairs, provided $E_\gamma \geq 2mc^2$ i.e. $E_\gamma \geq 1.02$ MeV, the threshold energy for e^+e^- pair production.
(b) Positrons are produced by artificial radioactivity, for example

$$^{30}\text{P} \rightarrow {}^{30}\text{Si} + \beta^+ + \nu_e$$

$$^{13}\text{N} \rightarrow {}^{13}\text{C} + \beta^+ + \nu_e$$

3.4.3 Positronium

When a positron is sufficiently slowed down in matter due to ionization, it may be captured by a free electron. Positron and electron form a hydrogen like atom with proton replaced by positron. This system called positronium differs from the hydrogen atom in that it exists for a short time before it annihilate into photons. Further, its energy levels are lowered by a factor of 2 because the reduced mass μ becomes half of that of hydrogen atom.

The spins of e^+ and e^- can be aligned antiparallel (with $l = s = 0$) for which mean lifetime is 1.25×10^{-10} sec before which they annihilate into two photons. In this case the three photon annihilations is forbidden. Alternatively e^+ and e^- can be in the triplet state (with $l = 0$, $s = 1$), for which the mean lifetime is 1.5×10^{-7} sec before which they annihilate into three photons. In this case the two-photon annihilation is forbidden. These results follow from C-invariance (Chap. 4). Annihilation into a single photon is forbidden because of linear momentum conservation.

Positronium is basically a loose structure as its ionization potential (6.8 V) is half of that of hydrogen atom. The wavelengths in the emission spectra of positronium are scaled up by a factor of 2.

3.5 Muon

3.5.1 Discovery

After the earlier discovery of the penetrating component of cosmic rays, C. Anderson and S.H. Neddermeyer [1] made energy loss measurements by placing l-cm platinum plate inside a cloud chamber at high mountain altitudes. By measuring the curvature of tracks on both sides of the plate, they were able to determine the loss in momentum for particles of momenta 100–500 MeV/c. The particles could be separated into two classes. For the first class, particles lost energy due to bremsstrahlung for which $dE/dx \propto E$ (Bethe and Heitler) and if sufficiently energetic may cause electromagnetic shower. The particles in the second class however lost practically no energy in the platinum plate. They were "penetrating". Further, they had to exclude the possibility that this component was due to protons.

About the same time Street and Stevenson confirmed the results obtained by Anderson and Neddermeyer and set to determine the mass of penetrating particles by ionization-momentum method. They used counters both in coincidence and anticoincidence to select events of interest. The counters fired only if a charged particle passed through them, the chamber was expanded to produce supersaturation and a picture taken only if the particle entered the chamber (coincidence) but was not detected while exiting (anticoincidence). This method of triggering the chamber was invented by Blackett and Occhialini. Also, a lead block was placed in front of the apparatus to screen out the shower particles. In late 1937, Street and Stevenson reported a track that ionized too much to be an electron with the measured momentum, but traveled too far to be a proton. They measured the mass crudely as 130 times the rest mass of the electron, a value smaller by a factor of 1.6 than the later improved value, but good enough to place it between the electron and the proton.

In 1940 Tomonaga and Araki [7] showed that positive and negative Yukawa particles should produce very different effects when they came to rest in matter. The negative particles would be captured into atomic-like orbits, but with very small radii. As a result they would overlap the nucleus substantially. Yukawa's particle supposed to explain nuclear forces, is expected to interact quickly with the nucleus, and get absorbed before decay. On the other hand the positive Yukawa particles would be repulsed by the nucleus and decay after coming to rest.

The mean lifetime of the penetrating particles was first measured by Franco Rassetti who found a values of about 1.5×10^{-6} sec. Improved results near 2.2×10^{-6} sec were obtained by Rossi et al. During World War II in Italy Conversi, Pancini and Piecioni investigated the decays of positive and negative penetrating particles from cosmic rays, using a magnetic focusing arrangement. They could select positive or negatively charged particles and then determine whether they decayed or not when stopped in matter. The positive particles did indeed decay as expected. When the absorber was iron the negative particles did not decay, but were absorbed by the nucleus, in accordance with the theory. However, when the absorber was carbon, the negative particles decayed. This implied that Tomonaga-Araki prediction was wrong by many orders of magnitude. These could not be the Yukawa particles. These particles, now called muons for brevity, in many ways resemble heavy electrons and belong to the Lepton family comprising particles that do not participate in nuclear interactions but are subject only to electromagnetic and weak interactions. Muons exist only in two charge states (μ^+, μ^-) and experiments show that they carry charge in magnitude equal to that of electron to an accuracy better than one part in 20,000. Neutral muons (μ^0) do not exist.

3.5.2 *Mass*

(a) Direct comparisons of the muon mass with those of the proton and pion were first made by using 184 inch cyclotron by the mass ratio method. Suppose the momenta and corresponding range of two kinds of particles are obtainable. When the ratio of their momenta is equal to the ratio of their ranges, their velocities are equal and their common ratio is equal to their mass ratio. Protons were used as comparison particles in the experiment to determine the μ^+/p mass ratio. Muon mass was found as $m_\mu = (105.7 \pm 0.1)$ MeV/c^2.
(b) The mass of μ^- particle has been determined from an accurate energy measurement of X-rays emitted in the phosphorus μ^- mesic atoms. When μ^- meson is captured by an atom, it is called mesic atom. The meson cascades down to lower energy levels. In this process either photons in X-ray region are emitted or Auger electrons are ejected. Owing to a large value of reduced mass, the

radii are shrunk by a factor of 200 compared to hydrogen-like atom. The 3D-2P transition energy for the phosphorus μ^- mesic atom is accidentally within the absorption edge of lead where absorption coefficient changes rapidly with X-ray energy. A measurement of attenuation coefficient permits X-ray energy to be determined. The measured X-ray energy is matched with that calculated from theory. In this way it is found that

$$m_\mu / m_e = 206.76 \pm 0.02$$

or

$$m_\mu = 105.654 \text{ MeV}/c^2$$

(c) The gyromagnetic ratio g_μ of muon. The gyromagnetic ratio is defined as the ratio of magnetic moment and the spin that is $g = \mu / I$. It is found that

$$g_\mu = 2\left[1 + (1.162 \pm 0.005) \times 10^{-3}\right]$$

and the ratio between the magnetic moments of the muon and proton.

$$\frac{\mu_\mu}{\mu_p} = 3.18334 \pm 0.00005$$

This result can be combined with the accurately known ratio between the magnetic moment of electron and that of proton.

$$\mu_e / \mu_p = 658.2107$$

$$\mu_e / \mu_\mu = \frac{g_e}{g_\mu} \cdot \frac{m_\mu}{m_e} = 206.767 \pm 0.003$$

Finally, using the accurately known value of the g-factor

$$g_e = 2\left[1 + (1.161 \pm 0.002) \times 10^{-3}\right]$$

and combining all these data

$$m_\mu = (105.653 \pm 0.002) \text{ MeV}/c^2$$

which is consistent with the values obtained by other methods.

The results for the value of g_μ are consistent with those for Dirac's particles. The absence of any anomalous part in the magnetic moment indicates that the muon is subject only to weak and electromagnetic interactions. The anomalous magnetic moments of proton and neutron are in fact attributed to the strong interactions to which they are subjected.

3.5.3 Lifetime

Lifetime of μ^+ was already known in cosmic ray work. But a more reliable value has been known for work with accelerator. The lifetime of μ^+ mesons has been determined by detecting decay of μ^+ arising from artificially produced π^+ mesons

Fig. 3.3 e^+ energy spectrum in μ^+ decay

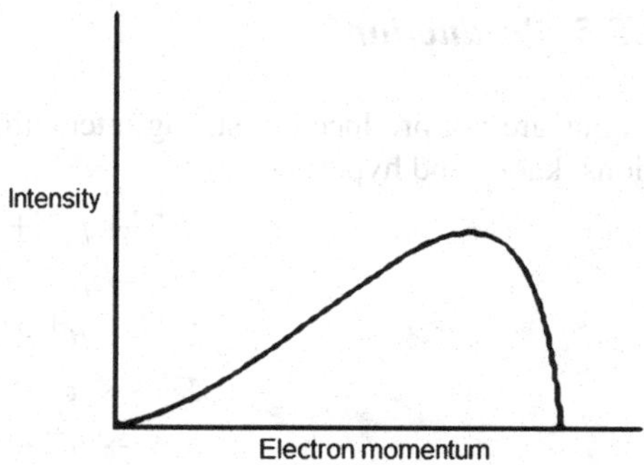

stopped in a carbon block in delayed coincidence using scintillator counters. The accepted value of the τ for μ^+ is $(2.197 \pm 0.015) \times 10^{-6}$ sec.

Free μ^- have the same decay constant as μ^+ with an accuracy of one part in thousand, a result which is anticipated from the CPT theorem for the identity of lifetimes for particle and antiparticle (Chap. 4). The lifetime of μ^- is modified when they stop in a substance and interact with nuclei. Observed mean lifetime is related to mean lifetime τ_μ for free decay and τ_c for capture as follows:

$$\frac{1}{\tau} = \frac{1}{\tau_\mu} + \frac{1}{\tau_c} \tag{3.14}$$

In very light elements τ is nearly the same as τ_μ but is considerably modified for heavier elements. For $Z = 11$, the two terms on the right side are almost equal.

A high value for τ_c implies a weaker interaction with nuclear matter. This important fact discovered by Piccioni et al. in 1947 [8], laid the foundation of discarding μ meson as Yukawa's particle.

3.5.4 *μ-Decay Spectrum*

As the e^+ spectrum for the decay of μ^+ is continuous, at least two other particles must be involved. μ^+ decays through the mode $\mu^+ \rightarrow e^+ + \overline{\nu}_\mu + \nu_e$, and $\mu^- \rightarrow e^- + \nu_\mu + \overline{\nu}_e$. The theoretical momentum spectrum of positrons for the decay of μ^+ is shown in Fig. 3.3. Excellent fit is obtained with observations. The decay proceeds via weak interaction and is similar to beta decay. Of the two neutrinos, one is believed to be of the kind found in beta decay, and the other of kind found in pion decay. The maximum momentum of the decay electrons is $P_{\max} = 52$ MeV/c. The energy of the electrons is greatest when the two neutrinos escape in the same direction opposite to electron.

3.5.5 Production

Muons are not produced in strong interactions, but result from the weak decays of pions, kaons and hyperons

$$\pi^+ \to \mu^+ + \nu_\mu$$
$$\pi^- \to \mu^- + \overline{\nu_\mu}$$
$$\Sigma^+ \to \mu^+ + \nu_\mu + n$$
$$\Sigma^- \to \mu^- + \overline{\nu_\mu} + n$$
$$\Lambda \to \mu^- + \overline{\nu_\mu} + p$$
$$K^+ \to \mu^+ + \nu_\mu \quad \text{etc.}$$

The above hyperon decays are the rare decays. Muons can also be produced in pairs by photo production processes. The formula which gives the production cross-section for muons is similar to that for electron pairs production except that the mass of muon replaces that of electron.

The cross-section for muon production from neutrino (ν_μ) interaction is small, being about 10^{-38} cm^2 at 1 GeV.

$$\nu_\mu + n \to \mu^- + p$$
$$\nu_\mu + p \to \mu^+ + n$$

Muon pairs may be produced in high energy hadron collisions. An example is

$$p + p \to \mu^+ + \mu^- + X$$

where X is any hadronic state. This is interpreted in terms of quark-parton model (Chap. 6).

3.5.6 Interactions

When μ^- reaches the ground state of a mesic atom of high Z element, the Bohr orbits may be well inside the nucleus itself. Instead of decaying μ^- would have a high probability of undergoing an elementary interaction of the type $\mu^- + p \to n + \nu_\mu$. The ν_μ resulting from the reaction carries most of the energy, and the neutron much less energy. Consequently, in many cases, the neutron may be trapped within the nucleus having suffered a collision. Neutron multiplicities from μ^- captures in Pb and Ag have been observed with cadmium loaded liquid scintillator tank. The observed distributions are in agreement with the theoretical distributions as given by Fermi gas model, the average multiplicity of neutrons in Mg, Na and Pb being 0.6, 1.0 and 1.7, respectively. When the neutron is unable to escape from the nucleus, the typical reactions observed are

$$\mu^- + {}^3\text{He} \to + {}^3\text{H} + \nu_\mu$$
$$\mu^- + {}^{12}\text{C} \to + {}^{12}\text{B} + \nu_\mu$$
$$\mu^- + {}^{16}\text{O} \to + {}^{16}\text{N} + \nu_\mu$$

3.5.7 Muon Scattering

Low energy muon-nucleon scattering is explained by Rutherford scattering from an extended charge distribution.

3.5.8 Catalytic Reactions

Alvarez discovered an unexpected nuclear reaction in which μ^- mesons stopping in a liquid hydrogen bubble chamber catalyzed the nuclear fusion reactions

$$H + D + \mu^- \rightarrow {}^3He + \mu^- + 5.3\ \text{MeV}$$

and

$$D + D \rightarrow {}^3H + {}^1H$$

In an experiment on K^- mesons, some μ^- got contaminated. Some of the μ^-'s were initially captured to form mesic atoms of hydrogen and this neutral system drifts until μ^- decays or encounters a deuteron. The reduced mass effect on the meson binding is 135 eV higher for $D^+ + \mu^-$ than for $H^+ + \mu^-$ and consequently the transfer of μ^- from H to D becomes feasible and this new system drifts until μ^- decays or a D–μ–H nuclear ion is formed in the next event. This mesic structure enables D and H to get closer compared to ordinary hydrogen molecular ion and vastly enhances the probability for the fusion reaction. A few events of the type $D + D \rightarrow {}^{+3}H + {}^{+1}H$ were also observed. An increased concentration of deuterium caused an increased occurrence of these events. If μ^- was a stable particle or its lifetime was several orders of magnitude longer, this phenomenon would have been a useful source of nuclear power.

3.5.9 Electric Dipole Moment

Experiments have been conducted to detect a possible electric dipole moment for the muon, with negative results (Chap. 6).

3.5.10 Spin

The spin assignment for muon is $1/2$. All the evidences point to this value.

a. The observation of zero cut-off at the high energy end of the electron spectrum in the muon decay excludes spin $3/2$ for the μ-meson. This is because all direct couplings of a spin $3/2$ μ-meson with a spin $1/2$ electron and two spin $1/2$ neutrinos lead to electron spectra with finite cutoffs. Spin 0 is also excluded on account of the observed general shape of the electron spectrum (Fig. 3.3).

b. Evidence in favour of spin 1/2 also comes from the observations on the bursts of ionization produced by very energetic μ mesons (the so-called "burst production" by the penetrating component of cosmic radiation). These bursts of ionization consist of showers of electrons and positrons similar to the showers produced in the lead or iron plates by high energy photons or electrons. Now the cross-sections for these electromagnetic processes are strongly spin-dependent at high energies. Data are consistent with spin 1/2 assignment.
c. Another argument comes from μ mesic atoms whose spectra show structures consistent with the assignment of spin 1/2 and $g \simeq 2$ for the muons.
d. A direct evidence comes from the study of muonium, a system formed by μ^+ and e^- resembling a hydrogen atom. The direction of emission of positrons from the decay of muons is observed. Under a magnetic field this direction precesses with a characteristic frequency and permits the spin and magnetic moment to be determined.
e. One other evidence for $s = 1/2$ comes from the experiments which show the non-conservation of parity in muon decay (Chap. 7).

Let a π^+ come to rest and decay emitting a neutrino and a μ^+ in opposite directions. The pion does not have any angular momentum. Here μ^+ has an angular momentum equal and opposite to that of the neutrino. If the neutrino is polarized with the spin antiparallel to the direction of motion, so is the muon μ^- a fact which has been verified experimentally. The μ meson comes to rest in a very short time compared with its mean lifetime. If the polarisation is not destroyed in the slowing down process then the positron and the two neutrinos are emitted from a polarized source. The positrons tend to travel preferentially parallel to the spin direction. Suppose now a magnetic field is applied to the muon at rest. The field causes a rotation of the direction of polarization with an angular velocity $\omega = \frac{g_e H}{2m_\mu c}$, where $g = \mu/I$ is the gyromagnetic ratio, i.e. the magnetic moment in natural units $e\hbar/2m_\mu c$ divided by angular momentum in units of $\hbar$. This rotation is observed as the change of direction of preferential emission of the positron and g was found to be $2 \times (1.0020 \pm 0.0005)$, very close to the value for electron. This result is a strong argument for spin 1/2 for muon. Indeed if the spin was 3/2 or larger the g value would have been completely different.

3.6 Pions

3.6.1 Discovery

Shortly afterwards D.H. Perkins used sensitive photographic emulsions and found an event which had a slow negative particle that came to rest in an atom, most likely from the CNO group, and was absorbed by the nucleus leading to its disintegration into three fragments. This single event apparently was consistent with the predictions of Yukawa.

The connection between the results of the Italian group and the observation of Perkins was found by the Bristol group of Lattes, Occhialini and Powell who established that there were indeed two different particles one of which decayed into the other. The pion (π) decayed into a muon, (μ) and very light particle probably neutrino. The π (probably seen by Perkins) was very much like Yukawa's particle, and μ (which Anderson and Neddermeyer had found) was like an election, only heavier.

The pion has three charge states π^+, π^0, π^-. The two charge states π^+ and π^- are charge conjugate of each other and yield μ^+ and μ^-, respectively in their decays.

While a negative pion would always be absorbed by a nucleus upon coming to rest, the absorption of negative muon was much like the well known radioactive phenomenon of K-capture in which an inner electron is captured by a nucleus while a proton is transformed into a neutron and a neutrino is emitted. In heavy atoms, the negative muon could be observed (because of large overlap with the nucleus) with small nuclear excitation and the emission of a neutrino, while in the light atoms it would usually decay, because there is insufficient overlap between the muon and the nucleus.

Yukawa's particle had been found and the only unanticipated particle was the muon, of which I.I. Rabi is said to have remarked "Who ordered that"? the question remains unanswered

The pion (π meson) was discovered in 1947 by C.F. Powell, Occhialini and Lattes [9], and was identified with the postulated particle of nuclear forces. Both π^+ and π^- were discovered using photographic emulsions exposed to cosmic rays at high mountains. The sequence of decays of $\pi^+ \rightarrow \mu^+ \rightarrow e^+$ were observed, and π^- was observed to produce a 'star' due to its capture by an atom of emulsion. The neutral counterpart π^0 was subsequently discovered through its electromagnetic decay $\pi^0 \rightarrow 2\gamma$.

The copious production of pions in nucleon-nucleon collisions and their strong interactions were in good qualitative agreement with the theory of nuclear forces.

The nuclear forces are believed to result from the mutual exchange of massive quanta of an intermediary meson field strongly coupled to neutrons and protons. The known range of these forces, given by $(\hbar/m_\pi c)$ indicates m_π to be of the order of $300 m_e$ which is in good agreement with the accepted mass of pion, $273 m_e$. Again, the artificial production of pions in N-N collisions and their large interaction cross-sections substantiated their expected behaviour.

3.6.2 Artificial Production of Pions

In 1948, a 184 inch synchrocyclotron was built at Berkley (California) which accelerated protons to 350 MeV. Energetic protons on hitting suitable targets produced pions. The charged pions (π^+, π^-) were observed by Lattes and Gardner using photographic emulsions. The neutral pions (π^0) produced in proton collisions in carbon and beryllium targets were observed by Bgorklund, Crandall, Moyer and York by

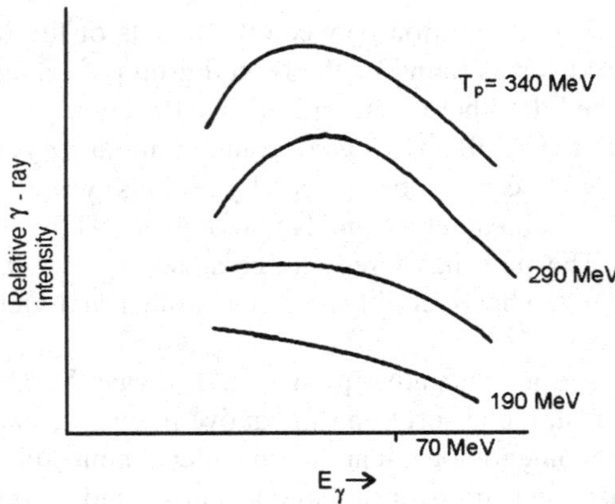

Fig. 3.4 Production of neutral pions and their decay into photons

recording e^+e^- pairs resulting from the decay $\pi^0 \to 2\gamma$, using a pair spectrometer. When proton energy increased from 175 MeV to 340 MeV, bremsstrahlung radiation dramatically increased, the increase in photon intensity being due to π^0 production.

At the higher energy, the photon spectra also drastically change their shape compared to the bremsstrahlung spectra. This was attributed to the production of neutral pions and their decays into photons, Fig. 3.4.

Evidence for π^0 was also provided in photographic emulsions by the Bristol group (Carlson, Hooper and King) in 1950, by the observation of conversion of photons into e^+–e^- pairs.

Direct evidence was provided by Steinberger, Panofsky and Stella [10] in the experiments with electron synchrotron. Electron beam was used to generate a beam of γ-rays, up to $E_\gamma = 330$ MeV. π^0's were produced when γ-rays struck Beryllium target. The γ-rays resulting from the decay of π^0's were accepted in coincidence by two photon detectors. The coincidence rate as a function of θ between planes of emitted photon and beam direction was consistent with the decay scheme $\pi^0 \to 2\gamma$.

Two photon decay proved that π^0 could not have spin 1. This is consistent with Yang's theorem which forbids the decay of spin-l particle into two photons. In 1951, Panofsky, Aamodt and Hadley studied reactions caused by π^- stopping in hydrogen and deuterium using a sophisticated pair spectrometer

$$\pi^- + p \to \pi^0 + n \tag{i}$$

$$\pi^- + p \to \gamma + n \tag{ii}$$

Reaction (ii) produced monochromatic photons and yielded a mass $m_{\pi^-} = 275.2 \pm 2.5 m_e$, in good agreement with the known values. Reaction (i) produced photons from π^0 decay which had Doppler broadening. It yielded the mass difference $m_{\pi^-} - m_{\pi^0} = 10.6 \pm 2.0 m_e$ (see Example 3.27 and Fig. 3.5).

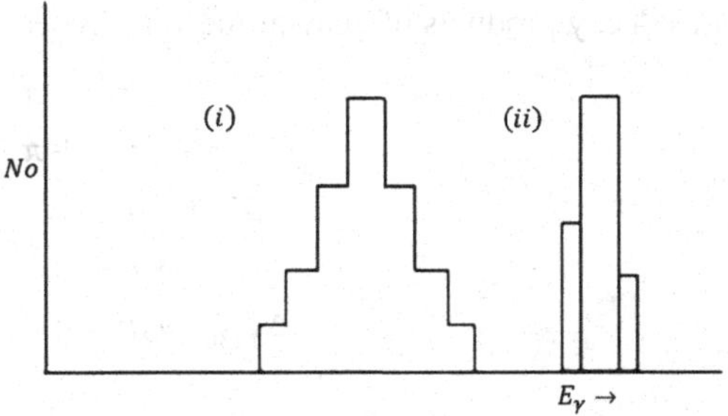

Fig. 3.5 Pair spectrum resulting from the absorption of π^- mesons in hydrogen

3.6.3 Threshold for Single Pion Production in N-N Collisions

Consider the interactions

$$p + p \rightarrow n + p + \pi^+$$
$$p + n \rightarrow p + p + \pi^-$$
$$p + n \rightarrow p + n + \pi^0$$

Calculations for the threshold for particle production have been done in Chap. 2

$$T_1(\text{threshold}) = \frac{m_1^2}{2m_2} + m_1\left(1 + \frac{m_2}{m_2}\right)$$
$$= \frac{(139.5)^2}{2 \times 938} + 139.5\left(1 + \frac{938}{938}\right)$$
$$= 289.37 \text{ MeV}$$

Note that this is slightly in excess of double the pion rest energy because of relativistic effects.

In case of a complex target nucleus, high energy collisions are expected to take place with individual nucleons, because of small de Broglie wavelength of the incident particle. However, because of the Fermi momentum of the target nucleons the threshold energy becomes much smaller. For a nucleon moving with maximum momentum of 218 MeV/c in the opposite direction to that of incidence corresponding to $T_F(\text{max}) = 25$ MeV, the threshold is lowered to about 160 MeV.

In case of photo-mesic production in hydrogen via

$$\gamma + p \rightarrow \pi^+ + n$$

the threshold energy is lowered to 150 MeV, but the production cross section is lower by two orders of magnitude because of electromagnetic interaction.

At higher incident energies, multiple meson production becomes possible.

3.6.3.1 Production from the Annihilation of Antiprotons

Pions are also produced in the annihilation of antinucleons

$$p^- + p \rightarrow n\pi$$

or as decay products of heavier mesons, hyperons or their resonances, such as

$$\begin{aligned} K^+ &\to \pi^+ + \pi^0 \\ &\to \pi^+ + \pi^+ + \pi^- \\ \Sigma^+ &\to p + \pi^0 \\ &\to n + \pi^+ \\ \Delta_{1236}^{++} &\to p + \pi^+ \end{aligned}$$

3.6.4 Decay

The dominant decay mode of charged pion is

$$\pi^+ \to \mu^+ + \nu$$
$$\quad \hookrightarrow e^+ + \nu_e + \overline{\nu}_\mu$$

The rare decay mode is

$$\pi \to e^+ + \nu$$

The ratio of the decay modes

$$R = \frac{\pi \to e\nu}{\pi \to \mu\nu}$$

provides a crucial test for the V-A theory (Chap. 6). The experimental ratio $R_{exp} = (1.267 \pm 0.023) \times 10^{-4}$ is in excellent agreement with the ratio $R = 1.275 \times 10^{-4}$ predicted for A (axial Vector) coupling and in violent disagreement with P (pseudo scalar) coupling which gives $R = 5.5$. The rare $\pi \to e\nu$ process produces positrons of unique energy, around 70 MeV. On the other hand the spectrum for μ^+ decay extends up to 53 MeV (Fig. 3.6).

In the experiment, the rare decays $\pi \to e^+\nu$ are distinguished from the more frequent scheme $\pi^+ \to \mu^+ \to e^+$ by the momentum, timing (the mean life time of the pion is 25 ns, that of muon 2200 ns), as well as the absence of muon pulse in the counters. Even rarer is the decay mode (10^{-8}), $\pi^+ \to \pi^0 + e^+ + \nu_e$ analogous to β decay.

Other decay modes are also possible for the charged pion such as

$$\pi \to \begin{cases} e\nu & \text{branching fraction } 1.27 \times 10^{-4} \\ \mu\nu\gamma & \text{branching fraction } 1.24 \times 10^{-4} \\ e\nu\gamma & \text{branching fraction } 5.6 \times 10^{-8} \\ \pi^0 e\nu & \text{branching fraction } 1.02 \times 10^{-8} \end{cases}$$

The neutral pions decay predominantly through the scheme

$$\pi^0 \to 2\gamma$$

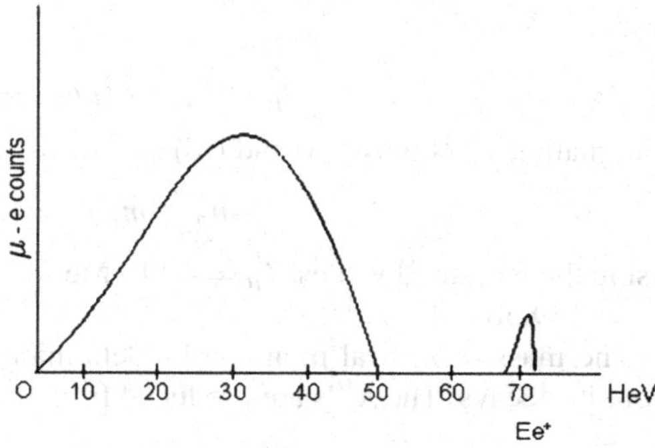

Fig. 3.6 Spectrum for μ^+ decay

The alternative rare decay mode is via Dalitz pair (1951)

$$\pi^0 \rightarrow e^+ + e^- + \gamma$$

This occurs at a frequency of two orders of magnitude lower (1.2 %). Still less frequent decay mode is via Double Dalitz Pair,

$$\pi^0 \rightarrow (e^+ + e^-) + (e^+ + e^-)$$

in which each photon internally converts into a pair, the process is down by four orders of magnitude.

3.6.5 Mass

If two different kinds of particles 1 and 2 having the same velocity are stopped in a photographic emulsion, then the momenta are in the same ratio as their ranges and their masses.

$$\frac{p_1}{p_2} = \frac{m_1}{m_2} = \frac{R_1}{R_2}$$

One of the particles could be some standard particle like proton. Barkas [11] measured the ratio of the pion-muon mass and of the pion-proton mass. The parameters measured were the range (in emulsion) and momentum. These measurements give the masses to an accuracy of about one part in a thousand. m_π is found to be equal to $273.12m_e$.

Pion mass can also be determined from carefully measured energy of μ^+ produced in the decay of π^+ at rest.

$$\pi^+ \rightarrow \mu^+ + \nu$$

Assuming

$$m_\nu = 0, \qquad T_\mu + T_\nu = m_\pi - m_\mu \quad \text{(Energy Conservation)} \tag{i}$$

$$p_\mu = p_\nu \quad \text{(Momentum Conservation)} \tag{ii}$$

or

$$p_\mu^2 = T_\mu^2 + 2T_\mu m_\mu = p_\nu^2 = T_\nu^2 \quad \text{(iii)}$$

Eliminating T_ν between (i) and (iii)

$$(m_\pi - m_\mu)^2 = 2T_\mu m_\pi \quad \text{(iv)}$$

Using the measured values, $T_\mu = 4.119$ MeV and $m_\mu = 105.659$ MeV/c^2, one finds $m_\pi = 139.569$ MeV/c^2.

The mass of neutral pion can be determined from the measurement of γ-rays from its decays. The π^0's are produced from the capture of π^- in hydrogen.

$$\pi^- + p \rightarrow \pi^0 + n \quad \text{(v)}$$
$$\pi^0 \rightarrow 2\gamma$$

The competitive reaction is

$$\pi^- + p \rightarrow n + \gamma \quad \text{(vi)}$$

The ratio of the cross-sections for the reaction (v) and (vi) is called Panofsky ratio and has a value of 1.56. Reaction (vi) gives monochromatic γ-rays of energy 130 MeV. In reaction (v), π^0 moves with constant speed. While in motion π^0 disintegrates into 2γ, each γ-ray carrying energy equal to the half of rest mass energy of π^0 in the frame of reference in which π^0 is at rest. For an isotropic angular distribution of γ-rays in the rest frame of π^0, the energy distribution of γ-rays in the Lab will be uniform, extending from the minimum value E_- to a maximum value E_+ and the mass of π^0 is obtained from $m_{\pi^0}c^2 = 2\sqrt{E+E^-}$ (see Example 3.27). The γ-ray energy was measured by a pair spectrometer. The accepted value for π^0 is $m_{\pi^0} = 134.9645$ MeV/c^2.

3.6.6 Mean Life Time

Direct measurement of mean life time of π^0 has been possible from observations of the Dalitz pairs. Suppose that π^0 emerges from a 'star' (nuclear disintegration) in the photographic emulsion and decays into a Dalitz pair and an invisible γ-ray. The time interval between emission and the decay of neutral pion can be determined from the distance between the star and the origin of Dalitz pair and from the momentum of the decaying pion. The momentum is also estimated from the study of the pair. The mean life time of π^0 is found to be 1.8×10^{-16} sec.

For positively charged pions the mean life is obtained by electronic measurements of the following sequence of events; arrival and stopping of the pion in a scintillator, emission of muon and much later decay of the muon into electron and neutrinos. The time interval between the arrival of the pion and its decay gives the mean life directly. The accepted value for the mean life time of π^+ is $(2.55 \pm 0.03) \times 10^{-8}$ sec.

3.6.7 Spin

The spin of charged pions can be determined from the principle of detailed balance for the reversible reaction

$$p + p \rightleftarrows \pi^+ + d$$

In general, consider the reaction

$$\underbrace{a + b}_{i} \rightarrow \underbrace{c + d}_{f}$$

with the particles a and b in the initial state i and c and d in the final state f. According to the time dependent perturbation theory, the number of transitions per sec is given by

$$W = \frac{2\pi}{h}|M_{if}|^2 \rho_f$$

where M_{if} is the matrix element describing the interaction and $\rho = dN/dE$ is the density of the final states. Now the detailed balance theorem implies that $M_{if} = M_{fi}$, that is

$$M_{\pi^+ d \rightarrow pp} = M_{pp \rightarrow \pi^+ d}$$

Summing over all the final spin states

$$W = \frac{2\pi}{\hbar}\sum_f |M_f|^2 \frac{dN}{dE}$$

As the transition takes place from all spin states which are equally probable, we must compute W by averaging over all the initial spin states which are $(2s_1 + 1) \cdot (2s_2 + 1)$ in number for the two particles with spins s_1 and s_2 in the initial state. Averaging over all states, the transition rate becomes

$$W = \frac{2\pi}{\hbar} \frac{1}{(2s_1 + 1)(2s_2 + 1)} \sum_f |M_f|^2 \frac{dN}{dE} \tag{3.15}$$

Now the cross-section σ is related to the transition rate as follows: Transition rate $(W) =$ Cross-section $(\sigma) \times$ (number of particles/unit volume) $\times$ relative velocity of interacting particles (v) so that

$$\sigma = \frac{1}{(2s_1 + 1)(2s_2 + 1)} \sum_f |M_f|^2 \frac{dN}{dE} \times \frac{\text{Unit volume}}{\text{Number of particles}} \times \frac{1}{v}$$

where the numerical constants are absorbed in $|M_f|^2$.

For a single particle in a volume V, we have

$$\sigma = \frac{1}{(2s_1 + 1)(2s_2 + 1)} \sum_f |M_f|^2 \frac{dN}{dE} \frac{V}{v_i}$$

where v_i is the relative velocity in the initial state. Now consider the density of states

$$dN = \frac{\text{Phase Space}}{\hbar^3} = \frac{(\text{momentum Space})}{h^3}(\text{volume})$$
$$= \frac{4\pi p^2 dp V}{\hbar^3}$$

so that

$$\frac{dN}{dp} = \frac{4\pi p^2 V}{h^3}$$

It ε_1 and ε_2 are the total particle energies, then

$$\frac{dE}{dp} = \frac{d\varepsilon_1}{dp} + \frac{d\varepsilon_2}{dp}$$

so that

$$\varepsilon\frac{d\varepsilon}{dp} = p \quad \text{or} \quad \frac{d\varepsilon}{dp} = \frac{p}{\varepsilon} = v$$

But

$$\varepsilon^2 = p^2 + m^2$$
$$\frac{dE}{dp} = v_1 + v_2 = v_f$$

where v_f is the relative velocity in the final state. Thus

$$\frac{dN}{dE} = \frac{dN}{dp} \cdot \frac{dp}{dE} = \frac{4\pi^2 V}{h^3 v_f}$$

Substituting, we get the expression

$$\sigma(ab \to cd) = \frac{1}{(2s_1+1)(2s_2+1)} \sum |M_f|^2 \frac{4\pi p_f^2 V^2}{\hbar^3 v_i v_f}$$

Now, at the same center-of-mass energy using the detailed balancing result that the matrix elements in both directions are equal,

$$M_{\pi^+ d \to pp} = M_{pp \to \pi^+ d}$$
$$\sigma_{\pi^+ d \to pp} \frac{(2S_{\pi^+}+1)(2S_d+1)}{p_p^2} v_{\pi^+ d} v_{pp} = \sigma_{pp \to \pi^+ d} \frac{(2s_p+1)^2 v_{\pi^+ d} v_{pp}}{p_{\pi^+}^2} \qquad (3.16)$$

Putting $S_d = 1$, $S_p = 1/2$, and introducing a factor $1/2$ to take into account the indistinguishability of protons in the final state in the $\pi^+ d$ absorption reaction,

$$2S_{\pi^+} + 1 = \frac{4}{3}\frac{\sigma_{pp \to \pi^+ d}}{\sigma_{\pi^+ d \to pp}} \cdot \frac{p_p^2}{p_\pi^2} \times \frac{1}{2}$$

The last equation can now be used to determine the spin of the π^+. Even approximate values of cross-sections are good enough to determine the spin of pion. Cross sections (0.18 mb) obtained by Cartwright et al. [12] for the reaction $pp \to \pi^+ d$ at proton energy of 340 MeV which corresponds to meson energy of 22.3 MeV in the CMS and $\sigma_{\pi^+d\to pp} = 3.1$ mb from the work of Durbin et al. [13] at pion energy of 29 MeV in CMS is quite close to that for the forward reaction in Cartwright's work. For a kinetic energy of 22.3 MeV for the pion and the corresponding energy of 85 MeV for each proton in the CMS, we obtain the ratio $p_\pi/p_p = 0.20$. Solving the equation for S_π, we find $S_\pi = -0.016$ or 0. Note that the above deduction is dependent on the assumption that in the π^+d absorption processes all the pion spin states are equally probable. In case π spin is non-zero and in case we are dealing with a polarized beam of pions then the above analyses would be invalidated.

A further evidence of zero spin comes from the observation that the decay of charged mesons is found to be isotropic in the CMS. Since π^- is the antiparticle of π^+, its spin is also assumed to be zero. Discussion for π^0 spin will be taken up after considering its parity (Chap. 3).

3.6.8 Isospin (T) of Pion

Since π^+, π^0 and π^- are regarded as the three charge states of pion, $2T + 1 = 3$, giving $T = 1$ for pion. Heizenberg's concept of isospin was extended to pion by Kemmer [14] who predicted the existence of neutral pion. Although π^0 mass (135 MeV/$c^{2)}$ is slightly smaller than charged pions (139 MeV/c^2 each), the approximate equality of the charged and neutral pions is reminiscent of the near equality of masses of neutron and proton.

3.6.9 Primakoff Effect

This refers to the process of π^0 production by the interaction of an incident γ-ray with the electromagnetic field (Coulomb's field) of the nucleus. Primakoff effect is basically a peripheral effect in which the classical impact parameters involved are large. The cross-section for the Primakoff effect is expected to be quite small, being of the order of 1 mb and very forward peaked. The Primakoff effect is identified by its characteristic angular distribution and must be distinguished from the nuclear production of π^0 which are produced at much wider angles. The determination of the cross-section for Primakoff effect affords the estimation of the lifetime of π^0. Primakoff cross-section is proportional to the reciprocal of the lifetime of π^0. The best fit to all the date yields a value of $(0.73 \pm 0.1) \times 10^{-16}$ sec. This leads to the value of $r_o = 1.2 \pm 0.07$ fm for the quantity in the formula $R = r_o A^{1/3}$ for the nuclear radius, which is in good agreement with the value obtained by other methods.

Figure 3.7 shows the angular distribution of π^0 due to Primakoff effect (solid line) and that due the nuclear production (dashed curve) for γ-rays interactions in lead at 1 GeV.

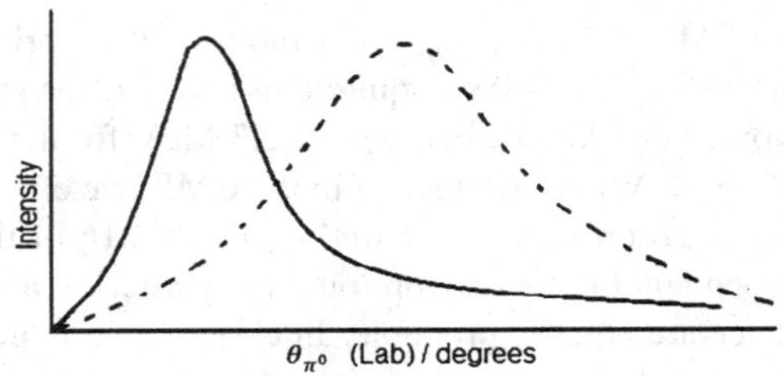

Fig. 3.7 Angular distribution of π^0 for γ-rays interactions in lead

3.7 Antiproton

3.7.1 Discovery

Following the discovery of positron in cosmic rays, it was natural to expect the observation of antinucleons. However, prior to 1955 no clear cut event of antiproton was observed.

Since the threshold energy for antiproton-proton pair production in the reaction $p + p \rightarrow p + p + p + p^-$ is $6mc^2$ or 5.64 GeV, a synchrotron called Bevatron was constructed at Berkeley, California (1955) and the p^- was discovered by Chamberlain et al. (Formerly 1 BeV = 1 GeV). The main problem was to find particles with charge $-e$ and the mass equal to that of proton. This was accomplished by determining the sign and the magnitude of charge, and the momentum and velocity of the particles. In the relativistic relation, $p = m\gamma\beta c$. If p and β (and hence γ) be known the mass m is determined.

The momentum is fixed by the trajectory of the particles provided the charge and magnetic fields are known. The latter are measured directly and the trajectory is fixed by the wire-orbit method. In this method, a flexible wire carrying current i and subjected to a mechanical tension T when placed in a magnetic field assumes exactly the form of the orbit of a particle of charge e and momentum p if the relation $T/i = p/e$ is satisfied.

The experimental layout is shown in Fig. 3.8, which essentially serves the purpose of a mass spectrograph. T = target, M_1, M_2 = bending magnets, Q_1, Q_2 = quadrupole magnets for focusing, S_1, S_2, S_3, S_4 = scintillation counters, C_1, C_2, C_3 = Cerenkov-counters. Particles in passing through the scintillation counters S_1, S_2, S_3 produce pulses of the same pulse height as those caused by protons of the same momentum showing there by the magnitude of charge is e and not greater. The sense of bending shows that charge is negative and that momentum is determined by the curvature measurement of the trajectory. In this experiment momentum of particles selected was found to be 1.19 GeV/c. This momentum corresponds to $\beta = 0.76$ for p^- whereas it is 0.99 for π^-. The measurement of velocity is the most difficult part of the experiment as p^-, s are masked out in the heavy background of pions mixed with some electrons and muons in the ratio of 50,000 pions to one p^-.

β is determined by the time of flight measurements between scintillators S_1 and S_2 and corroborated by the response of the special Cerenkov counter C_2 which

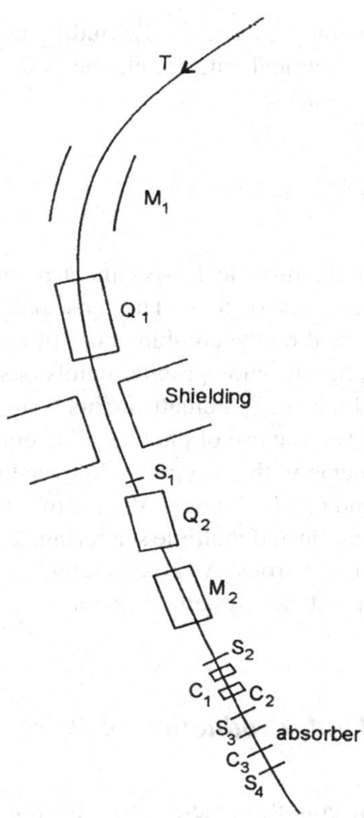

Fig. 3.8 The experimental layout of a mass spectrograph

responds only to particles with $0.75 < \beta < 0.78$. Cherenkov counter C_1 is in anticoincidence and responds to particles with $\beta > 0.79$ and serves to eliminate pions and lighter particles. Scintillator S_3 ensures that the p^- traverses the entire apparatus. The time of flight between S_1 and S_2 was 51 mμs for p^- and 40 mμs for π^- as their velocities are different. The time of flight and the response of C_2 provided independent velocity measurements and combined with the other counters as described above permitted the identification of p^- and its mass measurement within 5 % accuracy. The certified antiprotons were given by the signature $S_1 + S_2 + C_2 + S_3 - C_1$ (a signal for coincidence of the first four counters, not accompanied by signal from C_1) while $S_1 + S_2 + C_1 + S_3 - C_2$, corresponds to a signal for a pion. Further check was made on the antiproton yield by increasing the momenta of beam protons in the Bevatron. This resulted in an increased yield in the ratio of p^-/π^-.

3.7.2 Charge

The sign of the charge is determined by the curvature of the trajectory and its magnitude by the pulse size in the counter experiments and by grain density in photographic emulsions. Excluding the possibility of fractional charges, it is $-e$, which is identical with the charge of the electron.

3.7.3 Mass

In the historical experiment p^- mass was found to be equal to that of proton to an accuracy of 5 %. The most precise ratio of the p^- mass to that of proton is obtained by the combined use of measurement of momentum by the wire method and range in photographic emulsions. In this way, a value of 1.010 ± 0.006 has been obtained. Systematic errors from momentum measurements may be to the extent of 3 %. The use of photographic emulsions have provided independent mass measurements without separate knowledge of momentum by (i) Combination of ionization and residual range. A value of 1.009 ± 0.027 was obtained. (ii) Combination of ionization and multiple scattering. The emulsions were calibrated directly using protons or deuterons. A value of 0.999 ± 0.043 was obtained. Here, again the errors are only statistical. Systematic errors might be as high as 3 %.

3.7.4 Production in Pairs

In complex nuclei, the threshold for p^-–p pair production is lowered to about 4.0 GeV on account of Fermi momentum. The fact that no single antiproton has been observed is a proof that p^- is produced in pairs in proton-nucleon collisions that is, the reaction proceeds as $p + N \rightarrow p + N + p^- + p$.

3.7.5 Decay Constant

Antiprotons must be stable in a vacuum. However antineutrons must decay with a mean lifetime of free neutron through the scheme, $\overline{n} \rightarrow p^- + e^+ + \nu_e$.

3.7.6 Isotopic Spin

The isotopic spin T of an antinucleon is $\frac{1}{2}$ and the formula for the charge $\frac{Q}{e} = T_3 + \frac{B}{2}$, gives the assignment of $T_3 = -\frac{1}{2}$ to $p^-(B = Q/e = -1)$ and $T_3 = \frac{1}{2}$ to antineutron. p–p^- pair has $I_3 = 0$ but $I = 1$ or 0, where as a p–$\overline{n}$ pair or the p^-–n pair has $I = 1$.

3.7.7 Annihilation

A nucleon-antinucleon pair at rest annihilates with the release of energy equal to $2Mc^2 = 3877$ MeV. The energy manifests itself as mostly production of pions. Kaons may also be produced to the extent of ∼5 %. The average multiplicity is 4.8. Many of the pions are produced from the decay of p^-–p system through pion resonant states. In fact, ω^0(790 MeV) was discovered in the analysis of annihilation events in which the effective mass of three pions showed a distinct peak against the background of uncorrelated pions. The annihilation of $\overline{N}$–N system produces mesons while e^+–e^- system produces photons, the former being a nuclear phenomenon and the latter an electromagnetic one. As the p^- energy increases the average pion multiplicity also increases as larger energy will now be available. Thus at 5.7 GeV/c beam energy $\langle N_\pi \rangle = 7.3$ which is only 1.5 times greater than at rest, notwithstanding the fact that the C.M. energy has increased by factor of 2. The momentum spectrum of pions is found in accordance with the phase space considerations. At higher energies pionic resonances are not revealed so clearly as for annihilation at rest.

In the $\overline{N}$–N annihilation pions are mainly produced. Charged pions decay into muons which in turn decay into electrons, positrons and neutrinos, and neutral pions decay into γ-rays. Thus, the entire available energy is degraded. When p^- is incident on p or n with equal frequency as in deuterium, the exact consequence of charge independence is $\langle N_{\pi^+} \rangle + \langle N_{\pi^-} \rangle = 2\langle N_{\pi^0} \rangle$.

When the annihilation takes place in a complex nucleus the annihilation products would interact within the nucleus leading to knock-on nucleons and evaporation particles. Pions may be observed or their energy degraded. Annihilation of p^- may also be observed from the light emitted as the scintillation light or Cerenkov radiation by the charged particles produced directly or indirectly in the annihilation process.

3.7.8 Interaction Cross-Sections

The type of interactions in which antiprotons participate include elastic scattering, inelastic scattering, charge exchange scattering and annihilation. The corresponding cross-sections will be denoted by σ_e, σ_i, σ_c, σ_a, respectively. The reaction cross-section $\sigma_r = \sigma_i + \sigma_c + \sigma_a$, and the total cross-section $\sigma_t = \sigma_r + \sigma_e$. The term charge exchange scattering refers to the process $p^- + p \rightarrow n + \overline{n}$.

The annihilation process consists of $p^- + N \rightarrow n\pi$, where N is nucleon and n = number of pions produced. A genuine inelastic process is of the type $p^- + p \rightarrow \pi^0 + p + p^-$. At low energies, 0–100 MeV $\sigma \simeq 2\sigma_{geom}$.

As the beam energy increases σ_a decreases. Thus, $\sigma_a = 44$ mb at 900 MeV. Similarly, σ_e becomes smaller. σ_c stays at a constant value of 5 mb for the single pion production over a wide range of energies. The inelastic process ($p^- + p \rightarrow p^- + p + \pi^0$) obviously starts at 290 MeV. With the increase of beam energy, both the elastic and total annihilation cross sections decrease. The decrease of σ_t with

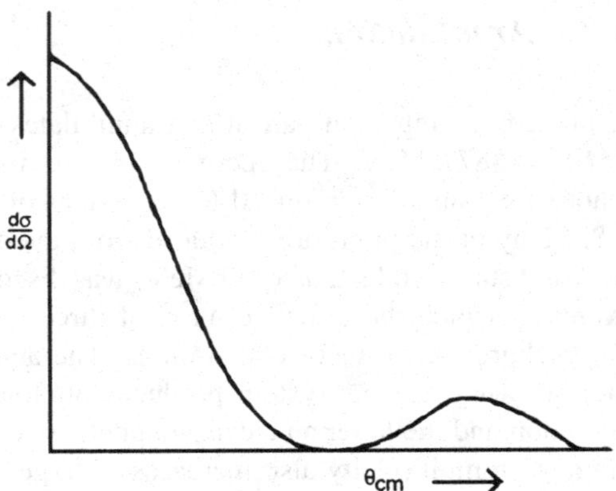

Fig. 3.9 Typical curve of diffraction scattering

increasing energy is mainly due to the drop in annihilation part of cross-section $\sigma_t = 167$ mb at 5 GeV/c and is 46 mb at 10 GeV/c.

The cross section for the non-annihilation inelastic channels is still smaller at 5 GeV than the value of the proton-proton inelastic cross section at the corresponding energy. The inelastic collisions with antiprotons have pronounced peripheral character. Typical nucleon isobars produced are N^*_{1236}, N^*_{1512}, N^*_{1688}, and their anti-isobars; thus $p^- + p \to \overline{N} + p\pi^+$ and for the isobars-anti-isobars pair production $p^- + p \to N^{*++}_{1238} + \overline{N}$ the cross section is enhanced to $\sigma = 1.05$ mb.

3.7.9 Elastic Scattering

The elastic scattering of antiproton has been studied at various energies. A typical curve is shown in Fig. 3.9 which is diffraction scattering. The angular distribution for p^-–p scattering is quite different from p–p scattering at similar energy.

At high energy, the angular distribution becomes narrowed. The large component of inelastic scattering together with the shape of the elastic scattering suggest that the proton-antiproton interaction conforms to "Black sphere" condition and $\sigma_{sc} = \sigma_r = \pi R^2$ with total cross section $\sigma_T = \sigma_{sc} + \sigma_r = 2\pi R^2$ and the differential cross section

$$\frac{d\sigma_{sc}}{d\Omega} = R^2 \left[\frac{J_1(KR\sin\theta)}{\sin\theta} \right]^2 \tag{3.17}$$

where R is the radius of interaction and J_1 is the Bessel function of order one. The value of R can be fixed from the measured total cross section. However, the parameter R is not constant but varies between 1.6 fm and 1.1 fm over the energy range 150–2000 MeV.

3.7.10 Strange Particle Production

About 4 % of the p–$\overline{p}$ annihilating at rest produce K-mesons. The production frequency increases to 8 % for p^- of 1 GeV/c. Production of hyperon-anti hyperon pairs becomes feasible at energies above the threshold. The cross-section for the reaction $p^- + p \to \Lambda^0 + \overline{\Lambda}$ at 1.6 GeV/c is 60 μb. In the interaction of antiprotons with complex nuclei, hyper-fragments have been observed.

3.7.11 Anti-neutrons

Anti-neutrons may be produced from the charge exchange of antiprotons and subsequently detected by their annihilation star in a bubble chamber or from the annihilation pulse produced in an annihilation counter.

Cork, Lambertson, Piccioni and Wenzel observed charge exchange process, $p^- p \to \overline{n}n$. An p^- beam was directed on a cube of liquid scintillator in which charge-exchange process occurred. The produced $\overline{n}$ continued forward into lead glass Cherenkov counter that detected annihilation of $\overline{n}$. To ensure that $\overline{n}$ and not p^- were responsible for annihilation, counters were placed in front of the Cerenkov counters and events with charged incident particles were rejected. The liquid scintillator was also monitored to ensure that for the reaction that took place there was indeed charge exchange rather their annihilation of incident p^-. The final annihilations occurring in the Cerenkov counter were compared with these produced directly by p^-. A great similarity in the annihilation of p^- and $\overline{n}$ has been observed.

3.7.12 Antibaryons

W. Powell and E. Segre et al. at Berkeley using 30 inch propane bubble chamber at Bevatron studied $\overline{n}$ annihilation stars. An p^- enters the bubble chamber at the top. Its track disappears as it charge exchanges, $pp^- \to n\overline{n}$. The $\overline{n}$ produces the star seen in the lower portion of the picture, Fig. 3.10. The energy released in the star was greater than 1500 MeV.

$\overline{\Lambda}$ production was first observed by D. Prowse M. Baldo-Ceolin [15] in photographic emulsions. $\Lambda\overline{\Lambda}$ production in a 72-inch hydrogen bubble chamber exposed to beam of p^- studied at Bevatron was observed by J. Button et al. [16], Fig. 3.11.

$\overline{\Sigma}^0$ *production* J. Button et al. provided 'evidence for the reaction $p^- p \to \overline{\Sigma}^0 \Lambda^-$ [17], and C. Baltay et al. [18]. H.N. Brown et al. observed the production Ξ^-–Ξ^+ pair [19]. Firestone et al. observed the production of $\overline{\Omega}^+$ [20]. Thus following the discovery of antiprotons all the antibaryons have been established.

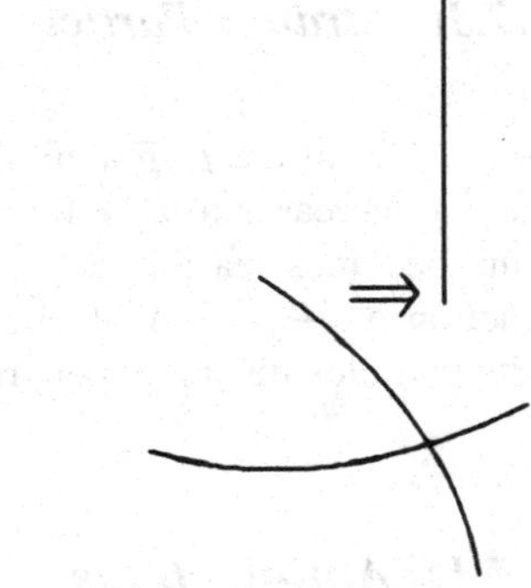

Fig. 3.10 Star produced by $\overline{n}$

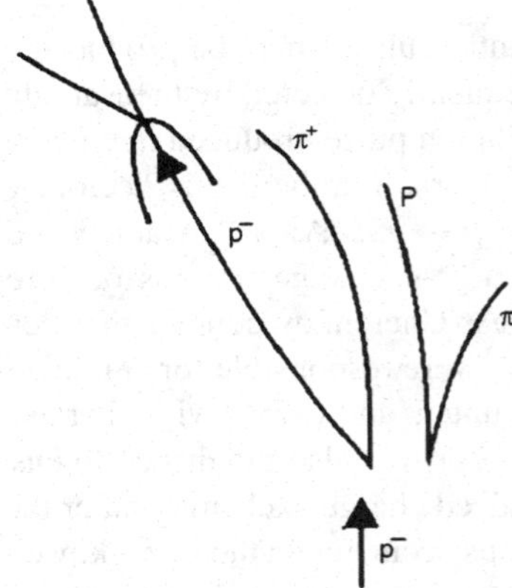

Fig. 3.11 Production of $\overline{\Lambda}$

3.8 Strange Particles

3.8.1 Production

Kaons and hyperons, known as strange particles are produced in high energy N–N or π–N collisions. One strange particle is always produced in association with another. This principle is known as the associated production of strange particles. Various types of production reaction with nucleons, pions and photons in which certain conservation laws are obeyed, are listed below. The corresponding production threshold energy are also indicated.

		Threshold energy
(i)	$p + p \rightarrow n + \Sigma^+ + K^+$	1782 MeV
(ii)	$p + p \rightarrow p + p + K^+ + K^-$	2106 MeV
(iii)	$\pi^- + p \rightarrow \Lambda + K^0$	755 MeV
(iv)	$\pi^- + p \rightarrow n + K^+ + K^-$	1361 MeV
(v)	$\gamma + p \rightarrow \Lambda + K^+$	911 MeV
(vi)	$p + p \rightarrow n + \Xi^- + K^+ + K^0$	3763 MeV

The above typical strong reactions conserve not only the charge Q/e, baryon number B and isospin I but also a new quantum number S, the strangeness. K^+ and K^0 which form an isospin doublet are assigned $S = +1$ and K^- and $\overline{K^0}$ which are also a doublet are assigned $S = -1$. Here K^0 and $\overline{K^0}$ are different particles. The sigma hyperons Σ^+, Σ^0 and Σ^- which constitute a triplet have $S = -1$. Λ hyperon which does not have any partner, that is, it is an isospin singlet has $S = -1$. The cascade hyperons Ξ^- and Ξ^0 form a doublet and are assigned $S = -2$. A more massive hyperon Ω^- (omega minus) is assigned $S = -3$, while ordinary particles $p, n, \pi^+, \pi^0, \pi^-$ have $S = 0$. In strong and electromagnetic reactions $\Delta S = 0$ and for weak decays $\Delta S = \pm 1$.

Note that the threshold for K^+ as in reaction (i) is less than that for K^- in reaction (ii), as K^+ can be produced in association with Σ or Λ hyperon where as K^- is produced in association with K^+.

In π–N interactions, (iii) and (iv), the threshold is lower than in N–N reactions. But obviously pions must be produced before they are made to collide with nucleons.

Reaction (v) also has lower threshold, but the process being electromagnetic has cross-section down by two orders of magnitude.

Reaction (vi) which involves the production of Ξ^- hyperon has a larger threshold not only because Ξ^- is more massive but also two K-mesons (K^+ and K^0) are to be produced to conserve strangeness.

In all the above reactions which proceed through strong interaction, strangeness is conserved. Note that the reaction

$$\begin{array}{lccccl} & \pi^- + & p \rightarrow & K^+ + & \Sigma^- & \\ s & 0 & 0 & +1 & -1 & \text{can proceed as } \Delta S = 0 \end{array}$$

but not

$$\begin{array}{lccccl} & \pi^- + & p \rightarrow & K^- + & \Sigma^+ & \text{since } \Delta S = -2, \\ s & 0 & 0 & -1 & -1 & \end{array}$$

although other conservations laws are obeyed.

3.8.2 K^+-Mesons (Kaons)

3.8.2.1 Mass

Masses of K^+-mesons have been directly measured using momentum-range method. When the momentum analysed beam of secondary particles is passed through emulsions, K^+-mesons are brought to rest within narrow band of ranges. Kaons with a given decay mode can be classified and their measured ranges permit their energies at the point of entry of the stack to be determined. The momentum of the particles is found from the geometry and the known magnetic field employed. The mass of individual particles is deduced from the combination of energy and momentum. The masses for various decay modes are found to be identical. The mean mass of K^+ mesons is found to be 493.67 MeV.

Independent mass measurements of kaons have been measured from the energy released in the decay mode, $K^+ \to \pi^+ + \pi^0$, $K^+ \to \mu^+ + \nu$, $K^+ \to \pi^+ + \pi^+ + \pi^-$. The masses thus deduced are in excellent agreement with those by the previous method.

It is also found that the relative frequency for various decay modes remains unaltered when the kaons undergo scattering or when different times of flight are involved. Thus the different decay modes are interpreted as the competitive decay modes of one and the same particle, K-meson.

3.8.2.2 Life Time

The life time of the two-body decay of K^+ into μ^+ or π^+ has been measured electronically with the use of scintillation counters and Cerenkov counters in coincidence and anti coincidence arrangement. The overall life time is found to be 1.237×10^{-8} sec.

3.8.2.3 Spin

a. All the evidences point to zero spin. No anisotropy has been found in $K_{\pi 2}$ decay $(K^+ \to \pi^+ + \pi^0)$ nor any asymmetry with respect to K-meson production plane. Such an anisotropy or asymmetry is expected if the K-mesons with non-zero spin are polarized at production.
b. The relative angular distribution of the three pions in the τ decay $(K^+ \to \pi^+ + \pi^+ + \pi^-)$ indicates spin 0 or 2 (Dalitz plot; Chap. 6).
c. The absence of the decay mode, $K^+ \to \pi^+ + \gamma$, is consistent with zero spin assignment. This mode is forbidden for spin 1 particle (Yang's theorem). Otherwise, for non-zero spin, it would strongly compete with the other decay modes.

3.8.2.4 Decay of Kaons

Similar decay modes are expected for K^- meson. The most frequent decay mode is $K_{\mu 2}$ (63.5 %) followed by $K_{\pi 2}$ (21.6 %). Some rare 2-body, 3-body and 4-body decay modes are also indicated in Table 3.1.

3.8.2.5 Nuclear Interactions

a. **Elastic Scattering.** At low energies, for angles up to 30° the scattering is mainly Coulomb scattering off complex nuclei. But the large angle scattering is observed to be larger than that expected for Rutherford scattering even after allowing for

Table 3.1 The percentage branching fractions of K^+ mesons

	Decay mode		Percentage branching fraction
1.	$\mu^+\nu_\mu$	$K_{\mu 2}$	63.5
2.	$\pi^+\pi^0$	$K_{\pi 2}$	21.6
3.	$\pi^+\pi^+\pi^-$	$K_{\pi 3}$	5.6
4.	$e^+\nu_e\pi^0$	$K_{\beta 3}$	4.5
5.	$\mu^+\nu_\mu\pi^0$	$K_{\mu 3}$	3.2
6.	$\pi^+\pi^0\pi^0$	$K_{\pi 3}$	1.7
7.	$\pi^+\pi^0\gamma$		2.2×10^{-4}
8.	$\pi^+\pi^+\pi^-\gamma$		1.0×10^{-4}
9.	$e^+\nu_e\pi^+\pi^-$		3.6×10^{-5}
10.	$e^+\nu_e\pi^0\pi^0$		
11.	$e^+\nu_e$		1.6×10^{-5}
12.	$\mu^+\pi^+\pi^-\nu_\mu$		7.7×10^{-6}
13.	$\pi^+\pi^+e^-\overline{\nu}_e$		$<1.0 \times 10^{-5}$
14.	$\pi^+\pi^+\mu^-\overline{\nu}$		
15.	$\pi^+e^+e^-$		
16.	$\pi^+\mu^+\mu^-$		

finite size of the nucleus and favours an additional nuclear scattering for K^+-meson, $K^+p \to K^+p$ or $K^+n \to K^+n$ without causing the excitation of the nucleus.

b. **Inelastic Scattering**. Here K^+ emerges with reduced energy generally causing a small 'star' in photographic emulsion.
c. **Charge Exchange Scattering.** Here the charge of K-meson and nucleon is exchanged, $K^+n \to K^0p$.

If a_1 and a_0 are the amplitudes for the $I = 1$ and $I = 0$ channels, respectively then we can express the cross-sections in terms of these amplitudes.

Elastic	$K^+p \to K^+p$	a_1^2
Inelastic	$K^+n \to K^+n$	$\frac{1}{4}(a_0 + a_1)^2$
Charge exchange	$K^+n \to K^0p$	$\frac{1}{4}(a_0 - a_1)^2$

We can find the branching ratio of charge exchange scattering to the total scattering by the use of iso-spin predictions. The system K^+p can proceed only through $I = 1$ state, its amplitude being a_1 and its total cross section $\sigma_1 = 4\pi a_1^2$ or proportional to a_1^2. The system K^+n can go through either $I = 1$ or $I = 0$ states. The cross sections for elastic scattering of K^+ with neutron would be proportional to $\frac{1}{4}(a_0 + a_1)^2$ and that of charge exchange proportional to $\frac{1}{4}(a_1 - a_0)^2$. If K^+ is incident with equal probability on p or n, then

$$\frac{\sigma_{ch.ex}}{\sigma_{total}} = \frac{\frac{1}{4}(a_1 - a_0)^2}{a_1^2 + \frac{1}{2}(a_1^2 + a_0^2)} \tag{3.18}$$

If $I = 1$ scattering dominates, i.e. $a_0 \to 0$, $\sigma_{ch.ex}/\sigma_{total} \to \frac{1}{6}$. On the other hand, if $I = 0$ dominates, i.e. $a_1 \to 0$, $\sigma_{ch.ex}/\sigma_{total} \to \frac{1}{2}$. Experiments show that the mean K^+-nuclear cross-section within nuclear matter is $\overline{\sigma} = 4$ mb of which $\sigma_{ch.ex}$ is 16 % showing thereby $a_0 \ll a_1$ for incident energies 40–160 MeV The angular distribution in the CMS is observed to be isotropic, which suggests S-wave scattering in the $I = 1$ channel.

The inelastic events are characterized by strong peaking in the backward direction. The total scattering cross-section remains at around 15 mb in the range of 40–60 MeV The K^+–n charge-exchange cross-section rises rapidly with energy, varying as cubic of the center of mass momentum. The enhancement in K^+–n charge exchange cross section sets in at about the threshold for π production in K–N collisions $K^+ + N \to K^+ + N\pi$ which is 220 MeV for scattering of a K^+ on free nucleon. Thus, at low energies $I = 1$ channel dominates and the scattering is mainly through S-waves. However, at higher energies $I = 0$ channel becomes important but S-wave interaction becomes relatively weak. A P-wave contribution is required to explain the backward peaking in the $K^+ + n$ interaction. The decrease in the K^+–n elastic cross-section is explained by an interference of P-waves in $I = 1$ and $I = 0$ channels.

3.8.3 K^--Mesons

3.8.3.1 Production

K^--mesons may be produced in high energy N–N or π–N collisions.

$$p + p \to K^+ + K^- + p + p \quad \text{Threshold} = 2106 \text{ MeV}$$

$$\pi^- + p \to n + K^+ + K^- \quad \text{Threshold} = 1360 \text{ MeV}$$

$$\to p + K^0 + K^-$$

3.8.3.2 Mass

a. Mass of K^--mesons has been measured by momentum-range method used for 0.998 ± 0.013 K^+-mesons. A direct comparison with the mass K^+ meson gives $m(K^-)/m(K^+) = 0.998 \pm 0.013$.
b. The mass of K^--meson can be found from the reaction, $K^- + p \to \Sigma^+ + \pi^-$, and the known masses of Σ^+ and π^-, and the measured O-value. In this way it is found that $m_K - = 493.7 \pm 0.3$ MeV which is to be compared with the value 493.8 MeV for K^+-meson obtained from $K_{\pi 3}$ decay.

3.8.3.3 Lifetime

The lifetime of the K^--meson can be measured only from observations of its decay in flight. The observed life time $(1.25 \pm 0.11) \times 10^{-8}$ sec is consistent with that of K^+ as expected for a charge conjugate particle.

3.8.3.4 Decay Modes

It is more difficult to observe the decays of K^- as they interact strongly with nuclei. Consequently, when K^- mesons are brought to rest, they preferably interact because of Coulomb attraction rather than decay. The modes $K^-_{\pi 2}$, $K^-_{\pi 3}$ have been observed in flight. However we should expect all the others charge conjugate modes to exist.

3.8.3.5 Interactions

When a K^--meson is sufficiently slowed down in a medium, it is captured in an outer Bohr's orbit and cascades down to lower orbits via radiative or Auger electron transitions. Owing to a strong nuclear interaction it is readily absorbed by a proton, neutron or even a pair of nucleons.

Many more reaction channels are open for K^- mesons than for K^+ mesons simply because the hyperon into which the nucleon is transformed has the same strangeness ($s = -1$) as the K^--meson.

Notice that in all the above strong interactions, strangeness is conserved. If K^- is incident with equal frequency on p and n, then in Reactions (1), (2), (3), (9) and (10) (see Table 3.2), principle of charge independence gives

$$n_{\Sigma^+} + n_{\Sigma^-} = 2n_{\Sigma^0}$$

where n is the number. This prediction is confirmed in deuterium bubble chamber

$$n_{\Sigma^+} + n_{\Sigma^-} = 150 \quad \text{and} \quad n_{\Sigma^0} = 86$$

Also,

$$\sigma_{15} = 2\sigma_{13}; \qquad \sigma_{20} = 2\sigma_{21}; \qquad \sigma_{22} = 2\sigma_{23}$$

The expected energy distribution of Λ from the Reactions (2), (4), (5) and (6) are shown in Fig. 3.12.

Reactions with a pair of nucleons, (13)–(18) are distinguished from captures in single nucleons, (1)–(12), by a much higher Q-value.

At energies above a few MeV, elastic as well as charge exchange scattering take place

$$K^- + p \to K^- + p \quad \text{Elastic} \tag{3.24}$$

$$\to \overline{K}^0 + n \quad \text{Charge exchange} \tag{3.25}$$

$$K^- + n \to K^- + n \quad \text{Elastic} \tag{3.26}$$

Table 3.2 Interactions of K^- mesons at rest

$K^- + p$	$\to \Sigma^+ + \pi^-$	+103 MeV	(1)
	$\to \Sigma^0 + \pi^0$	+105.6 MeV	(2)
	$\to \Sigma^- + \pi^+$	+96.6 MeV	(3)
	$\to \Lambda + \pi^0$	+182 MeV	(4)
	$\to \Lambda + \pi^+ + \pi^-$	+37.6 MeV	(5)
	$\to \Lambda + \pi^0 + \pi^0$	+47 MeV	(6)
	$\to \Lambda + \gamma$	+317 MeV	(7)
	$\to \Sigma^0 + \gamma$	+240 MeV	(8)
$K^- + n$	$\to \Sigma^- + \pi^0$	+101.5 MeV	(9)
	$\to \Sigma^0 + \pi^-$	+102 MeV	(10)
	$\to \Lambda + \pi^-$	+179 MeV	(11)
	$\to \Lambda + \pi^- + \pi^0$	+44 MeV	(12)
$K^- + P + n$	$\to \Sigma^0 + n$	+240 MeV	(13)
	$\to \Lambda + n$	+320 MeV	(14)
	$\to \Sigma^- + p$	+245 MeV	(15)
$K^- + p + p$	$\to \Sigma^+ + n$	+241 MeV	(16)
	$\to \Sigma^0 + p$	+240 MeV	(17)
	$\to \Lambda + p$	+317 MeV	(18)
$K^- + n + n$	$\to \Sigma^- + n$	+236 MeV	(19)
$K^- + {}^4\text{He}$	$\to \Sigma^- + {}^3\text{He}$	+195 MeV	(20)
	$\to \Sigma^0 + {}^3\text{H}$	+200 MeV	(21)
	$\to \Lambda + \pi^- + {}^3\text{He}$	+137 MeV	(22)
	$\to \Lambda + \pi^0 + {}^3\text{H}$	+142 MeV	(23)

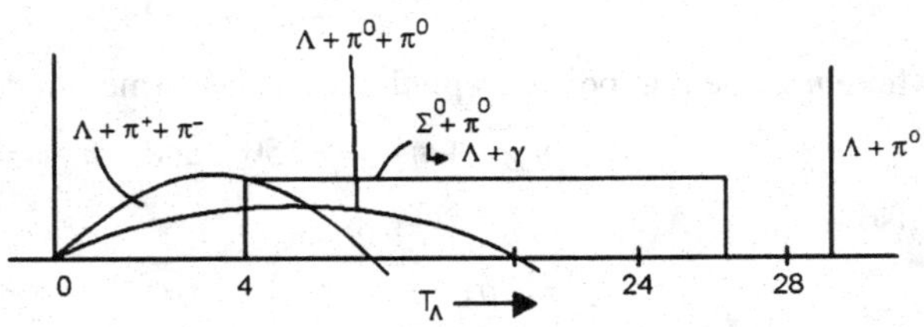

Fig. 3.12 Energy distribution of Λ

At higher energies, the production of the cascade hyperon and $\overline{\Omega}$ particle become feasible

$$K^- + p \to \Xi^- + K^+ \tag{3.27}$$

$$\to \Xi^0 + K^0 \tag{3.28}$$

Threshold = 614 MeV

$$K^- + p \to \overline{\Omega} + K^+ + K^0 \tag{3.29}$$

Threshold = 2698 MeV

Because of the large number of channels open for the production of hyperons ($S = -1$) by absorption of $K^-(S = -1)$ by nucleons, through strangeness conservation which is not possible for $K^+(S = +1)$ a basic asymmetry results in the behaviour of K^+ and K^- mesons. Consequently, the mean free path (λ) for K^+ and K^- in matter is vastly different. Thus, for example in photographic emulsions at 30–120 MeV, $\lambda \simeq 95$ cm for K^+ where as in the energy range 16–160 MeV, $\lambda \sim 30$ cm for K^-.

3.8.4 Σ Hyperons

3.8.4.1 Mass

Mass of Σ^+ is found from its decay at rest.

$$\Sigma^+ \rightarrow p + \pi^0 \quad \text{(i)}$$

$$\rightarrow n + \pi^+ \quad \text{(ii)}$$

Thus, from (ii), $m_\Sigma = \sqrt{p_\pi^2 + m_\pi^2} + \sqrt{p_\pi^2 + m_n^2}$ since $p_n = p_\pi$. Alternatively (i) may be used. The ranges of pion or proton are best measured in emulsions. It is found that $m_{\Sigma^+} = 1189.4 \pm 0.14$ MeV.

The mass of Σ^- hyperons has been determined for the measured ranges of π^+ in the reaction $K^- + p \rightarrow \Sigma^- + \pi^+$. This yields a value $m_{\Sigma^-} = (1197.3 \pm 0.14)$ MeV.

The mass of Σ^0 can be found from $\Sigma^- + p \rightarrow \Sigma^0 + n$, followed by the rare decay mode $\Sigma^0 \rightarrow \Lambda + e^+ + e^-$. A value of $m_{\Sigma^0} = (1192.0 \pm 0.2)$ MeV has been deduced.

3.8.4.2 Lifetime

The mean lifetime of Σ^+ is found to be $\mathcal{T}_{\Sigma^+} = (0.78 \pm 0.03) \times 10^{-10}$ sec, that of Σ^-, $\mathcal{T}_{\Sigma^-} = (1.6 \pm 0.06) \times 10^{-10}$ sec and that of Σ^0, $\mathcal{T}_{\Sigma^0} \sim 10^{-19}$ sec. For the neutral hyperon, the lifetime is necessarily short as the decay is electromagnetic ($\Sigma^0 \rightarrow \Lambda + \gamma$).

3.8.4.3 Decay Modes

$\Sigma^\pm$ hyperons decay into a system of nucleon and pion via weak decay, while Σ^0 decays through electromagnetic interaction. See Table 3.3.

Table 3.3 Decay modes of Σ hyperons

	Decays mode	Branching Fraction
Σ^+	$\rightarrow \pi^+ + n$	0.49 ± 0.025
	$\rightarrow \pi^0 + p$	0.51 ± 0.025
	$\rightarrow p + \gamma$	$(1.85 \pm 0.45) \times 10^{-3}$
	$\rightarrow e^+ + \nu_e + n$	$<10^{-4}$
	$\rightarrow \mu^+ + \nu_\mu + n$	$<10^{-4}$
	$\rightarrow \Lambda + e^+ + \nu_e$	$(0.4 \pm 0.2) \times 10^{-4}$
Σ^-	$\rightarrow \pi^- + n$	1.00
	$\rightarrow e^- + \overline{\nu} + n$	$(1.3 \pm 0.2) \times 10^{-3}$
	$\rightarrow \mu^- + \overline{\nu} + n$	$(0.66 \pm 0.15) \times 10^{-3}$
	$\rightarrow \Lambda + \overline{e} + \overline{\nu}$	$(0.74 \pm 0.2) \times 10^{-4}$
Σ^0	$\rightarrow \Lambda + \gamma$	1.0
	$\rightarrow \Lambda + e^+ + e^-$	

3.8.4.4 Interactions

Σ^+ hyperons can interact strongly with neutron through, $\Sigma^+ + n \rightarrow p + \Lambda$ at nominal energies. At rest, Coulomb's repulsion would suppress this reaction. The strong interactions of Σ^- particle with proton are

$$\Sigma^- + p \rightarrow \Lambda + n + 81 \text{ MeV} \tag{1}$$

$$\rightarrow \Sigma^0 + n + 4 \text{ MeV} \tag{2}$$

Phase space for (2) at rest, is much less than that for (1). Equation Reaction (1) can proceed only through $I = 1/2$ channel, where as in (2) both $I = 1/2$ and $I = 1/3$ can contribute to the reaction. Notice that the capture of Σ^- by neutron at rest cannot lead to a strong reaction since the final state $\Lambda + n + \pi^-$ is forbidden from energy considerations.

When Σ^- is captured by proton in complex nuclei Reaction (2) is suppressed for two reasons:

i. Some energy has to be absorbed in removing the proton bound to the nucleus.
ii. A high proportion of the neutron states which are energetically permissible are already filled.

Several possibilities exist in regard to the emergence of the product particles.

i. Both the neutron and Λ escape from the nucleus. The excitation energy is of the order of 20–30 MeV.
ii. The neutron is trapped within the nucleus but subsequently escapes, the excitation energy being about 40 MeV.
iii. The Λ-hyperon may collide with other nucleons or a group of nucleons to form a hyper-fragment which may come off either directly or through evaporation process.

iv. Λ may lose so much energy that it may not escape from the nucleus directly nor in the evaporation stage. In this case, it may form a hyper-nucleus of large Z. In this case, the predominant decay will be non-mesic with a release of 176 MeV.

3.8.5 Λ-Hyperon

The Λ-hyperon is produced in associations with K^+- or K^0-mesons in high energy N–N or π–N collisions (see Sect. 3.8.1).

3.8.5.1 Mass

The mass of Λ-hyperon has been determined from measurements of momentum of proton and π^- into which it decays, and the angle between them.

$$m_\Lambda^2 = m_p^2 + m_\pi^2 + 2(E_p E_\pi - p_p p_\pi \cos\theta)$$

The mass is found to be 1118 MeV.

3.8.5.2 Spin

The spin of Λ-hyperon has been determined from the observation of isotropic angular distribution of the decay products and is consistent with a value of $\frac{1}{2}$.

3.8.5.3 Lifetime

The lifetime has been determined from measurement of distances of decay points from the production points. The mean lifetime is found to be, $\tau = 2.63 \times 10^{-10}$ sec.

3.8.5.4 Decay Modes

Its principle decay modes are

$$\begin{aligned} \Lambda &\to p\pi^- \quad 67\,\% \\ &\to n\pi^0 \quad 33\,\% \end{aligned}$$

The rare decay mode is $\Lambda \to pe^-\overline{\nu}$.

3.8.5.5 Isospin

Λ-hyperon exists only in the neutral state. Hence it has $T = 0$.

3.8.6 Cascade Hyperon (Ξ)

3.8.6.1 Production

As the strangeness of Ξ-hyperon is -2, it is produced in association with two K-mesons of strangeness $+1$, in high energy N–N or π–N collisions (see Sect. 3.8.1).

3.8.6.2 Isospin

Ξ-hyperon exists in two charges states, Ξ^0 and Ξ^-. It therefore, has $T = \frac{1}{2}$.

3.8.6.3 Mass

The mass of Ξ^- has been determined by using K^- beam energy barely above threshold (35 MeV above threshold) so that additional pions are not produced in the collision with protons

$$K^- + p \rightarrow K^+ + \Xi^-$$
$$\Xi^- \rightarrow \Lambda\pi^-$$

Measurements of energy and momenta of K^-, K^+ and Ξ^- and the angle between K^+ and Ξ^- permit the mass of Ξ^- to be determined. Thus,

$$m_{\Xi^-} = (1321.3 \pm 0.1)\ \text{MeV}/c^2$$

Similar measurements give $m_{\Xi^0} = (1314.7 \pm 0.6)$ MeV.

It was a lot more difficult to establish the existence of Ξ^0 (1959) than that of Ξ^- (1953) as both the decay products of the former are neutral.

Since Ξ^0 has $S = -2$, its production by π^- is quite infrequent ($\pi^- + p \rightarrow K^0 + K^0 + \Xi^0$). The technique was to use mass separated beam of K^- at momentum 1 GeV/c at Bevatron and analyze the events in a bubble chamber

$$K^- + p \rightarrow K^0 + \Xi^0$$
$$\Xi^0 \rightarrow \Lambda\pi^0$$
$$K^0 \rightarrow \pi^+\pi^-$$

3.8.6.4 Decay Modes

$$\Xi^- \rightarrow \Lambda\pi^-$$
$$\Xi^0 \rightarrow \Lambda\pi^0$$

3.8.6.5 Lifetime

The mean lifetime for Ξ^- is found to be 1.8×10^{-10} sec and for Ξ^0, 3×10^{-10} sec.

Interaction A typical reaction with the cascade hyperon is

$$\Xi^- + p \to \Lambda + \Lambda \tag{i}$$

However, the Ξ^- neutron system is expected to be stable with respect to fast process.

Following the nuclear capture of Ξ^- by a proton as in (i), two Λ's are produced each of about 14 MeV. Several possibilities are open for the subsequent behaviour of Λ's.

i. Both Λ particles may escape directly or subsequently through evaporation process. The nuclear excitation energy is at most 28 MeV.
ii. One of the Λ particles escapes while the other one is trapped within the nucleus to form a hypernucleus which ultimately undergoes non-mesic decay releasing about 180 MeV energy.
iii. Both Λ particles may be trapped within the nucleus releasing about 30 MeV excitation energy. The nucleus reaches the ground state by emitting neutrons and gamma radiation forming a hypernucleus containing two Λ particles in the so-called double hyper-fragment. Its decay would be a two-stage process. The first Λ decay releases about 180 MeV energy and the second one may escape by evaporation either alone or as a part of a hyper-nucleus, which would then undergo decay close by or it may remain in the recoil nucleus which may decay after moving through an observable distance, leading to the appearance of a double-centred 'star', total energy released being about 350 MeV.

3.8.7 Hyperfragments

Danysz and Pncewski [21] observed a fragment emitted in a high energy cosmic ray collision with a nucleus in photographic emulsions which after separating from the parent nucleus had exploded. This fragment was interpreted as a nucleus in which a neutron is replaced by a Λ. Such fragments are known as hyper-fragments or hypernuclei and are denoted by the symbol ${}^{A}_{\Lambda}X$, where X is for the element and A the mass number. Thus ${}^{4}_{\Lambda}\mathrm{He}$ is a ${}^{4}\mathrm{He}$ nucleus in which a neutron has been replaced by a Λ. In order that a hyperfragment be formed, it is necessary that first a Λ be produced in the interaction of particles incident on complex nuclei. The hyper-fragment may result due to capture of K^-, Σ^-, Ξ^- or in the bombardment of high energy protons or pions. Since Λ is copiously produced in K^- interactions, the production of hyperfragments would also be greater compared to pion or proton interactions. When Λ is trapped within the nucleus, it cannot be transformed by nuclear interaction with the surrounding nucleon but must decay either by emitting nucleon and a pion (mesic decay) according to the scheme $\Lambda \to p + \pi^-$ or $\Lambda \to n + \pi^0$ or by emitting a pair of nucleons (non-mesic decay) according to the scheme $\Lambda + p \to p + n$ or $\Lambda + n \to n + n$. Examples of mesonic decays are ${}^4_\Lambda\text{H} \to {}^4\text{He} + \pi^-$; or $\to {}^3\text{He} + n + \pi^-$ or $\to {}^2\text{H} + {}^2\text{H} + \pi^-$. Examples of non-mesic decays are ${}^8_\Lambda\text{He} \to {}^4\text{He} + {}^4_\Lambda\text{H} + n$. For ${}^4_\Lambda\text{H}$ mean life time is found to be 1.2×10^{-10} sec, comparable with the lifetime of free Λ. Same order of magnitude for lifetime is found in other fragments. It is now found that the binding energy of Λ in light nuclei is approximately proportional to the mass number. This is in striking contrast with the binding energy of nucleons which fluctuate considerably in the light nuclei owing to the exclusion principle. However a single Λ is not subject to any such restriction. Thus, for example, whereas ordinary nuclei with $A = 5$ are unbound, there are hyper fragments with $A = 5$.

From the analysis of experimental data it is concluded that the Λ-nucleon force is comparable in magnitude to the nucleon-nucleon force and is spin dependent. Further evidence favours a spin $\frac{1}{2}$ for Λ. The mesonic decay of a hyperfragment is similar to the gamma emission of an excited nucleus. The non-mesonic decay mode is analogous to electron conversion in the deexcitation of nuclei. The ratio of mesonic to non-mesonic decays is then analogous to the internal conversion coefficient. Calculations show a strong dependence of the ratio on the orbital angular momentum l with which pion is emitted and suggests a value of $l = 0$ and a spin $\frac{1}{2}$ for the Λ.

The nonmesonic decay becomes increasingly important with the increasing atomic number of the nucleus ($Z \geq 3$).

The attraction of Λ to one or two protons is considerably weaker than the attraction for a neutron as evidenced from the non-existence of ${}^2_\Lambda\text{H}$ and ${}^3_\Lambda\text{He}$. Similarly, the data on ${}^3_\Lambda\text{H}$, ${}^4_\Lambda\text{He}$ and ${}^7_\Lambda\text{Li}$ show that the binding of Λ to a deuteron, ${}^3\text{He}$ or ${}^6\text{Li}$ is weaker than the binding of a neutron to these systems so that Λ-nucleon force appears less effective in binding than is the nucleon-nucleon force. On the other hand data on ${}^4_\Lambda\text{He}$, ${}^5_\Lambda\text{He}$ and ${}^9_\Lambda\text{Be}$ reveal that the Λ binding to ${}^3\text{H}$, ${}^4\text{He}$ or to ${}^8\text{Be}$ is stronger than that of neutron to these systems. In fact neutron binding is zero for ${}^4\text{H}$ and ${}^5\text{He}$ and only 1.67 MeV for ${}^9\text{Be}$. This is a consequence of the fact that the application of Pauli's principle to these systems which causes the binding energy of neutron to be zero or quite small has no effect on Λ as it is quite distinct from the rest of nucleons. Consequently Λ can occupy the lowest orbit in the potential well due to the attractive force of A nucleons to which it is attached. As A increases, the volume of this potential well increases, its depth remaining constant as the nuclear density is practically constant with A. Hence the Λ binding energy (B_Λ) is expected to increase monotonically with increasing A to some saturation value in very large hypernuclei. This is borne by experiments (Fig. 3.13).

The principle of charge independence applied to Λ requires that the nuclear interaction in the $\Lambda + p$ and $\Lambda + n$ systems should be identical. This is strongly supported by the experimental B_Λ values of 1.65 MeV and 1.60 MeV for mirror hyper-nuclei ${}^4_\Lambda\text{H}$ and ${}^4_\Lambda\text{He}$.

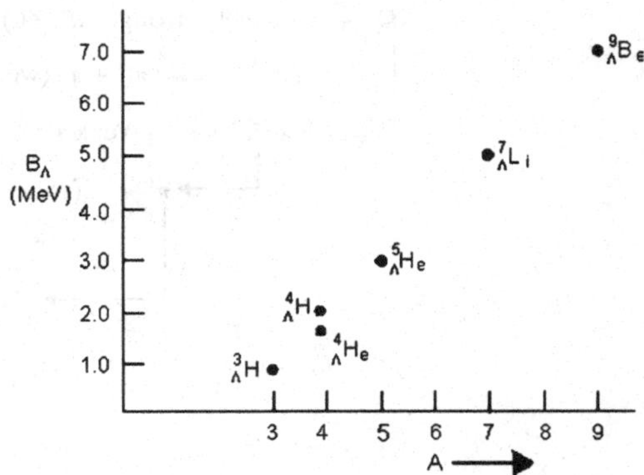

Fig. 3.13 Expected increase in binding energy (B_Λ) with increasing A

It is believed that the Λ-nucleon interaction is due to complicated processes

i. The exchange of a K^--meson, with a range of the order of $\hbar/m_K c \simeq 0.4$ fm; the K-meson Compton wavelength can produce an exchange force between Λ and nucleon.

$$\Lambda + p \rightarrow p + K^- + p \rightarrow p + \Lambda$$

ii. The exchange of several pions giving a non-exchange force with a range less than $\hbar/2m_\pi c \simeq 0.7$ fm.

$$\Lambda + p \rightarrow \Sigma^+ + \pi^- + p \rightarrow \Lambda + \left(\pi^+ + \pi^- + p\right) \rightarrow \Lambda + p$$

iii. More complicated processes involving both K-mesons and pions give interactions of still shorter range.

3.8.8 Ω^- (Omega Minus) Hyperon

3.8.8.1 Discovery

The existence of the Ω^- hyperon of strangeness $s = -3$, its mass as well as its decay modes were predicted within the framework of quark model and unitary symmetry before its discovery in 1964. The first Ω^- event was observed when about 10^5 pictures of 5 GeV/c K^- interactions in a 80-inch hydrogen bubble chamber were scanned at Brookhaven. The following sequence of production and decays were depicted in the historical event (1964).

K^+ was identified from momentum and gap-length measurements. The mass of Λ was determined from the measured kinematical quantities of proton and π^-. The event was unusual in that both the γ-rays from the decay of π^0 materialized within the chamber. Ξ^0 was identified from its decay products and shown not to be connected with the primary vertex. K^0 was identified from its decay into two charged pions. Also see Sect. 5.6.2.

$K^- + p \rightarrow \Omega^- + K^+ + K^0$ (strong, $\Delta S = 0$)

$K^0 \rightarrow \pi^+ + \pi^-$ (weak decay $\Delta S = 1$)

$\Omega^- \rightarrow \Xi^0 + \pi^-$ (Weak decay $\Delta S = 1$)

$\Xi^0 \rightarrow \pi^0 + \Lambda$ (weak decay, $\Delta S = 1$)

$\Lambda \rightarrow \pi^- + p$ (weak decay, $\Delta S = 1$)

$\pi^0 \rightarrow \gamma_1 + \gamma_2$ (e.m decay)

$\gamma_1 \rightarrow e^+e^-$, $\gamma_2 \rightarrow e^+e^-$

3.8.8.2 Mass

Its mass has been determined as 1672.5 MeV.

3.8.8.3 Spin

The angular distribution of the decay products in the rest frame of Ω^- suggests a value of $J = 3/2$ for its spin.

3.8.8.4 Isospin

As Ω^- is a singlet, its isospin $T = 0$.

3.8.8.5 Lifetime

The mean life is found to be $\tau = 0.82 \times 10^{-10}$ sec.

3.8.8.6 Decay Modes

The decay modes of Ω^- are

$$\begin{aligned}\Omega^- &\rightarrow \Xi^0 + \pi^- \\ &\rightarrow \Xi^- + \pi^0 \\ &\rightarrow \Lambda + K^-\end{aligned}$$

In each of the above modes, the rule $\Delta S = \pm 1$ for the weak decays is obeyed.

3.9 Neutrino

3.9.1 Introduction

The existence of neutrino, a light neutral particle, was postulated by Pauli [22] to save the conservation laws of energy, momentum and angular momentum from the observation of continuous energy spectrum in beta decay. The hypothesis that the continuous β energy spectrum may result from a given nuclear level to a variety of closely spaced levels was abandoned with the failure to observe the corresponding gamma rays.

Under the assumption that at least one neutral particle (ν) accompanies the decay not only the continuous character of β-ray spectrum was explained but also the linear momentum and angular momentum laws saved.

Although the existence of neutrino was taken for granted, its experimental discovery was delayed for almost thirty years owing to the feeble interaction which it suffers. Using Fermi's theory, Bethe and Pierls estimated the cross-section of neutrino inverting beta decay.

$$\overline{\nu} + p \rightarrow \beta^+ + n$$

The cross-section for the inverse β decay can be calculated using natural units ($\hbar = c = 1$)

$$\sigma\left(\overline{\nu}p \rightarrow ne^+\right) = \frac{W}{v_i} = \frac{G^2}{\pi}|M|^2 \frac{p^2}{v_i v_f}$$

where v_i, v_f are the relative velocities of the particles in the initial and final states ($v_i = v_f = c$) and p is the numerical value of the CMS momentum of the neutron and positron. Here we have the mixed transition with $M_F^2 = 1$ for the Fermi contribution ($\Delta J = 0$) and $M_{GT}^2 \simeq 3$ for the spin multiplicity factor for the Gamow-Teller contribution ($\Delta J = 1$), AAK1,3. Thus

$$\sigma = \frac{(M_F^2 + M_{GT}^2)}{\pi} G^2 p^2 \simeq \frac{4G^2 p^2}{\pi}$$

where $G = 1.16 \times 10^{-5}$ GeV^{-2} is the universal coupling constant.

For a few MeV neutrino energy E_ν above the threshold $Q = 1.8$ MeV, the CMS momentum $p \simeq (E_\nu - Q)/c$. For $cp \simeq 1$ MeV, we obtain

$$\begin{aligned}\sigma &= \frac{4}{\pi} \frac{(1.16 \times 10^{-5})^2 (10^{-9})^2}{(3 \times 10^8)^2} = 1.9 \times 10^{-47}\ \text{m}^2 \\ &\simeq 2 \times 10^{-43}\ \text{cm}^2\end{aligned}$$

For $\overline{\nu}$ of 2.3 MeV energy $\sigma \sim 10^{-44}$ cm^2. σ for the considered interaction increases with E_ν until it levels off at about 10^{-38} cm^2. A cross-section of 10^{-44} cm^2 is, in fact so small that it corresponds to a mean free path of $\sim$300 light years in water! It was therefore concluded that the observation of neutrino reaction was not possible with the facilities and techniques available at that time.

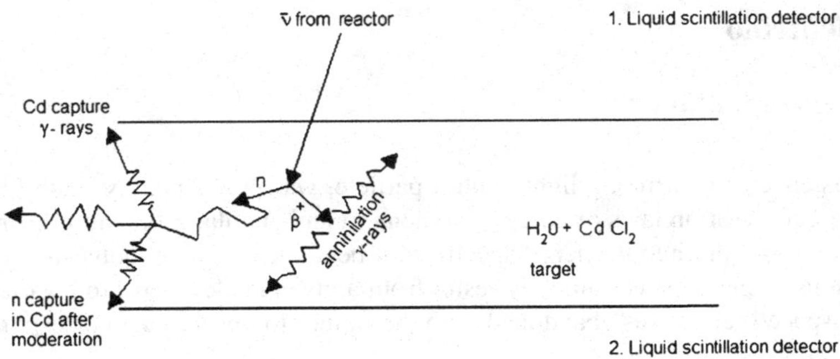

Fig. 3.14 Detection of antineutrino

3.9.2 The Experimental Discovery

With the introduction of power reactors large fluxes of antineutrinos became available and the detection of reaction with antineutrinos became feasible. The first measurement of $\overline{\nu}$ cross-section was made by Reins and Cowan in 1959 who employed a reactor as the source of uranium fission fragments which are neutron-rich and undergo β-decay emitting e^- and $\overline{\nu}$ averaging 6 per fission with average energy centered around 1 MeV. For a 1000 MW reactor the useful flux of antineutrinos is of the order of 10^{13} cm^{-2} s^{-1}.

Figure 3.14 shows schematically the detection system employed at Savannah river reactor. An antineutrino is incident on a water target containing dissolved $CdCl_2$. According to the reaction, a positron and a neutron are produced. The positron slows down and annihilates with an electron producing two 0.51 MeV γ-rays which penetrate the target and are detected in coincidence by two scintillation detector placed on either side of the target. The neutron is moderated and is captured by cadmium as signaled by the multiple γ-rays observed in coincidence by the two scintillation detectors. The antineutrino's signature is therefore a delayed coincidence between the prompt pulses produced by e^+ annihilation radiation and those produced microseconds later by the neutron captured in cadmium. Radiation other than $\overline{\nu}$ was ruled out as the cause of the signal. The small interaction cross-section necessitated the use of a detecting system which exclusive of lead shield weighed about 10 tons. A signal rate of $(3.0 \pm 2)/\text{hr}$ was observed. This implied a signal to total accidental background ratio 4/1. Measurements of positron and neutron detection efficiency and knowledge of the reactor flux showed this signal rate to be consistent with a cross section of $\overline{\sigma} = 1.2^{+0..7}_{-04} \times 10^{-43}$ cm^2, a value which is in agreement with the theoretical of expected value $\overline{\sigma} = (1.0 \pm 0.16) \times 10^{-43}$ cm^2. This corresponds to a mean free path for antineutrino absorption in water of 10^{20} cm or 100 light years.

3.9.3 Experiment to Demonstrate that Neutrino and Antineutrino Are Different Particles

From fission reactors one obtains flux of antineutrinos rather than neutrinos. Pontecarvo and Alvarez suggested that if neutrino is identical with antineutrino $\overline{\nu}$ from fission reaction it should be observable via the reaction

$$\overline{\nu} + {}^{37}_{17}\mathrm{Cl} \rightarrow {}^{37}_{18}A + \beta^- \tag{1}$$

This is equivalent to the conversion of a neutron of ^{37}Cl nucleus into a proton i.e.

$$\overline{\nu} + n \rightarrow p + \beta^- \tag{2}$$

But the inverse β decay is actually

$$\nu + n \rightarrow p + e^- \tag{3}$$

and so if $\overline{\nu} \equiv \nu$ then (2) and (1) should be observable. A radiochemical experiment was undertaken by Davies. The Experiment consisted of irradiation of a large volume of CCl_4 and then sweep out the product ^{37}A with He gas, separate it from He and then count it in a small low-background Geiger Muller detector. It was possible to quantitatively sweep out ^{37}A from a few thousand gallons of CCl_4 and then detect the ^{37}A by observing its K-capture. It was shown that the probability for the Reaction (1) to occur was less than $1/10$ of that expected for the assumption $\overline{\nu} \equiv \nu$. In view of the non-observation of Reaction (1) it was demonstrated that ν and $\overline{\nu}$ are not identical particles.

3.9.4 Muon Neutrino (ν_μ) is Different from Electron Neutrino (ν_e)

The layout of a neutrino beam is sketched in Fig. 3.15. First, secondary beams of pions and kaons are produced from high energy protons in the target. By employing bending magnets, quadrupole magnets and collimating slits secondaries of one charge in a small band of momentum can be selected. The pions and kaons enter a decay tunnel in which a fraction decays to muons and neutrinos ($\pi^+ \rightarrow \mu^+ + \nu_\mu$; $\pi^- \rightarrow \mu^- + \overline{\nu}$). Muons are stopped by a thick iron shield and the interactions of neutrinos are observed with neutrino detectors which consist of electronic and bubble chamber detectors. Initially, the beam mainly consists of muon neutrinos ν_μ with a very small admixture of electron neutrinos ν_e ($\sim$0.5 %) attributed to the three-body beta decay of kaons in flight $K^+ \rightarrow \pi^0 + e^+ + \nu_e$. Typical events of ν_μ and ν_e are:

$$\nu_\mu + n \rightarrow p + \mu^- \tag{1}$$

$$\nu_e + n \rightarrow p + e^- \tag{2}$$

The production of high energy secondary electron in (2) is recognized by the shower that is produced due to bremsstrahlung and pair production. Thus relative number of observed events in (1) and (2) are consistent with the calculated fluxes of ν_μ and ν_e in the beam, leading to the distinction between ν_μ and ν_e and thereby confirming the conservation of muon number (Chap. 4).

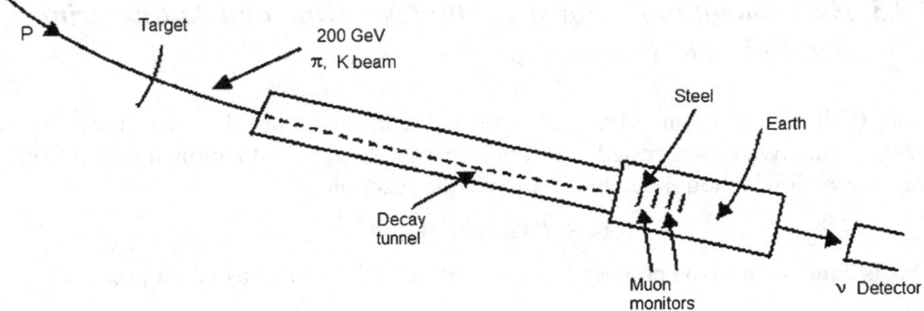

Fig. 3.15 Experiment to demonstrate that muon neutrino is different from electron neutrino

3.9.5 *Mass of Neutrino*

3.9.5.1 Mass of ν_e

The precise shape of an allowed β spectrum in the vicinity of the upper energy limit depends on the rest mass of neutrino (AAK 1, 3). The relativistic Fermi-Kurie plot of the beta spectrum for points not too near the high energy end gives information on ν mass.

The β spectrum emitted in the "super allowed" transition of ^{3}H to ^{3}He is amenable for the determination of neutrino mass as the transition energy is 18 keV. Very low energy β particles in the spectrum have been investigated with the use of proportional counters. For ^{35}S, the end point energy is 169 keV. Various investigations suggest a value of less the 65 eV for neutrino. Fritschi et al. [23] give a value less than 18 eV/c^2. It is difficult to improve the upper limit on neutrino's mass by this technique because of limitations of energy resolution and the uncertainty in the correct interpretation of the shape of the spectra near the end point. These experiments essentially determine the square of the mass m_ν^2. In some experiments it is embarrassing to obtain a negative value for m_ν.

3.9.5.2 Mass of ν_μ

The mass of muon neutrino is known with less accuracy. One way to determine its mass is to measure the ranges of muon in the decay of positive pions at rest. In this way it is found that $m_{\nu\mu} < 170$ keV.

3.9.5.3 Mass of ν_τ

The mass of ν_τ has been determined in the decay of τ particle, $\tau \to 5\pi + \nu_\tau$. An upper limit on its mass is set at 24 MeV/c^2.

On theoretical grounds, there is reason to believe that the mass of each neutrino should be zero. One of the original arguments of the two component neutrino theory was that it gives a natural reason for the vanishing of neutrino mass.

In the standard Model of particle physics neutrinos are assumed to be massless. Furthermore, they are completely polarized, and are left-handed with the component of spin vector against the momentum vector ($j_z = -\frac{1}{2}$), while antineutrinos are right-handed.

On the other hand, there is compelling reason to believe in the finite mass of neutrino to explain the apparent mismatch between the motional energy and the gravitational energy in the universe. Further, in order to understand the solar neutrino oscillations (Chap. 6) at least one of the flavours must be endowed with a finite mass. Thus, the problem of precise mass determination of neutrino has assumed paramount importance in recent years.

3.9.6 Charge

The charge of ν_e is known to be less than $4 \times 10^{-17}e$ and that of ν_μ less than $3 \times 10^{-5}e$.

3.9.7 Spin

The spin of neutrino is consistent with $\frac{1}{2}$.

3.9.8 Quantum Numbers for Leptons

The neutral particles, neutrinos and antineutrinos belonging to the family of leptons occur in three flavors ν_e, ν_μ and ν_τ with the subscript signifying their association with the respective charged leptons. The leptons are assigned a quantum number $L = +1$, and the antileptons $L = -1$. L is known as the lepton number. Other particles have $L = 0$. In any type of reaction L is conserved. Table 3.4 gives the lepton number for various members of lepton family.

In all the reactions, the flavour number L_e, L_μ, L_τ is also conserved additively. Thus, the μ^+ can decay as $\mu^+ \rightarrow e^+ + \nu_e + \overline{\nu_\mu}$, but not $\mu^+ \rightarrow e^+ + \overline{\nu_e} + \nu_\mu$. Similarly high energy γ-rays may produce e^+e^- or $\mu^+\mu^-$ pairs but not as μ^+e^- or $e^+\mu^-$. Again at energy above threshold $\overline{\nu}$ can be transformed to e^+ in the reaction $\overline{\nu_e} + p \rightarrow e^+ + n$ but not in $\overline{\nu_e} + p \rightarrow \mu^+ + p$.

Table 3.4 Quantum numbers L of leptons and antileptons

Q/e	Leptons (lepton number +1)		
−1	e^-	μ^-	τ^-
0	ν_e	ν_μ	ν_τ
Q/e	Antileptons (antilepton number −1)		
+1	e^+	μ^+	τ^+
0	$\overline{\nu_e}$	$\overline{\nu_\mu}$	$\overline{\nu_\tau}$

3.9.9 Helicity

The polarisation or Helicity H is defined as

$$H = \frac{I_+ - I_-}{I_+ + I_-} = \alpha \frac{v}{c}$$

where $I+$ and I_- represent the intensities for spin parallel and antiparallel to momentum p. Experimentally

$$\begin{aligned} \alpha &= +1 \quad \text{for } e^+, \ H = +v/c \\ &= -1 \quad \text{for } e^-, \ H = -v/c \end{aligned}$$

For $\overline{\nu}$, if rest mass $= 0$, $v = c$, $H = +1$ and for ν, $H = -1$.

3.9.10 Neutrino Sources

a. Radioactivity and decay of unstable particles

$$\begin{aligned} &n \to pe^- \overline{\nu_e} \\ &p \to ne^+ \nu_e \left(^{14}\text{O}\right) \\ &\Lambda \to pe^- \overline{\nu_e} \\ &\Sigma^- \to ne^- \overline{\nu_e} \\ &\pi^+ \to \mu^+ + \nu_\mu, \qquad \pi^- \to \mu^- \overline{\nu_\mu} \\ &\mu^+ \to e^+ \nu_e \overline{\nu_\mu}, \qquad \mu^- \to e^- \nu_\mu \overline{\nu_e} \end{aligned}$$

b. Photo-nuclear process $e^- + \gamma \to e^- + \overline{\nu_e} + \nu_e$
c. Annihilation of $e^- e^+$ pair $e^- + e^+ \to \nu_e + \overline{\nu}$
d. Fusion reactions: Hydrogen burning

***p–p* Cycle**

$$\begin{aligned} p + p &\to {}^2\text{H} + e^+ + \nu_e \\ {}^2\text{H} + p &\to {}^3\text{He} \\ {}^3\text{He} + {}^3\text{He} &\to {}^4\text{He} + 2p \end{aligned}$$

As helium is built at the centre of the star, the following bicycle may take place in helium burning

$$^{3}\mathrm{He} + {}^{4}\mathrm{He} \rightarrow {}^{7}\mathrm{Be} + \gamma$$
$$^{7}\mathrm{Be} + e^{-} \rightarrow {}^{7}\mathrm{Li} + \nu$$
$$^{7}\mathrm{Li} + p \rightarrow {}^{8}\mathrm{Be} + \gamma$$

$$^{8}\mathrm{Be} \rightarrow {}^{4}_{2}\mathrm{He}$$
$$^{7}\mathrm{Be} + p \rightarrow {}^{8}\mathrm{B} + \gamma \rightarrow {}^{8}\mathrm{Be} + e^{+} + \nu \rightarrow {}^{4}_{2}\mathrm{He}$$

CNO Cycle

$$^{12}\mathrm{C} + p \rightarrow {}^{13}\mathrm{N} + \gamma$$
$$^{13}\mathrm{N} \rightarrow {}^{13}\mathrm{C} + e^{+} + \nu_e$$
$$^{13}\mathrm{C} + p \rightarrow {}^{14}\mathrm{N} + \gamma$$
$$^{14}\mathrm{N} + p \rightarrow {}^{15}\mathrm{O} + \gamma$$
$$^{15}\mathrm{O} \rightarrow {}^{15}\mathrm{N} + e^{+} + \nu$$
$$^{15}\mathrm{N} + p \rightarrow {}^{12}\mathrm{C} + {}^{4}\mathrm{He}$$

About 4 % of stellar energy is dissipated in the form of neutrinos. When the star's hydrogen is exhausted, the next step is provided by burning helium. This occurs after the supply of hydrogen falls below a critical value and gravitation contracts the star and raises its temperature to about 1.5×10^{8} K. The following reactions take place:

$$^{4}\mathrm{He} + {}^{4}\mathrm{He} \rightleftarrows {}^{8}\mathrm{Be}$$
$$^{8}\mathrm{Be} + {}^{4}\mathrm{He} \rightleftarrows {}^{12}\mathrm{C}^{*} \rightarrow {}^{12}\mathrm{C} + \gamma$$
$$^{12}\mathrm{C} + {}^{4}\mathrm{He} \rightarrow {}^{16}\mathrm{O} + \gamma$$
$$^{16}\mathrm{O} + {}^{4}\mathrm{He} \rightarrow {}^{20}\mathrm{Ne} + \gamma$$
$$^{20}\mathrm{Ne} + {}^{4}\mathrm{He} \rightarrow {}^{24}\mathrm{Mg} + \gamma$$

As helium becomes exhausted the star again goes to higher temperature and densities by gravitational contraction. The reaction $e^{-} + \gamma \rightarrow e^{-} + \nu + \overline{\nu}$ becomes important, also $+\gamma \rightarrow e^{+} + e^{-} \rightarrow \nu + \overline{\nu}$. Pressure due to γ-rays and electrons which act along as a support to the star falls as these processes go on. The electron and photon energy is transferred to neutrinos which leak through the star. The star goes on contracting and the temperature increases to about 7×10^{9} K. Ultimately the star collapses and explodes. Before collapse, 20–30 % of the energy is released in neutrinos. When the collapse occurs practically all energy goes into neutrinos.

e. Cosmic Ray Secondary Neutrinos

When primary cosmic rays hit the upper atmosphere, nuclear reactions take place in which pions are produced and to some extent kaons. The charged pions decay with muons and muon-neutrinos ν_μ and $\overline{\nu}$.

3.9.11 Interactions of Neutrinos

While charged leptons participate in electromagnetic and weak interactions, the neutral ones, neutrinos are unique in that they can have only weak interactions with other particles. Neutrinos participate in weak scattering and reactions

$$\overline{\nu}_\mu + e^- \to \overline{\nu}_\mu + e^-$$
$$\nu_\mu + e^- \to \nu_\mu + e^-$$
$$\nu_\mu + n \to p + \mu^-$$
$$\nu_e + n \to p + e^-$$
$$\overline{\nu}_\mu + p \to n + \mu^+$$
$$\overline{\nu}_e + p \to n + e^+$$

The cross section for the absorption reactions is of the order of 10^{-38} cm^2 for 1–2 MeV energy. At lower energies the cross-sections are even smaller.

Other types of inelastic processes are:

$$\nu_\mu + n \to \Sigma^+ + \mu^-$$
$$\overline{\nu}_\mu + n \to \Sigma^- + \mu^+$$
$$\overline{\nu}_\mu + p \to \Lambda + \mu^+$$
$$\overline{\nu}_\mu + p \to \Sigma^0 + \mu^+$$
$$\nu_\mu + N \to N + \mu^- + K$$

The cross-section is reduced by a factor of 10 or so for the strange particles production compared to $\nu_\mu + n \to \mu^- + p$ or $\overline{\nu}_\mu + p \to \mu^+ + n$.

Example 3.1 In the decay of a charged pion at rest, the muon carries kinetic energy of 4.12 MeV. Calculate the mass of pion in terms of the electron mass (masses of muon and electron are 206.9 and 0.511 MeV/c^2, respectively).

Solution (a) $\pi^+ \to \mu^+ + \nu_\mu$

$$T_\mu = \frac{Q(Q + 2m_\nu)}{2(m_\mu + m_\nu + Q)} = \frac{(m_\pi - m_\mu)^2}{2m_\pi}$$
$$(\because Q = m_\pi - m_\mu - m_\nu = m_\pi - m_\mu \text{ and } m_\nu = 0)$$
$$m_\mu = 206.9 m_e = 206.9 \times 0.511 = 105.7 \text{ MeV}$$

Therefore,

$$T_\mu = 4.12 = \frac{(m_\pi - 105.7)^2}{2m_\pi}$$

Solving for m_π,

$$m_\pi = 141.39\ \text{MeV}$$

(b) $\mu^+ \to e^+ + \nu_e + \overline{\nu}_\mu$.

$T_{\max}$ for electron is obtained when ν_e and $\overline{\nu}_\mu$ fly together in opposite direction. Thus, the three-body problem is reduced to a two-body one.

$$m_\mu c^2 = 206.9 \times 0.511 = 105.72\ \text{MeV}$$

$$Q = 105.72 - 0.51 = 105.21\ \text{MeV}$$

$$T_e(\max) = \frac{Q^2}{2(m_e + Q)} = \frac{(105.72)^2}{2(0.51 + 105.72)} = 52.6\ \text{MeV}.$$

Example 3.2 A particle decays into two particles of mass m_1 and m_2 with a release of energy Q. Calculate relativistically the energy carried by the decay products in the rest frame of the decaying particle.

Solution Energy conservation gives

$$T_1 + T_2 = Q \tag{1}$$

Momentum conservation gives

$$\vec{P} + \vec{P} = 0$$

or

$$p_1^2 = p_2^2$$

or

$$T_1^2 + 2T_1 m_1 = T_2^2 + 2T_2 m_2 \tag{2}$$

Solving (1) and (2)

$$T_1 = \frac{Q(Q + 2m_2)}{2(m_1 + m_2 + Q)}; \qquad T_2 = \frac{Q(Q + 2m_1)}{2(m_1 + m_2 + Q)}$$

Example 3.3 Antiprotons are captured at rest in deuterium giving rise to the reaction

$$p^- + d \to n + \pi^0$$

Find the total energy of the π^0. The rest energies for p^-, d, n, π^0 are 938.2, 1875.5, 939.5 and 135.0 MeV, respectively.

Solution

$$Q = 938.2 + 1875.5 - (939.5 + 135.0) = 1739.2\ \text{MeV}$$

$$T_r = \frac{Q(Q + 2m_n)}{2(Q + m_\pi + m_n)} = \frac{1739.2(1739.2 + 2 \times 939.5)}{2(1739.2 + 135 + 939.5)} = 11.18\ \text{MeV}$$

Total energy $E_\pi = 1118 + 135 = 1253$ MeV.

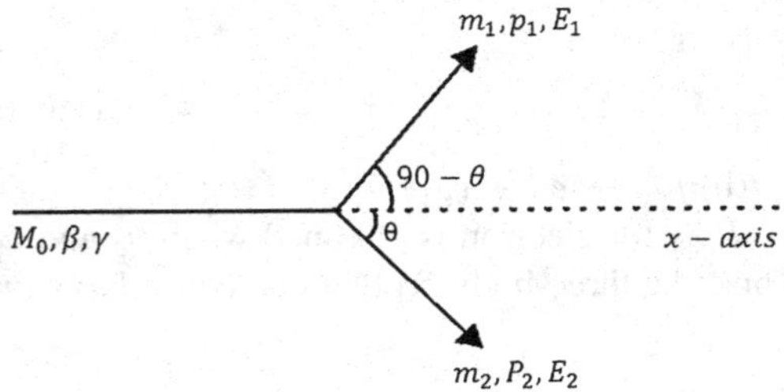

Fig. 3.16 Decay of an unstable heavy particle into two lighter particles

Example 3.4 An unstable heavy particle of relativistic energy decays into two lighter particles which are emitted with momenta $\vec{p}_1$ and $\vec{p}_2$ and total energies E_1 and E_2, respectively as shown in Fig. 3.16. If the decay products are emitted at right angles to each other, calculate the rest mass, velocity and the direction of motion of the primary particle.

Solution

$$M_0\beta\gamma = p_1 \sin\theta + p_2 \cos\theta \quad \text{(momentum conservation along } x\text{-axis)} \tag{i}$$

$$p_1 \cos\theta = p_2 \sin\theta \quad \text{(momentum conservation along } y\text{-axis)} \tag{ii}$$

$$M_0\gamma = E_1 + E_2 \quad \text{(energy conservation)} \tag{iii}$$

Solving (i), (ii) and (iii)

$$M_0 = \sqrt{m_1^2 + m_2^2 + 2E_1E_2}$$

$$\beta = \frac{\sqrt{p_1^2 + p_2^2}}{E_1 + E_2}$$

$$\theta = \tan^{-1}\left(\frac{p_1}{p_2}\right)$$

Example 3.5 A π^+ meson at rest decays into a μ^+ meson and a neutrino in 2.5×10^{-8} sec. Assuming that the π^+ meson has kinetic energy equal to its rest energy, what distance would the meson travel before decaying as seen by an observer at rest?

Solution

$$(\gamma - 1)M = M \quad \text{or} \quad \gamma = 2$$

$$\beta = \frac{\sqrt{\gamma^2 - 1}}{\gamma} = \frac{\sqrt{2^2 - 1}}{2} = \frac{\sqrt{3}}{2}$$

The dilated time $t_1 = \gamma t_0 = 2 \times 2.5 \times 10^{-8} = 5 \times 10^{-8}$ sec.

The distance traveled before decaying is

$$d = vt_1 = \beta c t_1 = \frac{\sqrt{3}}{2} \times 3 \times 10^8 \times 5 \times 10^{-8} = 13 \text{ metre.}$$

Example 3.6 If intense beams of charged pions be produced and allowed to decay in their flight in an evacuated tunnel, energetic beams of muons and muon neutrinos can be produced. Find the fraction of pions of momenta 100 GeV/c that will decay while traveling a distance of 200 m (mean lifetime of pions = 2.6×10^{-8} sec).

Solution Time $t = \frac{d}{v} = \frac{300}{3\times 10^8} = 1.0 \times 10^{-6}$ sec as $v \simeq c$ at ultra-relativistic velocity.

The proper lifetime is dilated

$$\begin{aligned}\tau &= \tau_0 \gamma \simeq \tau_0 \frac{E}{m} = 2.6 \times 10^{-8} \times \frac{(200 \times 10^3 + 140)}{140} \\ &= 3.71 \times 10^{-3}\end{aligned}$$

Fraction f of pions decaying is given by the radioactive law

$$\begin{aligned}f &= 1 - \exp(-t/\tau) \\ &= 1 - \exp(0.0269) \\ &= 0.027\end{aligned}$$

The pions and muons are subsequently stopped in thick walls of steel and concrete, pions through their nuclear interactions and muons through absorption by ionization.

Example 3.7 As a result of a nuclear interaction a K^* particle is created which decays to a K meson and a π^- meson with rest masses equal to $966m_e$ and $273m_e$, respectively. From the curvature of the resulting tracks in a magnetic field, it is concluded that the momentum of the secondary K and π mesons are 394 MeV/c and 254 MeV/c, respectively, their initial directions of motion being inclined to one another at 154°. Calculate the rest mass of the K^* particle.

Solution

$$M^2 = m_1^2 + m_2^2 + 2(E_1 E_2 - P_1 P_2 \cos\theta) \quad \text{(i)}$$

$$m_1 = 966m_e = 966 \times 0.511\ \text{MeV} = 493.6\ \text{MeV} \quad \text{(ii)}$$

$$m_2 = 273m_e = 273 \times 0.511\ \text{MeV} = 139.5\ \text{MeV} \quad \text{(iii)}$$

$$p_1 = 394\ \text{MeV}, \qquad p_2 = 254\ \text{MeV} \quad \text{(iv)}$$

$$E_1 = \sqrt{p_1^2 + m_1^2} = \sqrt{(394)^2 + (4936)^2} = 631.6 \quad \text{(v)}$$

$$E_2 = \sqrt{p_2^2 + m_2^2} = \sqrt{(254)^2 + (1395)^2} = 289.8 \quad \text{(vi)}$$

$$\cos\theta = \cos 154° = -0.898 \quad \text{(vii)}$$

Using (ii)–(vii) in (i),

$$M = 899.4\ \text{MeV}$$

Example 3.8 The kinetic energy and momentum of a particle are found to be 250 MeV and 368 MeV/c respectively. Determine the mass of the particle and identify it.

Solution

$$p^2 = T^2 + 2Tm$$

$$m = \frac{p^2 - T^2}{2T} = \frac{(368)^2 - (250)^2}{2 \times 250} = 145.85\text{ MeV}$$

$$= \frac{145.85}{0.511} = 285.4m_e$$

It is π meson, the accepted mass of π meson, being $273m_e$.

Example 3.9 Determine the mean free path of an energetic proton in lead using the concept of geometrical cross-section. Given density of lead = 11.3 g/cm^3, atomic weight of lead = 11.3 g/cm^3, $R = r_0 A^{1/3}$, with $r_0 = 1.3$ fm.

Solution Radius of lead, $R = r_0 A^{1/3} = 1.3 \times 10^{-13} \times (207)^{1/3} = 7.69 \times 10^{-13}$ cm. Geometrical cross-section $\sigma = \pi R^2 = 1.857 \times 10^{-24}$ cm^2. Mean free path $\lambda = \frac{1}{\sigma N}$

$$N = \frac{N_{av}\rho}{A} = \frac{6 \times 10^{23} \times 11.3}{207} = 3.275 \times 10^{22}$$

$$\lambda = \frac{1}{1.857 \times -24 \times 3.275 \times 10^{22}} = 16.4\text{ cm}$$

Example 3.10 The heavy boson W^+ is not produced in the reaction $\nu + p \rightarrow p + \mu^- + W^+$, where stationary protons are bombarded with neutrinos of energy $E_\nu = 100$ GeV. Estimate a lower limit for the mass of the W^+ (mass of proton 938 MeV/c^2, mass of muon 106 MeV/c^2, mass of neutrino zero).

Solution Using the invariance, $E^2 - |\Sigma \vec{p}|^2 = E^{*2} - |\Sigma p^*|^2$. At threshold: $(E_\nu + M_p)^2 - E_\nu^2 = (M_P + M_\mu + M_W)^2 - 0$

$$(100 + 0.938)^2 - (100)^2 = (0.938 + 0.106 + M_W)^2$$

$$M_W = 12.68\text{ GeV}$$

Since the reaction does not proceed, $M_W > 12.68$ GeV.

Example 3.11 Calculate the threshold energy for the reaction $p + p \rightarrow p + \Lambda + K^+$. Masses $m_p = 938$ MeV/c^2, $m_\Lambda = 1115$ MeV/c^2, $m_k = 494$ MeV/c^2.

Solution $T_{thr} = \frac{m^2}{2m_2} + m(1 + \frac{m_1}{m_2})$. Excess mass, $m = 1115 + 494 - 938 = 671$ MeV

$$T_{thr} = \frac{(671)^2}{2 \times 938} + 671\left(1 + \frac{938}{938}\right)$$

$$= 1582\text{ MeV}$$

Example 3.12 Antineutrinos of 2.3 MeV from the fission product decay in a reactor have a total cross-section with protons ($\vec{\nu}_e + p \rightarrow e^+ + n$) of 6×10^{-48} m^2. Calculate the mean free path of these neutrinos in water. Assume the antineutrinos are able to interact only with the free protons.

Solution $\lambda = \frac{1}{\Sigma} = \frac{1}{\sigma N}$, where N = number of H_2O molecules per cm^3

$$N = \frac{N_{av}\rho}{A} = 6 \times 10^{23} \times \frac{1}{18} \times 1 = 3.33 \times 10^{22}$$

Number of H atoms per cm$^3 = 2 \times 3.33 \times 10^{22} = 6.66 \times 10^{22}$

$$\Sigma = 6.66 \times 10^{22} \times 6 \times 10^{-44} = 4 \times 10^{-21}\ \text{cm}^2$$

$$\lambda = \frac{1}{\Sigma} = \frac{1}{4 \times 10^{-21}} = 2.5 \times 10^{20}\ \text{cm}$$

$$= \frac{2.5 \times 10^{20}}{945 \times 10^{15}} = 2.6 \times 10^4\ \text{light years.}$$

Example 3.13 Calculate threshold energy and the invariant mass of the system at the threshold energy. $p + p \rightarrow p + \Lambda + K^+$. The rest energies of p, Λ and K^+ are 938, 1115 and 494 MeV.

Solution $T_{thr} = \frac{M^2}{2m_2} + M(1 + \frac{m_1}{m_2})$, where m_1 and m_2 are the incident and target particles mass and M is the excess of mass.

$$m_1 = m_2 = 938, \qquad M = 1{,}115 + 494 - 938 = 671$$

$$T_{thr} = \frac{(671)^2}{2 \times 938} + 671\left(1 + \frac{1}{1}\right) = 1582\ \text{MeV}$$

Example 3.14 Calculate the threshold energy for the photo-meson production.

Solution $\gamma + p \rightarrow n + \pi^+$

$$T_{thr} = \frac{M^2}{2m_2} + M\left(1 + \frac{m_1}{m_2}\right)$$

$$m_1 = 0;\ m_2 = 938;\ M = 939 + 140 - 938 = 141$$

$$T_{thr} = \frac{(141)^2}{2 \times 938} + 141(1 + 0) = 152\ \text{MeV}$$

Example 3.15 Consider the production of Ω^- in the reaction $K^- + p \rightarrow K^0 + K^+ + \Omega^-$ with hydrogen target at threshold energy. Calculate the probability that Ω^- will survive after traversing a distance of 4 cm. The rest masses are $m_{k^-} = m_{k^+} = 494$ MeV/c^2, $m_{k^0} = 498$ MeV/c^2, $m_p = 938$ MeV/c^2, $m_{\Omega^-} = 1675$ MeV/c^2, mean lifetime of $\Omega^- = 1.3 \times 10^{-10}$ sec.

Solution

$$T_{thr} = \frac{m^2}{2m_2} + m\left(1 + \frac{m_1}{m_2}\right)$$

$$m_1 = 494;\ m_2 = 938$$

$$m = 498 + 1685 - 938 = 1245$$

$$T_{th} = \frac{(1245)^2}{2 \times 938} + 1245\left(1 + \frac{494}{938}\right) = 2727\ \text{MeV}$$

Minimum momentum

$$p_{K^-} = \sqrt{T^2 + 2Tm} = \sqrt{(2727)^2 + 2 \times 2727 \times 494} = 3183\ \text{MeV}$$

$$P_{thr} = 3183\ \text{MeV}/c$$

$$E_K = 2727 + 494 = 3221$$

$$\gamma_K = \frac{E_K}{m_K} = \frac{3221}{494} = 6.52$$

$$\gamma_c = \frac{\gamma + m_2/m_1}{\sqrt{1 + 2\gamma m_2/m_1 + (m_2/m_1)^2}}$$

$$\frac{m_2}{m_1} = \frac{938}{494} = 1.9$$

$$\gamma_c = \frac{6.52 + 1.9}{\sqrt{1 + 2 \times 652 \times 19 + (19)^2}} = 5.51$$

$$\gamma_\Omega = \gamma_c = 5.51; \qquad \beta_\Omega = \frac{\sqrt{\gamma_\Omega^2 - 1}}{\gamma_\Omega} = 0.983$$

Proper time

$$t_0 = \frac{d}{v} = \frac{d}{\beta c}$$

Observed time

$$t = \gamma t_0 = \frac{\gamma d}{\beta c} = \frac{5.51 \times 4 \times 10^{-2}}{0.983 \times 3 \times 10^8}$$

$$= 7.47 \times 10^{-10}\ \text{sec}$$

Probability that Ω^- will travel 4 cm before decay.

$$= \exp\left(-\frac{t}{\tau}\right)$$

$$= \exp\left(-\frac{7.47 \times 10^{-10}}{1.3 \times 10^{-10}}\right)$$

$$= 0.0032$$

Example 3.16 Show that in the collision of proton with hydrogen, the threshold for p^-–p production is $6mc^2$, where m is the mass of proton or antiproton.

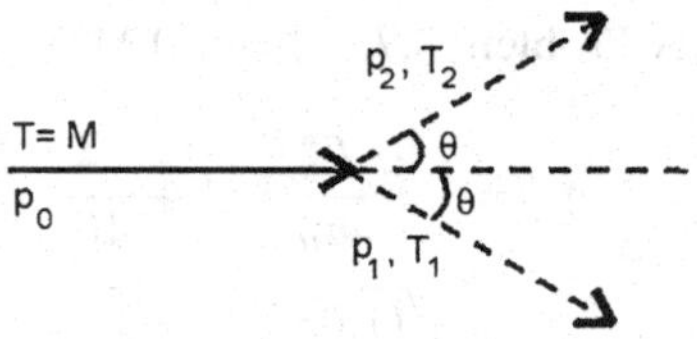

Fig. 3.17 Elastic collision between protons

Solution The reaction is

$$p + p \to p^- + p + p + p$$

Using the invariance $E^2 - p^2 = E^{*2} - p^{*2}$

$$(T + m + m)^2 - (T^2 + 2Tm) = (m + m + m + m)^2 - \text{zero}$$

or $T = 6m$ therefore $T_{thr} = 6mc^2$.

Example 3.17 A proton of kinetic energy 1876 MeV makes an elastic collision with a stationary proton in such a way that after collision, the protons are traveling at equal angles on either side of the incident proton. Calculate the angle between the direction of motion of the protons.

Solution With reference to Fig. 3.17

$$P_0 = p_1 \cos\theta + p_2 \cos\theta = 2p_1 \cos\theta \quad (\because p_2 = p_1 \text{ by symmetry})$$
$$P_0^2 = 4p_1^2 \cos^2\theta$$
$$T_0^2 + 2T_0 M = 4(T_1^2 + 2T_1 M)\cos^2\theta$$

But

$$T_0 = 938 \text{ MeV} = 2M \quad \text{and} \quad T_1 = \frac{T_0}{2} = M$$
$$4M^2 + 4M^2 = 4(M^2 + 2M^2)\cos^2\theta$$

whence

$$\cos\theta = 0.816$$
$$\theta = 35.3°$$
$$2\theta = 70.6°$$

Example 3.18 A π-meson with a kinetic energy of 280 MeV decays in flight into μ-meson and a neutrino. Calculate the maximum energy which (a) the μ-meson (b) the neutrino may have in the Laboratory system (mass of π-meson $= 140$ MeV/c^2, mass of μ-meson $= 106$ MeV/c^2, mass of neutrino $= 0$).

Solution

$$\gamma_\pi = 1 + \frac{T_\pi}{m_\pi} = 1 + \frac{280}{140} = 3$$
$$\beta_\pi = \frac{\sqrt{\gamma_\pi^2 - 1}}{\gamma_\pi} = \frac{\sqrt{3^2 - 1}}{3} = 0.9428$$

By Problem 3.7, $T^*_\mu = 4.0$ MeV therefore

$$\gamma^*_\mu = 1 + \frac{T^*_\mu}{m_\mu} = 1 + \frac{4}{106} = 1.038$$

$$\beta^*_\mu = \frac{\sqrt{(1.0377)^2 - 1}}{1.0377} = 0.267$$

$$\gamma_\mu = \gamma_\pi \gamma^*_\mu \left(1 + \beta_\pi \beta^*_\mu \cos\theta^*\right)$$

$$\gamma_\mu(\max) = \gamma_\pi \gamma^*_\mu \left(1 + \beta_\pi \beta^*_\mu\right) = 3 \times 1.038(1 + 0.9428 \times 0.267) = 3.898$$

$$(\because \theta^* = 0)$$

$$T_\mu(\max) = \left(\gamma_\mu(\max) - 1\right) m_\mu = 307 \text{ MeV}$$

Using the formula for optical Doppler effect

$$T_\nu(\max) = \gamma_\pi T^*_\nu (1 + \beta_\pi) = 3 \times 29.5(1 + 0.9428) = 172 \text{ MeV}$$

Example 3.19 Assume the decay $K^0 \to \pi^+ + \pi^-$ in flight. Calculate the mass of the primary particle if the momenta of each of the secondary particles is 360 MeV/c and the angle between the tracks is 70°.

Solution Let the mass of the primary particle be M, and that of secondary particles m_1 and m_2. Let the total energy of the secondary particles in the LS be E_1 and E_2, and momenta p_1 and p_2. Using the invariance of (total energy)^2 – (total momentum)^2

$$\begin{aligned}
M^2 &= (E_1 + E_2)^2 - \left(p_1^2 + p_2^2 + 2p_1 p_2 \cos\theta\right) \\
&= E_1^2 - p_1^2 + E_2^2 - p_2^2 + 2(E_1 E_2 - p_1 p_2 \cos\theta) \\
&= m_1^2 + m_2^2 + 2\left(E_1^2 - p_1^2 \cos\theta\right) \quad (\because E_1 = E_2, p_1 = p_2) \\
&= 2m_1^2 + 2\left(m_1^2 + p_1^2 - p_1^2 \cos\theta\right) \\
&= 4m_1^2 + 4p_1^2 \sin^2\frac{\theta}{2} \\
&= 4(139)^2 + 4(360)^2 \sin^2 35° \\
&= 247832 \\
M &= 498 \text{ MeV}/c^2
\end{aligned}$$

Example 3.20 A counter arrangement is set up for the identification of particles which originate from the bombardment of a target with 5.3 GeV protons. The negatively charged particles are subject to momentum analysis in a magnetic field which permits only those having momentum $p = 1.19$ GeV/c to pass through into a telescope system comprising two scintillation counters in coincidence with a separation of 12 m between the detectors.

Identify the particles whose time of flight in the telescope arrangement was determined as $t = 51 \pm 1$ ns. What would have been the time of flight of π^- mesons?

Solution

$$\beta = \frac{v}{c} = \frac{d}{ct} = \frac{12}{3 \times 10^8 \times 51 \times 10^{-9}} = 0.7843$$

$$\gamma = \frac{1}{\sqrt{1-\beta^2}} = 1.613$$

$$p = m\gamma\beta c$$

$$mc^2 = \frac{cp}{\gamma\beta} = \frac{1.19}{1.613 \times 0.7843} = 0.94 \text{ GeV}$$

The particles are antiprotons

For π^- mesons

$$\gamma\beta = \frac{\beta}{\sqrt{1-\beta^2}} = \frac{cp}{m_0c^2} = \frac{1.19}{0.14} = 8.9$$

whence $\beta = 0.9875$

Time of flight, $t = \frac{d}{\beta c} = \frac{12}{0.9875 \times 3 \times 10^8} = 4 \times 10^{-8}$ sec or 40 ns.

Example 3.21 A proton of momentum $\vec{p}$ large compared with its rest mass M, collides with a proton inside a target nucleus with Fermi momentum $\vec{p}_f$. Find the available kinetic energy in the collision, as compared with that for a free-nucleon target, when $\vec{p}$ and $\vec{p}_f$ are (a) parallel (b) anti parallel (c) orthogonal

Solution Using the invariance, $E^2 - |\sum \vec{p}|^2 = E^{*2}$

$$\begin{aligned} E^{*2} &= (E + E_f)^2 - \left(p^2 + p_f^2 + 2pp_f \cos\theta\right) \\ &= (E + E_f)^2 - \left(E^2 - M^2 + E_f^2 - M^2 + 2pp_f \cos\theta\right) \\ &= 2M^2 + 2(EE_f - \vec{p} \cdot \vec{p}_f) \end{aligned}$$

(a) for parallel momenta, $\theta = 0$, $\vec{p} \cdot \vec{p}_f = +pp_f$
(b) for anti-parallel momenta $\theta = \pi$, $\vec{p} \cdot \vec{p}_f = -pp_f$
(c) for orthogonal momenta $\theta = \frac{\pi}{2}$, $\vec{p} \cdot \vec{p}_f = 0$

Example 3.22 A positron of energy E_+, momentum $\vec{p}_+$ and an electron, energy E_-, momentum $\vec{p}$ are produced in a pair creation process

(a) what is the velocity of their CMS?
(b) what is the energy of either particle in the CMS.

Solution

(a) $\beta_c = \frac{|\vec{p}_+ + \vec{p}_-|}{E_+ + E_-}$

Using the invariance principle

$$(\text{total energy})^2 - (\text{total momentum})^2 = \text{invariant}$$

$$(E_+ + E_-)^2 - |\vec{p}_+ + \vec{p}_-|^2 = \left(E_1^* + E_2^*\right)^2 - \left|\vec{p}_1^* + \vec{p}_2^*\right|^2$$

But $E_1^* = E_2^*$ since the particles have equal masses. Also by definition of center of mass, $|\vec{p}_1^* + \vec{p}_2^*| = 0$.

Therefore $E_1^{*2} = E_2^{*2} = \frac{1}{4}[(E_+ + E_-)^2 - (p_+^2 + p_-^2 + 2p_+p_- \cos\theta)]$ where θ is the angle between e^+–e^- pair

$$\begin{aligned} E_1^{*2} = E_2^{*2} &= \frac{1}{4}\left[E_+^2 - p_+^2 + E_-^2 - p_-^2 + 2(E_+E_- - p_+p_- \cos\theta)\right] \\ &= \frac{1}{4}\left[m^2 + m^2 + 2(E_+E_- - p_+p_- \cos\theta)\right] \\ &= \frac{1}{2}\left(m^2 + E_+E_- - p_+p_- \cos\theta\right) \end{aligned}$$

or

$$E_1^* = E_2^* = \sqrt{\frac{1}{2}\left(m^2 + E_+E_- - p_+p_- \cos\theta\right)}$$

Example 3.23 Calculate the binding energy of Λ in the hyper fragment ${}^5_\Lambda$He if the rest mass energy of Λ, ^{4}He and ${}^5_\Lambda$He are 1115.58 MeV, 3727.32 MeV and 4839.82 MeV respectively.

Solution

$$\begin{aligned} B_\Lambda &= \left(M_{\Lambda}^{-} + M_{\text{He}}\right) - M_{\Lambda^5\text{He}} \\ &= (1115.58 + 3727.32) - 4839.82 \\ &= 3.08\ \text{MeV}. \end{aligned}$$

Example 3.24 A particle of mass m collides elastically with another identical particle at rest. Show that for a relativistic collision

$$\tan\theta \tan\phi = \frac{2}{\gamma + 1}$$

where θ, ϕ are the angles of the out-going particles with respect to the direction of the incident particle and γ is the Lorentz factor before the collision. Also, show that $\theta + \phi \le \frac{\pi}{2}$, where the equal sign is valid in the classical limit.

Solution

$$\tan\theta = \frac{1}{\gamma_c} \cdot \frac{\sin\theta^*}{(\cos\theta^* + \beta_c/\beta_1^*)} = \frac{1}{\gamma_c} \tan\frac{\theta^*}{2}$$
$$(\because \beta_c = \beta_1^*) \tag{1}$$

Also

$$\tan\phi = \frac{1}{\gamma_c} \tan\frac{\phi^*}{2} \tag{2}$$

where θ^* and ϕ^* are the corresponding angles in the CM system as shown in Fig. 3.18.

Multiply (1) and (2)

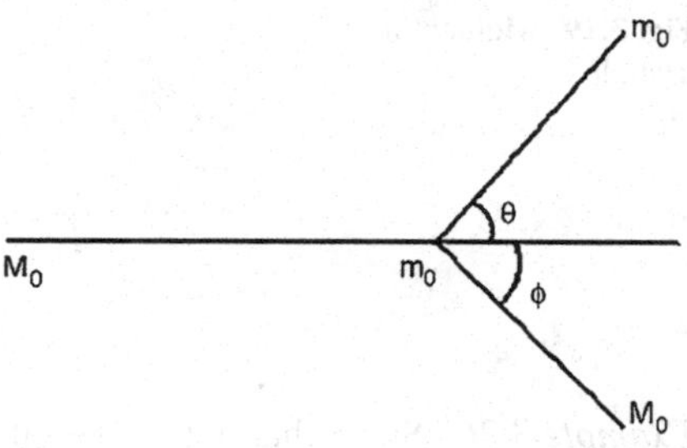

Fig. 3.18 Elastic collision between two identical particles

$$\tan\theta\tan\phi = \frac{1}{\gamma_c^2}\tan\frac{\theta^*}{2}\tan\frac{\phi^*}{2}$$

But $\phi^* = \pi - \theta^*$ and for $m_1 = m_2$, $\gamma_c = \sqrt{\frac{\gamma+1}{2}}$

$$\therefore \quad \tan\theta\tan\phi = \frac{2}{\gamma+1}$$

In the classical limit, $\gamma \to 1$ and $\tan\theta\tan\phi = 1$.
But since

$$\tan(\theta+\phi) = \frac{\tan\theta + \tan\phi}{1 - \tan\theta\tan\phi}$$
$$\tan(\theta+\phi) \to \infty$$

i.e.

$$\theta + \phi = \frac{\pi}{2}$$

For $\gamma > 1$, $\tan\theta\tan\phi < 1$. Hence $\tan(\theta+\phi)$ will be finite so that $(\theta+\phi) < \frac{\pi}{2}$. Hence, for $\gamma \geq 1; \theta + \phi \leq \frac{\pi}{2}$.

Example 3.25 An antiproton of momentum 5 MeV/*c* suffers a scattering. The angles of the recoil proton and scattered antiproton are found to be 82° and 2°30′ with respect to the incident direction. Show that the event is consistent with an elastic scattering of an antiproton with a free proton.

Solution By Example 3.24 it is sufficient to show that $\tan\theta\tan\phi = \frac{2}{\gamma+1}$

$$p = \gamma\beta m = \sqrt{\gamma^2 - 1}m$$

Therefore, $\gamma = \sqrt{1 + (\frac{p}{m})^2} = \sqrt{1 + (\frac{5}{0.9538})^2} = 5.41$

$$\tan\theta\tan\phi = \tan 82^\circ \tan 2^\circ 30' = 7.115 \times 0.04366 = 0.3106$$
$$\frac{2}{\gamma+1} = \frac{2}{5.41+1} = 0.3120$$

Hence the event is consistent with elastic scattering.

Fig. 3.19 Momentum triangle

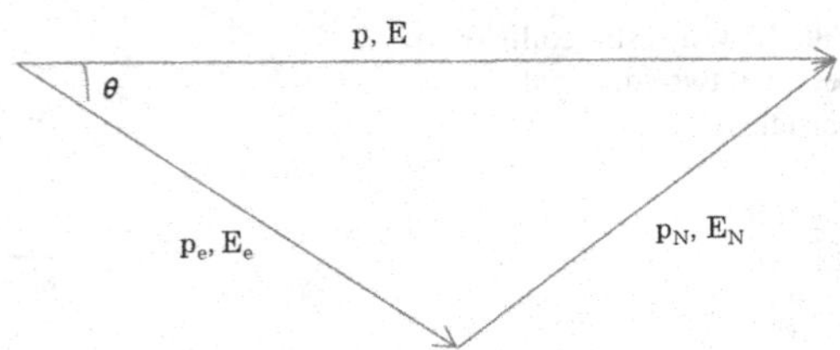

Example 3.26 Show that if E is the ultra-relativistic laboratory energy of electrons incident on a nucleus of mass M, the nucleus will acquire kinetic energy

$$E_N = \frac{E^2}{Mc^2} \frac{(1-\cos\theta)}{1+\frac{E}{Mc^2}(1-\cos\theta)}$$

where θ is the scattering angle.

Solution Let $\vec{p}$, $\vec{p}_e$, $\vec{p}_N$ be the momentum of the incident electron, scattered electron and recoil nucleus, respectively. From the momentum triangle, Fig. 3.19.

$$p_N^2 = p_e^2 + p^2 - 2pp_e\cos\theta = E_N^2 + 2E_N M \quad (1)$$

where we have put $c = 1$.

From energy conservation

$$E_N + E_e = E \quad (2)$$

$$\text{As} \quad E_e \simeq p_e \quad (3)$$

$$E \simeq p \quad (4)$$

Equation (2) can be written as p_e, E_e

$$E_N + p_e = E \quad (5)$$

Combining (1), (3), (4) and (5), we get

$$E_N = \frac{E^2(1-\cos\theta)}{M[1+\frac{E}{M}(1-\cos\theta)]}$$

Restoring c^2, we get the desired result.

Example 3.27 Neutral pions of fixed energy decay in flight into two γ-rays. Show that

a. the velocity of π^0 is given by

$$\beta = \frac{E_{\text{max}} - E_{\text{min}}}{E_{\text{max}} + E_{\text{min}}}$$

where E is the γ-ray energy in the laboratory

b. the rest mass energy of π^0 is given by

$$mc^2 = 2\sqrt{E_{\text{max}} E_{\text{min}}}$$

c. the energy distribution of γ-rays in the laboratory is uniform under the assumption that γ-rays are emitted isotropically in the rest system of π^0.
d. the angular distribution of γ-rays in the laboratory is given by

$$I(\theta) = \frac{1}{4\pi\gamma^2(1-\beta\cos\theta)^2}$$

e. the locus of the tip of the momentum vector is an ellipse.
f. in a given decay the angle ϕ between the two γ-rays is given by

$$\sin\frac{\phi}{2} = \frac{mc^2}{2\sqrt{E_1E_2}}$$

g. the minimum angle between the two γ-rays is given by

$$\theta_{\min} = \frac{2mc^2}{E_\pi}$$

Solution Consider one of the two γ-rays. From Lorentz transformation

$$cp_x = \gamma_c\left(cp_x^* + \beta_c E^*\right)$$

where the energy and momentum refer to one of the two γ-rays and the subscript c refers to π^0. Starred quantities refer to the rest system of π^0.

$$cp\cos\theta = \gamma\left(cp^*\cos\theta^* + \beta E^*\right)$$

where we have dropped off the subscript c. But for γ-rays $cp^* = E^*$ and $cp = E$, and because the two γ-rays share equal energy in the CMS, $E^* = mc^2/2$, where m is the rest mass of π^0.

Therefore

$$cp\cos\theta = \frac{\gamma mc^2}{2}(\beta + \cos\theta^*) \tag{1}$$

Also

$$\begin{aligned} E &= \gamma\left(E^* + \beta cp_x^*\right) = \gamma\left(E^* + \beta cp^*\cos\theta^*\right) \quad \text{or} \\ cp &= E = \gamma\frac{mc^2}{2}(1+\beta\cos\theta^*) \end{aligned} \tag{2}$$

Eliminating $\cos\theta^*$ between (1) and (2)

$$\begin{aligned} cp(1-\beta\cos\theta) &= \gamma\frac{mc^2}{2}(1-\beta^2) = \frac{mc^2}{2\gamma} \\ cp = \hbar\nu = E &= \frac{mc^2}{2\gamma(1-\beta\cos\theta)} \end{aligned} \tag{3}$$

When $\theta = 0$,

$$E_{\max} = \frac{mc^2}{2\gamma(1-\beta)} = \frac{mc^2(1+\beta)}{2\gamma(1-\beta^2)} = \frac{1}{2}E_\pi(1+\beta) \tag{4}$$

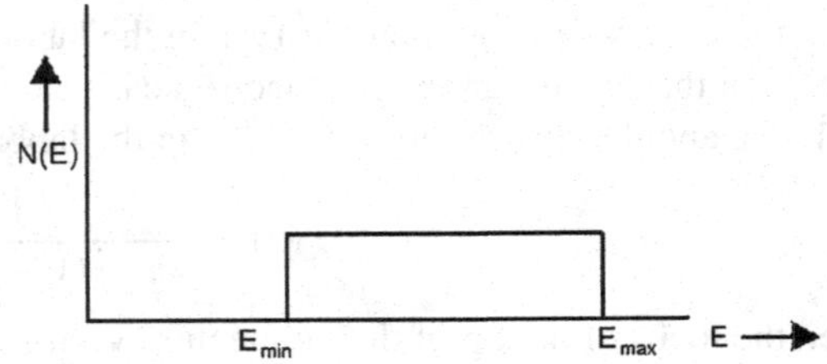

Fig. 3.20 Rectangular or uniform energy spectrum

When $\theta = \pi$,

$$E_{\min} = \frac{mc^2}{2\gamma(1+\beta)} = \frac{mc^2(1-\beta)}{2(1-\beta^2)\gamma} = \frac{1}{2}E_\pi(1-\beta) \tag{5}$$

a. From (4) and (5), $\frac{E_{\max}-E_{\min}}{E_{\max}+E_{\min}} = \beta$. From the measurement of $E_{\max}$ and $E_{\min}$, the velocity of π^0 can be determined.

b. Multiplying (4) and (5) and writing $E_\pi = \gamma m_\pi c^2$

$$m_\pi c^2 = 2\sqrt{E_{\max}E_{\min}}$$

From the measurement of $E_{\max}$ and $E_{\min}$, mass of π^0 can be determined.

If $E_{\max} = 75$ MeV and $E_{\min} = 60$ MeV, then $m_\pi c^2 = 2\sqrt{75 \times 60} =$ 134.16 MeV.

Hence the mass of π^0 is $\frac{134.16}{0.51} = 262.5 m_e$.

c.

$$\frac{dN}{dE} = \frac{dN}{d\Omega^*}\frac{d\Omega^*}{dE} = \frac{1}{4\pi}\frac{2\pi\sin\theta^* d\theta^*}{dE} = \frac{1}{2}\frac{d\cos\theta^*}{dE} \tag{6}$$

where we have put $d\Omega^* = 2\pi\sin\theta^* d\theta^*$ for the element of solid angle and $\frac{dN}{d\Omega^*} = \frac{1}{4\pi}$ under the assumption of isotropy.

Differentiating (2) with respect to $\cos\theta^*$

$$\frac{dE}{d\cos\theta^*} = \frac{\gamma\beta mc^2}{2}$$

or

$$\frac{1}{2}\frac{d\cos\theta^*}{dE} = \frac{1}{\gamma\beta mc^2} \tag{7}$$

Combining (6) and (7), the normalized distribution is

$$\frac{dN}{dE} = \frac{1}{\gamma\beta mc^2} = \text{constant} \tag{8}$$

This implies that the energy spectrum is rectangular or uniform. It extends from a minimum to maximum, Fig. 3.20.

From (4) and (5),

$$E_{\max} - E_{\min} = \beta E_\pi = \gamma\beta mc^2 \tag{9}$$

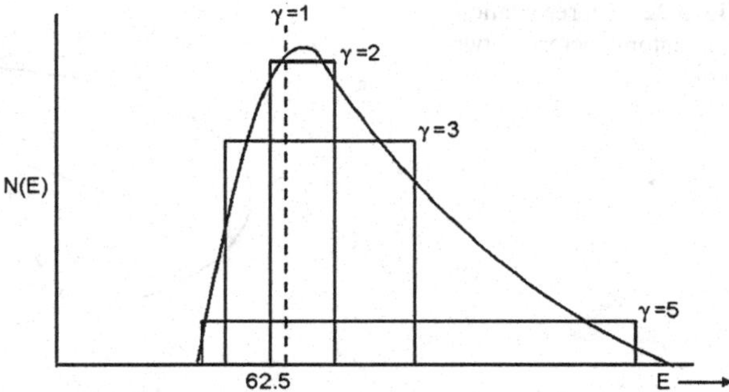

Fig. 3.21 Rectangular distribution of γ-ray energy

Note that the area of the rectangle is height × length

$$\left(\frac{dN}{dE}\right) \times (E_{\max} - E_{\min}) = \frac{1}{\gamma\beta mc^2} \times \gamma\beta mc^2 = 1$$

That is, the distribution is normalized as it should.

The higher the π^0 energy the larger is the spread in the γ-ray energy spectrum.

For a mono-energetic source of π^0's, we will have a rectangular distribution of γ-ray energy as in Fig. 3.21. But if the γ-rays are observed from π^0's of varying energy, as in cosmic ray events the rectangular distributions may be superimposed so that the resultant distributions may look like the solid curve, shown in Fig. 3.21.

Note that if the π^0's were to decay at rest ($\gamma = 1$) then the rectangle would have reduced to a spike at $E = 62.5$ MeV, half of rest energy of π^0.

d. The γ-rays of intensity $I(\theta^*)$ which are emitted in the solid angle $d\Omega^*$ in the CMS will appear in the solid angle $d\Omega$ in the LS with intensity $I(\theta)$. Therefore

$$I(\theta)d\Omega = I(\theta^*)d\Omega^* \quad \text{or}$$
$$I(\theta) = I(\theta^*)\frac{\sin\theta^* d\theta^*}{\sin\theta d\theta} \tag{10}$$

From the inverse Lorentz transformation

$$\begin{aligned} E^* &= \gamma E(1 - \beta\cos\theta) \\ &= \gamma E^* \gamma(1 + \beta\cos\theta^*)(1 - \beta\cos\theta) \\ \therefore \quad & \frac{1}{\gamma^2(1 - \beta\cos\theta)} = 1 + \beta\cos\theta^* \end{aligned}$$

Differentiating

$$-\frac{\beta\sin\theta d\theta}{\gamma^2(1 - \beta\cos\theta)^2} = -\beta\sin\theta^* d\theta^*$$

Therefore

$$\frac{\sin\theta^* d\theta^*}{\sin\theta d\theta} = \frac{1}{\gamma^2(1 - \beta\cos\theta)^2} \tag{11}$$

Also

$$I(\theta^*) = \frac{1}{4\pi} \tag{12}$$

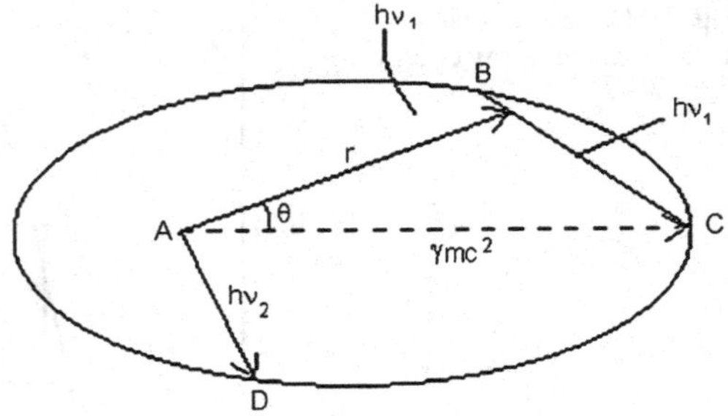

Fig. 3.22 Representation of momentum vectors of two photons

because of assumption of isotropy of photons in the rest frame of π^0.

Combining (10), (11) and (12)

$$I(\theta) = \frac{1}{4\pi\gamma^2(1-\beta\cos\theta)^2} \tag{13}$$

This shows that small emission angles of photons in the lab system are favoured.

e. In Fig. 3.22 $\vec{AB}$ and $\vec{AD}$ represent the momentum vectors of the two photons in the LS. $\vec{BC}$ is drawn parallel to $\vec{AD}$ so that ABC forms the momentum triangle, that is

$$\vec{AB} + \vec{BC} = \vec{AC}$$

From energy conservation

$$\hbar\nu_1 + \hbar\nu_2 = \frac{\gamma mc^2}{2}(1+\beta\cos\theta^*) + \frac{\gamma mc^2}{2}(1+\beta\cos(\pi-\theta^*)) = \gamma mc^2 = \text{const}$$

where we have used the fact that the angles of emission of the two photons in the rest frame of π^0, are supplementary.

Since momentum is given by $p = h\nu/c$, it follows that $AB + BC = \text{constant}$; which means that the locus of the tip of the momentum vector is an ellipse

$$E = \frac{mc^2}{2\gamma(1-\beta\cos\theta)} \tag{14}$$

Compare this with the standard equation for ellipse

$$r = \frac{a(1-\varepsilon^2)}{1-\varepsilon\cos\theta}$$

We find $\varepsilon = \beta$

$$a(1-\beta^2) = \frac{mc^2}{2\gamma} \quad \text{or}$$

$$a = \frac{\gamma mc^2}{2}$$

The larger the velocity of π^0, the greater will be the eccentricity ε.

f. The angle between the two γ-rays in the LS can be found from the formula

$$m^2c^4 = m_1^2c^4 + m_2^2c^4 + 2(E_1E_2 - c^2p_1p_2\cos\phi)$$

putting $m_1 = m_2 = 0$, $cp_1 = E_1$, $cp_2 = E_2$

$$\begin{aligned} m^2c^4 &= 2E_1E_2(1 - \cos\phi) = 4E_1E_2\sin^2\frac{\phi}{2} \\ \sin\frac{\phi}{2} &= \frac{mc^2}{2\sqrt{E_1E_2}} \end{aligned} \tag{15}$$

g. For small angle ϕ,

$$\phi = \frac{mc^2}{\sqrt{E_1E_2}} \tag{16}$$

set $E_2 = \gamma mc^2 - E_1$

$$\phi = \frac{mc^2}{\sqrt{E_1(\gamma mc^2 - E_1)}} \tag{17}$$

For minimum angle $\frac{d\phi}{dE_1} = 0$. This gives $E_1 = \frac{\gamma mc^2}{2}$.

Using this value of E_1 in (17), we obtain

$$\phi_{\min} = \frac{2}{\gamma} \tag{18}$$

Measurement of $\phi_{\min}$ affords the determination of E_π via γ.

Example 3.28 It is sometimes possible to differentiate between the tracks due to relativistic pions, protons and kaons in a bubble chamber by observing the high energy δ-rays which are produced. For a pion of momentum 8 GeV/c what is the minimum energy of a δ-ray which must be observed to prove it is not produced by a kaon or proton?

Solution Maximum energy transferred to the knock out electron (δ-ray) is given by

$$E'_{\max} = 2mc^2\beta^2\gamma^2$$

where mc^2 is the rest energy of electron, β and γ, refer to the incident particle. To identify a particle as a pion, it is necessary that

$$E' > \left(E'_{\max}\right)_K$$

For kaon of 8 GeV/c,

$$\begin{aligned} \beta &= \frac{p}{E} = \frac{1}{\sqrt{1 + (M/p)^2}} \\ &= \frac{1}{\sqrt{1 + (0.494/8)^2}} = 0.998 \\ \gamma &= \frac{1}{\sqrt{1 - \beta^2}} = 16.22 \\ \left(E'_{\max}\right) &= 2 \times 0.51 \times (0.998)^2(16.22)^2 \\ &= 268 \text{ MeV} \end{aligned}$$

Similarly

$$\left(E'_{\max}\right)_p = 8.6 \text{ MeV}$$
$$\left(E'_{\max}\right)_\pi = 1648 \text{ MeV}$$

Thus, the γ-ray energy must be greater than 268 MeV

3.10 Questions

3.1 How is the mass of π^- most accurately determined?

3.2 How is the mass of π^0 measured?

3.3 Use classical arguments to estimate the time required for a μ^- to fall from the radius of the lowest electron obit to the lowest μ orbit in iron. Assume the power is radiated continuously in accordance with the result of classical electrodynamics.

3.4 Suppose positive and negative kaon beams are available for an exposure of a hydrogen bubble chamber. For which beam is the threshold lowest for the production of $\overline{\Sigma}^-$, $\overline{\Sigma}^0$, $\overline{\Sigma}^+$, $\overline{\Xi}^+$, $\overline{\Xi}^0$ and $\overline{\Omega}^+$? Give the reaction that has the lowest threshold and the incident momentum at the threshold.

3.5 In the decay of π^+ at rest, the energy of μ^+ is unique, but in the decay of μ^+ at rest, the energy of e^+ varies from zero to a maximum value; why this is so?

3.6 Why are mesons thus called?

3.7 What was the name Leptons originally meant for?

3.8 Where does the name Baryon originate from?

3.9 Positronium is said to be a loose structure while a mesic atom is a relatively strongly bound structure. What is the physical reason?

3.10 What properties of neutrino and photon are (a) identical (b) different?

3.11 Describe the experimental discovery of neutrino.

3.12 How was it demonstrated that neutrino is different from its antiparticle antineutrino?

3.13 How was it demonstrated that ν_μ is different from ν_e?

3.14 The existence of neutrinos was predicted as early as 1928. But the experimental discovery came as late as 1956. Why was the delay?

3.15 What was the purpose of cadmium in the experimental set-up of Reins et al. for the discovery of $\overline{\nu}$?

3.16 Why mesons have to be massive in order to mediate nuclear forces?

3.17 When μ-meson was discovered initially it was suspected to be Yukawa's particle. But subsequently it was shown that it was not the correct candidate. What properties were in favour and what were against?

3.18 In the decay of π^+ meson at rest which particle carries more energy?

3.19 Is 2 MeV electron relativistic? Is 4 MeV proton relativistic?

3.20 In p–p collisions with hydrogen target, the threshold energy required for single pion production is 290 MeV, more than double the rest mass energy of pion which is about 140 MeV. Why is this necessary in a conventional accelerator?

3.21 In Yukawa's theory of nuclear forces massive mesons are assumed to provide the necessary glue between nucleons by jumping to and fro. Would not the energy conservation law break down when the meson is emitted or absorbed?

3.22 Can positronium annihilate into a single photon? If not, why not?

3.23 A relativistic Λ-hyperon decays in flight according to the scheme $\Lambda \rightarrow p + \pi^- + Q$. If E_p and E_π are the total energies in the Lab system for proton and pion, respectively and P_p and P_π, the corresponding momenta, show that the Q-value for the decay can be calculated from the formula

$$Q = \left[m_p^2 + m_\pi^2 + 2(E_p E_\pi - P_p P_\pi \cos\theta)\right]^{1/2}$$

where θ is the angle between and $\vec{p}_p$ and $\vec{p}_\pi$.

3.24 If the experiment in Question 3.20 is repeated in complex nuclei, then the threshold is brought down to nearly 150 MeV. Explain.

3.25 In the experimental discovery of antiproton, how p^-'s could be counted against a heavy background of π^-'s and μ^-'s?

3.26 By what process can an p^- be transformed into $\overline{n}$?

3.27 Why are strange particles thus called?

3.28 What was τ–θ puzzle?

3.29 A Σ^+-hyperon decays at rest ($\Sigma^+ \rightarrow p\pi^0$). Which particle is expected to carry more energy?

3.30 Why Ξ hyperon is called cascade hyperon?

3.31 Which of the following two reactions has lower threshold for the associated production?

$$p + p \rightarrow K^+ + \Sigma^+ + n$$
$$\pi^+ + p \rightarrow K^+ + \Sigma^+$$

3.32 At nominal energies, many more reaction channels are open for K^- mesons than for K^+ mesons. Explain.

3.33 What do you understand by the expression hyper-fragment?

3.34 Λ's may be produced from the K^- interactions at rest in

(a) $K^- + p \rightarrow \Lambda + \pi^0$

(b) $\rightarrow \Sigma^0 + \pi^0$, $\Sigma^0 \rightarrow \Lambda + \pi^0$

How are the energy distributions in (a) and (b) different?

3.35 Λ-hyperon has the same strangeness as Σ -hyperons. Why is not Λ associated with Σ^+ and Σ^- hyperons so as to form isospin triplet of Σ?

3.11 Problems

3.1 Consider the decay of muon at rest. If the energy released is divided equally among the final leptons, then show that the angle between the paths of any two leptons is approximately 120° (neglect the mass of leptons compared to the mass of muon mass.

3.2 Show that a Fermi energy of 25 MeV lowers the threshold incident kinetic energy for antiproton production by proton incident on nucleus to 4.3 GeV

3.3 In an event observed by Leprince-Ringuet and Lheritier a singly charged particle of mass $M \gg m_e$, scatters elastically from an electron. If the incident particle's momentum is p and the scattered electron's (relativistic) energy is E and Φ is the angle the electron makes with the incident particle (which is nearly undeflected) show that

$$M = p\left[\left[\frac{E + m_e}{E - m_e}\right]\cos^2\phi - 1\right]^{1/2}$$

In this event, the cloud chamber was in a magnetic field of about 2450 gauss. The incident particle had a radius of curvature of 7.0 m while that of the electron was

1.5 cm. If $\Phi = 20°$ and the scattering plane was perpendicular to the magnetic field, estimate M.
[Ans. $M = 555$ MeV/c^2]

3.4 Calculate the wavelengths of the first four lines of the Lyman series of the positronium on the basis of the simple Bohr's theory.
[Ans. 24.31 A, 20.53 A, 19.45 A, 19.00 A]

3.5 Calculate the threshold energy for the; production of e^+e^- pair in a high energy collision with an atomic electron.
[Ans. $6m_ec^2$]

3.6 One of the resonances in the K^--nucleon system has its maximum when the K^- meson momentum is about 1.1 GeV/c. Calculate it's mass.
[Ans. 842 MeV]

3.7 A π^+ meson decays at rest into π^+ and ν_μ. Estimate the energy carried by μ^+ and ν_μ, given that $m_\pi = 139.5$ MeV, $m_\mu = 106$ MeV, $m_\nu = 0$.
[Ans. $T_\mu = 4$ MeV, $E_\nu = 29.5$ MeV]

3.8 Consider the decay process $K^+ \to \pi^+\pi^0$ with the K^+ at rest. Find

(a) the total energy of the π^0 meson
(b) its relativistic kinetic energy

The rest mass energy is 494 MeV for K^+, 140 MeV for π^+, 135 MeV for π^0.
[Ans. (a) 245.6 MeV; (b) 110.6 MeV]

3.9 In the experiment of Curie and Joliot if the protons ejected from the hydrogen were due to Compton-like effect, $K_{\max} = 4.5$ MeV corresponds to the incident photon energy $h\nu = 50$ MeV. Photons of such energy would produce nitrogen recoil energy up to 377 keV. What would be the maximum energy of nitrogen nuclei under the neutron hypothesis?
[Ans. 1.12 MeV]

3.10 A neutral pion undergoes radioactive decay into two γ-rays. Obtain an expression for the laboratory angle between the direction of the γ-rays, and find the minimum value for this angle when the pion energy is 5 GeV ($m_{\pi^0} = 0.14$ GeV).
[Ans. $\theta_{\min} = \frac{2m}{E}$, 3.21°]

3.11 A particle of rest mass m_1 traveling in the positive x-direction with kinetic energy T collides with a stationary particle of rest mass m_2. After collision, two particles of rest mass m_3 and m_4 emerge. If particle 3 emerges with momentum p_3 along the negative z-axis, in what direction and with what momentum does particle 4 emerge after the collision
[Ans. $\theta = \tan^{-1} \frac{p_3}{\sqrt{T_2^2 + 2Tm_1}}$; $p_4 = \sqrt{p_3^2 + T^2 + 2Tm_1}$]

3.12 Find the threshold energy of the reaction

$$\gamma + p \rightarrow K^{+*} + \Lambda$$

The laboratory proton is at rest. The following rest energies may be assumed, for proton 940 MeV for K^*890 MeV, for Λ1110 MeV.
[Ans. 1562 MeV]

3.13 A bubble chamber event was identified in the reaction

$$p^- + p \rightarrow \pi^+ + \pi^- + \omega^0$$

The total energy available was 2.29 GeV while the kinetic energy of the residual particles was 1.22 GeV. What is the rest energy of ω^0 in MeV?
[Ans. 790 MeV]

3.14 A particle of rest mass m_1 and velocity v collides with a particle of mass m_2 at rest after which the two particles coalesce. Show that the mass M and velocity v of the composite particle are given by

$$M^2 = m_1^2 + m_2^2 + \frac{2m_1 m_2}{\sqrt{1 - v^2/c^2}}$$

3.15 If the mean track length of 100 MeV π mesons is 4.88 m up to the point of decay, calculate their mean lifetime.
[Ans. 2×10^{-8} sec]

3.16 A beam of π^+ masons of energy 2 GeV has an intensity of 10^6 particles per sec at the beginning of a 20 m flight path. Calculate the intensity of the neutrino flux at the end of the flight path (mass of π meson $= 139$ MeV/c^2, lifetime $=$ 2.56×10^{-8} sec).
[Ans. 8.34×10^5/s]

3.17 Find the threshold energy for the production of three pions by a pion incident on a hydrogen target, i.e. $\pi + p \rightarrow p + \pi + \pi + \pi$. Assume the rest masses of the pion and proton are $273m_e$ and $1837m_e$, where the rest energy of the electron $m_e c^2 = 0.51$ MeV.
[Ans. 362 MeV]

3.18 Σ^+-hyperon decays at rest via $\Sigma^+ \rightarrow p + \pi^0$. Find the energy of π^0, given $m_{\Sigma^+} = 1189$ MeV, $m_p = 938$ MeV, $m_{\pi^0} = 135$ MeV.
[Ans. 116 MeV]

3.19 One of the decay modes of K^+ mesons is $K^+ \rightarrow \pi^+ + \pi^+ + \pi^-$ what is the maximum kinetic energy that any of the pions can have, if the K^+ decays at rest? Given $m_k = 965m_e$ and $m_\pi = 273m_e$.
[Ans. 50 MeV]

3.20 Show that the energy E_n of positronium is given by

$$E_n = -\frac{\alpha^2 m_e c^2}{4n^2}$$

where m_e is the electron mass, n the principal quantum number and α the fine structure constant.

3.21 Calculate the threshold energy of the following reaction $\pi^- + p \rightarrow K^0 + \Lambda$. The mases for π^- and p, K^0 and Λ are, 140, 938, 498, 1115 MeV, respectively.
[Ans. 767 MeV]

3.22 Show that the threshold kinetic energy in the Laboratory for the production of n pions in the collision of protons with a hydrogen target is given by

$$T = 2nm_\pi \left(1 + \frac{nm_\pi}{4m_p}\right) \quad I$$

where m_π and m_p are respectively the pion and proton masses.

3.23 Show that for the decay in flight of a Λ-hyperon into a proton and a pion with laboratory momenta P_p and P_π, respectively, the Q value can be calculated from

$$Q = \left(m_p^2 + m_\pi^2 + 2E_p E_\pi - 2P_b P_\pi \cos\theta\right)^{1/2} - (m_p + m_\pi)$$

where θ is the angle between P_p and P_π in the Laboratory system and E is the total relativistic energy.

3.24 At what energy do the classical and relativistic kinetic energies of proton differ by 10 %?
[Ans. 61.9 MeV]

3.25 A gamma ray interacts with a stationary proton and produces a neutral pion according to the scheme

$$\gamma + p \rightarrow p + \pi^0$$

Calculate the threshold energy given $M_p = 940$ MeV and $M_\pi = 135$ MeV.
[Ans. 150 MeV]

3.26 The cascade hyperon Ξ^- can be produced in the reaction $\pi^- + p \rightarrow \Xi^- + K^+ + K^0$. If the threshold energy for this reaction is 2232 MeV and the masses for π^-, p, K^+, K^0 are respectively 140, 938, 494, 498 MeV/c^2, calculate the mass of Ξ^-.
[Ans: 1439 meV/c^2]

3.27 If a particle of mass M decays into m_1 and m_2, m_1 has momentum p_1 and total energy E_1, where as m_2 has momentum p_2 and total energy E_2, p_1 and p_2 subtend an angle θ, then show that

$$E_1 E_2 - p_1 p_2 \cos\theta = \text{invariant} = \frac{1}{2}\left(m^2 - m_1^2 - m_2^2\right)$$

3.28 If a proton of 10^9 eV collides with a stationary electron and knock it off at 3^0 with respect to the incident direction, what is the energy acquired by the electron? [Ans. 3.0 MeV]

References

1. Anderson, Neddermeyer (1937)
2. Chadwick (1932)
3. W. Pauli (1930)
4. Pais (1952)
5. Perl et al. (1975)
6. C.D. Anderson, Phys. Rev. **43**, 491 (1933)
7. Tomonaga, Araki (1940)
8. Piccioni et al. (1947)
9. C.F. Powell, Occhialini, Lattes (1947)
10. Steinberger, Panofsky, Stella, (1950)
11. Barkas, (1956)
12. Cartwright et al. (1953)
13. Durbin et al. (1951)
14. Kemmer (1939)
15. D. Prowse, M. Baldo-Ceolin, Phys. Rev. Lett. **1**, 179 (1958)
16. J. Button et al., Phys. Rev. **121**, 1788 (1961)
17. J. Button et al., Phys. Rev. Lett. **4**, 530 (1960)
18. C. Baltay et al., Antibaryon production in antiproton-proton reactions at 3.7 BeV/*c*. Phys. Rev. **140**, B1027 (1965)
19. H.N. Brown et al., Phys. Rev. Lett. **8**, 255 (1962)
20. Firestone et al., Phys. Rev. Lett. **26**, 410 (1971)
21. Danysz, Pncewski (1953)
22. Pauli (1927)
23. Fritschi et al. (1986)

Chapter 4
Conservation Laws and Invariance Principles

4.1 Fundamental Interactions

There are four types of interactions for elementary particles which sharply differ from one another, the gravitational, the electromagnetic, the strong and the weak. The gravitational interaction which results from the mass of the bodies has a small coupling constant which specifies the strength of interaction. This type of interaction is insignificant in particle physics in ordinary conditions simply because the masses of elementary particles themselves are very small. Thus, the gravitational energy between two protons which are a distance $r = 2$ fm apart, will be

$$V(r) = \frac{Gm_p^2}{r} = \frac{6.67 \times 10^{-11} \times (1.67 \times 10^{-27})^2}{2 \times 10^{-15} \times 16 \times 10^{-13}} \simeq 6 \times 10^{-37}\ \text{MeV}$$

This is much less than the rest mass energy of proton (938 MeV).

The electromagnetic interaction, that is the interaction of charged particles with photons and between charged particles is mediated by the exchange of photons. The strength of this type of interaction is characterized by the fine structure constant $\alpha = e^2/4\pi\varepsilon_0\hbar c = 1/137$. The electrostatic energy of two protons, a distance 2 fm apart is calculated from

$$V(r) = \frac{e^2}{4\pi\varepsilon_0 r} = \frac{1.44(\text{MeV} - \text{fm})}{r(\text{fm})} = \frac{1.44}{2} \simeq 0.7\ \text{MeV}$$

For strong interaction we can calculate the energy from the results of square well potential for the deuteron in the ground state (AAK 1,5).

$$V_0 a^2 \simeq \frac{\hbar^2}{16 m_p} \simeq 10^{-28}\ \text{MeV}\,\text{m}^2$$

which for $a = 2$ fm yields $V_0 = 25$ MeV.

For weak interactions, as in β-decay, the coupling constant in the dimensionless form has a value of the order of 10^{-5}, which is to be compared with a value of $1/137$ for the electromagnetic interaction. In what follows we will not be concerned

A. Kamal, *Particle Physics*, Graduate Texts in Physics,
DOI 10.1007/978-3-642-38661-9_4, © Springer-Verlag Berlin Heidelberg 2014

Table 4.1 Characteristics of the fundamental interactions

	Strong	Electromagnetic	Weak	Gravitational
Carrier of field	Gluon	Photon	$W^{\pm}$, Z^0	Graviton
Spin-parity (J^P) of quantum	1^-	1^-	$1^-, 1^+$	2^+
Coupling Constant	$\frac{f^2}{\hbar c}$	$\frac{e^2}{\hbar c}$	$\frac{g^2}{\hbar c}$	$\frac{G^2}{\hbar c}$
Mass	0	0	80–90 GeV	0
Relative strength	1	10^{-2}	$\leq 10^{-5}$	10^{-38}
Time scale	10^{-23} sec	10^{-18}–10^{-20} sec	$\leq 10^{-13}$sec	
Range	$\leq 10^{-15}$ m	∞	10^{-18} m	∞
Source	Colour Charge	Electric Charge	Weak Charge	Mass

with gravitational interactions which are too feeble even at very small distances of particles.

The contrast in the three types of interactions strong, electromagnetic and weak, is reflected by the magnitude of interaction cross-sections that are involved. For strong interactions σ is typically a few millibarns, while for electromagnetic interactions, it is down by two orders of magnitude. For weak interactions

$$\sigma \sim 10^{-38}\ \text{cm}^2 \quad \text{or} \quad 10^{-11}\ \text{mb}$$

These three types of interactions are also characterized by different range of life times of unstable particles. For strong interaction, $\tau \sim 10^{-23}$–10^{-25} sec as in the decay of resonances. For electromagnetic interactions, $\tau \sim 10^{-16}$–10^{-20} sec, as for Σ^0-decay and π^0-decay. For weak interactions, $\tau < 10^{-13}$ sec, as in charged pion decay and β-decay.

Table 4.1 summarizes the characteristics of the interactions.

We give below typical examples of the three types of interactions.

1. Strong (nuclear) interactions

   i. Nuclear scattering $n + p \rightarrow n + p$
   ii. Charge exchange scattering $\pi^- + p \rightarrow \pi^0 + n$
   iii. Meson production $n + p \rightarrow p + p + \pi^-$
   iv. Strange particles production $p + p \rightarrow p + \Sigma^+ + K^0$
   v. Annihilation of N–$\overline{N}$. $p + p^- \rightarrow \pi + \pi + \pi + \pi + \pi$
   vi. Decay of resonance $N^*(1236) \rightarrow N + \pi$
   vii. K^- Capture $K^- + p \rightarrow \Lambda + \pi^0$
   viii. Pion Capture $\pi^- + {}^2\text{H} \rightarrow n + n$

   Note that the absence of photons and leptons is not a sure test of strong interaction. For example, the decay $\Sigma^+ \rightarrow p + \pi^0$, is actually a weak interaction. One must also examine the lifetimes and interaction cross-sections which fit with the expectations of Table 4.1, and the conservation or otherwise of certain quantum numbers as shown in Table 4.2.

Table 4.2 Table of elementary particles

			Particle	Mass (MeV/c^2)	τ (sec)	Common decay mode
Hadrons	Mesons	Pions	π^-, π^+	139	2.5×10^{-8}	$\mu\nu$
			π^0	135	1.8×10^{-16}	$\gamma\gamma$
		Kaons	K^-, K^+	494	1.2×10^{-8}	$\mu\nu$
			K^0	498		$\pi^{\pm}\pi^0$
			Mixture of K_1, K_2		$\sim 10^{-10}$	$\pi^+\pi^-$
			K_1			$\pi^0\pi^0$
						$\pi^0\pi^0\pi^0$
					5.18×10^{-8}	$\pi^+\pi^-\pi^0$
			K_2			$\pi\mu\nu$
						$\pi e\overline{\nu}$
			η	550	10^{-18}	$\gamma + \gamma$
						$\pi^+ + \pi^- + \pi^0$
	Baryons	Nucleons	p	938.2	$>10^{11}$	stable
			n	939.5	10^3	$pe\nu$
		Hyperons	Λ	1115	2.6×10^{-10}	$p\pi^-, n\pi^0$
			Σ^+	1189	0.8×10^{-10}	$p\pi^0, n\pi^+$
			Σ^0	1192	10^{-14}	$\Lambda\gamma$
			Σ^-	1197	1.6×10^{-10}	$n\pi^-$
			Ξ^0	1314	3×10^{-10}	$\Lambda\pi^0$
			Ξ^-	1321	1.8×10^{-10}	$\Lambda\pi^-$
			Ω^-	1675	1.3×10^{-10}	$\Xi\pi$
						ΛK^-
Photon			γ	0	∞	stable
Leptons			τ^-	1784	3.4×10^{-13}	electrons and neutrinos
			μ^-	105	2×10^{-6}	$e^-\nu\overline{\nu}$
			e^-	0.51	∞	stable
			ν_e	0	∞	stable
			ν_μ	<0.5		stable
			ν_τ	<164	∞	stable
Graviton			?	0	–	stable

2. Electromagnetic interactions (EM)

   i. Compton scattering $\gamma + e^- \rightarrow \gamma + e^-$
   ii. Photoelectric effect $\gamma + e^- \rightarrow e^-$
   iii. Photodisintegration $\gamma + {}^2\text{H} \rightarrow p + n$
   iv. Photoproduction $\gamma + p \rightarrow p + \pi^0 \rightarrow n + \pi^+$

v. e^+–e^- annihilation $e^+ + e^- \rightarrow \gamma + \gamma \rightarrow \gamma + \gamma + \gamma$
vi. Radiative decay $\pi^0 \rightarrow \gamma + \gamma$, $\Sigma^0 \rightarrow \Lambda + \gamma$
vii. Bhabha scattering $e^+ + e^- \rightarrow e^+ + e^-$

Note that the involvement of photons is a sure test of electromagnetic interaction, but their absence is not, as in (vii) which occurs between charged particles.

3. Weak interactions

i. β-decay ${}^3\text{H} \rightarrow {}^3\text{He} + \beta^- + \overline{\nu}_e$
ii. Muon interaction $\mu^- + p \rightarrow n + \nu_\mu$
iii. Pion decay $\pi^+ \rightarrow \mu^+ + \nu_\mu$
iv. Kaon decay $K^+ \rightarrow \pi^+ + \pi^0$
v. Neutrino-electron scattering $\nu_\mu + e^- \rightarrow \nu_\mu + e^-$
vi. Antineutrino interaction $\overline{\nu}_e + p \rightarrow n + e^+$
vii. Electron capture $e^- + p \rightarrow n + \nu_e$

4.2 Classification of Elementary Particles

Elementary particles can be divided into various groups in different ways. One way is to group them according to masses, light (leptons) intermediate (mesons) and heavy (baryons). They can also be grouped according to their lifetimes (long, short and very short). One other possibility is to group them according to the statistics (Fermi-Dirac or Bose-Einstein) which they enjoy. Particles like e, ν, μ, N, Σ, Λ and their antiparticles are Fermions and π, K, γ and their antiparticles are bosons. The number of Fermions is conserved while that of bosons is not. It is necessary to distinguish between bosons and fermions since bosons can be absorbed or emitted singly but fermions only in pairs.

The most fruitful approach is to classify the particles in regard to their behaviour, the type of interaction (strong, e.m. or weak) in which they participate. Nuclear interaction is very much stronger than the electromagnetic interaction at short distances, their relative strength are characterized by the difference in the coupling constants. Strong interactions are responsible between hadrons. The mesons act as the quanta of the strong interactions on the nuclear scale. These interactions reflect the interaction between quarks due to the exchange of gluons on the sub-nuclear level, (Chap. 5).

The energy of the strong interaction between two hadrons at a distance larger than 5 to 10 fm is negligible but at small distance ($\sim$0.1 fm) the energy of strong interaction becomes of the same order of magnitude as the rest mass energy of the strongly interacting particles.

The EM interactions are responsible for the force between electrically charged particles and photons and are mediated by the exchange of photons.

Weak interactions are responsible for many particle decays such as radioactive decay, the basic process being, $n \rightarrow p + e^- + \overline{\nu}_e$, pion and muon decay and a number of other decay processes. The weak interaction is of very short range ($\sim 10^{-4}$ fm) which is substantially smaller than that of the strong interaction. The corresponding energy of the weak interaction is about five orders of magnitude smaller than the strong interaction for the same distance of separation. The particles which participate in weak interactions are known as Leptons.

Gravitational interactions exist between all particles having mass and are believed to be mediated by the gravitons which are not yet discovered. Although the gravitational interaction is familiar and important on the macroscopic scale for astronomical bodies, the force is unimportant for elementary particles under ordinary conditions.

The uncertainty principle argument demonstrates that the shorter the range of the force the heavier will be the associated exchange quantum. For the infinite range, as for the EM and gravitational forces, the quantum is massless. The heavy quantum associated with the weak interaction ($W^{\pm}$) reflects a very short range force. At the subnuclear level however, the gluons which are massless particles are an exception, their range being finite due to 'confinement' (Chap. 5).

The properties of familiar particles introduced in Chap. 3 are displayed in Table 4.2. The particles are divided into four families, hadrons, photon, leptons, and graviton, corresponding to strong, electromagnetic weak and gravitational interaction. Hadrons are further divided into Mesons (Bosons) and Baryons (Fermions), mesons into pions and kaons, baryons into nucleons and hyperons.

In the second family, photon stands out alone.

In the third family electrons, muons and tauons and their associated neutrinos constitute the leptons.

In the fourth family graviton, is a massless particle of spin 2, the quantum of gravitation is not yet discovered.

4.3 Conservation Laws

Certain conservation laws are universally valid as they are obeyed in all the three types of interaction, strong, em and weak. Such is the case of baryon number (B), charge Q/e etc. Some other quantities like parity (space inversion) are valid restrictively in that they are conserved in strong and e.m. interaction but violated in weak interactions. The knowledge of conservation of certain quantities in a given type of interaction puts severe constraints on the choice of Hamiltonian involving such quantities.

In some of the conservation laws, the quantum number like charge or Baryon number are additive. In some others, like parity or charge conjugation (replacing particle by antiparticle) they are multiplicative. In certain interactions more than one conservation law may be violated. Such is the case of the possible decay of proton, $p \to e^+ + \pi^0$. Here baryon number is violated as the baryon p disappears and lepton number is also violated as the lepton e^+ is produced singly.

4.3.1 Charge (Q/e) Conservation

Total charge is conserved in all the three types of interactions. Thus the annihilation of p^-, $p^- + p \rightarrow \pi^+ + \pi^+ + \pi^- + \pi^0 + \pi^0$, is forbidden by charge conservation; so also the decay $\pi^+ \rightarrow \mu^- + \overline{\nu_\mu}$.

4.3.2 Baryon Number (B) Conservation

Baryons are assigned baryon number $B = +1$, antibaryons, $B = -1$, others $B = 0$. Baryon number is conserved in all the interactions. Thus, the reaction $\pi^- + {}^2\mathrm{H} \rightarrow n + \pi^0$ is forbidden by baryon non-conservation.

If baryon number is not required to be conserved then antiprotons could be produced in the reaction,

$$p + p \rightarrow p + p^- + \pi^+ + \pi^+$$

the threshold for which is only 600 MeV. However, if baryon number is to be conserved then the energetically cheapest way of producing antiprotons is by means of the process

$$p + p \rightarrow p + p + p + p^-$$

for which the threshold energy of the incident proton is $6m_pc^2 = 5.6$ GeV. The fact that the antiprotons are produced in the second reaction and not the first one, is a proof for the baryon conservation. In the interactions with antiprotons the following are allowed to proceed.

$$p^- + p \rightarrow \overline{n} + n \quad \text{(charge exchange scattering)}$$
$$p^- + p \rightarrow \pi^+ + \pi^- \quad \text{(annihilation)}$$

The forbidden reactions are

$$p^- + p \rightarrow n + n$$
$$p^- + p \rightarrow p + \pi^-$$

It may be pointed out that the Grand unified theories (GUTS) (Chap. 6) do allow the decays

$$p \rightarrow \pi^+ + \overline{\nu}_e$$
$$\rightarrow e^+ + \pi^0$$

which violate both lepton and baryon number conservation.

4.3.3 Lepton Conservation

The fact that ν_e, ν_μ, and ν_τ are different means that they are to be distinguished by allotting a separate quantum number (L) to the members of each species. A subscript

Table 4.3 Lepton numbers

Q/e	$L_e = 1$	$L_\mu = 1$	$L_\tau = 1$
0 −1	$\begin{pmatrix} \nu_e \\ e^- \end{pmatrix}$	$\begin{pmatrix} \nu_\mu \\ \mu^- \end{pmatrix}$	$\begin{pmatrix} \nu_\tau \\ \tau^- \end{pmatrix}$
Q/e	$L_e = -1$	$L_\mu = -1$	$L_\tau = -1$
0 +1	$\begin{pmatrix} \overline{\nu}_e \\ e^+ \end{pmatrix}$	$\begin{pmatrix} \overline{\nu}_\mu \\ \mu^+ \end{pmatrix}$	$\begin{pmatrix} \overline{\nu}_\tau \\ \tau^+ \end{pmatrix}$

e, μ or τ is fixed to indicate the given species. The lepton number L_e, L_μ and L_τ assigned to the leptons are shown in Table 4.3.

The antiparticles are assigned opposite lepton numbers to those for particles. In any allowed process involving leptons the sum of the lepton numbers for each species must separately remain constant.

Thus,

$$\begin{aligned} &\pi^+ \to \mu^+ + \nu_\mu \quad \text{is allowed} \\ &L_\mu = [0]\ [-1]\ [+1] \end{aligned}$$

But

$$\begin{aligned} &\pi^+ \to \mu^+ + \nu_e \quad \text{is forbidden} \\ &L_\mu = [0]\ [-1]\ [0] \\ &L_e = [0]\ [0]\ [+1] \end{aligned}$$

Also,

$$\begin{aligned} &\mu^+ \to e^+ + \nu_e + \overline{\nu}_\mu \quad \text{is allowed} \\ &[L_\mu] = [-1]\ [0]\ [0]\ [-1] \\ &[L_e] = [0]\ [-1]\ [+1]\ [0] \end{aligned}$$

But

$$\begin{aligned} &\mu^+ \to e^+ + \overline{\nu}_e + \nu_\mu \quad \text{is forbidden} \\ &L_\mu = [-1]\ [0]\ [0]\ [+1] \\ &L_e = [0]\ [-1]\ [-1]\ [0] \end{aligned}$$

4.3.4 Energy (*E*) Conservation

The total energy refers to kinetic energy plus rest mass energy. Energy is conserved in all types of processes. In fact the apparent missing energy in β-decay motivated Pauli to postulate the existence of neutrino. The law is universal. The decay of a free proton into neutron

$$p \to n + e^+ + \nu_e$$

is forbidden simply because the rest mass energy of proton is smaller than that of neutron plus positron. Similarly, in the annihilation of antiprotons at rest, hyperon-antihyperon pairs are not produced for the same reason.

4.3.5 Linear Momentum Conservation

The law of linear momentum conservation is valid for both elastic as well as inelastic processes for all the three types of interactions. The law leads to the coplanarity of three pions in the decay of τ-meson ($K_{\pi 3}$ mode) at rest and the coplanarity of Λ-hyperon and the product particles p and π^- in the decay in flight.

4.3.6 Angular Momentum Conservation

The total angular momentum (orbital + spin) is universally conserved. In general, spin alone is not separately conserved.

4.3.7 Isotopic Spin (Isospin or Isobaric Spin)

The concept of isospin was discussed in (AAK1, 4) as well as in Chap. 3. It was pointed out that although there is no connection between ordinary spin and isospin, the rules for combining isospin are identical with those for ordinary angular momentum. The isospin space is associated with the isospin, in analogy with the ordinary space for the angular momentum. Isospin is so named because its mathematical description is entirely analogous to ordinary spin or angular momentum in quantum mechanics.

If neutron and proton which have nearly the same mass are not to be distinguished they are jointly called Nucleon. In what follows we shall use T for the isospin of a single particle and I for a system of particles. Similarly, T_3 and I_3 will refer to the third component of isospin. Thus, n and p are said to be the sub-charge states of the same particle nucleon. In general multiplicity of charge states n is related to the isospin T by

$$n = 2T + 1 \tag{4.1}$$

The projection of T along the z-axis (also called the third component T_3) in the isospin space takes on values $T_3 = +\frac{1}{2}$ for proton and $-\frac{1}{2}$ for neutron. If v_p and v_n denote the isospin wave functions for the proton state and the neutron state, respectively, then,

$$T_3 v_p = +\frac{1}{2} v_p \qquad T_3 v_n = -\frac{1}{2} v_n \tag{4.2}$$

The charge Q/e is related to T_3 through the formula

$$\frac{Q}{e} = T_3 + \frac{B}{2} \tag{4.3}$$

where B is the baryon number. Conservation of isospin is a more general statement than that of charge independence. The strong interactions thus depend on I and not on I_3 where I and I_3 refer to the total isospin and its third component of the system, respectively. We have no way to distinguish between neutron and proton in so far as strong interactions are concerned. Since Q/e and B are absolutely conserved, so also is I_3 in both strong and electromagnetic interactions. The *em* interactions however do not conserve I. In weak interactions both I and I_3 are violated. The rule is $\Delta I_3 = \pm 1/2$.

The merit of introducing isospin is that it vastly simplifies the problems of strong interactions. Thus for a system of two nucleons, instead of dealing with four charge states (pp, pn, np, nn) we are concerned with only two isospin states, $I = 1$ and $I = 0$.

The isospin concept tells us why certain strong interactions are forbidden on account of violation of isospin I. Furthermore, it is possible to obtain important relations for the branching ratios for reactions in the two-body problems. The formalism also permits the estimation of branching fractions of the two body modes of decay of unstable particles. However, it is not possible to obtain the absolute cross-section from the isospin formalism.

Charge independence for pion was first proposed by Kemmer [1]. Pion is assigned $T = 1$ as it has three charge states π^+, π^0, π^-, (4.3) is reduced to

$$\frac{Q}{e} = T_3 \tag{4.4}$$

since $B = 0$. The three charge states π^+, π^0, π^-, take on the values $T_3 = +1, 0$ and -1, respectively. The π–N system can exist in two states with total isospin $I = \frac{3}{2}$ or $\frac{1}{2}$; with $I_3 = T_3(1) + T_3(2)$.

Later the principle of charge independence was extended to the strange particles. In what follows we shall denote the isospin function of the system of particles by φ^I_{I3}.

4.3.7.1 Addition of Isospins

The rules for the addition of two isospins (I) are identical with those of the angular momentum (J). The third component I_3 takes the place of the magnetic quantum number m.

The number of charge states is $(2I + 1)$ instead of $(2J + 1)$.

4.3.7.2 Addition of $T(1) = \frac{1}{2}$ and $T(2) = \frac{1}{2}$; Two Nucleon System

The numbers in brackets refer to particles 1 and 2. The matrix representation of an isospin $1/2$ in the charge space is identical to that of a spin $1/2$ in spin space, namely the Pauli matrices. In the two-nucleon system we can have either $I = 1$ or 0 states. It we neglect the coulomb interaction the strong interaction depends on whether the system is in $I = 1$ or 0 state. It does not depend on the value of I_3. In particular the n–p and p–p strong interaction characteristics are different since the p–p system can only exist in the $I = 1$ state ($I_3 = \frac{1}{2} + \frac{1}{2} = +1$). Where as the n–p system can exist in the $I = 1$ state as well as $I = 0$ state ($I_3 = -\frac{1}{2} + \frac{1}{2} = 0$).

We use the ladder operators

$$L_-\varphi^I_{I3} = \left[(I + I_3)(I - I_3 + 1)\right]^{1/2}\varphi^I_{I3-1} \tag{4.5}$$

$$L_+\varphi^I_{I3} = \left[(I - I_3)(I + I_3 + 1)\right]^{1/2}\varphi^I_{I3+1} \tag{4.6}$$

Also we use the product rule for two functions

$$L_-(uv) = uL_-v + vL_-u \tag{4.7}$$

Using the notation u^T_{T3} and v^T_{T3} for the individual particles and φ^I_{I3} for the system, we start with

$$\varphi^1_1 = u^{1/2}_{1/2}v^{1/2}_{1/2}$$

Applying (4.5) and putting $I = I_3 = 1$, $L_-\varphi^1_1 = \sqrt{2}\varphi^1_0$.

Applying (4.7)

$$L_-\varphi^1_1 = L_-\left(u^{1/2}_{1/2}v^{1/2}_{1/2}\right) = u^{1/2}_{1/2}L_-v^{1/2}_{1/2} + v^{1/2}_{1/2}L_-u^{1/2}_{1/2}$$
$$= u^{1/2}_{1/2}v^{1/2}_{1/2} + v^{1/2}_{1/2}u^{1/2}_{1/2}$$
$$\varphi^1_0 = \frac{1}{\sqrt{2}}\left(u^{1/2}_{1/2}v^{1/2}_{-1/2} + v^{1/2}_{1/2}u^{1/2}_{-1/2}\right)$$

Repeating the procedure

$$L_-\varphi^1_0 = \sqrt{2}\varphi^1_{-1}$$
$$\sqrt{2}\varphi^1_{-1} = \frac{1}{\sqrt{2}}\left(u^{1/2}_{1/2}L_-v^{1/2}_{-1/2} + v^{1/2}_{-1/2}L_-u^{1/2}_{1/2} + v^{1/2}_{1/2}L_-u^{1/2}_{-1/2} + u^{1/2}_{-1/2}L_-v^{1/2}_{1/2}\right)$$
$$= \frac{1}{\sqrt{2}}2v^{1/2}_{-1/2}u^{1/2}_{-1/2} = \sqrt{2}u^{1/2}_{-1/2}v^{1/2}_{-1/2}$$
$$(\because L_-v^{1/2}_{-1/2} = L_-u^{1/2}_{-1/2} = 0)$$
$$\varphi^1_{-1} = u^{1/2}_{-1/2}v^{1/2}_{-1/2}$$

The function φ^0_0 can be found as follows. Let φ^0_0 be a linear combination of $u^{1/2}_{1/2}v^{1/2}_{-1/2}$ and $u^{1/2}_{-1/2}v^{1/2}_{1/2}$, that is

$$\varphi^0_0 = au^{1/2}_{1/2}v^{1/2}_{-1/2} + bv^{1/2}_{1/2}u^{1/2}_{-1/2}$$

Normalization condition gives

$$a^2 + b^2 = 1$$

Making φ_0^0 orthogonal on φ_0^1 gives

$$a + b = 0$$

Solving we find

$$a = \frac{1}{\sqrt{2}}, \qquad b = -\frac{1}{\sqrt{2}}$$

$$\varphi_0^0 = \frac{1}{\sqrt{2}}\left(u_{1/2}^{1/2} v_{-1/2}^{1/2} - v_{1/2}^{1/2} u_{-1/2}^{1/2}\right)$$

Collecting the four isospin functions with the change of notation $\varphi_{I,I3}$ for the two nucleon system.

$$I = 1 \begin{cases} \varphi_{1,1} = u_p(1)u_p(2) \\ \varphi_{1,0} = \frac{1}{\sqrt{2}}\left[u_p(1)u_n(2) + u_n(1)u_p(2)\right] \\ \varphi_{1,-1} = u_n(1)u_n(2) \end{cases}$$

$$I = 0 \quad \varphi_{0,0} = \frac{1}{\sqrt{2}}\left[u_p(1)u_n(2) - u_n(1)u_p(2)\right]$$

Clearly $I = 1$ state is symmetric and $I = 0$ is antisymmetric in the exchange of particles. Observe that the isospin functions for the two nucleons are exactly the same as for spins of two electrons.

4.3.7.3 Application to the Feasibility of a Strong Reaction

The conservation of the isospin forbids the following reaction to proceed

$$^2\mathrm{H} + {}^2\mathrm{H} \rightarrow {}^4\mathrm{He} + \pi^0$$

I	0	0	0	1
I_3	0	0	0	0

Both ^{2}H and ^{4}He have $I = 0$ while π^0 has $I = 1$ and $I_3 = 0$. This reaction can not proceed as a strong interaction. It may however proceed as an electromagnetic interaction with a corresponding smaller cross-section since I_3 is conserved.

Example 4.1 Find the branching ratio of the following reactions at the same energy.

$$p + p \rightarrow d + \pi^+ \tag{i}$$

$$n + p \rightarrow d + \pi^0 \tag{ii}$$

Solution For d, $T = 0$ as it is a singlet. The final state is, therefore, $I = 1$ since π is assigned $T = 1$. Conservation of I requires that the initial state can be only $I = 1$. However, the np state is an equal mixture of states $I = 0$ and $I = 1$. Denoting the amplitude of $I = 1$ state by a_1, ratio of the cross sections is given by

$$\frac{\sigma_1}{\sigma_2} = \frac{a_1^2}{(a_1/\sqrt{2})^2} = 2$$

4.3.7.4 Addition of $T(1) = \frac{1}{2}$ and $T(2) = 1$; Pion-Nucleon Scattering

Denote the isospin functions for $T(1) = \frac{1}{2}$ particle (N) by $u^{1/2}_{1/2}$ and $u^{1/2}_{-1/2}$ and for $T(2) = 1$ particle (π) by v^1_1, v^1_0 and v^1_{-1}, where the superscript refers to T value and subscript to T_3 value of the particles.

For $\pi^+ p$ system obviously

$$\varphi^{3/2}_{3/2} = u^{1/2}_{1/2} v^1_1 \tag{4.8}$$

In order to find the state $\varphi^{3/2}_{1/2}$, we apply the ladder operator as in (4.5) and obtain

$$L_- \varphi^{3/2}_{3/2} = \left[\left(\frac{3}{2} + \frac{3}{2}\right)\left(\frac{3}{2} - \frac{3}{2} + 1\right)\right]^{1/2} \varphi^{3/2}_{1/2}$$

or

$$L_- \varphi^{3/2}_{3/2} = \sqrt{3}\varphi^{3/2}_{1/2} \tag{4.9}$$

Using (4.7)

$$L_- \varphi^{3/2}_{3/2} = L_- u^{1/2}_{1/2} v^1_1 = u^{1/2}_{1/2} L_- v^1_1 + v^1_1 L_- u^{1/2}_{1/2} \tag{4.10}$$

Applying (4.5)

$$L_- v^1_1 = \sqrt{2} v^1_0 \tag{4.11}$$

$$L_- u^{1/2}_{1/2} = u^{1/2}_{-1/2} \tag{4.12}$$

Combining (4.9), (4.10), (4.11) and (4.12)

$$\varphi^{3/2}_{1/2} = \sqrt{2/3} u^{1/2}_{1/2} v^1_0 + \sqrt{1/3} u^{1/2}_{-1/2} v^1_1 \tag{4.13}$$

Proceeding along similar manner, application of the ladder operator to (4.13) yields

$$\varphi^{3/2}_{-1/2} = \sqrt{2/3} u^{1/2}_{-1/2} v^1_0 + \sqrt{1/3} u^{1/2}_{1/2} v^1_{-1} \tag{4.14}$$

and repeating this procedure to (4.14) gives

$$\varphi^{3/2}_{-1/2} = u^{1/2}_{-1/2} v^1_{-1} \tag{4.15}$$

Note that further application of the ladder operator L_- to (4.15) will make the state zero, as it should, because no more states are accessible. We arrange the four states belonging to $I = 3/2$ as follows

$$I = \frac{3}{2} \quad \begin{cases} \varphi^{3/2}_{3/2} = u^{1/2}_{1/2} v^1_1 \\ \varphi^{3/2}_{1/2} = \sqrt{\frac{2}{3}} u^{1/2}_{1/2} v^1_0 + \sqrt{\frac{1}{3}} u^{1/2}_{-1/2} v^1_1 \\ \varphi^{3/2}_{-1/2} = \sqrt{\frac{1}{3}} u^{1/2}_{-1/2} v^1_{-1} + \sqrt{\frac{2}{3}} u^{1/2}_{1/2} v^1_{-1} \\ \varphi^{3/2}_{-3/2} = u^{1/2}_{-1/2} v^1_{-1} \end{cases}$$

The remaining states belonging to $I = \frac{1}{2}$ may be obtained by requiring the vectors $\varphi^{1/2}_{\pm 1/2}$ to be normalized and be orthogonal to $\varphi^{3/2}_{\pm 1/2}$.

Let

$$\varphi^{1/2}_{1/2} = a u^{1/2}_{1/2} v^1_0 + b u^{1/2}_{-1/2} v^1_1 \tag{4.16}$$

Normalization gives

$$a^2 + b^2 = 1 \tag{4.17}$$

Orthogonality of $\varphi^{1/2}_{1/2}$ with $\varphi^{3/2}_{1/2}$ gives us

$$0 = \sqrt{2/3} a + \sqrt{1/3} b \tag{4.18}$$

Solving (4.17) and (4.18) gives a possible pair of values

$$a = \sqrt{1/3} b = -\sqrt{2/3} \tag{4.19}$$

The other solution with a change of sign of a and b, does not affect the final results. Inserting (4.19) in (4.16) we obtain $\varphi^{1/2}_{1/2}$. The vector $\varphi^{1/2}_{-1/2}$ is conveniently obtained by requiring it to be orthogonal to $\varphi^{3/2}_{-1/2}$

$$I = \frac{1}{2} \begin{cases} \varphi^{1/2}_{1/2} = \sqrt{1/3} u^{1/2}_{1/2} v^1_0 - \sqrt{2/3} u^{1/2}_{-1/2} v^1_1 & (4.20) \\ \varphi^{1/2}_{-1/2} = \sqrt{1/3} u^{1/2}_{-1/2} v^1_0 - \sqrt{2/3} u^{1/2}_{1/2} v^1_{-1} & (4.21) \end{cases}$$

The coefficients occurring in the addition of isospins are called Clebsch-Gordon coefficients or Wigner coefficients (Appendix B).

4.3.7.5 Application to Pion-Proton Interactions

Consider the interactions at the same bombarding energy

$$\pi^+ + p \to \pi^+ + p \quad \text{(elastic scattering)} \tag{a}$$

$$\pi^- + p \to \pi^- + p \quad \text{(elastic scattering)} \tag{b}$$

$$\pi^- + p \to \pi^0 + n \quad \text{(change exchange scattering)} \tag{c}$$

Assume that the bombarding energy is not very large, otherwise multiple pion production and strange particle production become important. Here we shall not be concerned with such processes. Let the amplitude $a_{3/2}$ be associated with $I = 3/2$ state and $a_{1/2}$ with the state $I = 1/2$.

Now reaction (a) can have amplitude $a_{3/2}$ only since $I_3 = 3/2$.

$$\sigma_a \propto |a_{3/2}|^2 \tag{4.22}$$

Reactions (b) and (c) are mixtures of $I = 3/2$ and $I = 1/2$ states since $I_3 = -1/2$ and therefore involves both the amplitudes $a_{1/2}$ and $a_{3/2}$. Solving (4.14) and (4.21)

$$\varphi^{1/2}_{1/2} v^1_{-1} = \sqrt{\frac{1}{3}} \varphi^{3/2}_{-1/2} - \sqrt{\frac{2}{3}} \varphi^{1/2}_{-1/2} \tag{4.23}$$

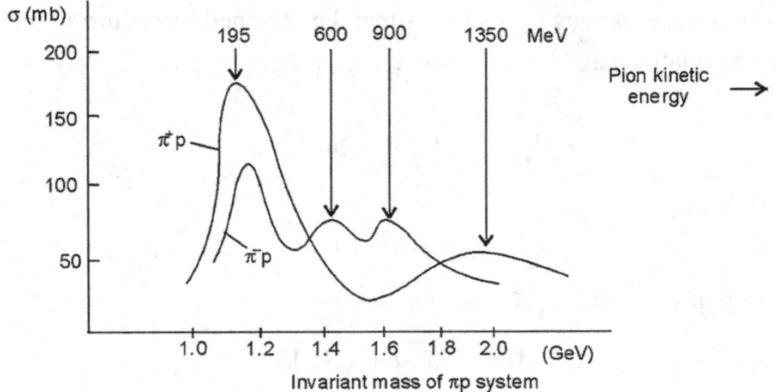

Fig. 4.1 Pion-nucleon resonances at higher energies

The $p\pi^-$ initial state, which is a mixture of $I = \frac{1}{2}$ and $I = 3/2$ states will be scattered into the final state to which contribute the original $I = 3/2$ component with amplitude $a_{3/2}$ and $I = \frac{1}{2}$ component with amplitude $a_1/2$, the total isospin being conserved. The final state is then

$$\varphi_f = \sqrt{\frac{1}{3}} a_{3/2} \varphi^{3/2}_{-1/2} - \sqrt{\frac{2}{3}} a_{1/2} \varphi^{1/2}_{-1/2} \tag{4.24}$$

Substituting (4.14) and (4.21) in (4.24) and writing u_p for $u^{1/2}_{1/2}$, u_n for $u^{1/2}_{-1/2}$, v_- for v^1_{-1} and v_0 for v^1_0, where $p, n, -, 0$ refer to proton, neutron, π^- and π^0, respectively,

$$\varphi_f = u_p v = \frac{1}{3}(a_{3/2} + 2a_{1/2}) + u_n v_0 \frac{\sqrt{2}}{3}(a_{3/2} - a_{1/2}) \tag{4.25}$$

The predicted relative cross-sections for the channels (a), (b) and (c) are

$$\sigma_a : \sigma_b : \sigma_c = |a_{3/2}|^2 : \frac{1}{9}|a_{3/2} + 2a_{1/2}|^2 : \frac{2}{9}|a_{3/2} - a_{1/2}|^2$$

At a given bombarding energy, one or the other amplitude may dominate. The special cases are

(i) $a_{3/2} \gg a_{1/2}$; $\sigma_a : \sigma_b : \sigma_c = 9 : 1 : 2$
(ii) $a_{3/2} \simeq a_{1/2}$; $\sigma_a : \sigma_b : \sigma_c = 1 : 1 : 0$
(iii) $a_{3/2} \ll a_{1/2}$; $\sigma_a : \sigma_b : \sigma_c = 0 : 2 : 1$

The pioneering experiments made by Fermi et al. [2] revealed that both for π^+ and π^- there is a strong peak in the total cross-section at pion kinetic energy of about 200 MeV. The ratio $\sigma_{\pi^+ p}/\sigma_{\pi^- p} = 3$. In our nomenclature this ratio is identical with $\sigma_a/(\sigma_b + \sigma_c)$, proving thereby the dominance of $I = 3/2$ amplitude, Fig. 4.1. The bump in the cross-section is referred to as a resonance, $N^*(1236)$ which was the forerunner of numerous resonances of nucleon, pion and kaon which were discovered a decade later. Figure 4.1 indicates more pion-nucleon resonances at higher energies. These resonances are interpreted as the excited states of nucleon which have very short lifetimes ($\sim 10^{-23}$–10^{-25} sec) and decay into nucleon and pion. In general, in any one region of pion-nucleon invariant mass several amplitudes will contribute, and the presence of a bump in cross-section need not necessarily imply a unique resonant state. Only for the first resonance $N^*(1236)$, also known as (3, 3) resonance, such an interpretation is unambiguous. This resonance is known to have four charge states, ++, +, 0, −, so that it is assigned $T = 3/2$. From the angular distribution of the decay products in the CMS, its spin J is found to be 3/2 (Chap. 5).

Example 4.2 Find the branching ratio of the reactions

$$p + d \to {}^3\mathrm{H} + \pi^+ \quad \text{(i)}$$
$$\to {}^3\mathrm{He} + \pi^0 \quad \text{(ii)}$$

Solution ^{3}He and ^{3}H form the isospin doublet with $T_3 = +\frac{1}{2}$ for ^{3}He and $T_3 = -\frac{1}{2}$ for ^{3}H. Deuteron is a singlet ($T = 0$) the initial state $\varphi^{1/2}_{1/2}$ is a pure $I = \frac{1}{2}$ state. But the final state is a mixture of $I = \frac{1}{2}$ and $I = 3/2$ (since $I_3 = \frac{1}{2}$). Conservation of I requires that the reaction can proceed only via $\varphi^{1/2}_{1/2}$. For the addition of $T(1) = \frac{1}{2}$ and $T(2) = 1$, we have from (4.20)

$$\varphi^{1/2}_{1/2} = \sqrt{\frac{1}{3}} u^{1/2}_{1/2} v^1_0 - \sqrt{\frac{2}{3}} u^{1/2}_{-1/2} v^1_1$$

Therefore

$$\frac{\sigma_1}{\sigma_2} = \frac{2/3a^2_{1/2}}{1/3a^2_{1/2}} = 2$$

Example 4.3 The $N^*(1236)$ resonance also known as Δ^+ is known to decay as

$$\Delta^+(1236) \to \pi^0 + p \quad \text{(i)}$$
$$\to \pi^+ + n \quad \text{(ii)}$$

Find the branching ratio of the two decay modes.

Solution The decay of Δ resonance proceeds through strong interaction as the lifetime is of the order of 10^{-25} sec. The initial state is $\varphi^{3/2}_{1/2}$, while the final state is a mixture of $I = \frac{3}{2}$ and $\frac{1}{2}$. Using (4.13)

$$\varphi^{3/2}_{1/2} = \sqrt{2/3} u^{1/2}_{1/2} v^1_0 + \sqrt{1/3} u^{1/2}_{-1/2} v^1_1$$

$$\frac{\sigma_1}{\sigma_2} = \frac{2/3a^2_{3/2}}{1/3a^2_{3/2}} = 2$$

4.4 Strangeness

The associated production of strange particles was confirmed at the Cosmotron of Brookhaven National Laboratory in 1953. The cosmic rays results on strange particles were confirmed. In order to explain the associated production Pais introduced a multiplicative quantum number to each particle analogous to parity. Pion and nucleon were assigned $+1$ and new particles, K, Λ etc. -1. The product of these number, in the initial and final state of strong interaction was expected to be identical. Thus the reaction $\pi^- + p \to K^0 + \Lambda$ was allowed but $\pi^- + p \to K^0 + n$ forbidden. Pais' parity was expected to be conserved in strong interactions. Weak decays like $\Lambda \to \pi^- + p$ were allowed as they need not obey this rule. Because the weak interactions are quite feeble, lifetimes of unstable particles would be much longer than those expected for strong interaction.

The associated-production proposal of Pais was only a partial explanation. The full solution is due to Gell-Mann who introduced new quantum numbers called Strangeness(s). The law was addition and not multiplication. The old particles n, p, π were assigned $S = 0$, for K^+, $S = +1$, for Λ and Σ'_S, $S = -1$. Thus, the interaction, $\pi^- + p \to K^+ + \Sigma^-$, is allowed, but not $\pi^- + p \to K^0 + n$. However, Gell-Mann's predictions differ from Pais, for example Gell-Mann's rule forbids $n + n \to \Lambda + \Lambda$ while Pais rule allows it. Similarly $\pi^- + p \to K^- + \Sigma^+$, is allowed by Pais' rule but forbidden by Gell-Mann's rule.

Strong interactions conserve I and S, EM interactions conserve S, but allow a unit change of isospin. Weak interactions violate I and allow a unit change of S, that is $\Delta S = \pm 1$.

Gell-Mann's classification places the K-meson into two isospin doublets (K^+, K^0) and $(\overline{K^0}, K^-)$. When this scheme was proposed serious objections were raised for the prediction that K^0 is not its own antiparticle. After all Kemmer [1] had shown that there is nothing wrong with this for π^0. However, Gell-Mann's scheme was ultimately accepted.

Gell-Mann's proposal, given independently by Nakano and Nishijima gives a modified relation between Q/e, T_3 and B through S

$$\frac{Q}{e} = T_3 + \frac{(B + S)}{2} \tag{4.26}$$

The quantity

$$Y = B + S \tag{4.27}$$

is known as hypercharge.

Since the mass of Σ^+ and Σ^- are not close enough to the mass of Λ, a new hyperon Σ^0 was predicted which decays through the scheme, $\Sigma^0 \to \Lambda + \gamma$.

Since the cascade hyperon decays weakly ($\Xi^- \to \Lambda + \pi^-$), Ξ^- was assigned $s = -2$ and Gell-Mann's scheme gives $T_3 = -\frac{1}{2}$. Thus Ξ^0 with $T_3 = +\frac{1}{2}$ was required as its doublet with the same strangeness $S = -2$. All these predictions have been verified.

4.5 Isotopic Spin and Strangeness Conservation

Guided by the overwhelming success of the charge independence theory of nuclear forces applied to nucleons and pions, Gell-Mann and Nishijima advanced independently the concept of charge independence to strange particles. Here, as in the case of pions and nucleons, the strong charge independent reactions conserve both I and I_3. The electromagnetic interactions again conserve I_3 but will, in general allow a change in I_3 equal to $\pm\frac{1}{2}$, provided the principle of charge independence is valid for the decayed products.

The Λ particle is assigned $T = 0$ since its charge counter-part does not exist. Thus, the decay $\Lambda \rightarrow p + \pi^-$, where both I_3 and I are violated must necessarily be weak. Here I is violated although Q/e and B conservation are satisfied. This is in contrast with a system of ordinary particles where the conservation of Q/e and that of B guarantees the conservation of I_3 through the relation.

$$\frac{Q}{e} = T_3 + \frac{B}{2} \tag{4.3}$$

We conclude that for strong interactions of strange particles the conservation of B, of I and of Q/e must be independently satisfied. Accordingly, the reaction of the type

$$\pi^- + p \rightarrow \Lambda + K^0$$

may be allowed to proceed through a strong charge independent channel. Since for the initial state $I_3 = -\frac{1}{2}$, and in the final state Λ has $T_3 = 0$, K^0 must be assigned $T_3 = -\frac{1}{2}$ so as to ensure the conservation of I_3. The initial state is a mixture of $I = \frac{1}{2}$ and $I = \frac{3}{2}$ states, but the final state has $I = \frac{1}{2}$ channel. The $I_3 = -\frac{1}{2}$ assignment for K^0 suggests that its counter-part K^+ has $I_3 = +\frac{1}{2}$ so that (K^+, K^0) form an isospin doublet, I_3 being $+\frac{1}{2}$ and $-\frac{1}{2}$, respectively.

The K^- meson may be interpreted as the antiparticle of K^+ meson. Since antiparticles are assigned opposite values, K^- has $T_3 = -\frac{1}{2}$ and its counter part $\overline{K^0}$ is assigned $T_3 = +\frac{1}{2}$ so that $(\overline{K^0}, K^-)$ form an isospin doublet, $\overline{K^0}$ being different from K^0.

We shall now consider the Σ hyperons. In order to ensure stability against a rapid decay to a system of pion and nucleon it is essential to assign an integral value to T. In the absence of doubly charge counterpart, $T = 1$ is an obvious choice. Strong interactions of the type $\Sigma \leftrightarrows \Lambda + \pi$ are then guaranteed which are I conserving. At the same time, rapid decay of Σ hyperon into a system of Λ and π is inhibited from energy conservation. The Σ triplet thus consists of Σ^+, Σ^0, Σ^- with $T_3 = +1, 0, -1$, respectively.

The radiative decay of Σ^0 hyperon, $\Sigma^0 \rightarrow \Lambda + \gamma$ deserves special attention. Since I_3 is conserved, the electromagnetic interaction is expected to be effected in a time of the order of 10^{-20} sec and the lack of evidence for its alternative mode of decay into a system of $p + \pi^-$ is explained since the latter is slower than the former by ten orders of magnitude.

Consider now a hypothetical reaction such as

$$K^+ + n \rightarrow \Sigma^+ + \pi^0$$

Q/e, B and I are conserved yet this reaction is not realized. The reason is that I_3 is violated. Hence, it can neither be a strong charge independent reaction nor an electromagnetic one. The criterion for an allowed charge independent reaction is the conservation of I_3 rather than I. Further, it was pointed out that the conservation of Q/e and B does not necessarily imply the conservation of I_3. This peculiarity is a natural consequence of the variation in the linear relation between Q/e and I_3 as we pass from multiplet to multiplet. We have thus

$$Q/e = I_3 \quad \text{for } \Sigma^+, \Sigma^0, \Sigma^-, \Lambda$$

$$Q/e = I_3 + \frac{1}{2} \quad \text{for } K^+, K^0$$

$$Q/e = I_3 - \frac{1}{2} \quad \text{for } K^-, \overline{K}$$

We are therefore, under the necessity of conserving the charge Q/e and the third component I_3 of the reaction, simultaneously and independently, if the reaction is to proceed in a strong charge independent manner. This is, indeed, the origin of the concept of "strangeness". We, therefore, introduce this new quantum number S, the "Strangeness" and assign appropriate value so that the relation (4.3) is generalized to (4.26).

Accordingly, we have $S = +1$ for (K^+, K^0), $S = -1$ for $K^-, \overline{K}, \Sigma^+, \Sigma^0, \Sigma^-, \Lambda$ and $S = 0$ for ordinary particles $p, n, \pi^+, \pi^0, \pi^-$.

It is now clear that the introduction of strangeness completely does away with the charge multiplicity; it is now enough to conserve S apart from conserving Q/e and B. A more realistic criterion for charge independent reactions is then $\Delta S = 0$, where S is the total strangeness of the reaction. This is of course not an entirely new conservation law. It stems from the better known conservation laws. But that it allows a unique classification of the fundamental particles is decidedly a great advantage. The strong reactions are allowed in so far as the strangeness is conserved, nevertheless they may be forbidden by other conservation laws, such as those of angular momentum and parity conservation.

For the weak interaction $\Delta S = \pm 1$ which is an immediate consequence of the fact that $\Delta I_3 = \pm\frac{1}{2}$ for such processes. For this reason the cascade hyperon Ξ^- which decays into a system of $\Lambda\pi^-$, is characterized by $S = -2$ and the existence of its neutral counterpart Ξ^0 was postulated. The observation of Ξ^0 is exceedingly difficult since its decay particles Λ and n are both neutral. Why the much more favourable decay $\Xi^- \rightarrow n + \pi^-$ does not exist is also clear, since the rule $\Delta S = \pm 1$ for weak decays would be violated.

A more massive hyperon Ω^- was predicted by Gell-Mann from a higher symmetry (SU_3). It is assigned $S = -3$. Its production and decay schemes are discussed in Chap. 5.

It must be pointed out that the selection rules provide only a qualitative frame work for the possible reactions and do not lead to any predictions of cross-sections.

Table 4.4 I and S assignments

I/S	-3	-2	-1	0	1	2	3
0	Ω^-		Λ		$\overline{\Lambda}$		$\overline{\Omega}^+$
1/2		Ξ^0 Ξ^-	K^- $\overline{K}^0$	p n $\overline{p}$ $\overline{n}$	K^+ K^0	$\overline{\Xi}^0$ $\overline{\Xi}^+$	
1			Σ^+ Σ^0 Σ^-	π^+ π^0 π^-	$\overline{\Sigma}^-$ $\overline{\Sigma}^0$ $\overline{\Sigma}^+$		

Nor do they embrace even a rudimentary theory resting on spin, parity or specific interaction. It is due, however, to the theory of charge independence that quantitative relations between the differential cross-section or total cross-sections at a give energy, of various reactions for a given charge multiplet can be explicitly formulated.

Table 4.4 summarizes the strangeness S for various hadron multiplets.

Gell-Mann's formula can also be written as

$$Q/e = I_3 + \frac{Y}{2} \tag{4.28}$$

where

$$Y = B + S \tag{4.27}$$

is called hypercharge.

In accordance with the $\Delta S = 0$, rule for strong interactions, following production process is allowed to proceed (above threshold energy of 906 MeV)

$$\begin{aligned} &\pi^- + p \to \Sigma^- + K^+; \quad \text{allowed} \\ &S = 0\,0\ \ -1\ \ +1 \end{aligned}$$

But the following reaction has never been observed

$$\begin{aligned} &\pi^- + p \to \Sigma^+ + K^-; \quad \text{forbidden} \\ &S = 0\,0\ -1\ -1 \end{aligned}$$

Example 4.4 Slow K^- mesons are incident equally on protons and neutrons. Show that the sum of charged Σ-hyperons produced is equal to twice the neutral Σ-hyperons.

Solution

$$\begin{aligned} K^- + p &\to \Sigma^+ + \pi^- \\ &\to \Sigma^0 + \pi^0 \\ &\to \Sigma^- + \pi^+ \\ K^- + n &\to \Sigma^- + \pi^0 \\ &\to \Sigma^0 + \pi^- \end{aligned}$$

For $K^- + p$ interaction the initial system of two $T = \frac{1}{2}$ particles has $I_3 = 0$ and consists equally of an $I = 0$ and an $I = 1$ state. Charge independence requires these amplitudes to depend only on I. The final $\Sigma + \pi$ state will be

$$\varphi_f = \frac{1}{\sqrt{2}}\left[a_0\varphi_0^0(\Sigma\pi) + a_1\varphi_0^1(\Sigma\pi)\right]$$

For

$$I(1) = I(2) = 1$$

$$\varphi_0^0 = \frac{1}{\sqrt{3}}(u_1v_{-1} - u_0v_0 + u_{-1}v_1)$$

$$\varphi_0^1 = \frac{1}{\sqrt{2}}(v_1u_{-1} - u_1v_{-1})$$

$$\therefore \quad \varphi_f = \frac{1}{\sqrt{2}}\left[\left(\frac{a_0}{\sqrt{3}} - \frac{a_1}{\sqrt{2}}\right)u_1v_{-1} - \frac{a_0}{\sqrt{3}}u_0v_0 + \left(\frac{a_0}{\sqrt{3}} + \frac{a_1}{\sqrt{2}}\right)u_{-1}v_1\right]$$

$$\therefore \quad (\Sigma^+\pi^-) : (\Sigma^0\pi^0) : (\Sigma^-\pi^+) = \frac{1}{2}\left(\frac{a_0}{\sqrt{3}} - \frac{a_1}{\sqrt{2}}\right)^2 : \frac{a_0^2}{6} : \frac{1}{2}\left(\frac{a_0}{\sqrt{3}} + \frac{a_1}{\sqrt{2}}\right)^2$$

The reaction $K^- + n$ goes through $I = 1$ only.

Since $I_3 = -1$, the final $\Sigma\pi$ state is

$$a_1\varphi_{-1}^1(\Sigma\pi) = \frac{a_1}{\sqrt{2}}[u_{-1}v_0 - u_0v_{-1}]$$

Thus

$$(\Sigma^-\pi^0) : (\Sigma^0\pi^-) = \frac{a_1^2}{2} : \frac{a_1^2}{2}$$

If K^- is incident with equal frequency on proton and neutron then

$$n_{\Sigma^-} + n_{\Sigma^+} = \frac{1}{2}\left(\frac{a_0}{\sqrt{3}} - \frac{a_1}{\sqrt{2}}\right)^2 + \frac{1}{2}\left(\frac{a_0}{\sqrt{3}} + \frac{a_1}{\sqrt{2}}\right)^2 + \frac{a_1^2}{2}$$

$$= \frac{a_0^2}{3} + a_1^2$$

$$n_{\Sigma^0} = \frac{a_0^2}{6} + \frac{a_1^2}{2}$$

It follows that

$$n_{\Sigma^-} + n_{\Sigma^+} = 2n_{\Sigma^0}$$

4.6 Invariance and Conservation Laws

The concept of symmetry or invariance is very important in physics in general and particle physics in particular. Newtonian equations are invariant (same form) with respect to Galilean transformations. Relativistic equations are invariant with respect to Lorentz transformations, that is they have the same form in all inertial systems under space-time transformations. Rotational invariance stems from the requirement that the field equations should be independent of the orientation of the spatial coordinate system.

The transformation considered can be either continuous as for translation or rotation in space, or discrete as for parity operation which is the spatial reflection through the origin. The associated conservation laws for the two cases are additive and multiplicative, respectively.

The importance of various invariance principles can be appreciated when we notice an intimate connection with the corresponding conservation laws. Thus, the law of conservation of momentum is a consequence of the invariance of the Hamiltonian under infinitesimal translations. The law of conservation of angular momentum is a consequence of the invariance of the Hamiltonian under infinitesimal rotations. The law of conservation of energy is a consequence of invariance of Hamiltonian under infinitesimal displacements in time. Gauge invariance of Maxwell's field equations are closely connected with the conservation of electric charge. These connections between conservation laws and invariance properties of the Hamiltonians are of considerable importance in selecting Hamiltonians for the description of physical systems which are subject to conservation laws. The invariance requirements act as severe restrictions on the possible choices of a Hamiltonian.

4.7 Invariance and Operators

The expectation value q of the operator is given by

$$q = \int \psi^* Q \psi d\tau \tag{4.29}$$

Heisenberg's equation of motion for the operator Q is

$$i\hbar \frac{dQ}{dt} = i\hbar \frac{\partial Q}{\partial t} + [Q, H] \tag{4.30}$$

where the commutator

$$[Q, H] = QH - HQ \tag{4.31}$$

and Q depends explicitly on time, so that $\partial Q/\partial t \neq 0$. When $\partial Q/\partial t = 0$, we have $dQ/dt = 0$ if $[Q, H] = 0$. Thus an operator, not depending explicitly on the time, will be a constant of the motion if it commutes with the Hamiltonian operator. In general, conserved quantum numbers are associated with operators commuting with the Hamiltonian.

4.7.1 Translation

The effect of an infinitesimal translation δr in space on a wave function ψ will be

$$\psi' = \psi(r+\delta r) = \psi(r) + \delta r \frac{\partial \psi(r)}{\partial r} = \left(1 + \delta r \frac{\partial}{\partial r}\right)\psi = D\psi \tag{4.32}$$

where

$$D = \left(1 + \delta r \frac{\partial}{\partial r}\right) \tag{4.33}$$

is an infinitesimal space translation operator.

Since $p = -i\hbar\partial/\partial r$, we can write

$$D = (1 + ip\delta r/\hbar) \tag{4.34}$$

A finite translation Δr can be obtained from n steps in succession ($\Delta r = n\delta r$), giving

$$D = \lim_{n\to\infty}\left(1 + \frac{i}{\hbar}p\delta r\right)^n = \exp\left(\frac{i}{\hbar}p\Delta r\right) \tag{4.35}$$

Thus D is a unitary operator, with $D^*D = D^{-1}D = 1$. The momentum operator p is called the generator of the operator D of space translations.

If the Hamiltonian H is independent of such space translations, then

$$[D, H] = 0 \tag{4.36}$$

From the form (4.34) it is seen that if D commutes with H, so also does the generator p

$$[p, H] = 0 \tag{4.37}$$

Thus, if the Hamiltonian is invariant under space translations, then the momentum operator p (which generates the translations) commutes with the Hamiltonian and the expectation value of p (the momentum of the system) is conserved. We may then state that momentum is conserved in an isolated system or the Hamiltonian is invariant under space translation or the momentum operator commutes with the Hamiltonian.

4.7.2 Rotations

In analogy with the operator D of space translation (4.33), the generator of infinitesimal rotations about some axis may be written

$$R = 1 + \delta\varphi \frac{\partial}{\partial \varphi} \tag{4.38}$$

The operator of the z-component of angular momentum is

$$J_z = -i\hbar \frac{\partial}{\partial \varphi} \tag{4.39}$$

where φ measures the azimuthal angle about the z-axis

$$R = 1 + \frac{i}{\hbar} J_z \delta\varphi \tag{4.40}$$

A finite rotation $\Delta\varphi$ is obtained by repeating the infinitesimal rotation n times: $\Delta\varphi = n\delta\varphi$. When $n \to \infty$ as $\delta\varphi \to 0$,

$$R = \lim_{n\to\infty} \left(1 + \frac{i}{\hbar} J_z \delta\varphi\right)^n = \exp\left(\frac{i}{\hbar} J_z \Delta\varphi\right) \tag{4.41}$$

Conservation of angular momentum about an axis corresponds to invariance of the Hamiltonian under rotations about that axis, and is also expressed by the commutation relation

$$[J_z, H] = 0 \tag{4.42}$$

Here the angular momentum J_z is the generator of the transformation.

Since the laws of physics must be independent of the orientation of the coordinate axis in space, the frames of reference linked by the rotation operators are equivalent and the rotations form a symmetry group. The conservation of angular momentum may be regarded as due to the isotropy of space. Alternatively, it may be viewed as a symmetry of the system; if the Hamiltonian of the system is invariant under rotation then angular momentum is conserved.

4.7.3 Isospin Symmetry

Symmetries have been recognized in particle physics, and have played an important role in our understanding. One such symmetry is the charge symmetry which says that n–n nuclear force is equal to p–p nuclear force and its generalization embodied in the principle of charge independance asserts that n–n = n–p = p–p nuclear force. The principle of charge independence which is more stringent than the charge symmetry, requires that for a given energy of relative motion, the nuclear force between a pair of nucleons depends only on the total angular momentum and parity of the two particles, regardless of the charge state. A new quantum number, isospin T was introduced to enunciate this principle. This principle was extended to pions and was adopted later to strange particles. The interactions between hadrons could be analyzed in terms of isospin amplitudes and the algebra which is relevant for isospin formalism is that of the group SU_2 and is similar to that used for ordinary angular momentum. For the pion-nucleon system, one need not distinguish between various charge states of pion and nucleon (six in all for pion triplet and nucleon doublet). One is concerned only with two amplitudes $a_{3/2}$ and $a_{1/2}$ for $I = 3/2$

and $I = 1/2$ states for the π–N system. Thus, the recognition of isospin multiplets simplifies the problem considerably.

The conditions of charge independence (conservation of I) and invariance under isospin rotations are equivalent. It is an empirical fact that the energies of two-nucleon states depend only on whether they are symmetric or antisymmetric ($I = 1$ or $I = 0$) under interchange of particles and not on whether the state contains two "equal" or two "unequal" nucleons. Thus, invariance under rotation in isospin space is the realization of the equivalence of neutron and proton. Proton and neutron may be considered as degenerate states of the particle nucleon. The small difference in masses (~1 part in 1000) is believed to be due to the difference in charges. If this is taken into account the degeneracy would be removed.

4.7.3.1 The Generalized Pauli Principle

The Pauli's principle is generalized when isospin is introduced. The proton and neutron are no longer different particles but the same particle (the nucleon) with different eigen values T_3 of a new intrinsic coordinate. The generalized principle states that the state consisting of a pair of nucleons must be antisymmetric under the exchange of two nucleons in space (l), spin (s) and isospin (I) variables. The behaviour of the overall wave function under the exchange of two particles comes from that of individual parts.

$$\begin{aligned}
&\text{Space} \sim (-1)^l \\
&\text{Spin} \sim (-1)^{s+1} \\
&\text{Isospin} \sim (-1)^{I+1}
\end{aligned}$$

The Pauli principle then yields

$$\begin{aligned}
&(-1)^l(-1)^{s+1}(-1)^{I+1} = -1 \quad \text{or} \\
&(-1)^{l+s+I} = -1 \qquad\qquad (4.43) \\
&\text{thus } (l + s + I) \text{ must be odd}
\end{aligned}$$

One consequence of this rule is that proton-neutron pair in an $I = 1$ state must have even $(l + s)$ and in an $I = 0$ state odd $(l + s)$. Thus deuteron which exists as an isospin singlet ($I = 0$) is in an orbital s-state (symmetric), as most of the ground states are, it must also be symmetric in spins. Hence the deuteron has a total spin of 1 in its ground state.

4.7.3.2 Angular Momentum $j = \frac{1}{2}$ Spinors

A nucleon is represented as a function of a dichotomic variable which can take one of the values $f = [\substack{a \\ b}]$. The proton is represented by a column vector $[\substack{1 \\ 0}]$ and neutron by $[\substack{0 \\ 1}]$. These are known as spinors. The standard linear operators which operate on such variables are the Pauli spin matrices and the unity operator. It is convenient to use the half values of the operators. These operators and their combinations are the only ones which act upon a dichotomic function.

The Pauli matrices used for spin $\frac{1}{2}$ particles are

$$\sigma_x = \begin{pmatrix} 0 & 1 \\ 1 & 0 \end{pmatrix} \qquad \sigma_y = \begin{pmatrix} 0 & -i \\ i & 0 \end{pmatrix} \qquad \sigma_z = \begin{pmatrix} 1 & 0 \\ 0 & -1 \end{pmatrix} \tag{4.44}$$

We form the operators

$$2\tau_1 = \sigma_x \qquad 2\tau_2 = \sigma_y \qquad 2\tau_3 = \sigma_z \qquad 2\tau_4 = \hat{1} \tag{4.45}$$

where

$$\hat{1} = \begin{pmatrix} 1 & 0 \\ 0 & 1 \end{pmatrix}$$

is the unit matrix. Then

$$2\tau_1 p = \begin{pmatrix} 0 & 1 \\ 1 & 0 \end{pmatrix}\begin{pmatrix} 1 \\ 0 \end{pmatrix} = \begin{bmatrix} 0 \\ 1 \end{bmatrix} = n \tag{4.46}$$

Similarly

$$2\tau_1 n = \begin{pmatrix} 0 & 1 \\ 1 & 0 \end{pmatrix}\begin{pmatrix} 0 \\ 1 \end{pmatrix} = \begin{bmatrix} 1 \\ 0 \end{bmatrix} = p \tag{4.47}$$

so that τ_1 transforms $n \leftrightarrow p$.

The effect of τ_2 and τ_3 can also be worked out.

$$2\tau_2 p = \begin{pmatrix} 0 & -i \\ i & 0 \end{pmatrix}\begin{pmatrix} 1 \\ 0 \end{pmatrix} = \begin{pmatrix} 0 \\ i \end{pmatrix} = i\begin{pmatrix} 0 \\ 1 \end{pmatrix} = in \tag{4.48}$$

$$2\tau_2 n = \begin{pmatrix} 0 & -i \\ i & 0 \end{pmatrix}\begin{pmatrix} 0 \\ 1 \end{pmatrix} = \begin{pmatrix} -i \\ 0 \end{pmatrix} = -i\begin{pmatrix} 1 \\ 0 \end{pmatrix} = -ip \tag{4.49}$$

The combination of τ_4 and τ_3 give

$$(\tau_4 + \tau_3) = \begin{pmatrix} \frac{1}{2} & 0 \\ 0 & \frac{1}{2} \end{pmatrix} + \begin{pmatrix} \frac{1}{2} & 0 \\ 0 & -\frac{1}{2} \end{pmatrix} = \begin{pmatrix} 1 & 0 \\ 0 & 0 \end{pmatrix} \tag{4.50}$$

$$(\tau_4 + \tau_3)p = p(\tau_4 + \tau_3)n = 0 \tag{4.51}$$

$$(\tau_4 - \tau_3)p = 0(\tau_4 - \tau_3)n = n \tag{4.52}$$

Thus these operators will project the proton as neutron parts, respectively out of a mixture of states.

Further

$$\tau_+ = (\tau_1 + i\tau_2) = \begin{pmatrix} 0 & 1 \\ 0 & 0 \end{pmatrix} \tag{4.53}$$

$$\tau_- = (\tau_1 - i\tau_2) = \begin{pmatrix} 0 & 0 \\ 1 & 0 \end{pmatrix} \tag{4.54}$$

They have the properties

$$(\tau_1 + i\tau_2)n = \tau_+ n = \begin{pmatrix} 0 & 1 \\ 0 & 0 \end{pmatrix} \begin{pmatrix} 0 \\ 1 \end{pmatrix} = \begin{pmatrix} 1 \\ 0 \end{pmatrix} = p \tag{4.55}$$

$$(\tau_1 - i\tau_2)p = \tau_- p = \begin{pmatrix} 0 & 0 \\ 1 & 0 \end{pmatrix} \begin{pmatrix} 1 \\ 0 \end{pmatrix} = \begin{pmatrix} 0 \\ 1 \end{pmatrix} = n \tag{4.56}$$

$$\tau_+ p = 0 \qquad \tau_- n = 0 \tag{4.57}$$

Thus, the operator τ_+ transforms a neutron into a proton state, and τ_- transforms a proton into a neutron. In other words, the operators $\tau\pm$ flip the sign of the third component of isospin of the nucleon.

The τ's obey the same commutation rules as the spin operators

$$\sigma_i \sigma_j = -\sigma_j \sigma_i (i \neq j) \tag{4.58}$$

$$= 1(i = j) \tag{4.59}$$

$$\sigma_x \sigma_y = i\sigma_z \quad \text{(and cyclic permutation)} \tag{4.60}$$

These properties are specific to $j = \frac{1}{2}$, and do not apply to $j > \frac{1}{2}$.

A properly constructed Hamiltonian involving forces which are charge independent is invariant under rotations in the I-spin space. Such a Hamiltonian then commutes with the corresponding rotation operators, and I-spin is a constant of motion. If the Coulomb force for, say a pair of protons, is included in the Hamiltonian then H no longer commutes with the total I-spin.

4.7.3.3 Isospin and SU(2) Group

In the framework of isotopic symmetry, proton and neutron are treated as upper and lower states of a spinor in the isospin space. The proton corresponds to the projection of isospin equal to $+\frac{1}{2}$ and the neutron to the projection of isospin equal to $-\frac{1}{2}$. (The projection is on to some axis in the isospin space, it is usually referred to as the z-axis.) Transformations of the isospin spinor under which the Lagrangian is invariant are realized by complex 2×2 matrices satisfying the unitary condition $U^{\uparrow}U = \hat{1}$, where $U^{\uparrow}$ is the Hermitian conjugate matrix, and $1^{\dagger}$ is the 2×2 identity matrix ($\det U = 1$). These 2×2 matrices give the simplest representation of the group SU(2). The letter S signifies that the transformations are special that is unimodular, and the letter U indicates that they are unitary, while the numeral 2 shows that the simplest representation of the group is formed by 2×2 matrices and the corresponding space is formed by the two component spinor.

SU(2) is a subgroup of U(2) which is isomorphic to the group of all unitary matrices of order 2. It is a 4-parameter, continuous, connected, compact Lie group.

The subgroup SU(2) contains all the unitary matrices of order 2 with determinant $+1$, the general elements being

$$\begin{pmatrix} a & -b^* \\ b & a^* \end{pmatrix}$$

with $aa^* + bb^* = 1$. Owing to the additional condition on the determinant, SU(2) is a three-parameter group.

The group SU(2) has one and only one inequivalent irreducible representation of every integral order.

4.7.3.4 Direct Product Representations of SU(2)

The direct product $D = D^{(j)} \otimes D^{(j')}$ of two irreducible representations of SU(2) can be reduced into linear combinations of the irreducible representations. Thus

$$D^{(j)} \otimes D^{(j')} = \sum_{J=|j-j|}^{j+j'} D^{(J)} \tag{4.61}$$

Each representation occurs at most once in the reduction of the direct product. Only those irreducible representations are contained in the reduction whose J-values satisfy the inequality

$$|j - j'| \leq J \leq |j + j'| \tag{4.62}$$

As an example consider two nucleons. Isospin of each nucleon transforms according to $D^{(\frac{1}{2})}$. The system of two nucleons would transform according to the direct product representation $D^{(\frac{1}{2})} \otimes D^{(\frac{1}{2})}$. This is a 4-dimensional representation of SU(2). Decomposing into irreducible representations gives

$$D^{(\frac{1}{2})} \otimes D^{(\frac{1}{2})} = \underset{\text{triplet}}{D^{(1)}} \otimes \underset{\text{singlet}}{D^{(0)}} \tag{4.63}$$

4.8 Parity

Parity is a rather subtle concept and has no classical analogue. It is concerned with the behaviour of wave function under space inversion

$$x \to -x; \qquad y \to -y; \qquad z \to -z$$

Consider the Schrodinger equation

$$\left[-\frac{\hbar^2\nabla^2}{2m} + V(r)\right]\psi(r) = E\psi(\underline{r})$$

Assume $V(r)$ is symmetrical. Now, Laplacian ∇^2 as well as $V(r)$ are invariant under parity operation

$$H\psi(r) = E\psi(r)$$

$$\therefore \quad H\psi(-r) = E\psi(-r)$$

Both $\psi(r)$ and $\psi(-r)$ satisfy the same equation

$$\therefore \quad \psi(-r) = K\psi(r)$$

where K is constant.

Replacing r by $-r$

$$\begin{aligned} \psi(r) &= K\psi(-r) \\ &= K^2\psi(r) \\ K^2 &= 1 \quad \text{or} \quad K = \pm 1 \\ P\psi &= \pm\psi \\ &\text{i.e. } P = \pm 1 \end{aligned} \tag{4.64}$$

P is also called the parity of the system. A wave function may or may not have a definite parity which can be even ($P = +1$) or odd ($P = -1$). For example, for $\psi(x) = \cos x$, $P(\psi) \to \cos(-x) = \cos x = +\psi$, $P = +1$ i.e. P is even.

For $\psi(x) = \sin x$, $P(\psi) \to \sin(-x) = -\sin x = -\psi$, $P = -1$; i.e. P is odd.

But for $\psi = e^x$, $P\psi = e^{-x} \neq \pm\psi$ so that it has no definite parity. However, it is easily verified that the parity of the function $e^x + e^{-x}$ is even.

Parity of a system will be a conserved quantum number if H and P commute i.e. $[H, P] = 0$, where H is the Hamiltonian of the system. Any spherically symmetric potential has the property that $H(-r) = H(r) = H(r)$ so that $[H, P] = 0$. The bound states of a system have a definite parity. Classification of these states according to their parity is quite useful. In the problem of energy levels of a particle trapped in a square well potential, the levels can be classified into even or odd, i.e. levels with a wave function which is either even or odd function of the coordinates. It is found that parity is conserved in strong and electromagnetic interactions but breaks down in weak interactions.

4.8.1 Parity of Particles

The absolute intrinsic parity of a particle can not be determined. Parity of a particle can be stated only relative to another particle. By convention baryons are allotted positive parity. Thus, the parity of proton or neutron is assumed to be positive. The parity of deuteron will then be also positive; that of antiproton is negative. In general all the antifermions have intrinsic parity opposite to the fermions. In fact the Dirac relativistic theory predicts parity for positron opposite to that of electron. On the other hand bosons have the same parity for particle and antiparticle.

4.8.2 Parity of a System of Particles

Parity is a multiplicative quantum number, so that the parity of a composite system is equal to the product of the parities of the parts.

We often deal with at least two particles. For such a system the parity of the total wave function can be considered as the product of the intrinsic parities of the

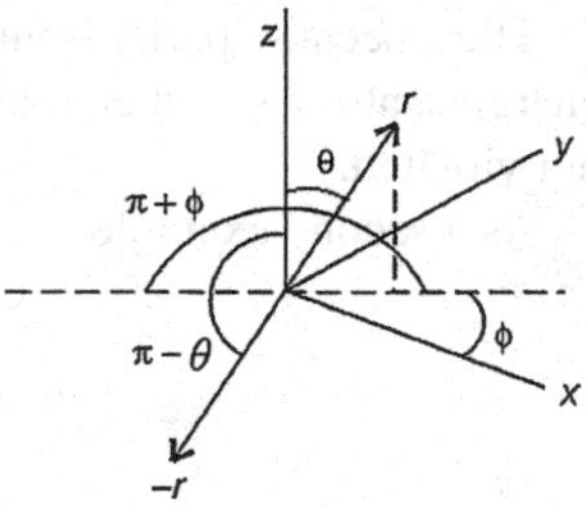

Fig. 4.2 Parity of a system of particles

individual parities and the parity of the orbital angular momentum part of the wave function. Thus for a system comprising particles A and B

$$P(AB) = P(A) \cdot P(B) \cdot P \quad \text{(orbital angular momentum wave function)} \tag{4.65}$$

Now, the orbital angular-momentum wave function has the form of spherical harmonics $Y_l^m(\theta, \varphi) \sim P_l^m(\cos\theta)e^{im\varphi}$, where l is the orbital angular momentum quantum number ($l = 0, 1, 2, \ldots$) and m is the magnetic quantum number $(-2, -1, 0, +1, +2)$, θ is the polar angle and φ is the azimuth angle. Space inversion, $r \to -r$ (Fig. 4.2) implies

$$\begin{aligned}
&r \to r, \qquad \theta \to \pi - \theta, \qquad \varphi \to \pi + \varphi \\
&e^{im\varphi} \to e^{im(\pi+\varphi)} = e^{im\pi}e^{im\varphi} = (-1)^m e^{im\varphi} \\
&P_l^m(\cos\theta) \to P_l^m\big[\cos(\pi - \theta)\big] = (-1)^{l+m} P_l^m(\cos\theta) \\
&Y_l^m(\theta, \varphi) \to Y_l^m(\pi - \theta, \pi + \varphi) = (-1)^{l+2m} Y_l^m(\theta, \varphi) \\
&\qquad\qquad\qquad\qquad\qquad\quad = (-1)^l Y_l^m(\theta, \varphi)
\end{aligned} \tag{4.66}$$

Thus, the spherical harmonies have parity $(-1)^l$ so that $s(l = 0), d(l = 2), g(l = 4), \ldots$ atomic states have even parity while $p(l = 1), f(l = 3), \ldots$ states have odd parity. Thus the parity is said to be conserved if the parity of the initial and final states are the same.

4.8.3 Conservation of Parity in EM and Strong Interactions

Consider a possible reaction with slow neutron absorption in ^{7}Be

$$n + {}^7\text{Be} \to {}^4\text{He} + {}^4\text{He} \tag{4.67}$$

The shell model assigns a parity $P_B = -1$ for ^{7}Be in the ground state. The intrinsic parity of neutron $P_n = +1$. Hence the parity for the initial state $p_i = p_n p_B(-1)^l$. For slow neutrons $l = 0$. Hence $P_i = (1)(-1)(1) = -1$. For the final state $P_f = P_\alpha P_\alpha (-1)^{l'}$, where $P_\alpha = +1$ for helium nucleus and l' is the relative angular momentum of ^{4}He nuclei. Now because ^{4}He are bosons, only $l' = 0, 2, 4, \ldots$ are allowed. Hence $p_f = +1$.

Thus, because parity is violated the above reaction can not proceed at low neutron energies, although other conservation laws (like baryon number, energy, charge) are not violated.

As a second example consider the γ-decay of polarized ^{19}F nuclei.

$$\begin{aligned} &^{19}\mathrm{F}^* \rightarrow {}^{19}\mathrm{F} + \gamma(110\,\mathrm{keV}) \\ &J^P = \frac{1}{2}^- \qquad J^P = \frac{1}{2}^+ \end{aligned} \tag{4.68}$$

This reaction can not proceed via strong or electromagnetic interactions as parity is violated. It can however go through via weak interaction. The observations are that in the decay of polarized ^{19}F nuclei, γ rays are emitted with a fore-aft asymmetry ascribed to weak interaction.

As a third example consider the α-decay of the excited state of ^{16}O at 8.87 MeV

$$\begin{aligned} &^{16}\mathrm{O}^* \rightarrow {}^{12}\mathrm{C} + \alpha \\ &J^P = 2^- \; 2^+ \; 0^+ \end{aligned} \tag{4.69}$$

where the initial state has odd parity, and the final state, even parity ($l = 0$). The extremely narrow partial width for this decay, $\Gamma_\alpha = 1.0 \times 10^{-10}$ eV is ascribed to parity violating contribution from weak interaction and may be contrasted with the width for γ-decay, $^{16}\mathrm{O}^* \rightarrow {}^{16}\mathrm{O} + \gamma$, of 3×10^{-3} eV.

Thus in experimental studies of both strong and electromagnetic interactions, extremely small parity violations are observed, not because of breakdown of parity conservation in these interactions as such, but because the Hamiltonian describing the interaction contains contributions as well from the weak interactions between the particles that are involved.

The first suspicion that there may be interactions (weak) in which parity is not conserved came from the study of two and three pion modes of kaon decay (Chap. 6).

The parity violation in strong interactions can be tested by investigating the effects due to pseudo-scalar terms in low energy nuclear processes. A pseudoscalar quantity is a scalar quantity which reverses its sign under parity operation. A direct test has been made by looking for non-conserving transitions in the decay of excited states of nuclei such as an excited state of ^{20}Ne with spin parity 1^+. This state may be formed by bombarding ^{19}F by protons. Such a state could decay to the ground state of ^{16}O by α-emission only if parity is conserved in the transition, since ^{16}O has spin parity 0^+ and the α-particle has spin 0. The orbital angular momentum for the final must be $l = 1$ in order to conserve the total angular momentum and the associated parity −1. The transition is not observed which sets a limit of about 10^{-6} on the relative size of the parity non-conserving amplitude.

Again, in the experiments on neutron-proton scattering from a hydrogen target with unpolarized neutrons, no up-down asymmetry is found, showing there by parity is conserved in strong interactions.

4.8.4 Conservation of Parity in Electromagnetic Interactions and Laporte Rule

For the dipole radiation emission, Laporte's selection rule states that allowed optical transitions occur only between states of opposite parity. Laporte's rule which was purely empirical rule is a special case of the law of parity conservation in electromagnetic interactions. Consider the integral

$$\int \psi_{l} * (\text{dipole})\psi_{l} d\tau$$

for the dipole radiation. Here l' and l are the orbital angular momentum quantum numbers for the initial and final atomic states. Since dipole has odd parity, $|l' - l|$ must be odd so that the overall parity of the integrand which is given by the product of these parities, is even and consequently we get non-zero intensity for the dipole radiation. Electric dipole transitions between bound states are therefore characterized by the selection rule $\Delta l = \pm 1$, so that as a result of the transition the parity of the atomic states must change (Laporte Rule).

It may be pointed out that although dipole-radiation is inhibited between even-even or odd-odd state, the quadrupole radiation is allowed. However, the intensity will be down by a factor of 10^8.

In the historical experiment made by Wu et al. to check non-conservation of parity in the β-decay of, ^{60}Co (Chap. 6) a by-product was the conservation of parity in electro-magnetic interactions. The ^{60}Co in the ground state with $J^P = 5^+$ β-decays through a Gamow-Teller transition to ^{60}Ni in the 4^+ state. The excited Ni state decays through two successive γ emission to 2^+ and then 0^+ states, with γ energy 1.173 and 1.332 MeV, respectively. The nuclei were aligned through adiabatic demagnetization. The angular distribution of γ rays was found out relative to the applied polarizing magnetic field. The γ rays appeared to be symmetrically emitted about the equatorial plane of the magnetic axis, showing thereby the conservation of parity in the electromagnetic interactions.

4.8.5 Determination of Parity of Negative Pions

Consider the following processes by which a π^- meson is absorbed at rest by a deuteron

$$\pi^- + d \rightarrow n + n \tag{4.70}$$

$$\pi^- + d \rightarrow n + n + \gamma \tag{4.71}$$

$$\pi^- + d \rightarrow n + n + \pi^0 \tag{4.72}$$

Reaction (4.70) is demonstrated to take place by direct observation of the neutrons which have a unique energy for this process. Reaction (4.71) proceeds with a cross-section lower by a factor of 0.42 compared to (4.70). Reaction (4.72) is not expected to go through as the phase space available is small. In the experiment nearly 70 MeV γ-rays from the decay of π^0 are not observed.

Let us focus on Reaction (4.70) and first examine the parity of the initial state. Since the parity of neutron and proton are $+1$, that of deuteron is also $+1$. Spin of deuteron is 1, and $l = 0$ mostly with 4% admixture of $l = 2$. The deuteron is in a state of total angular momentum $J = 1$. Thus, $J^P = 1^+$. Using the spectroscopic notation $^{2s+1}L_J$, deuteron's state is described by $^3S_1$ and $^3D_1$. Next we consider the parity of the angular momentum of the π^-–d system. It is shown that the time for a π^- to reach the K-orbit in the deuterium mesic atom is about 10^{-10} sec. Also, direct capture of π^- from $2p$ level is negligible compared to the transition from $2p$ to ls level. It is therefore concluded that all π^- sec will be captured from the s-state of the deuterium atom before they decay. Hence the parity arising from the angular momentum is $(-1)^l = (-1)^0 = +1$. If p_π is the intrinsic parity of π^- then the parity of the initial state will be

$$\pi_{in} = p_d \cdot p_\pi \cdot 1 = p_{\pi^-}$$

We will now examine the final state. The neutrons obey Fermi-Dirac statistics and must therefore be in antisymmetrical state. The two can be either in singlet spin state ($S = 0$, antisymmetric) or in triplet spin state ($S = 1$, symmetric). However, total wave function, product of space and spin parts, must be antisymmetric. Hence, $S = 0$ state must combine with orbital angular momentum states S and D, which are spatially symmetric and $S = 1$ spin state must combine with orbital angular momentum P state which is spatially antisymmetric. Initial state of $\pi^- d$ is characterized by $L = 0$ and $S = 1$ (since the deuteron spin $= 1$ and pion spin is zero). Hence the total angular momentum of initial state is 1. By conservation of angular momentum, final state must also have angular momentum 1. The possible final states are $^3S_1$ (symmetrical); $^3P_1$ (antisymmetrical); $^3D_1$ (symmetrical); $^1P_1$ (symmetrical).

Thus the correct state is only $^3P_1$ state. The symmetry of the final state can also be found out by the generalized Pauli's principle. Since $S_d = 1$ and $S_\pi = 0$, the initial total angular momentum is

$$J_i = S_d + S_\pi + l_i = 1 + 0 + 0 = 1$$

By angular momentum conservation $J_f = J_i = 1$. The neutrons can either be in the triplet state ($S = 1$) or singlet state ($S = 0$). According to Pauli's generalized principle the factor $(-1)^{L+S+I}$ must be negative. Here $I = 1$.

The total angular momentum $J = 1$ requires,

$$\begin{aligned}
l_f &= 0, \quad S = 1 \quad \text{(triplet)}\ ^3S_1 \\
l_f &= 1, \quad S = 1 \quad \text{(triplet)}\ ^3P_1 \\
l_f &= 2, \quad S = 1 \quad \text{(triplet)}\ ^3D_1 \\
l_f &= 1, \quad S = 0 \quad \text{(singlet)}\ ^1P_1
\end{aligned}$$

Of these only $l_f = S = 1$ has $L + S$ even. So the only possible state for two neutrons is $^3P_1$ state with parity $(-1)^L = -1$, which requires negative parity also for the initial state as parity is conserved in strong interactions. Neutrons and protons by

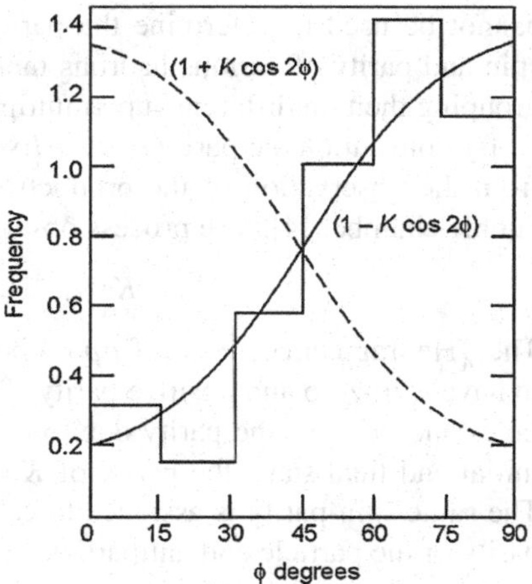

Fig. 4.3 Frequency distribution of angle ϕ between planes of polarization in the double Dalitz decay

convention have the intrinsic parity $+1$. The actual assignment is immaterial since baryon number is conserved, the baryon parities get cancelled in any reaction. It follows that the parity of deuteron is even. Hence, negative parity must be assigned to π^-.

Parity of π^0 has been established from the observations of the γ-ray polarization in the decay $\pi^0 \to 2\gamma$. Odd parity for pion would lead the plane polarisation vectors ($\vec{E}$ vectors) to be preferentially orthogonal.

In practice observations were made by Plano et al. [3] on the double Dalitz decay

$$\pi^0 \to \left(e^+ + e^-\right) + \left(e^+ + e^-\right)$$

Since the plane of each electron-positron pair lies predominantly in the plane of the E-vector, measurement of the angular distribution between the plane of the pairs permit the assignment of the odd parity to π^0 (Fig. 4.3). Such mesons are called pseudo scalar by spin-parity assignment $J^P = 0^-$. Particles with $J^P = 0^+$ are called scalar, $J^P = 1^-$ as vectors, and $J^P = 1^+$ as axial or pseudo vectors.

Figure 4.3 is the plot of frequency distribution of angle φ between planes of polarization in the double Dalitz decay

$$\pi^0 \to \left(e^+ + e^-\right) + \left(e^+ + e^-\right)$$

For a scalar π^0, the distribution should have the form $1 + K \cos 2\varphi$ and for a pseudo scalar π^0, $1 - K \cos 2\varphi$, where K is a constant.

4.8.6 Parity of Strange Particles

The parity of strange particles cannot be determined from the parity of decay products as conservation is known to be violated in weak decays. The decay modes

$$K^0 \to \pi^+ + \pi^- \quad \text{or} \quad \Lambda \to p + \pi^-$$

cannot be used to determine the parity of K^0 or Λ. It is important to know the spin and parity of various hadrons including resonances from the point of view of grouping them in different supermultiplets as in SU(3) (Chap. 5).

By convention the parity of Λ is fixed at $P = 1$. The parity of K^- is ascertained from the observation of the production of the hyperfragment ${}^4_\Lambda$He in the helium bubble chamber [4]. The process observed was

$$K^- + {}^4\text{He} \rightarrow \pi^- + {}^4_\Lambda\text{He}$$

The ${}^4_\Lambda$He fragment consists of $ppn\Lambda$ bound together. The hyperfragment is assumed to have spin-zero and positive parity. The reaction then involves only spin zero particles and because the parity due to orbital motion will have to be the same in the initial and final state, the parity of K^- is identical to that of π^-, i.e. $P_K - = -1$. The same spin-parity is assigned to K^+, K^0 and $\overline{K_0}$ (from the rule, for bosons the parity of the particle and antiparticle are the same).

The parity of the Σ-hyperon was determined from the study of the reaction

$$K^- + p \rightarrow \Sigma + \pi$$

at a centre of mass energy of 1520 MeV. At this energy a resonance is formed. The angular distribution of the produced particles indicated $J^p = 3/2^+$ for the resonances and the parity of Σ positive. The Ξ is assumed to have the same J^p value.

In Chap. 5 we will see that some of the meson resonances have even intrinsic parity and some of the baryon resonance have odd intrinsic parity.

4.8.7 The Isospin Versus Parity

The isospin and parity symmetries contrast is several respects. Parity is related to space time while isospin is not. For this reason, isospin is termed as an internal symmetry. Parity is a discrete symmetry while isospin is a continuous symmetry since it is possible to consider rotations in isospin space by an angle. Isospin is an approximate symmetry, since, for example, neutron and proton do not have exactly the same mass. Parity is an exact symmetry for strong and the electromagnetic interactions but breaks down in weak interactions. Isospin is conserved in strong interactions only. Isospin is the same for hadrons and antihadrons irrespective of the statistics they obey. Parity is the same for bosons and their antiparticles but for fermions and antifermions it is opposite.

4.9 Spin of Neutral Pion

In Chap. 3, we had elaborately discussed the method for the determination of spin of positive pion based on the detailed balance. Here, we shall focus on the neutral pion.

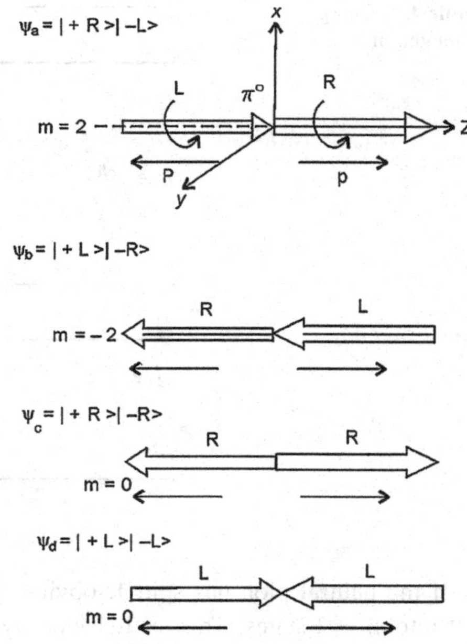

Fig. 4.4 Analysis of possible polarization of two γ rays produced by the decay of π^0

The fact that π^0 decays into 2γ rays suggests that π^0 is of integral spin and is a boson ($s_\gamma = 1$). Young [5] made a detailed analysis to exclude spin 1.

Consider a π^0 at the origin of a co-ordinate system emitting two γ rays one along $+z$ axis and the other along $-z$ axis, Fig. 4.4. Analyze the photons into circularly polarized components of helicity $+1$ or -1 (spin up and spin down). The photon wave function is denoted by $|+R\rangle, |+L\rangle, |-R\rangle, |-L\rangle$, where $+$ indicates a photon traveling in the positive z direction and a photon traveling in the negative z direction R a right-handed photon and L a left-handed photon.

For two photons the possible wave functions are:

$$\psi_a = |+R\rangle|-L\rangle \quad m = 2$$
$$\psi_b = |+L\rangle|-R\rangle \quad m = -2$$
$$\psi_c = |+R\rangle|-L\rangle \quad m = 0$$
$$\psi_d = |+L\rangle|-L\rangle \quad m = 0$$

Now rotate the system by 180° about the y-axis. This causes z-axis to change the direction but handedness remains the same

$$\psi'_a = |-R\rangle|+L\rangle = \psi_b$$
$$\psi'_b = |-L\rangle|+R\rangle = \psi_a$$
$$\psi'_c = |+R\rangle|+R\rangle = \psi_c$$
$$\psi'_d = |-L\rangle|+L\rangle = \psi_d$$

Thus, only ψ_c and ψ_d are eigen states of this rotation, with eigen values 1.

Table 4.5 Charge conjugation

	Proton	Antiproton
Q/e	+1	−1
B	+1	−1
$\mu = \frac{e\hbar}{2MC}$	+2.79	−2.79
	Parallel to spin	Antiparallel to spin
σ	$\frac{1}{2}\hbar$	$\frac{1}{2}\hbar$
	Electron	**Positron**
Q/e	−1	+1
L_e	$-\frac{e\hbar}{2mc}$	$+\frac{e\hbar}{2mc}$
σ	$\frac{1}{2}\hbar$	$\frac{1}{2}\hbar$

If the neutral pion has spin 1, obviously it could decay into $m = 0$ states, but not into $m = 2$ states. This excludes decay into ψ_a or ψ_b states. Suppose $S_\pi = 1$, then only $m = 0$ is possible. In this case the two-photon amplitude must behave under spatial rotation like the polynomial $P_l^m(\cos\theta)$ with $m = 0$, where θ is the angle relative to the z-axis. Under a 180° rotation about the y-axis, $\theta \to \pi - \theta$, since $P_1^0 \propto \cos\theta$, it therefore changes sign. For $m = 0$, this corresponds to two right (left) circularly polarized photons travelling in opposite directions. The above rotation is then circularly polarized photons travelling in opposite directions. The above rotation is then equivalent to interchange of the two photons for which the wave function must be symmetric. Here $S_\pi = 1$ is ruled out. It follows that $S_\pi = 0$ or ≥ 2. In high-energy N–N collisions, it is observed that neutral, positive and negative pions are produced in equal numbers, showing thereby the neutral pion has spin zero by analogy with π^+.

4.10 Charge Conjugation (C)

Charge conjugation is the operation of reversing the sign of the charge and magnetic moment of a particle (leaving all other coordinates unchanged). Effectively it changes a particle into antiparticle and vice versa. It also implies reversal of baryon number or lepton number. For electron and proton the effect of C is indicated below in Table 4.5.

Strong and Electromagnetic interactions are invariant under C-operation but weak interactions are not.

4.10.1 Invariance of C in Strong Interactions

The experimental test consists of comparing the characteristics of particles and their antiparticles in strong interactions. In the annihilation of antiprotons with protons the rates and spectra of π^+ and π^-, and K^+ and K^- are found to be identical.

$$
\begin{aligned}
p + p^- &\rightarrow \pi^+ + \pi^- + \cdots \\
&\rightarrow K^+ + K^- + \cdots
\end{aligned}
$$

4.10.2 Invariance of C in Electromagnetic Interactions

Similarly in the electromagnetic decays of η^0(549 MeV/c^2) the energy spectra of π^+ and π^- are found to be identical.

4.10.3 Eigen States of the Charge Conjugation Operator

Consider the operation of charge conjugation performed on a charged pion wave function $|\pi^+\rangle$.

$$
\begin{aligned}
C|\pi^+\rangle &\rightarrow |\pi^-\rangle \\
C|\pi^-\rangle &\rightarrow |\pi^+\rangle
\end{aligned}
$$

Hence $|\pi^+\rangle$ and $|\pi^-\rangle$ are not C-eigen states. However, for a neutral system, the C-operator may have a definite eigen value.

$$C|\pi^0\rangle = \eta|\pi^0\rangle$$

since π^0 transforms into itself and η is a constant.

Repeating the operation

$$C^2|\pi^0\rangle = C\eta|\pi^0\rangle = \eta C|\pi^0\rangle$$

or

$$
\begin{aligned}
|\pi^0\rangle &= \eta^2|\pi^0\rangle \\
\eta^2 = 1 \quad &\text{or} \quad \eta = \pm 1 \\
\therefore \quad C|\pi^0\rangle &= \pm|\pi^0\rangle
\end{aligned}
$$

In general a system whose charge is not zero cannot be an eigen function of C. However, if $Q = B = S = 0$, the effect of C is just to produce eigen value ± 1.

4.10.4 C-Parity of Photons

E.M. fields are produced by moving charges (currents) which change sign under charge conjugation. Consequently, the photon has $C = -1$. Since the charge-conjugation quantum number is multiplicative, this means that a system of n photons has C-eigen value $(-1)^n$.

Consider the decay

$$\pi^0 \to 2\gamma \quad \text{(even } C\text{-parity)}$$
$$C|\pi^0\rangle = +|\pi^0\rangle$$

It follows that the decay $\pi^0 \to 3\gamma$ is forbidden if the E.M. interactions are invariant under charge conjugation. Experiments have established that the ratio

$$\pi^0 \to \frac{\pi^0 \to 3\gamma}{\pi^0 \to 2\gamma} < 5 \times 10^{-6}$$

4.10.5 Positronium Annihilation

Positronium is a bound system of e^+ and e^-. Restrictions are imposed by C-invariance on the states of positronium which undergoes annihilation in the modes

$$e^+ + e^- \to 2\gamma$$
$$\to 3\gamma$$

The bound states of e^+e^- system possess energy levels similar to the hydrogen atom, but with about half the spacing because of the factor $\frac{1}{2}$ for the reduced mass instead of 1. The total wave function for this system may be written as

$$\psi(\text{total}) = \Phi(\text{space}) \propto (\text{spin}) \chi(\text{charge})$$

Consider the behaviour of these functions under the particle exchange.

The spin functions for a combination of two spin $\frac{1}{2}$ particles are written as Triplet

$$S = 1 \begin{cases} \propto(1,1) = \psi_1\left(\frac{1}{2}, \frac{1}{2}\right)\psi_2\left(\frac{1}{2}, \frac{1}{2}\right) \\ \propto(1,0) = \frac{1}{\sqrt{2}}\left[\psi_1\left(\frac{1}{2}, \frac{1}{2}\right)\psi_2\left(\frac{1}{2}, -\frac{1}{2}\right) + \psi_2\left(\frac{1}{2}, \frac{1}{2}\right)\psi_1\left(\frac{1}{2}, -\frac{1}{2}\right)\right] \\ \propto(1,-1) = \psi_1\left(\frac{1}{2}, -\frac{1}{2}\right)\psi_2\left(\frac{1}{2}, -\frac{1}{2}\right) \end{cases}$$

$S_z = 1, 0, -1$

Singlet

$$\propto (0,0) = \frac{1}{\sqrt{2}}\left[\psi_1\left(\frac{1}{2},\frac{1}{2}\right)\psi_2\left(\frac{1}{2},-\frac{1}{2}\right) - \psi_2\left(\frac{1}{2},\frac{1}{2}\right)\psi_1\left(\frac{1}{2},\frac{1}{2}\right)\right]$$

$$S = 0$$

$$S_z = 0$$

Neither e^+ nor e^- is an eigen state of C. However, the system e^+–e^- in a definite (l, s) state is an eigen state of C. According to the generalized Pauli principle under the total exchange of particles consisting of changing Q, $\vec{r}$ and S labels, ψ must change its sign.

Space exchange gives a factor $(-1)^l$ as this involves parity operation.
Spin exchange gives a factor $(-1)^{s+1}$.
Charge exchange gives a factor C.
The condition becomes

$$(-1)^l(-1)^{S+1}C = -1$$

or

$$C = (-1)^{l+S}$$

where S is total spin.

We shall now show how C invariance restricts the annihilation into photons. Let n be the number of photons in the final state.

Conservation of C-parity gives

$$(-1)^{l+S} = (-1)^n$$

(i) Singlet S state ${}^1S_0$; $l = S = 0$ (para-positronium)

$$e^+e^- \to 2\gamma \quad \text{allowed with lifetime } 1.25 \times 10^{-10} \text{ sec}$$
$$\to 3\gamma \quad \text{forbidden}$$

(ii) Triplet S state ${}^3S_1$; $l = 0$, $S = 1$ (Ortho-Positronium)

$$e^+e^- \to 3\gamma \quad \text{allowed with lifetime } 1.5 \times 10^{-7} \text{ sec}$$
$$\to 2\gamma \quad \text{forbidden}$$

4.11 G-Parity

Sometimes it is possible to obtain a useful conservation law by combining two known conservation laws. The new conservation law may reveal features not obvious in the original ones. Lee and Yang pointed out that charge conjugation can be combined with rotation in isospin space to yield a useful conservation law.

It was also pointed out that I-spin invariance is valid only for strong interactions. While charge-conjugation invariance holds good both for strong interactions as well as for electromagnetic interactions. This then means that only for strong interactions can we combine charge conjugation and I-spin invariance to obtain a useful selection rule.

It was also pointed out that the charge conjugation operator C can act on system with $Q = B = S = 0$ ($S =$ strangeness) for example π^0, γ, η, e^+e^-, $p^-{-}p$ systems. For non-zero values, Q, B and S change. In order that eigen values may exist it is important to formulate selection rules for charged systems. This can be accomplished for strong interactions by combining the operation of charge conjugation (C) with an isotopic rotation (R) about the y-axis or z-axis.

A rotation θ about an axis defined by a unit vector $\hat{n}$ in isospin space is given by

$$R(\theta) = \exp(-i\tau \cdot \hat{n}\theta)$$

where τ is the isospin operator. For example for an isospin doublet, τ_1, τ_2 and τ_3 are simply the Pauli matrices.

The operation G consists of rotation R of 180° about the y-axis in isospin space followed by charge conjugation

$$G = C\exp(-i\tau_2\pi) \tag{4.73}$$

Now, from the commutation relations, it is known that the isospin operations have the same algebraic properties as the ordinary angular momentum operators. Under a rotation π about the y axis an angular momentum state (j, m) transforms as

$$\exp(-i\tau_y\pi)|j, m\rangle = (-1)^{l-m}|j, -m\rangle$$

By analogy, for rotations in isospin space we have

$$\exp(-i\tau_y\pi)|I, I_3\rangle = (-1)^{I-I_3}|I, -I_3\rangle$$

Thus, for a rotation π about y-axis in isospin space we have

$$R_2(\pi)\left|\pi^+\right\rangle = (-1)^{1-1}|1, -1\rangle = \left|\pi^-\right\rangle$$

$$R_2(\pi)\left|\pi^-\right\rangle = (-1)^{1+1}|1, +1\rangle = \left|\pi^+\right\rangle$$

$$R_2(\pi)\left|\pi^0\right\rangle = (-1)^{1-0}|1, 0\rangle = -\left|\pi^0\right\rangle$$

Consider the isotopic triplet of pions, π^+, π^0, π^-. Application of charge conjugation has the effect of transforming $\pi^+ \to \pi^-, \pi^- \to \pi^+$ and $\pi^0 \to \pi^0$. The π^0 is an eigen state of C with eigen value $+1$

$$C\left|\pi^0\right\rangle = +1\left|\pi^0\right\rangle$$

For the charged pions we have

$$\begin{aligned} C\left|\pi^+\right\rangle &= \pm 1\left|\pi^-\right\rangle \quad \text{and} \\ C\left|\pi^-\right\rangle &= \pm 1\left|\pi^+\right\rangle \end{aligned} \tag{4.74}$$

The $\pm$ sign is introduced as there is an arbitrary phase. For neutral π^0 the G-parity is unambiguously equal to -1. Since the strong interactions conserve isospin and are invariant under the charge conjugation operation, it is plausible to expect that the

G-parity of the charged pions is identical with that of the neutral pion. We therefore choose the phases in the charge conjugation operation (4.74) in line with neutral pion so that

$$C|\pi^{\pm}\rangle = (-1)|\pi^{\mp}\rangle$$

Then under the G-transformation

$$G|\pi^{+,-0}\rangle = (-1)|\pi^{+,-0}\rangle \tag{4.75}$$

so that G-parity of the pion is -1. Like parity, G parity is also a multiplicative quantum number. Since the C-operation reverses the sign of the baryon number, it is obvious that G-parity has baryon number zero. As the charge-conjugation quantum number is multiplicative and isospin additive, G-parity is multiplicative.

Thus, the G-parity of a system of n pions will have G-parity $(-1)^n$. G-parity is conserved in strong interactions and is a good quantum number for non-strange mesons. Its conservation results in the selection rules for example in nucleon antinucleon annihilation into pions, and the decay of vector mesons ρ, ω, Φ and f mesons through strong interactions into pions.

Consider the nucleon-antinucleon system of total spin s and orbital angular momentum l, the effect of C-operation is to introduce a factor $(-1)^{l+s}$, similar to the case of positronium. Thus, the effect of the operation $G = CR$ on a neutral nucleon-antinucleon system $|\psi\rangle$ will be

$$G|\psi\rangle = (-1)^{l+s+I}|\psi\rangle \tag{4.76}$$

Suppose an antiproton annihilates with a neutron.

$$p^- + n \rightarrow \pi^0 + \pi^- \tag{4.77}$$

Using the formula

$$Q/e = T_3 + B/2$$

for p^-, $Q/e = -1$ and $B = -1$ so that $T_3 = -\frac{1}{2}$ (for $\overline{n}$, $T_3 = +\frac{1}{2}$). Also for n, $T_3 = -\frac{1}{2}$. Hence total $T_3 = -1$ for the initial state of reaction. The reaction can proceed through $I = 1$ isospin channel. For the two pion annihilation

$$G = (-1)^{l+s+I} = (-1)^2 = +1 \tag{4.78}$$

because $G(n\pi) = (-1)^n$. It follows that $l + s$ must be odd. Two possibilities exist. The annihilation may occur either in a singlet state or triplet state.

Singlet state $(s = 0)$ If $l = J = 1$, the parity of the left-hand side of (4.77) is $-(-1)^l$ or $(-1)^{l+1}$, where we have taken into account the fact that for the p^-–N system the intrinsic parity will be -1. This becomes $+1$ for $l = 1$. Now parity of the two-pion $l = 1$ state is $(-1)^l = (-1)^1 = -1$ or odd. Thus the ${}^1P_1$ state is forbidden by parity conservation. It also follows that both the singlet states ${}^1P_1$ and ${}^1S_0$ are forbidden.

Triplet state $(s = 1)$ The states ${}^3P_{0,1,2}$ are forbidden, since $l + s$ must be odd. However, an S-state is allowed, since $l = 0$, $J = 1$ gives negative parity in both

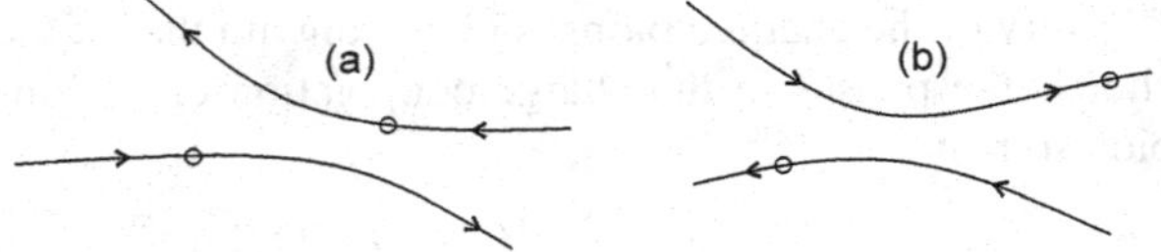

Fig. 4.5 (**a**) Collision between two molecules. (**b**) Time reversed collision

initial and final states. We conclude that the annihilation (4.77) can only proceed in the $I = 1, {}^3S_1$, state.

It is instructive to compare the selection rules for the positronium and nucleon-antinucleon system

$$e^+e^- \rightarrow n\gamma$$
$$p\text{-parity} \quad (-1)^{l+1}$$
$$C\text{-parity} \quad (-1)^{l+s} = (-1)^n$$
$$N\overline{N} \rightarrow n\pi$$
$$p\text{-parity} \quad (-1)^{l+1}$$
$$G\text{-parity} \quad (-1)^{l+s+I} = (-1)^n$$

It must be stressed that the concept of G-parity is not new but merely stems from the more familiar concepts, viz the charge conjugation and isospin invariance. However, its importance in the selection rules, notably in the decays of meson resonances can hardly be doubted (Chap. 5).

4.12 Time Reversal

By time reversal is meant the reflection of the time coordinate i.e., changing the sign just as the parity operation involves changing the sign of the three space coordinates. This means that a film of any elementary particle process running backward should still show a physically allowable process.

Both invariance and non-invariance under time reversal are familiar in classical physics. For example, Newton's law, $F = m\frac{d^2x}{dt^2}$ is invariant under the change of sign of the time coordinate. A film of the trajectory of a projectile in the earth's gravitational field looks equally realistic whether we run it forward or backward if we neglect the air resistance.

The situation is different for the laws of heat conduction or diffusion, which depend only on the first derivative of the time coordinate.

In the case of elementary collisions between molecules, the principle of microscopic reversibility is obeyed. This principle was used for the determination of spin of π^+ mesons (Chap. 3). For each collision, there exists a time reversed collision as shown in Fig. 4.5.

For a gas in equilibrium the two types of collisions occur with equal probability and the entropy is constant.

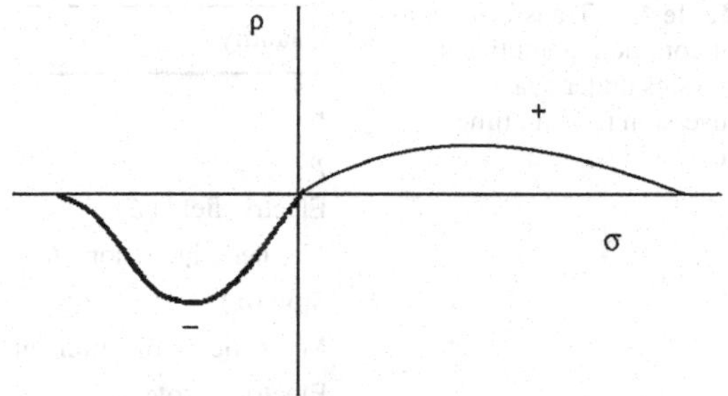

Fig. 4.6 Electric Dipole Moment (EDM) and Time Invariance

4.12.1 *T*-Invariance in Strong Interactions

There is no direct experimental evidence for the non-invariance of T-reversal in strong interactions. For example, a comparison of the forward and backward rates of the nuclear reaction

$$\mathrm{P} + {}^{27}\mathrm{Al} \leftrightarrows \alpha + {}^{24}\mathrm{Mg} \tag{4.79}$$

shows that T violation is less than $\simeq 5 \times 10^{-4}$.

4.12.2 Electric Dipole Moment (EDM) and Time Invariance

The existence of an electric dipole moment of neutron implies violation of both T- and P-invariance. Although neutron is uncharged, an asymmetric charge density of positive and negative clouds in the neutron with respect to the spin direction $\vec{\sigma}$ could result in a non-zero EDM, Fig. 4.6. This would provide a sensitive test for T-invariance since a dipole moment can be measured with great precision. In the experiment of Dress et al. [6], a resonance technique was used for the polarized thermal neutrons obtained from a reactor. The result sets a limit on the electric dipole moment of $(0.4 \pm 1.5) \times 10^{-24} e$ cm. The results obtained by Altarev et al. [7] set a limit on the electric dipole moment of $(2.3 \pm 2.3) \times 10^{-25} e$ cm and by Pendlebury et al. [8] $(0.3 \pm 4.8) \times 10^{-25} e$ cm. The existence of non-zero EDM for neutron is not yet established. The result of T-violation would explain the predominance of matter over antimatter with the evolution of time with no initial asymmetry (Chap. 6).

The transformations of common quantities in physics under space inversion (p) and time reversal (T) are given in Table 4.6.

Notice that an elementary particle of spin σ is not expected to possess a static electric dipole moment if the interaction of the particle with the electromagnetic field is to be invariant under the T or P operation.

Table 4.7 gives the conservation (Yes) and nonconservation (No) for various physical quantities in the three types of interactions.

Table 4.6 Transformations of common quantities in physics under space inversion (p) and time reversal (T)

Quantity	T	P
r	R	$-r$
p	$-p$	$-p$
Electric field (E)	E	$-E$
Magnetic induction (B)	$-B$	B
Spin (σ)	$-\sigma$	σ
Magnetic dipole moment ($\sigma \cdot B$)	$\sigma \cdot B$	$\sigma \cdot B$
Electric dipole moment ($\sigma \cdot E$)	$-\sigma \cdot E$	$-\sigma \cdot E$

4.13 Charge Conservation and Gauge Invariance

Of various types of continuous translations in particle physics an important one is that of gauge transformation. It is closely connected with the charge conservation. Charge is known to be conserved to an accuracy of better than one part in 10^{22}.

The classical electromagnetic field is described by Maxwell's equation in terms of the vectors $\vec{E}$, the electric component of the field and $\vec{B}$, the magnetic component along with ρ, the charge density and $\vec{j}$, the current density. The fact that charge is conserved is given by the equation of continuity

$$\frac{\partial \rho}{\partial t} + \vec{\nabla} \cdot \vec{j} = 0 \tag{4.80}$$

The vectors $\vec{B}$, and $\vec{E}$, may be derived from the vector potential $\vec{A}$, and scalar potential φ satisfying the equations

$$\vec{B} = \vec{\nabla} \times \vec{A} \tag{4.81}$$

Table 4.7 Conversation laws for the three types of interactions

Quantity	Strong	Electromagnetic	Weak
1. Q (charge)	Yes	Yes	Yes
2. B (Baryon number)	Yes	Yes	Yes
3. J (Angular momentum)	Yes	Yes	Yes
4. Mass + Energy	Yes	Yes	Yes
5. Linear momentum	Yes	Yes	Yes
6. I (Isospin)	Yes	No	No
7. I_3 (Third component of I)	Yes	Yes	No $\Delta I_3 = \pm 1/2$
8. S (Strangeness)	Yes	Yes	No $\Delta S = \pm 1$
9. p (Parity)	Yes	Yes	No
10. C (Charge conjugation)	Yes	Yes	No
11. G (G-parity)	Yes	No	No
12. L (Lepton number)	Yes	Yes	Yes

$$\vec{E} = -\vec{\nabla}\varphi - \frac{1}{c}\frac{\partial \vec{A}}{\partial t} \tag{4.82}$$

$\vec{B}$ and $\vec{E}$ satisfy Maxwell's equations which incorporate conservation of charge. These equations, however, do not uniquely define the potentials and it can be shown that $\vec{B}$ and $\vec{E}$ are invariant under gauge transformation.

$$\vec{A} \to \vec{A}' = \vec{A} + \vec{\nabla}\psi \tag{4.83}$$

$$\varphi \to \varphi' = \varphi - \frac{1}{c}\frac{\partial \psi}{\partial t} \tag{4.84}$$

where ψ is an arbitrary scalar function of space and time. The freedom to choose ψ is the essence of gauge invariance. In other words $\vec{A}$ and φ can be chosen in many ways. For example, we may further specify $\vec{A}$ and φ by the use of the Lorentz gauge

$$\nabla \cdot \vec{A} = -\frac{\partial \varphi}{\partial t} \tag{4.85}$$

In this case both $\vec{A}$ and φ obey the equations

$$\nabla^2 \vec{A} - \frac{1}{c^2}\frac{\partial^2 \vec{A}}{\partial t^2} = -\vec{j} \tag{4.86}$$

$$\nabla^2 \varphi - \frac{1}{c^2}\frac{\partial^2 \varphi}{\partial t^2} = -\rho \tag{4.87}$$

It can be shown that both ρ and $\vec{j}$, and φ and $\vec{A}$, are four-vectors which can be written as

$$\vec{j}_\mu = (\rho, \vec{j}) \tag{4.88}$$

$$\vec{A}_\mu = (\varphi, \vec{A}) \tag{4.89}$$

The wave equations (4.86) and (4.87) become

$$\left(\frac{1}{c^2}\frac{\partial^2}{\partial t^2} - \nabla^2\right)\vec{A}_\mu = \vec{j}_\mu \tag{4.90}$$

Since $(\frac{1}{c^2}\frac{\partial^2}{\partial t^2} - \nabla^2)$ is an invariant and A_μ and j_μ are four-vectors, the Maxwell's equations expressed in the form (4.90) are the same in any frame of reference. A four-vector is a quantity $\sum_{i=1}^{4} x_i^2$ which is covariant (same form) under Lorentz transformation.

4.14 The CPT Theorem

The CPT theorem, derived in different forms by Schwinger [9], Luders [10] and Pauli [11], states that for locally interacting fields, a Lagrangian which in invariant under proper Lorentz transformation is invariant with respect to the combined operation CPT. That is the observables are invariant under the combined operation CPT. Here the operation CPT stands for charge conjugation, parity transformation and the time reversal, taken in any order. The proof is valid only for the combined set of operations although the observables may not be invariant under the individual operations C, P and T.

Consequently, if an interaction is invariant under one of the operations then it would be invariant under the product of the other two. The theorem is based on only two assumptions (i) Lorentz invariance (ii) microscopic causality (signals cannot propagate faster than the velocity of light even over microscopic distances). Apart from its theoretical basis, CPT invariance rests on a very firm experimental footing. Some of its predictions are

(i) The existence of an antiparticle for every particle
(ii) The equality of masses, lifetimes, spins, magnetic moments of particles and antiparticles.

Equality of life times have been checked with an accuracy of about 0.1 to 0.2 percent for μ's, π's and K's.

$$(\tau_{\pi^+} - \tau_{\pi^-})/\tau_\pi < 10^{-3}$$
$$(\tau_{\mu^+} - \tau_{\mu^-})/\tau_\mu < 2 \times 10^{-3}$$
$$(\tau_{K^+} - \tau_{K^-})/\tau_K < 10^{-3}$$

Masses have been checked with varying degree of accuracy

$$(M_{\pi^+} - M_{\pi^-})/M_\pi < 10^{-3}$$
$$(M_{p^-} - M_p)/M_p < 8 \times 10^{-3}$$
$$(M_{K^+} - M_{K^-})/M_K < 10^{-3}$$
$$(M_{K^0} - M_{\overline{K^0}})/M_{K^0} < 10^{-14}$$

The magnetic moments have also been verified to a high degree of accuracy

$$(|\mu_{\mu^+}| - |\mu_{\mu^-}|)/|\mu_\mu| < 3 \times 10^{-9}$$
$$(|\mu_{e^+}| - |\mu_{e^-}|)/|\mu_e| < 10^{-5}$$

In order that CPT invariance may hold, if any of C, P or T is violated, one of the others may also be violated in a complimentary manner. Thus T violation implies CP violation and vice versa.

Example 4.5 Explain that charged pions, π^+ and π^- are of equal mass, but charged sigma hyperons Σ^+ and Σ^- have masses which differ by 8 MeV/c^2.

Solution π^+ and π^- mesons are particle-antiparticle pair and by CPT theorem they are expected to have identical mass, although they are the members of the same isospin triplet. The Σ^+ and Σ^- hyperons are the members of the Σ iso-spin triplet but they need not have the same mass. Actually the masses of Σ^+, Σ^0 and Σ^- differ slightly.

Example 4.6 Explain that the mean life of the neutral sigma hyperons is many orders of magnitude smaller than those of Lambda or neutral hyperons.

Solution Σ^0 baryon decays through the scheme $\Sigma^0 \to \Lambda + \gamma$, which is electromagnetic since $\Delta I = -1$, $\Delta S = 0$. On the other hand the decays $\Lambda \to p + \pi^-$ or $\Lambda \to n + \pi^0$ are weak decays as $\Delta S = 1$ and $\Delta I = \pm\frac{1}{2}$. Also $\Xi^0 \to \Lambda + \pi^0$ which is again a weak decay as $\Delta S = 1$ and $\Delta I = -\frac{1}{2}$. As the electromagnetic interactions are expected to be 10^{11} stronger than the weak interactions, the corresponding life times are expected to be longer by a similar factor.

Example 4.7 Explain why the reaction $\pi^- + p \to \Sigma^+ + K^-$ has never been observed at any energy.

Solution The strangeness in the initial state is $S_i = 0 + 0 = 0$, while in the final state $S_f = -1 - 1 = -2$; thus $\Delta S = -2$. So strangeness is violated in the strong interaction and the reaction does not take place at any energy.

Example 4.8 Consider the elastic scattering of charged pions

$$\begin{aligned} \pi^+ + p &\to \pi^+ + p \\ \pi^- + p &\to \pi^- + p \\ &\to \pi^0 + n \end{aligned}$$

Show that $\sqrt{\sigma_{\pi^+ p}} + \sqrt{\sigma_{\pi^- p}} - \sqrt{2\sigma_{\pi^0 n}} \geq 0$.

Solution Now from Sect. 4.3.7.5

$$\begin{aligned} \sqrt{\sigma_{\pi^+ p}} &= |a_{3/2}| \\ \sqrt{\sigma_{\pi^- p}} &= \frac{1}{\sqrt{3}}|a_{3/2} + 2a_{1/2}| \\ \sqrt{\sigma_{\pi^0 n}} &= \frac{\sqrt{2}}{3}|a_{3/2} - a_{1/2}| \end{aligned}$$

or

$$\sqrt{2\sigma_{\pi^0 n}} = \frac{2}{3}|a_{3/2} - a_{1/2}|$$

We can draw the triangle of the amplitudes in the complex plane as shown in Fig. 4.7. From the triangle, the required inequality follows from the fact that the sum of two sides is equal or greater than the third side.

Example 4.9 The spherical harmonies $Y_l^m(\theta, \varphi)$ are defined in terms of the associated Legendre polynomials $P_l^m(\cos\theta)$ through the relation

$$Y_l^m(\theta, \varphi) = \left[\frac{2l+1}{4\pi}\frac{(l-m)!}{(l+m)!}\right]^{1/2} P_l^m(\cos\theta)\exp(im\varphi)$$

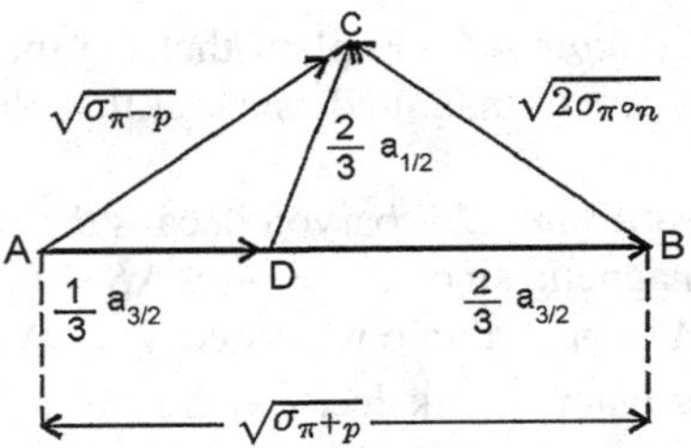

Fig. 4.7 Triangle of the amplitudes in the complex plane

where the associated Legendre polynomials are in even (odd) powers of $\cos\theta$ if $l-|m|$ are even (odd). Show that the parities of the spherical harmonics are $(-1)^l$.

Solution Under the parity operation $\theta \to \pi - \theta$ and $\varphi = \pi + \varphi$.
Then

$$\exp(im\varphi) \to \exp(im\pi)\exp(im\varphi) = (-1)^{|m|}\exp(im\varphi)$$

and

$$p_l^m(\cos\theta) \to p_l^m[\cos(\pi-\theta)] = p_l^m(-\cos\theta) = (-1)^{l-|m|}p_l^m(\cos\theta)$$

Hence,

$$Y_l^m(\theta,\varphi) \to Y_l^m(\pi-\theta,\pi+\varphi) = (-1)^P Y_l^m(\theta,\varphi)$$

and the parity of the spherical harmonium $Y_l^m(\theta,\varphi)$ is $(-1)^P$, independent of m.

Example 4.10 The matrices

$$\tau_1 = \frac{1}{2}\begin{pmatrix} 0 & 1 \\ 1 & 0 \end{pmatrix}; \qquad \tau_2 = \frac{1}{2}\begin{pmatrix} 0 & -i \\ i & 0 \end{pmatrix}; \qquad \tau_3 = \frac{1}{2}\begin{pmatrix} 1 & 0 \\ 0 & -1 \end{pmatrix}$$

apart from the factor $\frac{1}{2}$, are the Pauli spin matrices. Show that these matrices satisfy the well known commutation relations for angular momenta. Construct raising and lowering operators $\tau\pm = \tau_1 \pm i\tau_2$ and show that they have the desired properties by examining their effect on the I spin states $\binom{1}{0}$ and $\binom{0}{1}$, representing proton and neutron respectively.

Solution

$$[\tau_1,\tau_2] = \frac{1}{4}\begin{pmatrix} 0 & 1 \\ 1 & 0 \end{pmatrix}\begin{pmatrix} 0 & -i \\ i & 0 \end{pmatrix} - \frac{1}{4}\begin{pmatrix} 0 & -i \\ i & 0 \end{pmatrix}\begin{pmatrix} 0 & 1 \\ 1 & 0 \end{pmatrix} = \frac{i}{2}\begin{pmatrix} 1 & 0 \\ 0 & -1 \end{pmatrix}$$
$$= i\tau_3 \quad \text{etc.}$$

$$\tau_+ = \tau_1 + i\tau_2 = \begin{pmatrix} 0 & 1 \\ 0 & 0 \end{pmatrix}$$

Now $|p\rangle = \binom{1}{0}$ and $|n\rangle = \binom{0}{1}$

$$\tau_+|n\rangle = \begin{pmatrix} 0 & 1 \\ 0 & 0 \end{pmatrix}\begin{pmatrix} 0 \\ 1 \end{pmatrix} = \begin{pmatrix} 1 \\ 0 \end{pmatrix} = |p\rangle$$

Also $\tau_- = \tau_1 - i\tau_2 = \begin{pmatrix} 0 & 0 \\ 1 & 0 \end{pmatrix}$

$$\tau_-|p\rangle = \begin{pmatrix} 0 & 0 \\ 1 & 0 \end{pmatrix}\begin{pmatrix} 1 \\ 0 \end{pmatrix} = \begin{pmatrix} 0 \\ 1 \end{pmatrix} = |n\rangle$$

Example 4.11 Express the ratio of the cross-sections for the reactions $K^- + p \to \pi^- + \Sigma^+$ and $K^- + p \to \pi^+ + \Sigma^-$ in terms of two possible I spin amplitudes.

Solution Since $I = \frac{1}{2}$ for both K^- and p, the resultant I spin is either 0 or 1. In the final state $I = 1$ for both π and Σ so that the resultant is 0, 1 or 2. Contribution can be made only from $I = 0$ and $I = 1$ channels, with amplitudes a_0 and a_1, respectively, as conservation of I spin rules out $I = 2$ channel.

Using Clebsch-Gordon coefficients for $\frac{1}{2} \times \frac{1}{2}$, we have

$$K^- p = \frac{1}{\sqrt{2}}\big(|1,0\rangle - |0,0\rangle\big)$$

Using the Clebsch-Gordan coefficients for 1×1, we have

$$\pi^-\Sigma^+ = -\frac{1}{\sqrt{2}}|1,0\rangle + \frac{1}{\sqrt{3}}|0.0\rangle$$

$$\pi^+\Sigma^- = \frac{1}{\sqrt{2}}|1,0\rangle + \frac{1}{\sqrt{3}}|0.0\rangle$$

The probability amplitude for the process

$$K^- p \to \pi^-\Sigma^+$$

$$\begin{aligned}
\langle \pi^-\Sigma^+|H|K^-p\rangle &= \left(-\frac{1}{\sqrt{2}}\langle 1,0| + \frac{1}{\sqrt{3}}\langle 0,0|\right) H \frac{1}{\sqrt{2}}\big(|1,0\rangle - |0,0\rangle\big) \\
&= -\frac{1}{2}\langle 1,0|H|1,0\rangle - \frac{1}{\sqrt{6}}\langle 0,0|H|0,0\rangle \\
&= -\left(\frac{1}{2}A_1 + \frac{1}{\sqrt{6}}A_0\right)
\end{aligned}$$

Similarly,

$$\langle \pi^+\Sigma^-|H|K^-p\rangle = \left(\frac{1}{2}A_1 - \frac{1}{\sqrt{6}}A_0\right)$$

Hence,

$$\frac{\sigma(K^-p \to \pi^-\Sigma^+)}{\sigma(K^-p \to \pi^+\Sigma^-)} = \frac{|\frac{1}{2}A_1 + \frac{1}{\sqrt{6}}A_0|^2}{|\frac{1}{2}A_1 - \frac{1}{\sqrt{6}}A_0|^2}$$

Example 4.12 Calculate the ratio of the cross-section for the reaction $\pi^- p \to \pi^- p$ and $\pi^- p \to \pi^0 n$ on the assumption that the two I spin amplitudes are equal in magnitude but differ in phase by 45°.

Solution Since pions have $T = 1$ and nucleons $T = \frac{1}{2}$, the resultant I spin both in the initial state and final states can be $I = \frac{1}{2}$ or $\frac{3}{2}$.

We have

$$\pi^- p \to \sqrt{\frac{1}{3}}\left|\frac{3}{2}, -\frac{1}{2}\right\rangle - \sqrt{\frac{2}{3}}\left|\frac{1}{2}, -\frac{1}{2}\right\rangle$$

$$\pi^0 n \to \sqrt{\frac{2}{3}}\left|\frac{3}{2}, -\frac{1}{2}\right\rangle + \sqrt{\frac{1}{3}}\left|\frac{1}{2}, -\frac{1}{2}\right\rangle$$

Hence

$$\langle\pi^- p|H|\pi^- p\rangle = \frac{1}{3}a_3 + \frac{2}{3}a_1 \quad \text{and}$$

$$\langle\pi^0 n|H|\pi^- p\rangle = \frac{\sqrt{2}}{3}a_3 - \frac{\sqrt{2}}{3}a_1$$

and the ratio of the cross-sections is

$$\frac{\sigma_1}{\sigma_2} = \frac{|A_3 + 2A_1|^2}{|\sqrt{2}(A_3 - A_1)|^2}$$

Put $A_3 = A_1 e^{i\varphi}$ with $\varphi = \pm 45°$.

$$\frac{\sigma_1}{\sigma_2} = \frac{1}{2}\frac{|e^{i\varphi} + 2|^2}{|e^{i\varphi} - 1|^2} = \frac{1}{2}\frac{(e^{i\varphi} + 2)(e^{-i\varphi} + 2)}{(e^{i\varphi} - 1)(e^{-i\varphi} - 1)}$$

$$= \frac{1}{2}\frac{(5 + 4\cos\varphi)}{(2 - 2\cos\varphi)} = 6.68$$

Observe that the ratio is independent of the sign of φ.

Example 4.13 Use the conservation of angular momentum and parity to determine the possible angular momentum states involved in the annihilation process $pp^- \to \pi^+\pi^-$. Does charge conjugation invariance impose any further restriction on the allowed angular momentum states?

Solution For a $\pi^+\pi^-$ system, the charge conjugation operation gives $C = (-1)^P$. Let L be the relative orbital angular momentum in the p^-p system. The spins of the p and p^- can couple to give a total spin of 0 or 1, hence the total angular momentum in the initial state is $J_i = L, L \pm 1$. Since the pions have zero spin, the total angular momentum is also $J_f = L$. Conservation of angular momentum ($J_i = J_f$) gives $l = L, L \pm 1$. As proton and antiproton have opposite parity, conservation of parity rules out $l = L$. Therefore, the only possibility is $\ell = L \pm 1$. Hence the possible transitions are ${}^3S_1 \to P_1$, ${}^3P_0 \to S_0$ etc.

Charge conjugation invariance gives $(-1)^{l+s} = (-1)^l$, where S is the total spin of the p^-p system, and therefore $L + S = P$. Since S can be 0 or 1, this is not as stringent a condition as that arising from conservation of angular momentum and parity. Charge conjugation invariance therefore imposes no further restriction.

Example 4.14 Explain why absence of the decay $K^+ \to \pi^+ + \gamma$ can be considered an argument in favour of spin zero for K^+.

Solution The occurrence of this decay would violate the conservation of angular momentum since photon has spin 1 and pion zero and K is assumed to have zero spin.

Example 4.15 A hyper-nucleus is formed when a neutron is replaced by bound Λ-hyperon. ${}^4_\Lambda$He and ${}^4_\Lambda$H are a doublet of mirror hyper-nuclei. Deduce the ratio of the reaction rates.

$$K^- + {}^4\text{He} \rightarrow {}^4_\Lambda\text{He} + \pi^- \quad (1)$$
$$\rightarrow {}^4_\Lambda\text{H} + \pi^0 \quad (2)$$

Solution ^{4}He is a singlet and so $I = 0$. The initial state is pure $I = \frac{1}{2}$ since K^- has $= \frac{1}{2}$. Also $I_3 = -\frac{1}{2}$. The final state therefore has $I = \frac{1}{2}$, $I_3 = -\frac{1}{2}$.

Final state $\sim \frac{1}{\sqrt{3}}|\pi^0\text{H}_\Lambda\rangle - \sqrt{\frac{2}{3}}|\pi^-\text{H}_\Lambda\rangle$

$$\therefore \quad \frac{\sigma_1}{\sigma_2} = \frac{(\sqrt{2/3})^2}{(1/\sqrt{3})^2} = \frac{2}{1}$$

Example 4.16 Discuss, the possible decay modes of the Ω^--hyperon allowed by the conservation laws, and show that the weak decay is the only possibility.

Solution $M_{\Omega^-} = 1672$ MeV.

Strong strangeness-conserving decays are forbidden by energy conservation, e.g.

$$\Omega^- \nrightarrow p + 2K^- + \overline{K^0}$$
(938) (988) (498) (1675 MeV)

$$\nrightarrow \Lambda + K^- + \overline{K^0}$$
(1115) (494) (498)

$$\nrightarrow \Xi^0 + K^-$$
(1314) (494)

$$\nrightarrow \Sigma^+ + K^- + K^-$$
(1189) (494) (494)

Also, the strangeness conserving electromagnetic decays are forbidden by energy conservation.

Thus,

$$\Omega^- \nrightarrow \Xi^- + \Lambda + \gamma$$
(1675) (1321) (1115)

Weak decays are the only possibility

$$\Omega^- \rightarrow \Xi^0 + \pi^-$$
$$\rightarrow \Xi^- + \pi^0$$
$$\rightarrow \Lambda + K^-$$

Example 4.17 Capture of negative kaons in helium sometimes leads to the formation of the hyper-nucleus (a nucleus in which a neutron is replaced by a Λ-hyperon) according to the reaction $K^- + {}^4\text{He} \to {}^4_\Lambda\text{He} + \pi^0$. Study of the decay branching ratios of ${}^4_\Lambda$H, and the isotropy of decay products establishes that $J({}^4_\Lambda\text{H}) = 0$. Show that this implies negative parity for the K^-, independent of the orbital angular momentum of the state from which the K^- is captured.

Solution If l is the orbital angular momentum and the spins of all the particles is zero, it follows that equating parities in the initial and final states,

$$(-1)^l p_K = (-1)^l p_\pi \quad \text{and so } p_K = p_\pi = -1$$

Example 4.18 Show that the reaction $\pi^- + d \to n + n + \pi^0$ cannot occur for pions at rest.

Solution $J_d = 1$ and capture takes place from s-state. Hence in the final state $J = 1$. Since the Q-value is only 0.5 MeV, the final $nn\pi^0$ must be an s-state. Then the two neutrons must be in a triplet spin state (symmetric) forbidden by the Pauli principle.

Example 4.19 What restrictions does the decay mode $K_1^0 \to 2\pi^0$ place on (a) the kaon spin (b) the kaon parity?

Solution

a. J_K is even by Bose symmetry.
b. None, parity is not conserved in weak interactions.

Example 4.20 State which of the following combinations can or cannot exist in a state of $I = 1$, and give the reasons: (a) $\pi^0\pi^0$ (b) $\pi^+\pi^-$ (c) $\pi^+\pi^+$ (d) $\Sigma^0\pi^0$ (e) $\Lambda\pi^0$.

Solution Bose statistics holds for pions, and the wave function of a system of pions must be symmetric upon interchange of space, isospin coordinates of any two pions. For each pion $I = 1$. For a system of two pions, total isospin, $I = 2, 1, 0$. The isospin states are given by

$$|2, \pm 2\rangle, \qquad |2, \pm 1\rangle, \qquad |2, 0\rangle \quad \text{(symmetric)}$$
$$|1, \pm 1\rangle, \qquad |1, 0\rangle \quad \text{(antisymmetric)}$$
$$|0, 0\rangle \quad \text{(symmetric)}$$

The requirement of Bose statistics on the two pion states will impose some restrictions on the properties of the space part of the wave function

$$\psi(\text{total}) = \psi(\text{space})\psi(\text{isospin})$$

Therefore, for symmetric isospin states like $|2, \pm 1\rangle$ or $|0, 0\rangle$, ψ(space) must be symmetric. But for two pions, the interchange of the spatial positions introduces a factory $(-1)^l$ so that only even l states are allowed. Thus, (π^+, π^+), (π^-, π^-), (π^0, π^0) in (a) and (c) can be found in the states of $L = 0, 2, \ldots$ On the other hand, the states $|1, \pm 1\rangle, |1, 0\rangle$ are antisymmetric and this requires, $L = \text{odd} = 1, 3$ for the relative motion of the two pions. Thus (a) and (c) can exist in $s(L = 0)$, $d(L = 2)$-states. (b) being antisymmetric can have odd values of L viz $p(L = 1)$ etc. (d) and (e) can exist in $I = 1$ state regardless of L value since the two particles for each case are Fermion and Boson so that the previous considerations are of no consequence.

Example 4.21 The partial width for the decay

$$\begin{array}{lll} & {}^{16}\text{O}^* \rightarrow & {}^{12}\text{C} + \alpha \\ J^P & 2^- & 0^- \end{array}$$

is of the order of 10^{-10} eV. What information do you get for the strong force?

Solution If angular momentum is conserved, the relative orbital angular momentum, l in the final state must have $l = 2$. But such a state has even parity. The decay is parity changing. The observed width 10^{-10} eV implies a mean lifetime of about 10^{-7} sec. In such light nuclei Coulomb barrier effects are negligible, and the mean lifetime is very long compared with the mean lifetime expected for a decay through strong interaction, namely $< 10^{-23}$. It is concluded that strong interaction cannot significantly violate parity conservation. The decay is probably consistent with weak interaction effects.

Example 4.22 Given that the ρ meson has a width of 158 MeV/c^2 in its mass, how would you classify the interaction for its decay?

Solution Using the uncertainty principle the mean lifetime is estimated from

$$\tau \sim \frac{\hbar}{\Delta E} = \frac{6.2 \times 10^{-20}\ \text{MeV} - \text{sec}}{158} \simeq 4 \times 10^{-24}\ \text{sec}$$

Hence the decay is a strong interaction.

Example 4.23 At 600 MeV the cross section for the reactions $p + p \rightarrow d + \pi^+$ and $p + n \rightarrow d + \pi^0$ are $\sigma^+ = 3.15$ mb and $\sigma^0 = 1.5$ mb. Show that the ratio of the observed cross sections is in accordance with the iso-spin predictions.

Solution In the table below we are given the nucleon-nucleon state. 1 refers to the particle 1 and 2 particle 2, + for proton and − for neutron

I	I_3	State	System
	1	$\|1+\rangle\|2+\rangle$	p-p symmetric
1	0	$\frac{1}{\sqrt{2}}[\|1+\rangle\|2-\rangle + \|1-\rangle\|2+\rangle]$	p-n symmetric
	−1	$\|1-\rangle\|2-\rangle$	nn symmetric
0	0	$\frac{1}{\sqrt{2}}[\|1 + \|2-\rangle - \|1-\rangle\|2+\rangle\|]$	p-n antisymmetric

The p–p state is

$$|p\rangle|p\rangle \equiv |1+\rangle|2+\rangle = |1, 1\rangle$$

The p–n state is

$$|p\rangle|n\rangle \equiv |1+\rangle|2-\rangle = \frac{1}{\sqrt{2}}[|1, 0\rangle + |0, 0\rangle]$$

where the extreme right hand side has the notation $|I, I_3\rangle$. These have been obtained by reversing the equations of the above table.

It is seen that the initial $(p + p)$ system is a pure $I = 1$ state but the $(p + n)$ system is linear superposition with equal statistical weight of $I = I$ and $I = 0$ states. Therefore if isospin is conserved only the $I = 1$ state can contribute, since the final state is pure $I = 1$ state. The expected ratio of the cross section at all energies is

$$\frac{\sigma^+}{\sigma^0} = \frac{\sigma(p + p \to d + \pi^+)}{\sigma(p + n \to d + \pi^0)} = \frac{1}{(1/\sqrt{2})^2} = 2$$

The observed ratio $3.15/1.5 = 2.1$ is consistent with the expected ratio of 2.0.

4.15 Questions

4.1 Name the antiparticles of the following $e^+, p, \nu_e, \pi^0, K^0, n, \Lambda$.

4.2 What type of interactions are the following (strong, em or weak)?

(i) $\Lambda \to p + \pi^-$
(ii) $p + p \to p + n + \pi^+$
(iii) $\Sigma^0 \to \Lambda + \gamma$
(iv) $K^- + p \to \Sigma^- + \pi^+$
(v) $\gamma + p \to \pi^+ + n$
(vi) $\pi^0 \to \gamma + \gamma$
(vii) $\pi^- + d \to n + n$
(viii) $K^+ \to \pi^+ + \pi^+ + \pi^-$
(ix) $\mu^+ \to e^+ + \nu_e + \overline{\nu_\mu}$
(x) $N^{*++}_{(1236)} \to p + \pi^+$

4.3 Which of the following particles are hadrons?

(a) μ
(b) π
(c) η
(d) γ
(e) p^-

4.4 Which of the following particles are leptons?

(a) Σ^+
(b) K
(c) μ^-
(d) π
(e) e^+
(f) ν_e

4.5 Give the isospin of the following particles

(a) pion
(b) kaon
(c) nucleon
(d) Λ hyperon

4.6 Which of the following are unstable particles?

(a) e^+
(b) μ^-
(c) p^-
(d) free neutron

4.7 Why are the following processes forbidden?

(a) $n \to p + e^-$
(b) $\pi^- + p \to \Sigma^+ + K^-$
(c) $p \to e^+ + \gamma$
(d) $\Lambda \to n + K^0$
(e) $\mu^+ \to e^+ + \overline{\nu_e} + \nu_\mu$
(f) $d + d \to \alpha + \pi^0$

4.8 State whether the following are fermions or bosons?

(a) μ^-
(b) π^-
(c) d
(d) $\propto$
(e) Λ

4.9 Which of the following are examples of strong, electromagnetic, weak and gravitational interaction.

(i) binding forces in atoms
(ii) binding forces in molecules
(iii) binding forces in solar system
(iv) binding forces in the nuclei
(v) the decay $K^0 \to \pi^+ + \pi^-$

[Ans. (i) Em; (ii) Em; (iii) gravitational; (iv) strong; (v) weak]

4.10 Verify $\Xi^- + p \rightarrow \Lambda + \Lambda$ is allowed by conservation laws.

4.11 With the help of certain characteristic numbers in appropriate dimensionless units indicate (without proof) the relative strength of strong electromagnetic weak and gravitational interactions.

4.12 Which of the following processes are allowed by the conservation laws

(a) $p \rightarrow e^+ + \gamma$
(b) $p \rightarrow \pi^+ + \gamma$
(c) $n \rightarrow p + \gamma$
(d) $p^- + n \rightarrow \pi^- + \pi^0$

[Ans. only (d) is allowed]

4.13 Investigate the change in isospin, ΔI in the weak decays of Λ, Σ, Ξ and K particles. What is the common feature?
[Ans. $\Delta I = \pm\frac{1}{2}$]

4.14 The Σ-hyperon exists in three charge states +, 0 and −. What is the strangeness of Σ?
[Ans. −1]

4.15 The Ξ baryons form an isospin doublet with $S = -2$. What charge states are possible?
[Ans. 0, −]

4.16 Which of the following quantum numbers are additive and which ones are multiplicative?

(i) Strangeness
(ii) parity
(iii) Isospin
(iv) G-parity
(v) C-parity
(vi) Baryon number
(vii) Lepton number
(viii) charge.

[Ans. (i) A; (ii) M; (iii) A; (iv) M; (v) M; (vi) A; (vii) A; (viii) A]

4.17 A μ^+ meson may capture an electron through Coulomb attraction and form a system called muonium. Which of the following decays can occur?

(a) $(\mu^+ e^-) \rightarrow \nu_e + \overline{\nu_\mu}$
(b) $(\mu^+ e^-) \rightarrow \gamma + \gamma$

(c) $(\mu^+ e^-) \rightarrow e^+ + e^- + \nu_e + \overline{\nu_\mu}$

[Ans. (a) and (c) allowed but (b) is forbidden because of non-conservation of lepton number]

4.18 Explain qualitatively why the mean life time of the π^+ meson is 2.6×10^{-8} sec where as that of the π^0 meson is 0.8×10^{-16} sec.

4.19 When negative pions are slowed down, they are captured into atomic orbits of deuterium. How does one know that they cascade down into ls orbital?

4.20 Experiment cannot define intrinsic parity of a baryon. Only relative parity has meaning. Why?

4.21 A pionic deuterium atom in the ground state decays through the strong interaction $\pi^- + d \rightarrow n + n$. In which $^{2s+1}L_J$ state may the two neutron system be?

4.22 Which the following reactions are allowed by the conservation laws and which are forbidden and give the reason in each case.

(a) $\pi^0 \rightarrow e^+ + e^-$
(b) $p \rightarrow n + e^+ + \nu_e$
(c) $\mu^+ \rightarrow e^+ + e^- + e^+$
(d) $K^+ + n \rightarrow \Sigma^+ + \pi^0$

4.23 Assign the lepton generation subscript $\nu_e, \overline{\nu_e}, \nu_\mu, \overline{\nu_\mu}, \nu_\tau, \overline{\nu_\tau}$

(a) $\pi^+ \rightarrow \pi^0 + e^+ + \nu$
(b) $\nu^- \rightarrow e^- + \nu + \nu$
(c) $\mu^+ \rightarrow e^+ + \nu + \nu$
(d) $\Sigma^- \rightarrow n + \mu^- + \nu$
(e) $\overline{K^0} \rightarrow \pi^0 + e^- + \nu$
(f) $K^+ \rightarrow \pi^0 + e^+ + \nu$
(g) $\nu + p \rightarrow n + e^+$
(h) $D^0 \rightarrow K^- + \pi^0 + e^+ + \nu$
(i) ${}^3_1\mathrm{H} \rightarrow {}^3_2\mathrm{He} + e^- + \bar{\nu}$
(j) $\nu + n \rightarrow e^- + p$
(k) $\nu + p \rightarrow \mu^- + p + \pi^+$
(l) $\nu + {}^{37}_{17}\mathrm{Cl} \rightarrow {}^{37}_{18}\mathrm{Ar} + e^-$
(m) $\pi^- \rightarrow e^- + \nu$
(n) $\pi^+ \rightarrow \mu^+ + \nu$
(o) $\tau^- \rightarrow \pi^- + \pi^0 + \nu$

[Ans. (a) ν_e; (b) $\overline{\nu_e}$, ν_μ; (c) ν_e, $\overline{\nu_\mu}$; (d) $\overline{\nu_\mu}$; (e) $\overline{\nu_e}$; (f) ν_e; (g) $\overline{\nu_e}$; (h) ν_e; (i) $\overline{\nu_e}$; (j) ν_e; (k) ν_μ; (1) ν_e; (m) $\overline{\nu_e}$; (n) ν_μ; (o) ν_τ]

4.24 Explain why the following processes are never observed.

(a) $\pi^+ + p \rightarrow \Sigma^- + K^+$

(b) $p + n \to p + K^+ + K^-$
(c) $\pi^- + p \to \Lambda + \pi^0$
(d) $p^- + p \to \Sigma^- + \Lambda + \pi^+$
(e) $p^- \to n + \pi^-$
(f) $\Sigma^0 \to \Lambda$
(g) $n \to p + e^-$
(h) $\Sigma^0 \to \Lambda + \pi^0$

[Ans. (a) $\Delta \frac{Q}{e} = -2$; (b) $\Delta B = -1$; (c) $\Delta s = -1$; (d) $\Delta S = -2$; (e) $\Delta B = 2$; (f) momentum violation; (g) $\Delta L = 1$; (h) energy violation]

4.25 Which conservation law or laws require that at least two photons be emitted when a positron and electron annihilate? Explain. Is it possible for three photons to be emitted? Again, Explain.
[Ans. Momentum conservation, Yes]

4.26 The following fusion reaction has been proposed, $^2\text{H} + {}^2\text{H} \to {}^4\text{He} + \pi^0$. Explain why, even when the ^{2}H nuclei have sufficient energy, the reaction is not detected?
[Ans. I is not conserved]

4.27 Explain which of the following reactions is not allowed?

(a) $K^+ \to \pi^+ + \pi^0 + \pi^0$
(b) $K^+ \to \mu^+ + \nu_\mu + \overline{\nu_e}$
(c) $\Lambda \to p + e^- + \nu_e$
(d) $\Lambda \to n + \pi^0$
(e) $\pi^- + p \to \Sigma^+ + K^-$
(f) $K^- + p \to \Xi^0 + K^+ + K^-$
(g) $\pi^- + p \to K^0 + \Lambda$

[Ans. (b) $\Delta L = -1$; (c) $\Delta L = 2$; (e) $\Delta s = -2$; (f) $\Delta S = -1$]

4.28 Briefly point out the feasibility of the following decay processes:

(a) $n \to p + e^- + \overline{\nu_e}$
(b) $p \to e^+ + \gamma$
(c) $\pi^+ \to \mu^+ + \nu_\mu$
(d) $\Lambda \to n + K^0$

[Ans. (a) and (c) allowed]

4.29 Find the allowed values of the total isospin for the final states in the following interactions:

(a) $\pi^+ + p \to n + \pi^+ + \pi^+$
(b) $p + d \to n + p + p$
(c) $\pi^- + p \to n + \pi^0$

[Ans. (a) $I = \frac{3}{2}$; (b) $I = \frac{1}{2}$; (c) $I = \frac{1}{2}, \frac{3}{2}$]

4.30 Determine the parameters a, b and c in the expression

$$Q = aI_3 + bB + cS$$

where Q is the charge in the units of e, I_3 is the third component of the isotopic spin, and S is the strangeness for the particles Σ^+, K^0 and Ξ^-
[Ans. $a = 1; b = c = \frac{1}{2}$]

4.31 A particle X decays weakly as follows:

$$X \to \pi^0 + \mu^+$$

Determine the following properties of X:

(a) baryon number
(b) charge
(c) lepton number
(d) statistics
(e) A lower limit for its mass (in MeV/c^2)

[Ans. (a) $B = 0$; (b) $Q = e$; (c) $L_\mu = -1$, $L_e = 0$; (d) Fermi-Dirac; (e) $M \geq$ 240.6 MeV/c^2]

4.32 How are K^0 and $\overline{K^0}$ distinguished?
[Ans. By their interactions]

4.16 Problems

4.1 At a given centre of mass energy what is the ratio of the cross-section for the reactions, $P + d \to {}^3\text{He} + \pi^0$ and $P + d \to {}^3\text{H} + \pi^+$.
[Ans. 1:2]

4.2 Show that the branching ratio of reactions

$$\begin{aligned} K^- + {}^4\text{He} &\to \Sigma^- + {}^3\text{He} \\ &\to \Sigma^0 + {}^3\text{H} \end{aligned}$$

is 2:1.

4.3 Show that the branching ratio of the reactions

$$\begin{aligned} K^- + {}^4\text{He} &\to \Lambda + \pi^- + {}^3\text{He} \\ &\to \Lambda + \pi^0 + {}^3\text{H} \end{aligned}$$

is 2:1.

4.4 Find the ratio of the gravitational potential energy to the Coulomb potential energy between two electrons at the same distance of separation.
[Ans. 2.4×10^{-43}]

4.5 How can you determine the parity of Ξ^- from the study of the reaction $\Xi^- + p \to \Lambda + \Lambda$?

4.6 In the $\pi^+ - p$ elastic scattering a maximum is observed in the total cross-section at pion lab energy of 196 MeV. Deduce the mass and spin of the nucleon isobar formed.
[Ans. 1236 MeV, 3/2]

4.7 Show that the parity operation is equivalent to a reflection followed by a rotation through 180°.

4.8 Determine the parity of ortho-positronium and para-positronium. Demonstrate that parity conservation in EM interactions rules out two quanta annihilation of ortho positronium.

4.9 Show that the decay $\pi^0 \to \gamma + \gamma + \gamma$ is forbidden.

4.10 If the p^-–p annihilation at rest proceeds via s-states, explain why the reaction $p^- + p \to \pi^0 + \pi^0$ cannot be a strong interaction.

4.11 Show that a meson which decays to $\pi^+\pi^-$ pair by the strong interaction must have $C = P = (-1)^J$, where J is the spin of the meson.

4.12 Show that p^- annihilation is possible into an even (odd) number of pions if $I + L + s$ is even (odd).

4.13 The following processes are not found to occur in nature.

(a) $e^- \to e^- + \gamma$ in vacuum
(b) $K^+ \to \pi^+ + \gamma$

Explain briefly how a conservation law is violated which is the reason why the processes do not occur.
[Ans. (a) Both energy and momentum cannot be conserved simultaneously; (b) Angular momentum is violated]

4.14 Deduce through which isospin channels the following reactions may proceed

$$K^- + p \to \Sigma^0 + \pi^0$$
$$\to \Sigma^+ + \pi$$

Find the ratio of cross-sections assuming that one or other channel dominates.
[Ans. If $a_0 \gg a_1$, ratio is $1:1$; if $a_0 \ll a_1$, the ratio is $0 : a_1^2/4$]

4.15 Which of the following reactions are allowed by the conservation laws and which are forbidden, and give reasons in each case?

(a) $\pi^0 \to e^+ + e^-$
(b) $p \to n + e^+ + \nu_e$
(c) $\mu^+ \to e^+ + e^- + e^+$
(d) $K^+ + n \to \Sigma^+ + \pi^0$

[Ans. (a) Allowed; (b) Forbidden for free proton because of energy conservation but allowed for bound proton if Q-value is positive; (c) Forbidden by muon number conservation; (d) Forbidden by energy conservation]

4.16 Write down the quantum numbers (G, I, J^P) of the S- and P-states of the pp^- system which can decay to

(a) $\pi^+\pi^-$
(b) $\pi^0\pi^0$
(c) $\pi^0\pi^0\pi^0$.

[Ans. (a) $^3S_1, (+, 1, 1^-)$; $^3P_0(+, 0, 0^+)$; $^3P_2(+, 0, 2^+)$; (b) $^3P_0$, $(+, 0, 0^+)$; $^3P_2, (+, 0, 2^+)$; (c) $^1S_0, (-, 1, 0^-)$; $^3P_1, (-, 1, 1^+)$; $^3P_2, (-, 1, 2^+)$]

4.17 If antiprotons are incident with equal frequency on protons and neutrons in deuterium atoms, show that the expected number of charged pions must be equal to twice the number of neutral pions produced in the annihilation process.

References

1. Kemmer (1938)
2. Fermi et al. (1952)
3. Plano et al. (1959)
4. M. Block et al., Phys. Rev. Lett. **3**, 291 (1959)
5. Young (1951)
6. Dress et al. (1978)
7. Altarev et al. (1981)
8. Pendlebury et al. (1984)
9. Schwinger (1953)
10. Luders (1954)
11. Pauli (1955)

Chapter 5
Strong Interactions

5.1 Resonances

In Chaps. 3 and 4 we have studied stable particles as well as unstable particles. The stable particles are neutrinos (ν), photon (γ), electron (e), proton (p) and their antiparticles. The experimental lower limit on the lifetime of electron is $\sim 10^{22}$ years and that of proton $\sim 10^{31}$ years-much longer than the lifetime of the universe, 10^{10} years. The unstable particles are those which decay by weak or electromagnetic processes. The weak decays like $\pi^+ \to \mu^+ \nu_\mu$, $K^+ \to \pi^+ \pi^0$, $\Lambda \to p\pi^-$, $\mu^+ \to e^+ \nu_e \overline{\nu}_\mu$ etc. have lifetimes $< 10^{-13}$ sec. The electromagnetic decays like $\pi^0 \to 2\gamma$ or $\Sigma^0 \to \Lambda\gamma$ have lifetimes in the range 10^{-16}–10^{-20} sec.

In 1960s a new class of particles were discovered in high energy collisions of hadrons. They are so short-lived ($\sim 10^{-23}$–10^{-25} sec) that it is impossible to observe a track in a bubble chamber between the point of production and the point of decay. Their existence must be inferred only indirectly. Such short-lived particles are called resonant or excited states of familiar baryons and mesons. They may be simply called Resonances and may be likened to the excited states of atoms. Applying the uncertainty principle in the form

$$\Delta E \cdot \Delta t \sim \hbar \tag{5.1}$$

for the resonance with lifetime $\tau = \Delta t \sim 10^{-23}$ sec, the uncertainty ΔE in its mass energy will be appreciable, about 100 MeV, so that the measured mass has a spread of 100 MeV/c^2, of the order of 10 %. This is in contrast with the weak decays which have a mass spread of a few keV/c^2 or less. The fast decay of resonances also means that the decay occurs through strong interaction.

The first evidence of the nucleon resonance actually dates back to 1952 when Anderson, Fermi, Long and Nagle [1] discovered a broad peak in the total cross-section in pion-proton scattering experiments around 1230 MeV in the pion-proton CMS (Fig. 4.1). Later more massive resonances of nucleons were discovered. Similarly the resonances of pions, kaons and hyperons were established. It became apparent that the resonances had to be given the same status as the familiar particles.

A. Kamal, *Particle Physics*, Graduate Texts in Physics,
DOI 10.1007/978-3-642-38661-9_5, © Springer-Verlag Berlin Heidelberg 2014

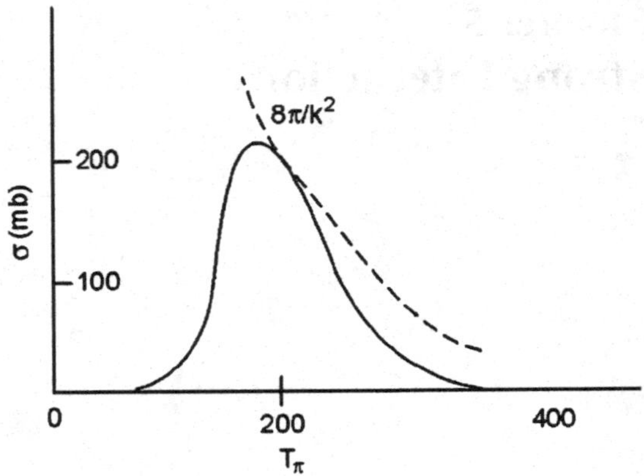

Fig. 5.1 Occurrence of the resonance is indicated by the peak in the total cross-section

According to Abdus Salam "all these are elementary particles, some are more elementary than others". In order to understand the particle physics it was necessary to establish patterns in their properties. To this end the mass, spin, parity, isospin and lifetimes of countless resonances began to be studied systematically so that an order may emerge from the chaos. There are two types of experiments which enable us to establish the occurrence of resonances. (1) Formation Experiments (2) Production Experiments. In the first type the cross-section for a given process such as πp elastically scattering is studied as a function of incident particle momentum p_1, or kinetic energy. The occurrence of the resonance is indicated by the peak in the total cross-section, Fig. 5.1. The mass of the resonance is given by

$$M^* = \left[\left(\sum E\right)^2 - \left(\left|\sum \vec{p}\right|\right)^2\right]^{1/2} \tag{5.2}$$

For a two-body collision when the incident particle of mass m_1 has momentum p_1 and the target particle of mass m_2 is at rest (5.2) becomes

$$M^* = \left[(m_1 + T + m_2)^2 - p_1^2\right]^{1/2} \tag{5.3}$$

where we have used natural units ($\hbar = c = 1$).

The second type is concerned with the study of two-body decay of the resonance plus the third particle. In such a case the primary reaction is in fact a two-body process. Suppose the resonance decays into particles 1 and 2. Since the resonance M^* and particle 3, m_3 share momentum in a unique way in the C.M.S of the colliding particles, the distribution of momentum of m_3 is no longer governed by phase-space considerations for a three-body decay but would exhibit a peak against a background, Fig. 5.2.

From the peak value of p_3 which is identical with that of the resonance, one can determine the mass of the resonance.

An alternative way of determining the mass of the resonance in the production experiments is to rely on the measurement of energy and momentum of a group of particles. Using the general relativistic relation between total energy E, momentum

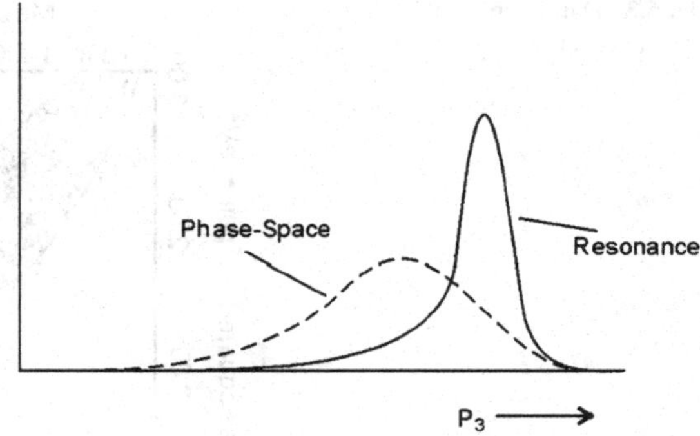

Fig. 5.2 Momentum distribution

$\vec{p}$ and mass m for a single particle, $m^2 = E^2 - p^2$, the effective mass for a group of particles becomes

$$m_1^2 \ldots n = \left(\sum_{n=1}^{i} E_n\right)^2 - \left(\left|\sum_{n=1}^{i} \vec{p}_n\right|\right)^2 \tag{5.4}$$

If the interaction products result without the resonance then the mass combination squared ($m_1^2 \ldots m_i^2$) will be a continuous distribution as expected from phase space. However, if a certain group of particles comes from the short-lived resonance then a narrow peak in the mass combination $m_1 \ldots m_i$ will show up at this mass. Calculations for the phase-space factors can be done analytically, up to four particles. For a larger number of particles recursion relations can be used to relate the phase space for n bodies to that for $n - 1$ bodies. Alternatively, Monte Carlo calculations may be used in which artificial events are generated randomly subject to momentum and energy conservation.

5.1.1 Dalitz Plot

One other technique conveniently used in the study of resonances in the production experiments involving three particles in the final state is the Dalitz plot. The Dalitz plot was originally applied to the analysis of τ-meson ($K_{\pi 3}$) decay into 3 pions. The Dalitz plot is discussed in detail in Appendix E and is applicable only to three particles which are similar or dissimilar. The Dalitz plot enables one to find out whether any pair of final-state particles ($\pi p, \pi\pi$ etc.), result from the decay of a resonance. The observed distributions of momentum and angle of final state particles are assessed. In case of a constant matrix element, the distributions will be governed by phase space. Any departure of the distribution from that expected for phase-space factors, is an indication of the occurrence of a resonance.

As an example, consider the case of hyperon resonance $Y_1^*(1385)$, also known as $\Sigma(1385)$ formed in the interactions of 1.15 GeV/c K^--mesons in a liquid hydrogen bubble chamber [2].

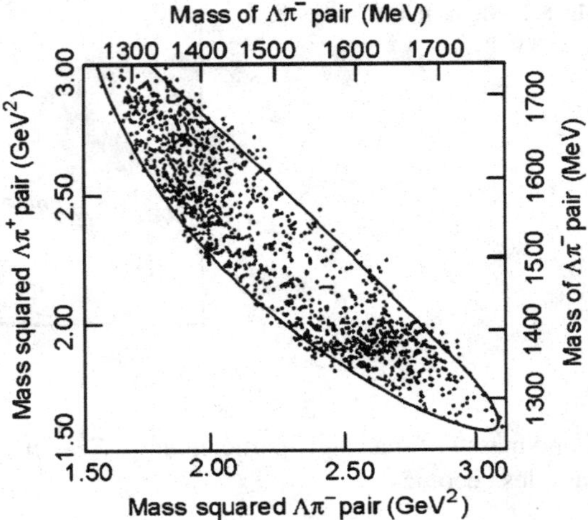

Fig. 5.3 Dalitz plot of the $\Lambda\pi^+\pi^-$ events

The reaction studied was

$$K^- + p \rightarrow \pi^+ + \pi^- + \Lambda \tag{5.5}$$

For each event, the momenta of both pions were recorded and the kinetic energies of π^+ and π^--mesons as computed in the overall centre-of-mass system, are plotted along the x- and y-axes respectively.

If Q is the total available kinetic energy of the three final-state particles in this system then

$$T_\Lambda = Q - (T_{\pi^+} + T_{\pi^-}) \tag{5.6}$$

In a three-body breakup there are infinitely numerous ways in which the total energy released (Q) can be distributed between the three particles subject to momentum conservation.

For $T_\Lambda =$ constant the sum $T_{\pi^+} + T_{\pi^-}$ would be constant. In other words, lines of constant T_Λ are inclined at 45° to the axes.

The points representing individual events are constrained to lie inside the distorted ellipse due to momentum and energy conservation, Fig. 5.3. In the absence of any strong correlations in the final state of (5.5), the density of points should be uniform.

Figure 5.3 is the Dalitz plot of the $\Lambda\pi^+\pi^-$ events from reaction (5.5) as measured by Shafer et al. [3] for 1.22 GeV/c incident momentum. The plot shows strong departure from uniform density as revealed by horizontal and vertical bands on the plot due to favoured values of $T_{\pi^+} + T_{\pi^-}$. This implies that reaction (5.5) is proceeding as a two-body reaction, one body consisting of the resonant state of Λ with unique mass $m_{\Lambda\pi} = 1385$ MeV and the second body a pion. The formation and break-up of reaction (5.5) can be written as

$$K^- + p \rightarrow \Sigma^\pm(1385) + \pi^\mp \tag{5.7}$$

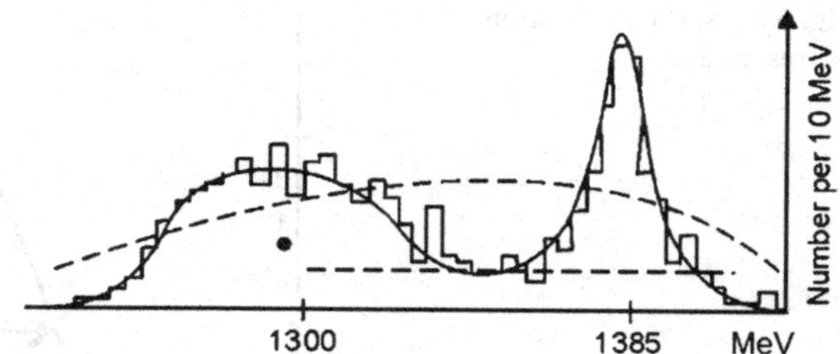

Fig. 5.4 The $\Lambda\pi^+$ mass spectrum

$$\Sigma^{+}_{I=1}(1385) \rightarrow \underset{0}{\Lambda} + \underset{1}{\pi^{\mp}} \tag{5.8}$$

If p is the CMS momentum of each, the total CMS energy will be

$$W = \sqrt{M^2_{\Lambda\pi} + p^2} + \sqrt{m^2_{\pi} + p^2} \tag{5.9}$$

so that for a fixed $M_{\Lambda\pi}$, the momentum is fixed.

If the resonance was sharp the points on the Dalitz plot should fall along two straight lines as in Fig. 5.3. The distribution of the points along these lines gives the width of the resonance. For a broad resonance the kinetic energy T_π will have a spread in values, resulting in a band rather than a line on the Dalitz plot. Since either π^+ or π^- can resonate with the Λ-hyperon, both vertical and horizontal bands are formed. This resonance must have isospin 1 since it decays by strong interactions for which isospin is conserved.

In the Dalitz plot it is usual to display $M^2_{\Lambda\pi^-}$ instead of T_{π^+} along the x-axis and $M^2_{\Lambda\pi^+}$ along the y-axis so that the $\Lambda\pi$ invariant mass can be read off directly. From (5.9) one can easily obtain

$$M^2_{\Lambda\pi^-} = W^2 + m^2_{\pi^+} - 2WE_{\pi^+} = a + bT_{\pi^+} \tag{5.10}$$

where a and b are constants. Figure 5.4 shows the $\Lambda\pi^+$ mass spectrum. The pure phase-space distribution as well as the curve for the assumption that Λ and π may resonate (Breit-Wigner formula) are shown.

5.2 Baryon Resonances

5.2.1 Nucleon Resonances

$\Delta(1232)$ resonance: The experimental observation on pion-proton scattering cross-sections were analysed in great detail in Sect. 4.3.7.5 in terms of isospin amplitudes $a_{3/2}$ and $a_{1/2}$. Referring to Fig. 4.1 it was shown that the peak for π^+p and π^-p scattering is due to the dominance of $I = 3/2$ evidenced by the ratio $\sigma_{\pi^+p}/\sigma_{\pi^-p} = 3$. Figure 4.1 shows total cross-section for a π^+p and π^-p interaction as a function of the incident pion energy. There is a very obvious $I = \frac{3}{2}$ resonance at $T_\pi = 195$ MeV corresponding to a pion-proton mass of 1232 MeV. This

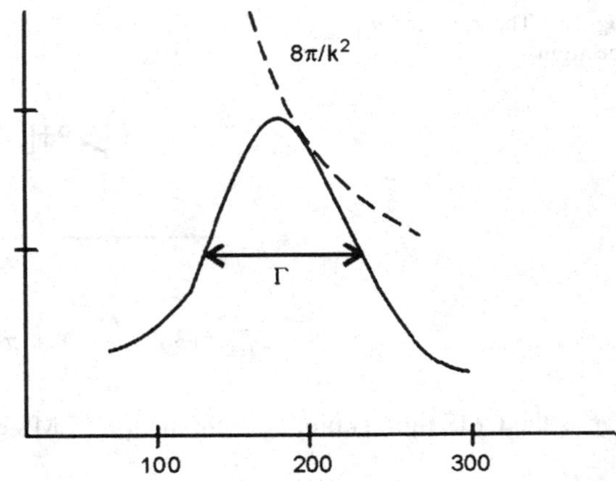

Fig. 5.5 Mass distribution of the resonance

was discovered by Fermi and Anderson in 1949 and is designated as $P_{33}(1232)$ showing that it is a p-wave ($\ell = 1$) pion-nucleon resonance of $I = \frac{3}{2}$ and $J = \frac{3}{2}$. At higher energies other humps and bumps in σ_{total} are observed such as $D_{13}(1520)$, $F_{15}(1688)$ and $F_{37}(1950)$.

5.2.1.1 Mass

The mass m of the Δ resonance is calculated by requiring the total energy E^* in the centre of mass system to be equal to m.

$$m = E^* = \sqrt{(E^2 - p^2)} = \sqrt{(m_\pi + m_p)^2 + 2Tm_p} \tag{5.11}$$

where we have used $E = T + m_\pi + m_p$ and $p^2 = T^2 + 2Tm_\pi$. Inserting $m_\pi = 0.139567$ GeV, $m_p = 0.938279$ and $T_\pi = 0.195$ GeV, we find $m = 1.236$ GeV.

The mass distribution of the resonance has the width $\Gamma = 120$ MeV. Using the uncertainty principle the lifetime is estimated as $\tau \sim \frac{\hbar}{\Gamma} = \frac{105.\times 10^{-21}}{120} \simeq 10^{-23}$ sec, characteristic of strong decay, see Fig. 5.5.

5.2.2 The Spin and Parity

The spin of the resonance can be determined by the study of $\pi^+ p$ differential cross-sections at the resonant energy. Take the incident pion direction along z-axis as the quantization axis. Since the pion is spin less, in the initial state the orbital angular momentum l in units of $\hbar$ must be combined with the proton spin to obtain the total angular momentum J. In the initial state the proton spin can be 'up' or 'down' so that $J = l \pm \frac{1}{2}$ for the state. The angular-momentum wave function for the pion in the p-state will be $\phi(l, m) = \phi(1, 0)$ since the pion is spinless. For the proton, the spin function is $\chi(S, S_z) = \chi(\frac{1}{2}, \pm\frac{1}{2})$ corresponding to the two orientations of the spin. Note that the proton spin may be 'flipped' or may not be flipped in the interaction.

In case of spin flip since J is conserved, a change $\Delta S = \pm 1$ in the z-component of the proton spin implies $\Delta l_z = 1$ as $\Delta j_z = 0$. Choosing the z-direction along the line of flight of the pion or proton in the CMS so that $l_z = 0$ in the initial state, for spin flip $l_z(\text{final}) = 1$. We arbitrarily choose the initial proton spin 'up', $S_z = +\frac{1}{2}$.

This choice is irrelevant since what is significant is flip or no flip. For this choice the initial state will be proportional to $\phi(l, 0)\chi(\frac{1}{2}, +\frac{1}{2})$. For the final state wave function we write χ' and ϕ'.

The final state with $J = \frac{3}{2}$, $J_z = \frac{1}{2}$, is then

$$\psi\left(\frac{3}{2}, \frac{1}{2}\right) = \sqrt{\frac{1}{3}}\phi'(1,1)\chi'\left(\frac{1}{2}, -\frac{1}{2}\right) + \sqrt{\frac{2}{3}}\phi'(1,0)\chi'\left(\frac{1}{2}, \frac{1}{2}\right)$$

where we have used the Clebsch-Gordon coefficient for combining angular momenta 1 and $\frac{1}{2}$. Note that, since the pion is scattered at some angle θ to the z-direction it may have the projected value 1 or 0 on the original axis. The ϕ' are the spherical harmonics

$$\phi'(1,1) = Y_1^1 = -\sqrt{\frac{3}{4\pi}}\sin\theta\frac{e^{i\phi}}{\sqrt{2}}$$

$$\phi'(1,0) = Y_1^0 = \sqrt{\frac{3}{4\pi}}\cos\theta$$

where θ is the polar angle which is identical with the scattering angle, and ϕ is the azimuthal angle. The angular distribution of the scattered pions is then

$$I(\theta) = \psi^*\psi = \frac{1}{3}\left(Y_1^1\right)^2 + \frac{2}{3}\left(Y_1^0\right)^2$$

since the cross-terms are zero as Y_1^1 and Y_1^0 as well as $\chi'(\frac{1}{2}, -\frac{1}{2})$ and $\chi'(\frac{1}{2}, \frac{1}{2})$ are orthonormal.

Hence

$$I(\theta) \propto \sin^2\theta + 4\cos^2\theta = 1 + 3\cos^2\theta \tag{5.12}$$

Figure 5.6 shows the angular distribution of the scattered pions in the CMS $(dN/d\cos\theta)$ for three different values of T_π in the region of resonance. Clearly for the resonant energy ($T_\pi = 195$ MeV) the curve is in excellent fit with the distribution of the form $1 + 3\cos^2\theta$ but deviates considerably for T_π above or below the resonant energy. Thus the Δ resonance occurs for the pure $P_{\frac{3}{2}}$ state. Note that for $P_{\frac{1}{2}}$ state we would have written

$$\psi\left(\frac{1}{2}, \frac{1}{2}\right) = \sqrt{\frac{2}{3}}\phi'(1,1)\chi'\left(\frac{1}{2}, -\frac{1}{2}\right) - \sqrt{\frac{1}{3}}\phi'(1,0)\chi'\left(\frac{1}{2}, \frac{1}{2}\right) \tag{5.13}$$

and $I(\theta) \propto \sin^2\theta + \cos^2\theta =$ constant, i.e. isotropic.

If $l = 0$ (s-wave scattering) so that the intermediate state (resonance) has spin $\frac{1}{2}$, then the spherical harmonic $Y_0^0 = \frac{1}{\sqrt{4\pi}}$ would come in the calculations and once again the angular distribution would be isotropic.

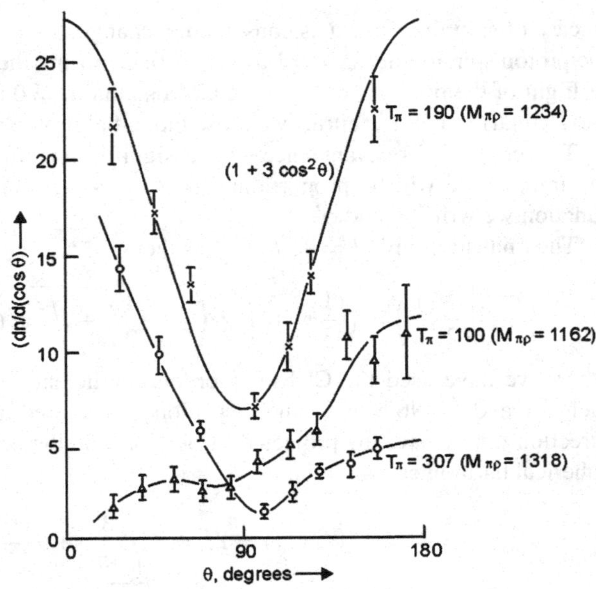

Fig. 5.6 The angular distribution of the scattered pion, in $\pi^+ p$ elastic scattering, as measured in the center-of-mass frame. In the region of the Δ-resonance of mass 1236 MeV ($T_\pi = 190$ MeV), the distribution has the form $1 + 3\cos^2\theta$, as in (5.12) [4]

5.2.3 Resonance Spin from Total Cross-Section

The partial-wave analysis leads to an expression for the total elastic cross-section (AAK 1, 5)

$$\sigma_T = \frac{2\pi}{k^2}\sum_l (2j+1)\sin^2\delta_l \tag{5.14}$$

where $\hbar k$ is the momentum of the incoming particle, $J\hbar$ is the total angular momentum and δ_l is the 'phase shift' of the wave with orbital angular momentum $l\hbar$. In case the interaction takes place predominantly by a single partial wave then only one term in the summation will be important giving

$$\sigma_T = \frac{2\pi}{k^2}(2j+1)\sin^2\delta_l \tag{5.15}$$

The maximum value of σ_T occurs for $\delta_l = \frac{\pi}{2}$. Thus for $J = (3/2)$ maximum upper limit becomes $(8\pi/k^2)$ for the elastic scattering through an intermediate state of spin $\frac{3}{2}$. The limit $(8\pi/k^2)$ is drawn in Fig. 5.5. It is seen that the total cross-section for elastic scattering does indeed just pass through this limit at the resonance energy.

5.2.4 Parity

Since the Δ resonance state is $P_{\frac{3}{2}}$, the parity is

$$P = P(\pi)P(p)(-1)^l = (-1)(+1)(-1)^1 = +1$$

5.2.5 Isospin

Figure 4.1 shows that there is a strong peak in the total elastic cross section for both π^+ as well as π^- mesons at ~200 MeV with proton; the experimental value for the ratio $(\sigma_{\pi^+p})/(\sigma_{\pi^-p}) = 3$. It was shown (Chap. 4) that this unequivocally establishes that $I = \frac{3}{2}$ dominates over $I = \frac{1}{2}$ amplitude. This bump is referred to as a resonance, the $\Delta(1236)$, 1236 MeV being the invariant pion-nucleon mass. It is also referred to as (3, 3) resonance because the spin-parity $J^P = \frac{3}{2}^+$ and $I = \frac{3}{2}$. The isospin $I = \frac{3}{2}$ implies that there are four charge states, Δ^{++}, Δ^+, Δ^0 and Δ^-. These states are found both in formation experiments as well as production experiments. The masses of Δ's are not exactly equal. They range from 1232–1236 MeV, the central mass being 1232 MeV.

The resonances at higher energy occur as for 1525 MeV, $I = \frac{1}{2}$. In general the existence of higher mass resonances is not unambiguous an account of the presence of tails of those of lower mass. For such resonances the angular distributions are analysed to obtain amplitudes of partial waves by means of phase-shift analyses. The $I = \frac{1}{2}$ resonances are called N^*'s and $I = \frac{3}{2}$ nucleon resonances are called Δ's. So far resonances with $I > \frac{3}{2}$ are not found. Δ's have been identified with spins up to $\frac{11}{2}$ and masses up to 3230 GeV/c^2. The simplest experiments for the determination of total cross-section involve attenuation of pions in traversing liquid hydrogen.

5.3 Hyperon Resonances

The study of hyperon resonances have been carried out mainly by the K^- beams because of copious production of hyperons in K^- interactions than for any other incident particles. Both production and formation experiments have been done with hydrogen bubble chambers. Compared to nucleon resonances, the study of hyperon resonances is fraught with difficulties. First separated K^- beams are much less intense than the pion beams, second a large number of photographs are required for meaningful statistics.

The first resonance $Y^*(1385)$ is produced copiously in the K^-p interactions over a wide range of momenta. The simplest process in which this resonance is detected is

$$K^- + p \rightarrow \Lambda + \pi^+ + \pi^- \tag{5.16}$$

which appears in a bubble chamber as a two prong event and a neutral Vee if Λ decays via $P\pi^-$. A similar type of event from $K^-p \rightarrow K^0 p\pi^-$ is usually distinguished from the kinematics of production and decay combined with the track density. A Dalitz plot for reaction (5.16) with $M^2(\Lambda\pi^+)$ against $M^2(\Lambda\pi^-)$ shows two bands at masses of 1382 MeV/c^2 corresponding to formation of $Y_1^*(1385)$ (this resonance is known from the old mass measurements at 1385 MeV/c^2).

The subscript denotes that the isospin of the resonance is $I = 1$. This is because $Y^*(1385)$ which is also known as $\Sigma(1385)$, decays into $\Lambda\pi$. Now for Λ, $I = 0$

and for π, $I = 1$. From isospin conservation, for $\Sigma(1385)$, $I = 1$. The three charge states and decay modes are

$$Y^{*+} \to \Lambda + \pi^+, \qquad Y^{*0} \to \Lambda\pi^0, \qquad Y^{*-} \to \Lambda\pi^-$$

The spin of the $Y^{*\pm}(1385)$ has been measured to be $\frac{3}{2}$ from the angular distribution of decay products.

The parity of $Y_1^*(1385)$ is $+1$. There are also higher mass resonances such as $\Sigma(1660)$ etc.

5.3.1 Cascade Hyperon Resonance

The lowest mass resonance of cascade hyperon is $\Xi^*(1530)$ [5, 6]. This was found in the production process

$$K^- + p \to \Xi^* + K \quad (\Xi^* \to \Xi + \pi) \tag{5.17}$$

The isospin is determined from the ratio of Ξ^* decays to $\Xi^0\pi^-$, $\Xi^-\pi^0$, $\Xi^-\pi^+$ and $\Xi^0\pi^0$,

$$\frac{\Xi^0\pi^-}{\overline{\Xi}\pi^0} = \frac{\overline{\Xi}\pi^+}{\Xi^0\pi^0} = \begin{cases} 2 & \text{for } I_{\Xi^*} = \frac{1}{2} \\ \frac{1}{2} & \text{for } I_{\Xi^*} = \frac{3}{2} \end{cases} \tag{5.18}$$

The results are in agreement with $I = \frac{1}{2}$. Also the ratio

$$\frac{K^- + p \to \Xi^0 + \pi^- + K^+}{K^- + P \to \Xi^- + \pi^0 + K^+} = 2 \tag{5.19}$$

establishes that for Ξ^*, $I = \frac{1}{2}$. Its spin parity is established as $J^P = \frac{3^+}{2}$ from the decay angular distributions.

An unusual feature of this resonance is its narrow width of 9.1 MeV/c^2. Other resonances such as $\Xi(1820)$ and $\Xi(2030)$ are also found.

5.4 Meson Resonances

5.4.1 Pion Resonances

5.4.1.1 ω^0 Meson

ω^0 was first observed by Maglic et al. [7] in the study of annihilation process

$$p^- + p \to \pi^+ + \pi^+ + \pi^- + \pi^- + \pi^0 \tag{5.20}$$

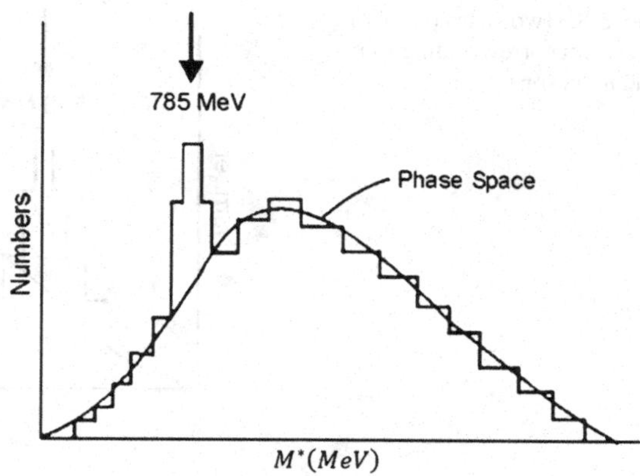

Fig. 5.7 The invariant mass distribution of pion combination $(\pi^+\pi^-\pi^0)$

The invariant mass m^* for all combinations of pions $(\pi^+\pi^-\pi^0)$ in Fig. 5.7 shows shows a very sharp spike at $M^* = 785$ MeV. Corresponding distributions for the combinations $(\pi^+\pi^+\pi^-)$, $(\pi^+\pi^+\pi^0)$, $(\pi^-\pi^-\pi^+)$ etc. did not show any corresponding peak. In such cases the distributions were consistent with that expected from phase-space considerations. It was therefore concluded that the meson is an $I = 0$ state.

The half-width of ω^0 is ≤ 12 MeV M^* (MeV) implying a life time $\tau_{\omega^0} \geq 4 \times 10^{-23}$ sec. From the study of Dalitz plot it was inferred that the spin-parity assignment for ω^0 meson is $J^P = 1^-$. Thus, the ω^0 meson is a 'vector' meson. The G-parity of ω^0 meson is -1, and its decay into 3π is expected by G-parity conservation for strong interaction. The decay modes are

$$\begin{aligned} \omega(785) &\to \pi^+\pi^-\pi^0 \quad 90\,\% \\ &\left.\begin{aligned} &\to \pi^+\pi^- \\ &\to \pi^0\gamma \end{aligned}\right\} \quad 10\,\% \end{aligned} \tag{5.21}$$

5.4.1.2 η-Meson

η-meson was first observed by Pevsner et al. [8] as a peak in the M^* distribution of $\pi^+\pi^-\pi^0$ combination in the interaction

$$\pi^+ + d \to p + p + \pi^+ + \pi^- + \pi^0 \tag{5.22}$$

at incident momentum of 1.23 GeV/c. Instead of the smooth phase-space distribution the spectrum shows two sharp peaks at masses 550 and 780 MeV/c^2, corresponding to η and ω meson respectively, as in Fig. 5.8. The dominant decay modes are

$$\begin{aligned} &\left.\begin{aligned} &\eta \to \gamma\gamma \\ &\eta \to \pi^0\pi^0\pi^0 \end{aligned}\right\} \quad 71\,\% \\ &\eta \to \pi^+\pi^-\pi^0 \quad 24\,\% \\ &\eta \to \pi^+\pi^-\gamma \quad 4.9\,\% \end{aligned} \tag{5.23}$$

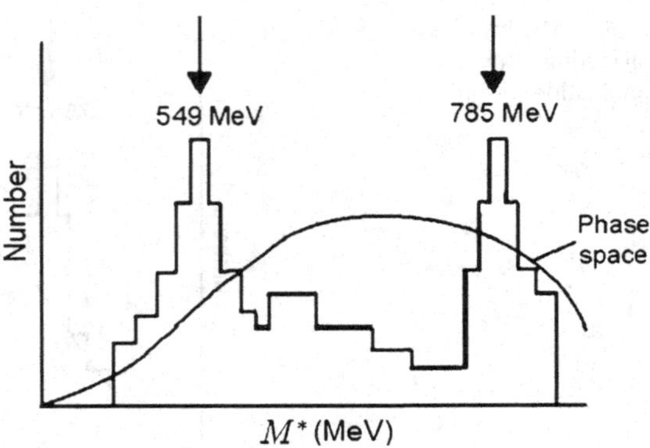

Fig. 5.8 Two sharp peaks of spectrum corresponding to η and ω mesons

No charged counterpart was observed indicating that $I = 0$. The decay to $\gamma\gamma$ limits the spin to 0 or 2. The Dalitz plot gives the spin-parity assignment for η-meson as $J^P = 0^-$. Thus the η-meson is a 'pseudo scalar' meson. The existence of the $\gamma\gamma$-decay mode of η-meson proves that this meson decays by electromagnetic transition with a total width of 0.9 keV only since $I = 0$ and $C(2\gamma) = +1$, it must have $G = +1$. The strong decay into two pions is forbidden by parity conservation, leaving the three-pion G-violating electromagnetic decay as the only possibility.

5.4.1.3 η'-Meson

Its mass is 958 MeV/c^2 with a total width of 0.3 MeV and shares common features with η-meson. It has $G = +1$ and decays electromagnetically into 2γ-rays. The dominant pionic decay mode is $\eta' \to 5\pi$.

5.4.1.4 ρ-Meson

The ρ-meson was first detected from a study of the reactions

$$\pi^- + p \to \pi^- + \pi^0 + p \tag{5.24}$$

and $\pi^- + p \to \pi^- + \pi^+ + n$, for π^- incident momentum of 1.9 GeV/c. The invariant mass m^* of the 2π system is plotted for selected event in which the momentum transfer to the proton is small (<400 MeV/c). The mass distribution shows a marked peak at 765 MeV as shown in Fig. 5.9, obviously deviating from a phase-space distribution. The half-width of the peak is ~75 MeV, the peak occurring for both $(\pi^-\pi^0)$ and $(\pi^+\pi^-)$ pairs so that $I = 0$ is excluded, leaving the option to $I = 1$ or $I = 2$. The ratio $\frac{\pi^-\pi^+n}{\pi^-\pi^0p}$ is expected to be 2/9 for $I = 2$ and 2 for $I = 1$. The observed ratio of 1.8 ± 0.3 favours the value $I = 1$ for the resonance. Since $I = 1$, Bose-Einstein statistics requires the two pions value be in a state of odd parity and therefore $J =$ odd for the ρ-meson. The angular distribution in the CM system relative to the direction of motion of the ρ-meson shows a dominant $\cos^2\theta$ term in the

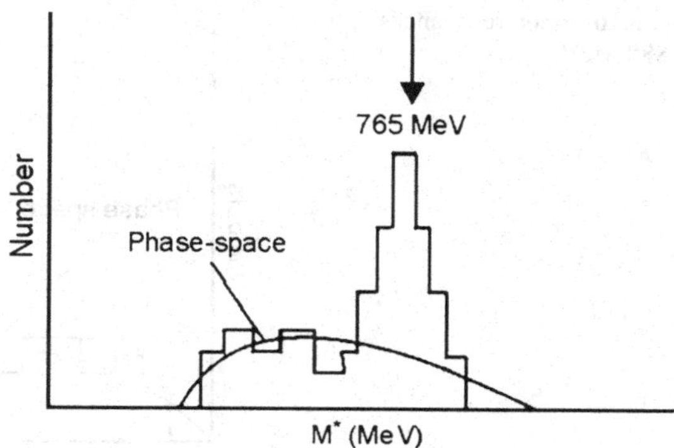

Fig. 5.9 The mass distribution showing a marked peak at 765 MeV for ρ-meson

resonant region. This shows that the two pions are in a p-state of relative motion and that the ρ-meson has spin-parity 1^-. Thus ρ-meson is a 'vector' meson and decays by strong interaction. Its G-parity is $+1$.

Events with small momentum transfer correspond to peripheral collisions which involve large classical impact parameters. Such interactions may be described by virtual exchange of the lightest particle available, pion. The interaction is simply $\pi\pi$ scattering, dominated by a spin-1 resonance near 765 MeV.

5.4.1.5 ϕ-Meson

It has mass 1020 MeV. It is also a non-strange vector meson with $J^P = 1^-$, isospin $I = 0$ and $G = -1$. It decays strongly into K^+K^-, $K_L K_S, \pi^+\pi^-\pi^0$. The kaon mode constitutes 84 per cent while pion mode 15 per cent.

5.4.1.6 f-Meson

It has mass 1270 MeV with $J^P = 2^+$, $I = 0$ and $G = +1$. It strongly decays into 2π.

5.4.2 Kaon Resonances

These are the K–π resonant states or K^*. Alston et al. [9] found a clear evidence for a resonance $(\overline{K}^0\pi^-)$ at about 888 MeV in the study of the reaction

$$\begin{aligned} K^- + P &\rightarrow K^- + \pi^0 + p \\ &\rightarrow \overline{K}^0 + \pi^- + p \end{aligned} \tag{5.25}$$

at K^- incident momentum of 1150 MeV/c, see Fig. 5.10. The width is not very narrow, being 50 MeV. A similar resonance was found in the $(K^-\pi^0)$ system. The isospin of K^* is deduced from the branching ratio $\overline{K}^0\pi^-/K^-\pi^0$ which is expected

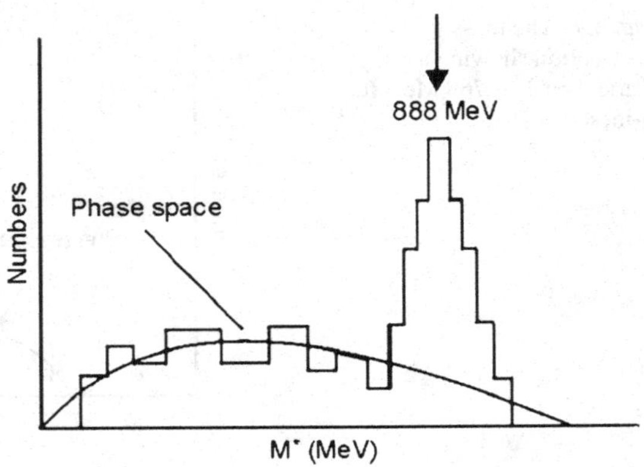

Fig. 5.10 Kaon resonances at 888 MeV

to be 0.5 for $I(\overline{K}^*) = 3/2$ and 2.0 for $I(\overline{K}^*) = \frac{1}{2}$. The observed ratio of 1.4 ± 0.4 favours the $I = \frac{1}{2}$ assignment.

The charge invariance implies the existence of a doublet of resonant states of $(K^0\pi^+)$ and $(K^+\pi^0)$ which have been observed in reactions

$$\begin{aligned} \pi^- + p &\to \Sigma^- + K^0 + \pi^+ \\ &\to \Sigma^- + K^+ + \pi^0 \end{aligned} \tag{5.26}$$

We thus have

$$\begin{aligned} (K^*)^{+}_{0} &\to K^0 + \pi^{+}_{0} \\ &\to K^+ + \pi^{0}_{-} \end{aligned} \tag{5.27}$$

$$\begin{aligned} (K^*)^{0}_{-} &\to K^- + \pi^{+}_{0} \\ &\to \overline{K}^0 + \pi^{0}_{-} \end{aligned} \tag{5.28}$$

The spin-parity assignment for K^* is $J^P = 1^-$.

5.5 Static Quark Model

The concept that the hadrons are elementary particles received a serious jolt with the discovery of resonant states. Familiar hadrons like nucleons and pions are elementary in the sense that none of them can be broken into constituents. Nevertheless, they are not structure less.

5.5.1 *Predecessors to the Quark Model*

In 1949 Fermi and Yang [10] proposed a model in which a pion is considered as a bound state of a nucleon and antinucleon. The model can be generalized to accommodate a large number of other non-strange mesons. The nucleon isospin doublet and the antinucleon doublet are the fundamental building blocks out of which non-strange states can be constructed. In this model n and p constitute the fundamental doublet, $\overline{n}$ and $\overline{p}$ form the second doublet. The meson isospin multiplicities are obtained from the Clebsch-Gordon series of the two doublets. $2 \otimes \overline{2} = 3 \oplus 1$. The model predicts that all non-strange mesons should be either isospin triplets or singlets. This prediction is in agreement with the known facts.

The model can be generalised to include nucleon excited states. Such excited nucleon states are considered as bound states of two nucleons and an antinucleon. The Clebsch-Gordon series being $2 \otimes 2 \otimes \overline{2} = 2 \oplus 2 \oplus 4$. The model predicts that all non-strange baryons should be isospin doublets or quartets. This is also in accord with the known facts. Thus the nucleon and antinucleon doublets are the building blocks of all non-strange hadrons.

Inspite of the merits this model is inadequate to account for strange particles. Sakata [11] in an attempt to overcome this shortcoming introduced Λ-hyperon as an additional state to the fundamental building blocks. In Sakata's model all hadrons can be constructed from the triplet p, n, Λ and their antiparticles.

Ikeda et al. regarded p, n and Λ to be a fundamental triplet of $U(3)$. As in Fermi-yang model, the meson is considered as the bound state of a baryon and an antibaryon.

The Clebsch-Gordan series for this scheme is

$$3 \times \overline{3} = 8 \oplus 1$$

Thus, the Sakata model leads to the successful prediction that mesons should occur in octets and singlets. Furthermore, the octet and singlet are expected not to differ appreciably in mass. Under this assumption mesons are expected to occur in nonets in agreement with the known fact.

Although Sakata model successfully explained the meson multiplicities, there was no way to reconcile with baryon multiplicities. For example, according to this model Σ-hyperon is composed of $N\overline{N}\Lambda$ (in particular $\Sigma^+ = p\overline{n}\Lambda$) in mutual s-state so that $J^P = \frac{1}{2}^-$ and not $\frac{1}{2}^+$ as is experimentally found.

5.5.2 The Quark Model

The breakthrough came with the work of Gell-Mann and Zweig [12] who independently proposed a model in which baryons and mesons are composites of a fundamental triplet of $U(3)$. Murray Gell-Mann named these particles "Quarks". It is believed that the complexity of hadronic phenomena results from the fact that hadrons, like atoms or nuclei are composite structures built up from structureless objects called quarks. In the beginning it was thought that quarks are merely mathematical objects to aid the classifications of hadrons. It must be stressed that $SU(3)$ classifications of supermultiplets does not depend on the actual existence of quarks, but, later it was found that there is an overwhelming evidence for the quark model as it explains virtually all of hadronic physics, hadronic spectroscopy, weak and

Table 5.1 Quantum number for light quarks

Flavour	B	J	I	I_3	S	Q/e
u	$\frac{1}{3}$	$\frac{1}{2}$	$\frac{1}{2}$	$+\frac{1}{2}$	0	$+\frac{2}{3}$
d	$\frac{1}{3}$	$\frac{1}{2}$	$\frac{1}{2}$	$-\frac{1}{2}$	0	$-\frac{1}{3}$
s	$\frac{1}{3}$	$\frac{1}{2}$	0	0	-1	$-\frac{1}{3}$

electromagnetic decays, collisions with electrons or neutrinos, hadron production in e^+e^- annihilation, prediction of Ω^-, magnetic moments of baryons and relations in cross-sections of hadrons belonging to different supermultiplets etc.

In the quark model, quarks denoted by q are Fermions, the most economical value for spin being $J = \frac{1}{2}$.

The baryons are bound states of three quarks written as qqq. The baryon spectrum corresponds to various quantum states of the qqq system. Since the baryon number B is additive, all the quarks carry baryon number $B = \frac{1}{3}$. The antiquarks $\overline{q}$ with $B = -\frac{1}{3}$ are distinct from quarks.

The meson is composed of a quark and an antiquark, $q\overline{q}$ and therefore has $B = 0$. The meson spectrum corresponds to different quantum states of the $q\overline{q}$ system. As its constituents can annihilate the meson states are unstable against decay.

These quarks consist of non-strange doublet, labeled u (for up with $I_3 = +\frac{1}{2}$) and d (for down with $I_3 = -\frac{1}{2}$), and an $S = -1$ isosinglet, labeled S (for strange). The assignments u, d, s which distinguish the types of quark are called the flavour of the quark.

The u and d quarks are approximately degenerate in mass and constitute an $SU(2)$ isospin doublet. The u and d quarks can therefore, account for the additive quantum number I_3 in hadrons. The third quark s enlarges the group from $SU(2)$ to $SU(3)$ and carries the additive quantum number S. These three quarks are called light quarks.

From the relation for strangeness

$$Q/e = \frac{1}{2}(B + S) + I_3 \tag{5.29}$$

it is obvious that the charge of the quarks is also fractional. In Table 5.1 are listed the properties of three flavours of quarks which the model needs for the description of commonly found hadrons.

The charge of the u quark is $+\frac{2}{3}$ and that of d quark is $-\frac{1}{3}$. For antiquarks the B, I_3, S and Q/e values are reversed.

Using the relation (5.29) we obtain the quark charges

$$\frac{Q_u}{e} = \frac{1}{2} \cdot \frac{1}{3} + \frac{1}{2} = \frac{2}{3}$$

$$\frac{Q_d}{e} = \frac{1}{2} \cdot \frac{1}{3} - \frac{1}{2} = -\frac{1}{3}$$

$$\frac{Q_s}{e} = \frac{1}{2}\left(\frac{1}{3} - 1\right) + 0 = -\frac{1}{3}$$

Thus quarks have fractional charges, whereas by construction all baryons q_i, q_j, q_k and all mesons $q_i\overline{q_j}$ have integer eigen values of Q/e. Since only the quarks u and d carry isospin, the isospin properties of hadrons must come from these quarks. Similarly, the strange particles must contain the s quark as one of its constituents. Thus, the proton is built of two u quarks and one d quark $p = uud$. The neutron is built from two d quarks and one u quark.

According to the non-relativistic quark model, quarks have zero orbital angular momenta within nucleons. The total spin of two u quarks in the proton is unity.

When this is added vectorially to the spin of the d quark we get for proton a spin of $1/2$ which is the required value. Similarly, the neutron is constructed by interchanging u and d quarks.

The third components I_3 is also correctly given, $I_3(p) = \frac{1}{2} + \frac{1}{2} - \frac{1}{2} = +\frac{1}{2}$, so also for the neutron, $I_3(n) = -\frac{1}{2}$.

A whole lot of other hadrons can be constructed as though they were toy cubes. Thus the quartet of Δ baryons are obtained from three quarks if their spins are parallel to give $J = \frac{3}{2}$.

$$\Delta^{++} = uuu, \qquad \Delta^{+} = uud, \qquad \Delta^{0} = udd, \qquad \Delta^{-} = ddd \tag{5.30}$$

The three quarks that determine the quantum numbers of the nucleons are called valence quarks. Apart from these so-called sea quarks, virtual quark-antiquark pairs, also exist in the nucleon. Their effective quantum numbers average out to zero and do not alter those of the nucleon.

5.5.2.1 Mass of Quarks

Free quarks have not been discovered. Quarks in hadrons are assumed to be confined. This leads to a difficulty in estimating the mass of the quark. The mass of a particle can be precisely defined by its energy only when it is a free particle. Since isolated quarks are not observed it is difficult to talk about the precise value of its mass. At the best one can refer to the 'effective mass' m_{eff} rather than the true mass m. The value of m_{eff} would depend upon the interaction of one quark with another quark. The situation is analogous to the phenomenon in condensed matter physics where a moving electron in a solid behaves as if it had an 'effective mass' that can be quite different from the true mass.

It is more meaningful to refer to the mass difference between quarks of different flavours. The mass difference between u and d quarks is expected to be small as they form an isospin doublet and the splitting is expected to be small. The splitting is expected to be of the order of a couple of MeV, due to electromagnetic interaction. The mass of the s-quark is believed to be higher than that of u or d quark by about 150 MeV.

We can estimate the approximate mass difference between the d quark and u quark as follows.

$$m_n - n_p = m_{ddu} - m_{uud} = m_d - m_u = +1.3 \text{ MeV}$$

Similarly, $m_{\Sigma^-} - m_{\Sigma^+} = +8.0$ MeV, $m_{\Xi^-} - m_{\Xi^0} = +6.4$ MeV. This shows that $m_d > m_u$.

Now the coulomb energy difference associated with the electrical energy between pairs of quarks is of the order of $\frac{e^2}{4\pi\varepsilon_0 R_0} = \frac{e^2}{4\pi\varepsilon_0 \hbar c} \times \frac{\hbar c}{R_0}$, where R_0 is the size of a baryon. With $\hbar c = 197$ MeV fm, $R_0 \cong 0.8$ fm, $\frac{e^2}{4\pi\varepsilon_0 \hbar c} = \frac{1}{137}$, we find $\frac{e^2}{4\pi\varepsilon_0 R_0} \cong$ 1.4 MeV.

Thus, $m_d - m_u \cong 2$ MeV. At the present the origin of quark masses and the near equality of u and d quark masses is not understood.

5.6 Supermultiplets

Various multiplets are described as the irreducible representations (I.R) of $SU(3)$ group, i.e. with 3 degrees of freedom corresponding to the three flavours of quarks, u, d, s under transformation in "unitary spin space".

Supermultiplets have much greater symmetry than in isospin. Multiplets, having the same J^P values are grouped under the same supermultiplets. The members of each isospin multiplets are plotted with Y (or s) versus I_3 values, Fig. 5.11. In a given supermultiplet each isospin multiplet has essentially the same central mass, differing only by a few MeV, characteristic of E.M. mass splittings in isospin multiplets. The states of different strangeness or hyper-charge (Y) differ considerably in mass. In the case of Baryon decuplet the mass difference for each increment of strangeness is roughly the same. In order to explain the regularities in the decuplet three types of Fermion constituents in a Baryon, called Quarks were postulated. Each has quantum numbers shown in Table 5.1.

The inequality of masses of the decuplet members indicates that the unitary symmetry is in fact badly broken.

By combining 3 quarks, we obtain baryons distributed in a singlet, two octets and a decuplet according to

$$3 \otimes 3 \otimes 3 = 1 \oplus 8 \oplus 8 \oplus 10 \tag{5.31}$$

This decomposition is similar to the decomposition $2 \otimes 2 = 3 \oplus 1$ used for the combination of two nucleons or $3 \otimes 2 = 4 \oplus 2$ for the πN scattering.

The baryon octets and the decuplet are shown in Fig. 5.11 with S versus I_3. The members in brackets are the mean masses (in MeV/c^2) of the isospin multiplets. Such diagrams are called weight diagrams. Here, with the change of notation we have written I for T.

5.6.1 Baryon Decuplet

First consider the baryon decuplet, Fig. 5.11(c) whose quark structure is given in Fig. 5.12. The diagram shows evidence for a repeated simple structure of three particles u, d and s. These three particles are the three basic quarks from which the hadrons are composed.

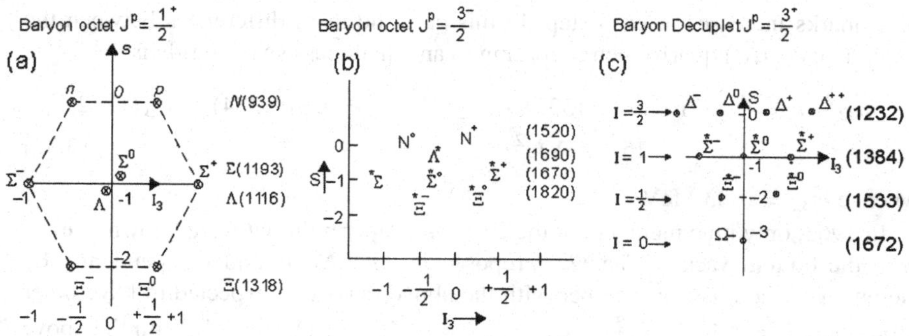

Fig. 5.11 The baryon octets and the decuplet

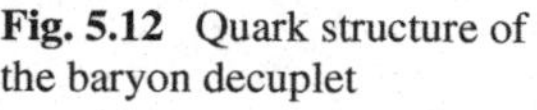
Fig. 5.12 Quark structure of the baryon decuplet

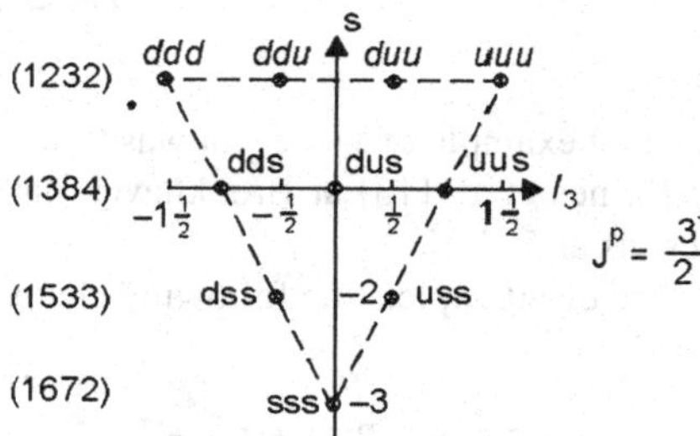

Figure 5.12 shows the baryon states of lowest mass and of spin-parity $J^P = \frac{3}{2}^+$, where we plot the strangeness S against the third component of isospin I_3, for each of the ten members. The first row consists of an $S = 0$, $I = \frac{3}{2}$ isospin quadruplet, the $\Delta 1232$ existing in the charge substates Δ^{++}, Δ^{+}, Δ^{0}, Δ^{-}. The number 1232 in parenthesis indicates the central resonance mass in MeV. Next row corresponds to $I = 1$ isospin triplet of $S = -1$, the $\Sigma^*(1384)$, resonance followed by an $S = -2$, $I = \frac{1}{2}$ isospin doublet, the $\Xi^*(1533)$ and finally an $I = 0$ singlet of $S = -3$, the $\Omega^-(1672)$ which is a particle and not a resonance. The members of each isospin multiplet essentially have the same central mass differing by a few MeV characteristic of electromagnetic mass splitting.

5.6.2 The Ω^--Hyperon

The symmetry breaking of $SU(3)$ in a given super-multiplet is due to large difference in the effective masses of s quark and (u, d) quark. We note that no space on the diagram is occupied by more than one particle, so that possible 'mixing' problems are avoided. Also as we move down from one I-spin multiplet to another the number of s-quarks increases at each step. In this case the mass differences between the s-quark and (u, d) quarks then according to an equal mass spacing rule is

$$m_{\Omega} - m_{\Xi^*}(1533) = m_{\Sigma^*}(1533) - m_{\Sigma^*}(1384) = m_{\Sigma^*}(1384) - m_{\Delta}(1232)$$
$$= 150\ \text{MeV}/c^2 \tag{5.32}$$

whence $m_{\Omega} = 1683\ \text{MeV}/c^2$

In 1962, only nine members of the $J^P = \frac{3}{2}^+$ supermultiplet were known to exist with the bottom vacant. The Ω^- proposed by Gell-Mann and independently by Ne'eman as the missing member with the above mass was expected to have other properties, $B = +1$, $J^P = \frac{3}{2}^+$, $S = -3$, $I = 0$, $I_3 = 0$, $Q/e = -1$. With the above properties the only $\Delta S = 1$ decays which are kinematically allowed are

$$\Omega^- \rightarrow \begin{cases} \Lambda K^- \\ \Xi\pi^0 \\ \Xi^0\pi^- \end{cases} \tag{5.33}$$

The first example of Ω^- event was found in a hydrogen bubble chamber photograph by Barnes et al. [13] at Brookhaven national laboratory in the interactions of K^- mesons at 5 GeV/c.

The event depicts the following chain of events

$K^- + p \rightarrow \Omega^- + K^+ + K^0$

$\quad\hookrightarrow \Xi^0 + \pi^-$ **($\Delta S = 1$ weak decay)**

$\qquad\hookrightarrow \pi^0 + \Lambda$ **($\Delta S = 1$ weak decay)**

$\qquad\qquad\hookrightarrow \pi^- + p$ **($\Delta S = 1$ weak decay)** (from Λ)

$\qquad\qquad\hookrightarrow \gamma + \gamma$ **(em decay)** (from π^0)

$\qquad\qquad\qquad\downarrow \quad \downarrow$

$\qquad\qquad\qquad e^+e^- \quad e^+e^-$

The K^0 is not observed to decay in the chamber but the event could be fully analysed because both γ-rays from the π^0 produced electron pairs in the bubble chamber. The line diagram is given in Fig. 5.13 (Courtesy Brookhaven National Laboratory).

Since then a number of examples of Ω^- particles decaying into each of the three mode of decays have been recorded and relativistic beams of strange baryons including Ω^- has been developed at CERN. The best values of the mass and lifetime are

$$M_{\Omega^-} = 1672.45 \pm 0.32\ \text{MeV}/c^2;$$
$$\tau_{\Omega^-} = (0.819 \pm 0.027) \times 10^{-10}\ \text{sec}$$

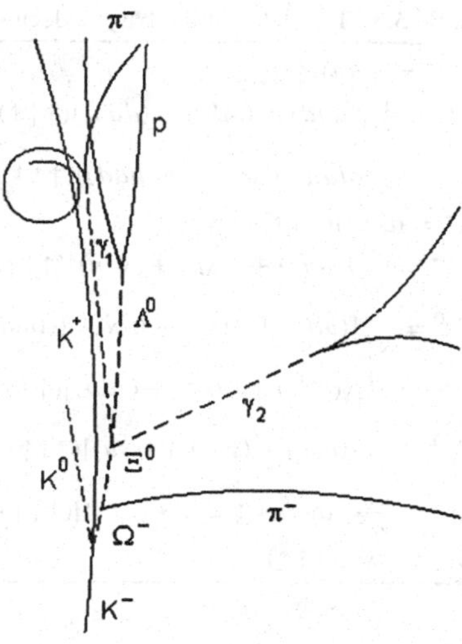

Fig. 5.13 The line diagram of the analysis of γ-rays from the π^0 producing electron pairs in the bubble chamber

Table 5.2 Principal decay modes of members of the baryon decuplet

Symbol	Mass (MeV)	Full width (MeV) or mean life time	Principal decay modes
Δ^{++}	1231	120 MeV	$N\pi$
Δ^{+}	1231		
Δ^{0}	1232		
Δ^{-}	1239		
Σ^{*+}	1382	35 MeV	$\Lambda\pi$ (90 %)
Σ^{*0}	1381	39 MeV	$\Xi\pi$ (10 %)
Σ^{*-}	1386	42 MeV	
Ξ^{*0}	1532	9 MeV	$\Sigma\pi$ (100 %)
Ξ^{*-}	1535	10 MeV	
Ω^{-}	1672	0.82×10^{-10}	$\Xi^0\pi^-$ $\Xi^-\pi^0$ ΛK^-

The expected spin of $\frac{3}{2}$ has been confirmed from measurements of angular distribution of decay products. Thus the discovery of Ω^- three years after its prediction has been a singular triumph of $SU(3)$ symmetry.

The mass full width at half maximum and principal decay modes of the members of the decuplet are shown in Table 5.2.

Table 5.3 Ten states of the baryon decuplet

$\Delta^{++} = (uuu)(\uparrow\uparrow\uparrow)$
$\Delta^{+} = \frac{1}{\sqrt{3}}[(uud) + (udu) + (duu)](\uparrow\uparrow\uparrow)$
$\Delta^{0} = \frac{1}{\sqrt{3}}[(ddu) + (dud) + (udd)](\uparrow\uparrow\uparrow)$
$\Delta^{-} = (ddd)(\uparrow\uparrow\uparrow)$
$\Sigma^{*+} = \frac{1}{\sqrt{3}}[(uus) + (usu) + (suu)](\uparrow\uparrow\uparrow)$
$\Sigma^{*0} = \frac{1}{\sqrt{6}}[(uds) + (usd) + (dus) + (dsu) + (sud) + (sdu)](\uparrow\uparrow\uparrow)$
$\Sigma^{*-} = \frac{1}{\sqrt{3}}[(dds) + (dsd) + (sdd)](\uparrow\uparrow\uparrow)$
$\Xi^{*0} = \frac{1}{\sqrt{3}}[(uss) + (sus) + (ssu)](\uparrow\uparrow\uparrow)$
$\Xi^{*-} = \frac{1}{\sqrt{3}}[(dss) + (sds) + (ssd)](\uparrow\uparrow\uparrow)$
$\Omega^{-} = (sss)(\uparrow\uparrow\uparrow)$

5.6.3 States of the Decuplet

For all the members of the decuplet, the quark spin $\frac{1}{2}$ must all be aligned in the same direction to give $J = \frac{3}{2}$ as all of them are assumed to be in the orbital state $l = 0$. As such states are symmetric in space, the symmetry character of u, d and s requires their states to be symmetric under interchange of flavour and spin of any quark pair. The particles at the three corners (uuu), (ddd), (sss) are already symmetrical in spin and flavor. It is natural to require the same symmetry for the other states. The state Δ^{++} may be written as

$$\Delta^{++} = \psi(u\uparrow u\uparrow u\uparrow) \quad \text{or} \quad (uuu)(\uparrow\uparrow\uparrow)$$

Different states in the multiplet can be obtained from each other horizontally, by applying the isospin ladder shift operators successively. Thus,

$$\begin{aligned} I^{-}(uuu) &= \left[I^{-}(u)\right](uu) + u\left[I^{-}(u)\right](u) + uu\left[I^{-}(u)\right] \\ &= duu + udu + uud \end{aligned}$$

Using the normalization factor

$$\Delta^{+} = \frac{1}{\sqrt{3}}(duu + udu + uud)(\uparrow\uparrow\uparrow)$$

In the given multiplet the states with negative I_3 are obtained by exchanging d with u.

Also, successive rows are constructed by replacing u-quarks with s quarks.

We summarized ten states of the decuplet in full in Table 5.3.

5.6.4 Baryon Octet ($J^P = \frac{1^+}{2}$)

In the decomposition $3 \otimes 3 \otimes 3 = 1 \oplus 10 \oplus 8 \oplus 8$, ten is already accounted for. Of the remaining combinations one is completely antisymmetric and is of the form

$$\frac{1}{\sqrt{6}}(uds + dus + sud - usd - dsu - sdu)$$

The remaining 16 consist of two octets of mixed symmetry.

Another possibility for symmetric three-quark combinations is to consider members with two spins parallel and one antiparallel to give $J = \frac{1}{2}, \uparrow\uparrow\downarrow$. This combination is not symmetric in the spins and therefore cannot be symmetric in flavour either. The asymmetry in spin must be compensated by an asymmetry in flavour if the state is to be completely symmetric.

First we examine the states with non-strange quarks u and d. Starting with proton (uud), put two quarks in a spin singlet state

$$\frac{1}{\sqrt{2}}(\uparrow\downarrow - \downarrow\uparrow)$$

which is antisymmetric. In order to make the overall state symmetric we need a flavour antisymmetric combination of u and d which is a singlet

$$\frac{1}{\sqrt{2}}(ud - du)$$

We then add the third quark u, with spin up to obtain,

$$(u\uparrow d\downarrow - u\downarrow d\uparrow - d\uparrow u\downarrow + d\downarrow u\uparrow)u\uparrow$$

Although the expression in brackets is symmetric under the interchange of the first and second quark (flavour and spin), the whole expression has to be symmetrized by making permutations to give the final result

$$\psi(p) = \frac{1}{\sqrt{18}}[2(u\uparrow u\uparrow d\downarrow) + 2(d\downarrow u\uparrow u\uparrow) + 2u\uparrow d\downarrow u\uparrow - u\downarrow d\uparrow u\uparrow$$
$$- u\uparrow u\downarrow d\uparrow - u\downarrow u\uparrow d\uparrow - d\uparrow u\downarrow u\uparrow - u\uparrow d\uparrow u\downarrow - d\uparrow u\uparrow u\downarrow]$$

The neutron ($I_3 = -\frac{1}{2}$) is obtained by exchanging u and d. We get corresponding functions for the other states of the $J = \frac{1}{2}$ octet in a similar fashion. There are altogether eight states which constitute the $J = \frac{1}{2}$ baryon octet.

This octet is shown on an (I_3S) diagram, Fig. 5.11(a). But in contrast to the mesons, no member of a baryon multiplet is the antiparticle of another member.

5.6.4.1 Spins in Octet

As in the case of decuplet, the mass increases with the number of s-quarks, but unlike the decuplet the splittings are not equal to each other. Combinations of u, d and s-quarks of Baryon Octet spins are shown in Table 5.4.

Table 5.4 Baryon octet: combinations of u, d and s-quarks

$\psi(u\uparrow u\uparrow d\downarrow) = N^+ = p$
$\psi(d\uparrow d\uparrow u\downarrow) = N^0 = n$
$\psi(u\downarrow s\uparrow s\uparrow) = \Xi^0$
$\psi(d\downarrow s\uparrow s\uparrow) = \Xi^-$
$\psi(u\uparrow u\uparrow s\downarrow) = \Sigma^+$
$\psi(u\uparrow d\uparrow s\downarrow) = \Sigma^0$
$\psi(u\uparrow d\downarrow s\uparrow) = \Lambda$
$\psi(d\uparrow d\uparrow s\downarrow) = \Sigma^-$

5.6.4.2 Masses of Isospin Multiplets

If we assume that the difference in multiplets arises due to the strange quark content as for baryon decuplet then this hypothesis gives the following results.

For the baryon octet, Fig. 5.11(a)

$$M_\Sigma = M_\Lambda$$
$$1193\text{ MeV} \quad 1116\text{ MeV}$$
$$M_\Lambda - M_N = M_\Xi - M_\Lambda$$
$$177\text{ MeV} \quad 203\text{ MeV}$$

For the Decuplet

$$\Sigma^*(1385) - \Delta(1232) = \Xi^*(1530) - \Sigma^*(1385) = \Omega^-(1672) - \Xi^*(1530)$$
$$153\text{ MeV} \qquad 145\text{ MeV} \qquad 142\text{ MeV}$$

For the octet the agreement is far from satisfactory.

5.6.5 Gell-Mann-Okubo Mass Formula

Making certain assumptions about the nature of the breaking of unitary symmetry, Gell-Mann and Okubo gave the mass formula for baryon supermultiplets

$$M = M_0 + M_1 Y + M_2\left[I(I+1) - \frac{Y^2}{4}\right] \tag{5.34}$$

where $Y = B + S$ is the hypercharge, I is the isospin value of the isospin multiplet, M_0, M_1 and M_2 are constants.

In the decuplet the states are related by

$$Y = B + S = 2(I - 1)$$

so that

$$M = A + BY$$

We therefore get linear S-dependence of mass.

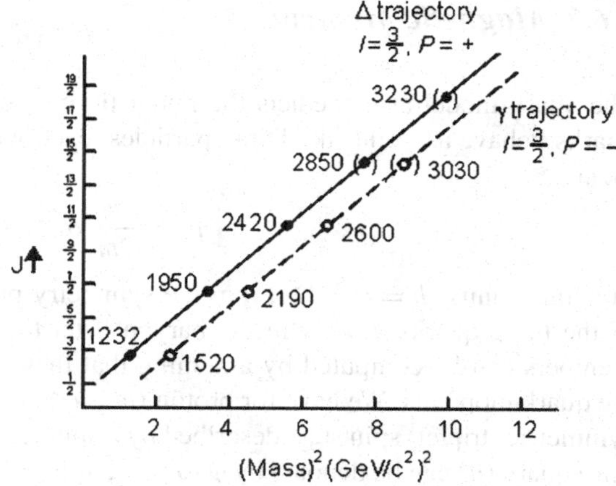

Fig. 5.14 The trajectories for the plots of mass-squared as a function of spin J for some of the non-strange baryon resonances

For the octet we find the relation

$$\begin{array}{cc} 3M_\Lambda + M_\Sigma = & 2(M_\Xi + M_N) \\ 4541\ \text{MeV} & 4514\ \text{MeV} \end{array} \tag{5.35}$$

which is correct to 1 per cent. The physical reasons on which the formula is based are not clear.

Summing up, the low-lying baryons fall elegantly into two multiplets, the $J = \frac{1}{2}$ octet and $J = \frac{3}{2}$ decuplet. These are in accordance with the expectations of symmetric space spin-flavour state in the quark model.

More baryons in the higher excited states can be classified as states for which the quarks have non-zero orbital angular momentum. Among these excited states there are also rotational "band" which reveal the same $m^2 - J$ relationship as the rotational "bands" in meson spectroscopy.

5.6.6 Regge Trajectories

In $SU(3)$ particles of the same spin parity and baryon number are grouped together. A more general relationship between multiplets with different spins is provided by the so-called 'Regge trajectories'. The trajectory consists of spin-mass^2 plot.

Figure 5.14 shows the trajectories for the plots of mass-squared as a function of spin J for some of the non-strange baryon resonances. There is a remarkable straight line fitbracketed states correspond to unmeasured spin values. The linearity for the Regge trajectories is not satisfactorily explained.

5.6.7 Magnetic Moments

The quark model also predicts the magnetic moments of baryons. Assuming that quarks behave as point like Dirac particles, each will have a magnetic dipole moment

$$\mu_i = \frac{e_i}{2m_i}\sigma_i \tag{5.36}$$

in natural units ($\hbar = c = 1$). From the symmetry properties of the wave functions of the three quarks u, d, s in the baryon octet the magnetic moments of various members can be computed by assuming that they are equal to the vector sums of the quark moments. We have for proton ($u \uparrow u \uparrow d \downarrow$), the two u-quarks will be in a symmetric (triplet) spin state described by a spin function χ ($J = 1$; $m = 0 \pm 1$), the third quark (d) can be denoted by ϕ ($J = \frac{1}{2}, m = \pm\frac{1}{2}$). The total angular momentum function for a spin-up proton will be ψ ($J = \frac{1}{2}$, $m = \frac{1}{2}$) and the Clebsch-Gordon coefficients for combining $J = 1$ and $J = \frac{1}{2}$ give us

$$\psi\left(\frac{1}{2}, \frac{1}{2}\right) = \sqrt{\frac{2}{3}}\chi(1,1)\phi\left(\frac{1}{2}, -\frac{1}{2}\right) - \sqrt{\frac{1}{3}}\chi(1,0)\phi\left(\frac{1}{2}, \frac{1}{2}\right) \tag{5.37}$$

For the first combination the moment will be $\mu_u + \mu_u - \mu_d$ and for the second just μ_d.

Hence the proton moment is given by

$$\mu_p = \frac{2}{3}(2\mu_u - \mu_d) + \frac{1}{3}\mu_d = \frac{4}{3}\mu_u - \frac{1}{3}\mu_d \tag{5.38}$$

where we have used the charge for μ and d-quark.

The result for neutron can be immediately written down with the labels u and d interchanged

$$\mu_n = \frac{4}{3}\mu_d - \frac{1}{3}\mu_u \tag{5.39}$$

For Σ^+, μ_d is replaced by μ_s in (5.38) and for Σ^-, μ_d by μ_s, and μ_u by μ_d. Λ-hyperon, which is a uds combination with $I = 0$ (i.e. antisymmetric) isospin state must be in antisymmetric spin state ($J = 0$).

Consequently, the u and d in the Λ make no contribution to the moment, and $\mu_\Lambda = \mu_s$. Similarly, values of μ for Σ^0, Ξ^0 and Ξ^- can be worked out.

If we replace $d \to s$ in (5.38) we get μ_{Σ^+}, $u \to s$ in (5.39) we get μ_{Σ^-}, by $u \to s$ in (5.38) we get μ_{Ξ^-}, $d \to s$ in (5.39) we get μ_{Ξ^0}.

Table 5.5 shows the observed and predicted values of baryons using quark masses. The values are given in terms of nuclear magnetons (n.m)

$$\mu_N = e\hbar/2Mc \tag{5.40}$$

where M is the proton mass.

We can express the quark magnetic moments in terms of the magnetic moments of only three baryons.

Table 5.5 Magnetic moments in the quark model

Baryon	Magnetic moment in quark model	Prediction (n.m)	Observed (n.m)
p	$\frac{4}{3}\mu_u - \frac{1}{3}\mu_d$		2.793
n	$\frac{4}{3}\mu_d - \frac{1}{3}\mu_u$	input −1.86	−1.913
Λ	μ_s		−0.614
Σ^+	$\frac{4}{3}\mu_u - \frac{1}{3}\mu_s$	input 2.68	2.33
Σ^0	$\frac{2}{3}(\mu_u + \mu_d) - \frac{\mu_s}{3}$	1.6	1.8
Σ^-	$\frac{4}{3}\mu_d - \frac{1}{3}\mu_s$	−1.05	−1.1
Ξ^0	$\frac{4}{3}\mu_s - \frac{\mu_u}{3}$	−1.4	−1.25
Ξ^-	$\frac{4}{3}\mu_s - \frac{\mu_d}{3}$	−0.47	−1.85

Choosing p, n and Λ, we obtain

$$\mu_u = \left(\frac{\mu_n + 4\mu_p}{5}\right) \tag{5.41}$$

$$\mu_d = \left(\frac{\mu_p + 4\mu_n}{5}\right) \tag{5.42}$$

$$\mu_S = \mu_\Lambda \tag{5.43}$$

Using $\mu_p = 2.79$, $\mu_n = -1.913$ and $\mu_\Lambda = -0.614$, we find $\mu_u = 1.85$ n.m, $\mu_d = -0.97$ n.m, $\mu_s = -0.61$ n.m.

A reasonable approximation is that $m_d \simeq m_u$.

With this approximation $m_n - m_p = m(ddu) - m(uud) \simeq m_d - m_u \sim 1$ MeV

$$\mu_p \propto \frac{2}{3}e \quad \text{and} \quad \mu_n \propto \frac{1}{3}e \quad \text{so that}$$
$$\mu_d = -\frac{1}{2}\mu_u \tag{5.44}$$

using (5.43) in (5.38) and (5.39) we find

$$\mu_p = \frac{3}{2}\mu_u, \qquad \mu_n = -\mu_u \tag{5.45}$$

whence

$$\frac{\mu_n}{\mu_p} = -\frac{2}{3} \tag{5.46}$$

which is in excellent agreement with the experimental value of $-\frac{1.913}{2.793}$ or -0.68.

Further predictions of $SU(3)$ are: $\mu_p = \mu_{\Sigma^+}$; $\mu_{\Sigma^-} = \mu_{\Xi^-}$; $\mu_n = \mu_{\Xi^0}$. The experimental values are 2.79 n.m for μ_p and 2.59 n.m for Σ^+.

The close agreement between the simple model and experiment is considered a triumph for the symmetric quark model with symmetric baryon functions.

For anti-symmetric wave functions there would be a serious disagreement with the experiments.

5.6.8 Mass of Quarks from Magnetic Moments

In analogy with (5.40)

$$\mu_q = \frac{Q_q e\hbar}{2m_q c} \tag{5.47}$$

For the u quark, 1.85 n.m $= \frac{2}{3}(\frac{e\hbar}{2m_p c}) \cdot \frac{m_p}{m_u} = \frac{2}{3}\frac{m_p}{m_u}$ n.m

$$m_u = \frac{2}{3} \times \frac{938}{1.85} = 338 \text{ MeV} \tag{46a}$$

Similarly, we find

$$m_d = 322 \text{ MeV} \quad \text{and} \quad m_s = 510 \text{ MeV} \tag{5.48}$$

These calculations give m_s heavier by about 180 MeV than the mean of m_u and m_d and that m_u and m_d are close to each other. However, one undesirable feature is that $m_u > m_d$ which contradicts the evidences obtained from other sources. This discrepancy is attributed to a small amount of orbital angular momentum in the wave functions as well as our neglect of relativistic effects such as exchange currents.

A crude mass of the u and d quarks can be found from the structure of nucleons.

$$m_u c^2 \simeq m_d c^2 = \frac{1}{3} m_{Nc2} \simeq 310 \text{ MeV} \tag{5.49}$$

As ϕ meson $= s\bar{s}$, the mass

$$m_s c^2 = \frac{1}{2} m_\phi C^2 = \frac{1}{2} \times 1019 \simeq 510 \text{ MeV} \tag{5.50}$$

These calculations are in agreement with those obtained from magnetic moments.

5.6.9 Meson Multiplets

When mesons and their resonances are classified according to the spin and parity then these multiplets which are nonets are found to obey $SU(3)$ symmetry group. This is the pseudo-scalar meson nonet with the quark antiquark spins antiparallel ($J^P = 0^-$), the vector mesons ($J^P = 1^-$) (with the spins parallel) and tensor mesons ($J^P = 2^+$).

A scalar particle has $J^P = 0^+$, a pseudoscalar has $J^P = 0^-$, a vector $J^P = 1^-$, an axial vector $J^P = 1^+$, a tensor particle ($J^P = 2^+$) and a pseudotensor ($J^P = 2^-$).

5.6.9.1 Pseudo-Scalar Mesons ($J^P = 0^-$)

The isospin triplet of pions (π^+, π^0, π^-) and the isospin doublets of K-mesons (K^+, K^0) and ($\overline{K}^0$, K^-) have $J^P = 0^-$. The other members of the nonet are η and

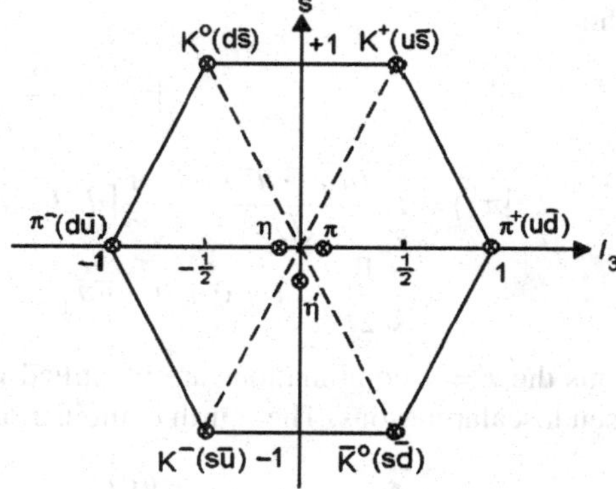

Fig. 5.15 Figure showing other members of the nonet η and η' meson

η' meson, Fig. 5.15. The wave functions for the pion triplet can be generated by using the ladder operators which were used in Chap. 4.

$$I_+\left|\psi^I_{I3}\right\rangle = \sqrt{(I - I_3)(I + I_3 + 1)}\left|\psi^I_{I3+1}\right\rangle \tag{5.51}$$

$$I_-\left|\psi^I_{I3}\right\rangle = \sqrt{(I + I_3)(I - I_3 + 1)}\left|\psi^I_{I3-1}\right\rangle \tag{5.52}$$

Applying (5.51) and (5.52) to single quark states and remembering that

$$I(u) = I(d) = \frac{1}{2}, \qquad I_3(u) = +\frac{1}{2}, \qquad I_3(d) = -\frac{1}{2}$$
$$I(\overline{u}) = I(\overline{d}) = \frac{1}{2}, \qquad I_3(\overline{u}) = -\frac{1}{2}, \qquad I_3(\overline{d}) = +\frac{1}{2}$$
$$I_-|u\rangle = \sqrt{\left(\frac{1}{2}+\frac{1}{2}\right)\left(\frac{1}{2}-\frac{1}{2}+1\right)}|d\rangle = |d\rangle$$

Similarly

$$I_+|u\rangle = I_+|\overline{d}\rangle = 0 \qquad I_-|d\rangle = I_-|\overline{u}\rangle = 0$$
$$I|\overline{d}\rangle = |-\overline{u}\rangle$$

The minus sign preceding $\overline{u}$ is introduced in accordance with the phase convention under C-conjugation.

Further

$$I_-\left|\psi^1_1\right\rangle = \sqrt{2}\left|\psi^1_0\right\rangle; \qquad I\left|\psi^1_{-1}\right\rangle = 0$$
$$I_+\left|\psi^1_0\right\rangle = \sqrt{2}\left|\psi^1_1\right\rangle; \qquad I_+\left|\psi^1_1\right\rangle = 0$$

Applying these results to quark-antiquark combinations we find

$$I_-\left|\pi^+\right\rangle = I_-|u\overline{d}\rangle = uI_-|\overline{d}\rangle + \overline{d}I_-|u\rangle$$
$$= -u\overline{u} + \overline{d}d = \sqrt{2}\left|\pi^0\right\rangle$$

Thus

$$|\pi^0\rangle = \frac{d\overline{d} - u\overline{u}}{\sqrt{2}} \tag{5.53}$$

$$\begin{aligned} I_-|\pi^0\rangle &= I_- \frac{(d\overline{d} - u\overline{u})}{\sqrt{2}} = \frac{-1}{\sqrt{2}}\left[d_- I_-|\overline{d}\rangle + \overline{d} I_-|d\rangle - u I_-|\overline{u}\rangle - \overline{u} I_-|u\rangle\right] \\ &= \frac{-1}{\sqrt{2}}\left[-d\overline{u} + 0 + 0 - \overline{u}d\right] = -\sqrt{2}d\overline{u} = \sqrt{2}|\pi^-\rangle \end{aligned}$$

Thus the $I = 1$ combinations are identified with π^+, π^0 and π^-, the lowest mass pseudoscalar mesons. The fourth combination has the property

$$\begin{aligned} I_+|\eta\rangle &= I_+ \frac{|d\overline{d} + u\overline{u}\rangle}{\sqrt{2}} \\ &= \frac{1}{\sqrt{2}}\left[d I_+|\overline{d}\rangle + \overline{d} I_+|d\rangle + u I_+|\overline{u}\rangle + \overline{u} I_+|u\rangle\right] \\ &= \frac{1}{\sqrt{2}}[0 + \overline{d}u - u\overline{d} + 0] = 0 \end{aligned}$$

Similarly $I_-|\eta\rangle = 0$.

Thus $|\eta\rangle$ is an isospin singlet which does not undergo an isospin transformation into another state. It is orthogonal to $|\pi^0\rangle$ so that $\langle\eta|\pi^0\rangle = 0$. The singlet state is symmetric at the quark level ($u \to d, \overline{u} \to \overline{d}$) while π^+, π^0, π^- states are antisymmetric as they change sign. η', which treats u, d and s symmetrically, is unaffected by $SU(3)$ transformation. It is a singlet under $SU(3)$ in exactly the same sense that the π^0 is a singlet under $SU(2)$ (isospin). On the other hand the η transforms as part of an $SU(3)$ octet whose numbers are the three pions and the four kaons.

The η' meson is also narrow (0.28 MeV/c^2) with mass 958 MeV/c^2 and has decay modes

$$\begin{aligned} \eta' &\to \pi^+\pi^-\eta \\ &\to \pi^+\pi^-\gamma \end{aligned}$$

The quantum numbers are I, $J^{pcG} = 0, 0^{-++}$. Thus the eighth particle of the octet is formed by η, and η' forms an isospin singlet.

The three quarks u, d and s give a total of $3^2 = 9$ which break into an octect and a singlet η is the eighth member of the octet, its state being $(d\overline{d} + u\overline{u} - 2s\overline{s})/\sqrt{6}$. The pseudoscalar nonet comprises the pion isotriplet the K and $\overline{K}$ iso-doublets, and the two iso-singlets η and η'. The eight states of the octet and the symmetric combinations of $q\overline{q}$ for $\eta' = (d\overline{d} + u\overline{u} + s\overline{s})/\sqrt{3}$ transform into one another by the interchange of u, d and s. Observe that η is orthogonal to both η' and π^0. The square roots are written to normalise the states, Table 5.6.

For mesons Gell-Mann-Okuba mass formula (5.34) for baryons is not satisfied. It works better if m^2 is used instead of m.

$$\begin{aligned} 2\left(M^2_{K^0} + M^2_{\overline{K}^0}\right) = \quad & 4M^2_{K^0} \quad = M^2_{\pi^0} + 3M^2_\eta \\ & 0.988\ \text{GeV}^2 \qquad\quad 0.924^2 \end{aligned} \tag{5.54}$$

Table 5.6 Pseudoscalar meson state as $q\overline{q}$ combinations

	Meson	I	I_3	S	Quark combination	Decay	Mass MeV/c^2
Octet	π^+	1	1	0	$u\overline{d}$	$\pi^{\pm} \to \mu\nu$	140
	π^-	1	-1	0	$d\overline{u}$		
	π^0	1	0	0	$\frac{u\overline{u}-d\overline{d}}{\sqrt{2}}$	$\pi^0 \to 2\gamma$	135
	K^+	$\frac{1}{2}$	$\frac{1}{2}$	$+1$	$u\overline{s}$	$K^+ \to \mu\nu$	494
	K^0	$\frac{1}{2}$	$-\frac{1}{2}$	$+1$	$d\overline{s}$	$K^0 \to \pi^+\pi^-$	498
	K^-	$\frac{1}{2}$	$-\frac{1}{2}$	-1	$\overline{u}s$	$K^- \to \mu\nu$	494
	$\overline{K}^0$	$\frac{1}{2}$	$\frac{1}{2}$	-1	$\overline{d}s$	$\overline{K}^0 \to \pi^+\pi^-$	498
	η	0	0	0	$(d\overline{d}+u\overline{u}-2s\overline{s})/\sqrt{6}$	$\eta \to 2\gamma$	549
Singlet	η'	0	0	0	$(d\overline{d}+u\overline{u}-s\overline{s})/\sqrt{3}$	$\eta' \to \eta\pi\pi$ $\to 2\gamma$	958

5.6.9.2 Vector Mesons ($J^P = 1^-$)

The discovery of resonances of K-mesons, ω and ρ meson have already been discussed. No charged ω has been seen, so we expect $I = 0$, $G = -1$.

Therefore ω can decay into three pions, while ρ decays into two pions. $\phi = s\overline{s}$ is one of the eigen states which decays as

$$\phi(1020) \to K^+K^-, K^0\overline{K}^0, \pi^+\pi^-\pi^0$$

and w decays as

$$\omega(783) \to \pi^+\pi^-\pi^0, \pi^+\pi^-\pi^0\gamma$$

The isospin triplet of ρ mesons, the two doublets of K^* and the singlets ω and ϕ constitute the vector mesons nonet, ϕ is the eighth member of the octet.

The vector meson nonets are displayed on S–I_3 plot in Fig. 5.16 and the corresponding quark structure is shown in Fig. 5.17 as well as in Table 5.7.

The diagrams for the two meson nonets have several interesting features. They are hexagons, i.e. have three-fold symmetry, reflecting the equivalence of the three quarks of which the mesons are composed. Isomultiplets lie on the horizontal lines, e.g. the K^+K^0 doublet and $\pi^+\pi^0\pi^-$ triplet. Horizontal displacement by one unit corresponds to substituting a u quark by a d quark (or vice versa). Also, displacement along a line at 120° with the I_3 axis causes (u, s) and (d, s) to interchange. Mesons of the same charge Q/e (say K^+ and π^+) lie on the same line at 120° to the I_3 axis. At the centre of the hexagon ($I_3 = S = 0$) there is three fold degeneracy as the particles π^0, η, η', and ρ^0, ω and ϕ occupy the same point. The mesons at the centre are their own antiparticles.

Substitution of S quarks raises the mass. Strong spin-spin forces must exist since ρ mesons are so much heavier than π-meson.

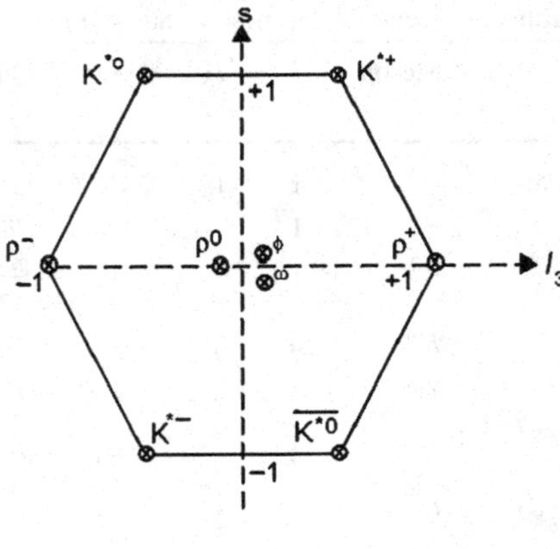

Fig. 5.16 The vector meson nonets displayed on S–I_3 plot

Fig. 5.17 Corresponding quark structure of the vector meson nonets

$d\bar{s}$ $u\bar{s}$

$d\bar{u}$ $\frac{d\bar{d}-u\bar{u}}{\sqrt{2}}$ $u\bar{d}$

$s\bar{u}$ $s\bar{d}$

There are also higher states of the $q\overline{q}$ system with non-zero orbital angular momentum L. Each given orbital angular momentum must be represented by the same nine combinations of u, d and s quarks as for $L = 0$. When the two spins are paral-

Table 5.7 Vector meson states as quark-antiquark combinations

	Meson	Quark combination	Mass MeV/c^2	Decay	I	I_3	S
Octet	ρ^+	$u\overline{d}$	770	$\rho^+ \to \pi^+\pi^0$	1	+1	0
	ρ^-	$d\overline{u}$		$\rho^- \to \pi^-\pi^0$	1	−1	0
	ρ^0	$\frac{1}{\sqrt{2}}(u\overline{u} - d\overline{d})$		$\rho^0 \to \pi^+\pi^-$	1	0	0
	K^{+*}	$u\overline{s}$	892	$K^{+*} \to K^+\pi^0$	$\frac{1}{2}$	$+\frac{1}{2}$	+1
	K^{0*}	$d\overline{s}$	892	$K^{0*} \to K^0\pi^0$	$\frac{1}{2}$	$-\frac{1}{2}$	+1
	$\overline{K}^{0*}$	$\overline{d}s$	892	$\overline{K}^{0*} \to \overline{K}^0\pi^0$ $\to K^-\pi^+$	$\frac{1}{2}$	$+\frac{1}{2}$	−1
	K^{-*}	$\overline{u}s$	892	$K^{-*} \to K^-\pi^0$ $\to \overline{K}^0\pi^-$	$\frac{1}{2}$	$-\frac{1}{2}$	−1
	ϕ	$s\overline{s}$	1019	$\phi \to K^+K^-$ $\to K^0K^-$ $\to \pi^+\pi^-\pi^0$	0	0	0
Singlet	ω	$(u\overline{u} + d\overline{d})/\sqrt{2}$	783	$\omega \to \pi^+\pi^-\pi^0$	0	0	0

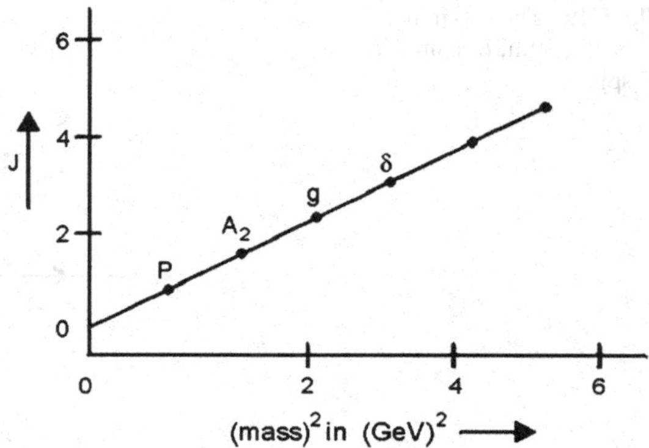

Fig. 5.18 A plot between m^2 and J

lel, their total spin $s = 1$ combines with L to $J = L \pm 1$ and $J = L$. In the case of $L = 1$, $J = 2^+$ nine mesons are identified for the nonet.

When meson states of different L are grouped corresponding to the rotational excitations of mesons they bear a similarity to the rotational bands of molecular or nuclear spectroscopy.

A relation $m^2 \propto J$ displayed in Fig. 5.18 is similar to the plot shown for baryons, Fig. 5.14.

5.6.10 U-Spin

Along the I-spin axis the hyper-charge is constant but the charge varies. The step operators for the basic multiplet change $d \leftrightarrow u$. There is a hexagonal symmetry for all multiplet diagrams. Instead of the single I_3 axis in $SU(2)$ there are three symmetry axes at 120° to each other.

These are I-spin, U-spin and V-spin (Fig. 5.19). Along the U-spin axis, $d \leftrightarrow s$. If U invariance holds then properties of the system would be independent of hyper-charge just as for I-invariance properties are independent of the charge. The third axis known as V-spin axis corresponds to transformation $u \leftrightarrow s$ has no such simple interpretation.

5.6.10.1 Tests of U-Spin Invariance

Although the e.m interaction does not conserve isospin, it is expected to conserve U-spin, since all members of a given U-spin multiplet have the same electric charge.

Assuming that U-spin conservation holds for e.m interactions, the e.m. properties of the baryons can be predicted. Since p and Σ^+ belong to the same U-spin

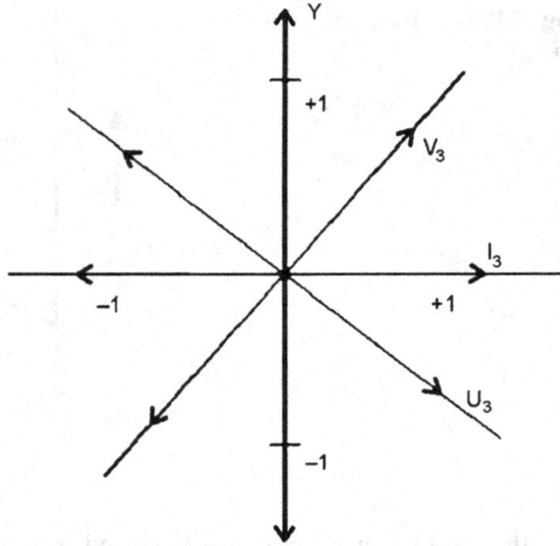

Fig. 5.19 Three symmetry axes of I-spin, U-spin and V-spin

multiplet they must have to same magnetic moment μ. Thus, the prediction is

$$\mu(\Sigma^+) = \mu(p)$$
2.6 n.m   2.79 n.m
$$\mu(\Sigma^-) = \mu(\Xi^-)$$
−1.5 n.m   −1.8 n.m

U-spin invariance can also be used to obtain relations among the electromagnetic mass splittings of baryons octet

$$M_n \quad - M_p + \quad M_{\Xi^-} \quad - M_{\Xi^0} = \quad M_{\Sigma^-} \quad - M_{\Sigma^+}$$
1.3 MeV   6.4 MeV   7.98 MeV

The predicted mass splitting is well satisfied.

5.6.11 Mixing Angle

The Gell-Mann-Okuba mass formula for baryons is satisfactory both for the octet as well as the decuplet. The mass operator used for baryons is linear. When this formula is extended for mesons, square of masses for the operator appears to be preferable. However, it turns out that the Gell-Mann-Okuba mass formula does not hold for mesons, independently of whether masses or squares of masses are used.

Consider a meson octet. It contains two isospin doublets, one with $S = 1$ and the other with $S = -1$. Both these doublets have the same mass as they are antiparticles of each other. Then there is a triplet with $S = 0$. Finally there are two neutral singlets, one belonging to the octet and the other independent of it, lying at the centre with $S = 0$.

If the mass of the isospin doublet, triplet and singlet mesons are M_2, M_3 and M_1, the Gell-Mann-Okuba mass formula becomes

Table 5.8 Comparison of Gell-Mann-Okuba mass formula with experimental masses of mesons

M_1 (MeV) predicted			
J^P	Observed mesons and mass	Linear mass formula	Quadratic mass formula
0^-	$\eta(549), \eta'(958)$	619	570
1^-	$\omega(783), \phi(1020)$	938	934
2^+	$f(1271), f'(1516)$	1458	1456

Table 5.9 Linear and quadratic $SU(3)$ mixing angles

J^P	Nonet member	θ (linear)	θ (quadratic)
0^-	$\pi K \eta, \eta'$	$-24°$	$-11°$
1^-	$\rho K^* \phi, \omega$	$36°$	$39°$
2^+	$a_2 K^* f', f$	$29°$	$31°$

$$\text{Linear:} \quad M_2 = \frac{1}{4} M_3 + \frac{3}{4} M_1 \tag{5.55}$$

$$\text{Quadratic:} \quad M_2^2 = \frac{1}{4} M_3^2 + \frac{3}{4} M_1^2 \tag{5.56}$$

We wish to compare these formulas with the experimental masses of pseudoscalar, vector and tensor meson octets. There is, however, one complication in identifying the singlets, that is which one belongs to the octet and which one to $SU(3)$ singlet. The procedure is to use the experimental values of the masses of isospin doublet and singlet to obtain that of the $SU(3)$ singlet and compare it with the observed masses of the singlets. The results are shown in Table 5.8.

We observe that the masses of the isospin singlets calculated from either the linear or quadratic Gell-Mann-Okuba mass formula do not agree with any of the experimental masses. However we see that in every case the predicted mass lies between the values of the mass of two mesons with the right quantum numbers. We assume that the $SU(3)$ symmetry breaking interaction mixes the isosinglet members of the octet with the $SU(3)$ singlet. A mixing angle θ can be defined as follows. Let M_1 be the mass of the isosinglet member of the octet as predicted by the Gell-Mann-Okuba mass formula and M be the masses of the two physically observed isosinglet mesons. Then we define θ by the formula

$$\text{Linear:} \quad M_1 = m \cos^2\theta + M \sin^2\theta \tag{5.57}$$

$$\text{Quadratic:} \quad M_1^2 = m^2 \cos^2\theta + M^2 \sin^2\theta \tag{5.58}$$

If we assume that m is mostly octet, then $|\theta| \leq 45°$. The sign of θ is taken to be positive if the heavier meson is mostly octet, otherwise negative. The values of the linear and quadratic mixing angles for the pseudoscalar, vector and tensor nonets are given in Table 5.9.

It is clear that by introducing the angle θ, we are able to obtain agreement with the experimental masses. The importance of the mixing angle θ lies in the predictions of other processes. For example, we can make a prediction about the decay modes of the ϕ and ω vector mesons, using $SU(3)$ mixing (Chap. 6).

5.6.12 Change of Flavour of Quarks

The mesons are constructed from the combination of quark and antiquark. Thus

$$\pi^+ = u\bar{d}\pi^0 = \frac{1}{\sqrt{2}}(u\bar{u} - d\bar{d})\pi^- = d\bar{u} \tag{5.59}$$

The negative sign comes from the requirement that the wave function for π^0 be orthogonal to that of the ϕ meson in the octet multiplet of mesons.

The quark and antiquark in the π meson are in a state with zero orbital angular momentum and oppositely directed spins so that the total spin of the pion is zero.

If a quark and antiquark with zero orbital angular momentum have parallel spins then they form mesons with spin 1: ρ^+, ρ^0, ρ^-. These vector mesons are resonances which decay into two pions, $\rho \to 2\pi$ in a time of the order of 10^{-23} sec. The ρ mesons are lightest among meson resonances. A large number of heavy meson resonances are known in which the quark-antiquark pairs are in excited states.

In strong and electromagnetic interactions all flavours of quarks are conserved. For example, the strangeness conserving decays $\Sigma^0 \to \Lambda^0 + \gamma$ and $K^* \to K + \gamma$ are observed but not the $\Delta S = 1$ transitions $\Sigma \to N + \gamma$ and $K \to \pi + \gamma$. This means that at quark level, the transitions $s \to u + \gamma$ or $s \to d + \gamma$ are strictly forbidden even though they violate no principle of electrodynamics. However, in the case of weak interactions the flavour carried by hadrons sometimes, but not always, changes in a weak process. In other words the quark may change its flavour in reactions caused by the weak interactions.

It was pointed out (Chap. 3) that the strange particles with the copious production but long life times posed a puzzle. The puzzle was resolved by suggesting that the strange particles are produced in pairs in strong interactions (characteristic time $\sim 10^{-23}$ sec) but decayed into non-strange hadrons slowly via weak interaction ($\sim 10^{-8}$–10^{-10} sec). This can be explained by the quark model. Consider an example for their strong production in which all the flavours including S are conserved.

$$\begin{array}{cccc} \pi^- + & p & \to K^+ + & \Sigma^- \\ d\bar{u} & uud & u\bar{s} & dds \end{array} \tag{5.60}$$

The typical weak decay of the above strange particles show:

$$\begin{array}{cccc} K^+ \to & \pi^+ + & \pi^+ + & \pi^- \\ u\bar{s} & u\bar{d} & u\bar{d} & d\bar{u} \end{array} \tag{5.61}$$

Here the flavour $\bar{s}$ changes to d. Similarly in the weak decay

$$\begin{array}{ccc} \Sigma^- \to & n & +\pi^- \\ sdd & udd & d\bar{u} \end{array} \tag{5.62}$$

the flavour s changes to d.

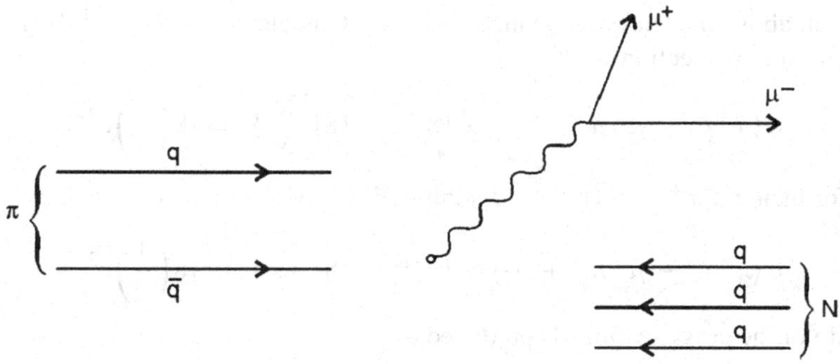

Fig. 5.20 Feynman diagram representing an interaction of particles

Strong interactions between hadrons are assumed to result from strong forces (gluons) between quarks inside the hadrons as in neutron or proton. This is similar to the electrical force between electrons and nuclei producing the chemical binding.

Furthermore the strong forces between quarks are assumed to be independent of the flavours. In particular the members of the isodoublet (u, d) therefore have the same strong interactions. Since the u–d mass difference is small compared to all hadron masses there is virtually complete symmetry between the members of this doublet. The known isospin invariance of the strong interaction is a direct consequence of this symmetry. However the small u–d mass difference is not understood at the present.

There is another approximate symmetry which is of considerable interest in hadronic spectroscopy. The mass splitting between the quarks s on one hand and (u, d) on the other is about only 150 MeV which is small compared to typical hadron masses. To this approximation the triplet (u, d, s) may be considered as degenerate. This leads to the $SU(3)$ symmetry.

5.6.13 Quark Charge Assignment Test in Drell-Yan Process

The process of lepton pair production by charged pions on nucleons provides a test for the quark charge assignment. This process, known as Drell-Yan mechanism (Fig. 5.20), consists of the annihilation of an antiquark from the pion with a quark from nucleon to produce a virtual photon decaying to a muon pair. Figure 5.20 is a Feynman diagram representing an interaction of particles. Time flows horizontally from left to right and space vertically. The arrows indicate the direction of motion of particles entering or leaving the vertices. Incoming particle (momentum $\vec{p}$) can be replaced by outgoing antiparticle (momentum $-\vec{p}$) without change of matrix element. More details are given in Chap. 6. The cross-section is proportional to the square of quark and antiquark charges. Consider a high energy $\pi^-(=\overline{u}d)$ collision much above the heavy resonances, with a ^{12}C nucleus ($= 18u + 18d$). The $\overline{u}u$ annihilation cross-section is

$$\sigma\left(\pi^- C \to \mu^+\mu^- + \cdots\right) \propto 18Q_u^2 = 18\left(\frac{2e}{3}\right)^2 = 18\left(\frac{4}{9}\right)e^2$$

while for incident $\pi^+(= u\overline{d})$, the $\overline{d}d$ annihilation cross-section is

$$\sigma\left(\pi^+ C \to \mu^+\mu^- + \cdots\right) \propto 18Q_d^2 = 18\left(\frac{1}{3}e\right)^2 = 18\left(\frac{1}{9}\right)e^2$$

The ratio of the cross-sections is predicted as

$$\sigma\left(\pi^- C\right)/\sigma\left(\pi^+ C\right) = 4$$

which is borne out by experiments.

5.7 Colour

Consider the three baryons at the corners of the diagram for the decuplet (Fig. 5.11) with quark structure $ddd = \Delta^-$, $uuu = \Delta^{++}$ and $sss = \Omega^-$. These are systems composed of three identical quarks all with the same quantum numbers. Since the orbital angular momentum of the quarks is assumed to be zero, in each of these configurations the total spin $\frac{3}{2}$ must be obtained from individual spins $\frac{1}{2}$ aligned in the same direction. This will violate Pauli's principle as all the quantum numbers in each case are identical. Mathematically this difficulty is expressed by writing down the total wave function for each of the composite particles.

$$\psi = \psi_{spin}\psi_{space}\psi_{flavour} \tag{5.63}$$

For the particles Δ^{++}, $\Delta^-$and Ω^-, each of the three factors is symmetric under the exchange of any two quarks. This means that the total wave function ψ is symmetric which contradicts Pauli's principle that ψ must be antisymmetric under the exchange of any two identical Fermions.

The concept of colour was first introduced by Greenberg [14] to save the spin-statistics theorem for certain members of the baryon decuplet. Greenberg suggested that the quarks of each flavour were endowed with a hidden degree of freedom which distinguished the quarks and permitted the total wave function to be antisymmetrical. He called this degree of freedom as 'colour'. This has nothing to do with the ordinary optical colour. Accordingly, there are three colours, say red (R), blue (B), and green (G) for the quarks. The antiquarks have the opposite or complimentary colours. The antired ($\overline{R}$) is cyan, antigreen ($\overline{G}$) is magenta and antiblue ($\overline{B}$) is yellow. The real particles are assumed to contain all the three colours in baryons or colour-anticolour combinations in mesons such that on the whole they are 'white' or colourless (colour singlets). Here the analogy with the ordinary colours is strong, the overlap of red, green and blue light falling on a screen gives white, as does the overlap of a colour and its compliment.

For baryon the wave function is written as the product of a symmetric space-spin-flavour wave function and an antisymmetric colour wavefunction

$$\psi = \psi_{spin}\psi_{space}\psi_{flavour}\psi_{colour} \tag{5.64}$$

where ψ_{colour} is the total antisymmetric combination of wavefunctions symbolically represented by R, G and B.

Baryons:

$$\psi_{colour} = \frac{1}{\sqrt{6}}[RGB + BRG + GBR - RBG - BGR - GRB] \tag{5.65}$$

Notice that the interchange of any two quark labels (for example, the first and second) has the effect $\psi_{colour} \rightarrow -\psi_{colour}$, as it should for an antisymmetric wavefunction.

For mesons which are bosons Pauli's principle does not apply. The colourless quark-antiquark combination for meson is as shown below.

Mesons:

$$\psi_{colour} = \frac{1}{\sqrt{3}}[R\overline{R} + G\overline{G} + B\overline{B}] \tag{5.66}$$

This is a symmetric wavefunction, as we expect for integral spin mesons.

Any combination of colour which produces net colour is unobservable.

The colour hypothesis has been verified in those experiments which require the counting of the number of possible quark states.

5.7.1 SU(3) Colour Group

The three states of a quark are regarded as a three dimensional representation of an $SU(3)$ colour group. Note that with $SU(3)$ symmetry combining three quarks having three different colour states leads to a singlet state

$$3 \otimes 3 \otimes 3 = 1 \oplus 8 \oplus 8 \oplus 10 \tag{5.67}$$

also does the combination of quark and antiquark

$$3 \otimes \overline{3} = 1 \oplus 8 \tag{5.68}$$

Other combinations of quark and antiquark such as $qq\overline{q}, qq \ldots$ can be shown not to lead to a colour singlet.

5.7.2 Colour Singlet

The requirement that all hadrons are colour singlets restricts the combination of quarks observed in nature, just to two structures qqq and $q\overline{q}$. Also, the $SU(3)$ colour symmetry is only coincidentally the same as the flavor (u, d, s) $SU(3)$ symmetry.

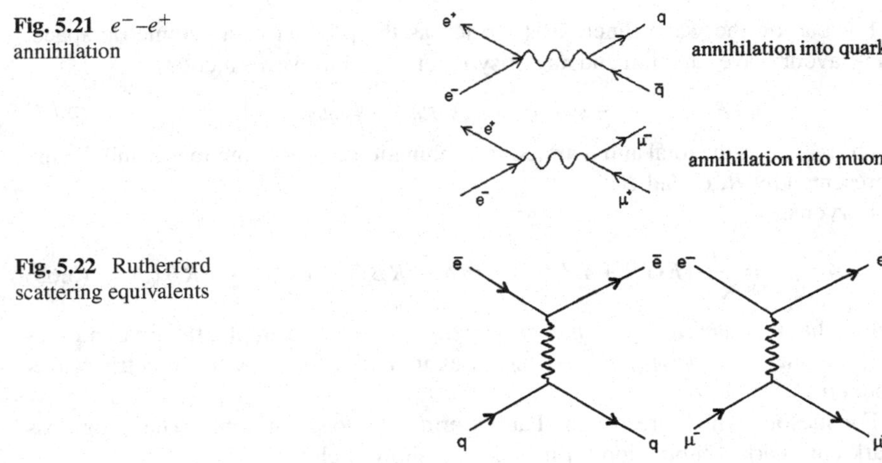

Fig. 5.21 e^{-}–e^{+} annihilation

Fig. 5.22 Rutherford scattering equivalents

Both symmetries happen to be based on the $SU(3)$ group but are physically quite different. The flavour symmetry because of the different quark masses is only approximate where as colour $SU(3)$ symmetry is believed to be exact.

Colour singlets (hadrons) are invariant under unitary transformations among the three quark colours. Hadrons can exist in colour singlet states which have zero values for all colour charges. While quarks which have non-zero colour charges can only exist confined within them. This explains why hadrons have integral electric charges while fractionally charged combinations like qq and $q\overline{q}q$ are forbidden.

5.7.3 Experimental Confirmation of Colour Hypothesis

Consider e^{-}–e^{+} annihilation, Fig. 5.21.

The high energy reaction can be visualised as

$$e^{+}e^{-} \rightarrow \gamma \rightarrow q + \overline{q} \tag{5.69}$$

The energetic quark (q) and antiquark ($\overline{q}$) which are not observed directly, then interact to form numerous mesons or baryons. The Feynman diagrams shown are identical with the diagram for electron-quark Rutherford scattering. The number of different diagrams of this sort that can be drawn is exactly equal to the number of quarks that can be produced, and the total cross-section for all possible $q\overline{q}$ productions is the sum over all such final states:

$$\sigma\left(e^{+}e^{-} \rightarrow \text{hadrons}\right) = \sum_{i} \sigma\left(e^{+}e^{-} \rightarrow q_i\overline{q_i}\right) \tag{5.70}$$

Each of these diagrams is similar to the diagram for $e^{+}e^{-}$ annihilation into $\mu^{+} + \mu^{-}$ which is also in turn identical with the diagram for $e^{-}\mu^{-}$ Rutherford scattering, Fig. 5.22.

We can therefore write the ratio of cross-sections at a given energy

$$\frac{\sigma(e^+e^- \to \text{hadrons})}{\sigma(e^+e^- \to \mu^+\mu^-)} = R = \frac{\sum_i Q_{qi}^2}{Q_\mu^2} \tag{5.71}$$

because the Rutherford cross-section is proportional to the square of the electric charge of the scatterer.

All the kinematical factors in the respective cross sections will cancel. If there are three types of quarks (u, d, s) that can be produced the ratio R is

$$R = \frac{(\frac{2}{3})^2 + (\frac{1}{3})^2 + (\frac{1}{3})^2}{1^2} = \frac{2}{3} \tag{5.72}$$

But if there are nine types $(U_R, U_B, U_G, d_R, d_B, d_G, S_R, S_B, S_G)$ the ratio

$$R = 3 \times \frac{2}{3} = 2.0 \tag{5.73}$$

The ratio R in the range 1–3 GeV centre-of mass energy of the e^+–e^- system clusters about $R = 2$, rather than $R = 2/3$.

Other evidence for the colour comes from the decay $\pi^0 \to 2\gamma$ for which counting procedure similar to the above may be employed. The decay rate calculated without colour gives $1/3$ of the observed rate, while with colours experimental values agree with the theoretical value within 10 per cent.

5.8 Gluons

The force between quarks is an exchange force mediated by the exchange of massless spin-l bosons called gluons. The field that binds the quarks is a colour field. Colour is to the strong interaction between quarks as electric charge is to the electromagnetic interaction between electrons. The gluons are massless and carry their colour-anticolour properties just as other particles may carry electric charge. It is the fundamental strong "charge" and is carried by both quarks and gluons, which must therefore be represented as combinations of a colour and a possibly different anticolour.

Figure 5.23 shows a gluon $R\overline{B}$ being exchanged by the red and blue quarks. In effect the red quark emits redness into a gluon and acquires blueness. The blue quark, on the other hand absorbs the $R\overline{B}$ gluon, canceling its blueness and acquiring a red colour in the process. Note that the gluon flavor does not change the flavor of the quark. We expect nine possible gluons $R\overline{R}$, $R\overline{B}$, $R\overline{G}$, $B\overline{R}$, $B\overline{G}$, $B\overline{B}$, $G\overline{R}$, $G\overline{B}$, $G\overline{G}$. The three net "colourless" combinations can be constructed for the requirements of the symmetry properties of the colour fields. They can be coupled in three different ways.

$$\frac{1}{\sqrt{2}}(R\overline{R} - G\overline{G}); \qquad \frac{1}{\sqrt{6}}(R\overline{R} + G\overline{G} - 2B\overline{B}); \qquad \frac{1}{\sqrt{3}}(R\overline{R} + G\overline{G} + B\overline{B}) \tag{5.74}$$

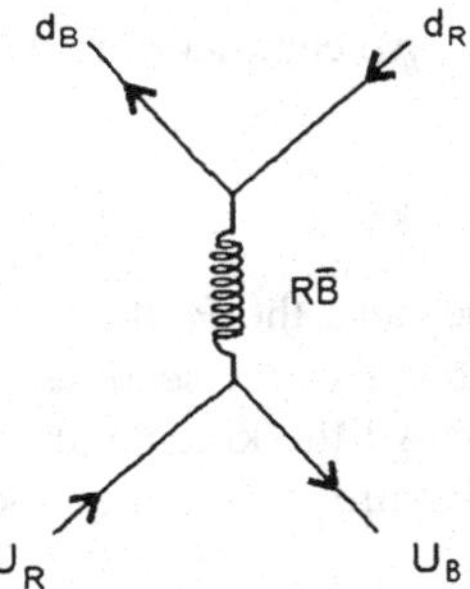

Fig. 5.23 A gluon RB being exchanged by the red and blue quarks

The first and second of these colourless combinations can transmit colour, but the third being a colour singlet cannot and it therefore must be excluded as an agent of the colour singlet cannot and it therefore must be excluded as an agent of the colour field as it does not carry net colour charge. This leaves 8 colour gluons as the source of quark-quark interaction. The main difference between gluons and photons is that there is only one photon which is electrically neutral but gluons are eight and all carry colour charges. Gluons strongly interact with one another and emit one another. Just as charge couples the carrier of the interquark interaction (the gluon) to a quark. Evidence for existence of gluons comes from two experiments.

i. Deep inelastic scattering experiments of electrons with protons show that only 50 per cent of the internal momentum is carried by non-quark systems, that is gluons. High energy (15–200 GeV) events in which electrons, muons and neutrinos collide with nucleons and transfer large momentum and energy correspond to deep inelastic scattering (Chap. 6).
ii. Second piece of evidence comes from e^+e^- annihilation experiments in which hadrons are produced in the final state. The mechanism $e^+ + e^- \to \gamma \to q + \overline{q}$ produces a highly energetic quark antiquark pair travelling in opposite directions. In a time of the order of strong interaction time, the quark and antiquark are transformed into a "shower" of hadrons, but the momentum of the original quark-antiquark pair is preserved. Thus are observed two jets as in Fig. 5.24. The visible tracks represent the trajectories of the hadrons bent into arcs by a magnetic field. A few of the reactions, however produce 3 (or more) jets as in Fig. 5.25. Since it is not possible for $e^+ + e^-$ to produce any combination of quarks and antiquarks totaling 3, the interpretation is that one of the quarks radiates a gluon, which then forms its own hadron jet. The observed angles between gluon and quark jets are consistent with the spin-1 assignment to the gluon.

The creation of hadrons in the e^+e^- annihilation proceeds in two steps. First is the electromagnetic process resulting in the virtual photon. Second is the fate of the $q\overline{q}$ pair. But for the strong interaction the primary pair would escape to infinity, Fig. 5.26, where the pair $s\overline{s}$ is considered. As the separation increases, the strong interaction between the pair increases without bound to confine the quarks. This prevents the quarks to run away to infinity. Instead it is energetically favourable for the strong field to produce a secondary quark pair, say $u\overline{u}$. Then u bonds to $\overline{s}$

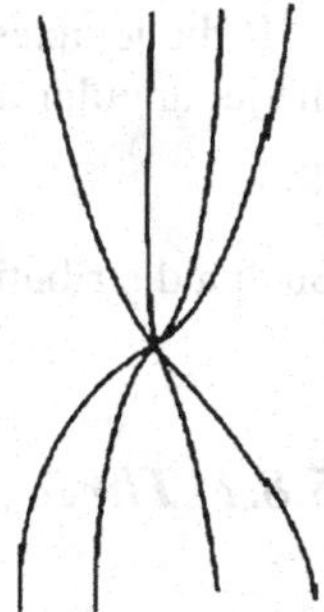

Fig. 5.24 The visible tracks represent the trajectories of the hadrons bent into arcs by a magnetic field

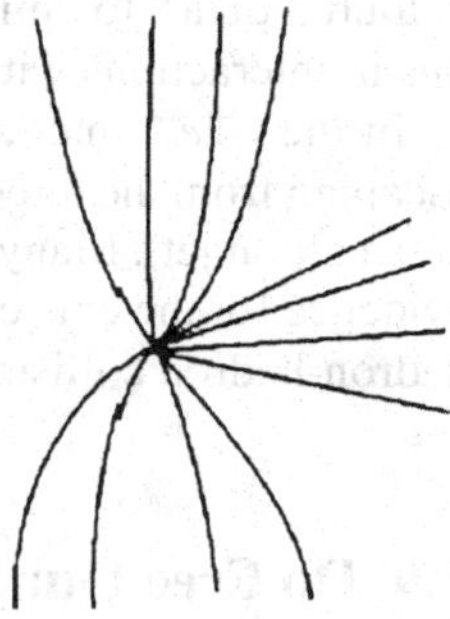

Fig. 5.25 Three or more jets produced by few reactions

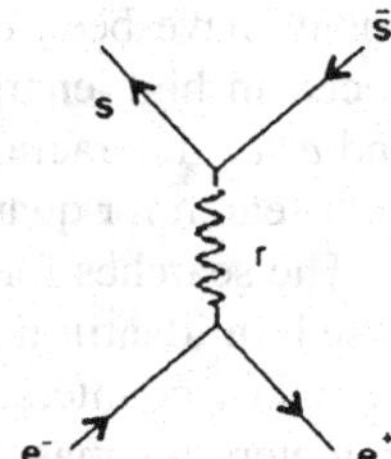

Fig. 5.26 The primary pair escapes to infinity for strong interactions where the pair ss is considered

and $\overline{u}$ to s to form colour singlets K^+ and K^-. There are no long-range confining forces between colour singlets, and therefore $e^+e^- \to K^+K^-$ can actually occur. At higher energy K^+ or K^- may be formed in an excited state K^* which upon decay $K^* \to K\pi$ can produce more mesons. Thus the production process would be $e^+e^- \to K^*K^- \to K\pi K^-$.

When the energy of the e^+e^- pair is very large compared to meson masses, the primary and secondary quark can not bond into stable mesons because of the large relative momenta between the primaries and the secondaries. Therefore, pair production must occur repeatedly until there are enough pairs with low relative momenta to form a stable multihadron state. The number of hadrons increases indefinitely with the increasing e^+e^- energy.

If the jet axes represent the directions of parent quarks (which are not seen) then the jet angular distribution for spin $\frac{1}{2}$ quarks should be

$$I(\theta) \propto 1 + \cos^2 \theta$$

Such a distribution is indeed verified in experiments.

5.8.1 Three Jet Events

Just as an electrically-charged particle will radiate photon if subject to acceleration when it passes near another charge (bremsstrahlung) in the same way we expect a coloured quark to emit QCD (quantum chromodynamics) bremsstrahlung, i.e. gluons in interactions with other coloured particles.

In the e^+e^- interaction producing a $q\overline{q}$ pair, one of which radiates a gluon in escaping from the interaction, we expect to see three jets (a quark jet, an antiquark jet and a gluon jet). Many examples of 3-jet events have been seen. This provides direct evidence for the existence of gluons. Jets have also been seen in other processes like hadron-hadron collisions, muons and neutrinos collisions with nucleons.

5.9 Do Free Quarks Exist?

Experiments have been extensively carried out to discover free quarks. The experiments have been of two types; one is concerned with the search of quarks produced in high-energy collisions at accelerators, in hadron-hadron, lepton-hadron, and e^+e^- interactions, and among cosmic-ray secondaries; the other one is tied up with search for quarks already existing in matter.

The searches for quarks among the secondaries produced at the accelerators are based on identifying particles of charge $2e/3$ and $e/3$ through the combination of Cerenkov counters, pulse height and time of flight techniques and magnetic spectrometers for masses up to 10 GeV. In the second type, searches for pre-existing quarks have extended to samples of water and sediments from deep oceans, moon rocks, meteorites and even oyster shells. The experiments are prototype of Millikan's oil drop method. The general conclusions drawn from all evidences, is that free quarks do not exist, even if they do then it is at the level of $\leq 10^{-20}$ of bound quarks. Thus free quarks have not yet been discovered. They are believed to be confined to hadrons and cannot be isolated.

5.10 Quark Dynamics—Confinement

No free quark has been observed (even in Fermi Lab's 1000 GeV collider), it suggests that quarks are permanently confined in hadrons and that no amount of energy can liberate a quark from its hadronic environment. On the other hand at very short distances, as revealed by deep inelastic scattering experiments with e^- and ν (Chap. 6), the quarks exhibit a different behaviour, viz they are found to move almost freely as if they were unbound. These two proprieties are called infrared slavery (confinement of quarks to regions of the size of hadrons with large, perhaps infinite energies to liberate them to larger distances but asymptotic freedom, that is free movement at short distances). Any successful theory of quark interactions must be able to explain these apparently contradictory properties.

In analogy with quantum eletrodynamics (QED), the quantum theory of electromagnetic field, we have quantum chromodynamics (QCD), the quantum theory of the colour field. We will not go into the complexities of the theory but will merely point out its main features by comparison with QED.

In QED, electric charges interact through the exchange of real or virtual photons. In QCD, quarks interact by exchanging gluons. The photons are the carriers of the electromagnetic field while gluons are the carriers of the strong colour field. The main difference is this, photons themselves carry no electric charge and are therefore unaffected by electric fields. On the contrary gluons carry a net colour, and therefore interact directly with the quarks, that is, a quark can emit a gluon and then interact with it and create additional gluons; a photon cannot by itself exchange photons with nearby charge. This property of gluons forces QCD into a considerable level of mathematical complexity.

The emission of coloured gluons provides a clue for the operation of asymptotic freedom. An electron emitting virtual photons still remains an electron with a charge of $-e$, but a quark emitting a virtual gluon must change its colour charge. The colour charge of a quark is therefore spread out over a sphere of radius of the order of a hadron (0.5–1 fm). If another quark were to penetrate the sphere, this "smeared out" color field would cause a considerably reduced quark-quark interaction. If we sample the quark's interactions over a radius small compared with 1 fm, we observe only a small fraction of its colour charge, and it appears only very weakly bound or nearly free.

The behaviour of quark interactions as the separation is increased can be explained by a comparison with QED. Figure 5.27(a) and (b) shows the difference between the electric field of two charges and the colour field of two quarks, respectively.

As the distance between two point charges increases, the electric field (corresponding to the density of electric field lines crossing a unit surface area) decreases. The colour field remains constant as the distance increases. Eventually, attempts to separate the quarks to large enough distance results in the production of a new $q\overline{q}$ pair.

The density of electric field lines through any surface is proportional to the electric field at that location; thus as the charges are separated, the density of electric field lines through any surface is proportional to the electric field at that location; so as charges are separated the density of the field lines and the electric field between the charges decreases. Regarding the electric field lines as representative of virtual photon exchange between the charges, it is immediately apparent why the QCD field lines behave differently—the exchanged photons do not interact, while the exchanged gluons do. Thus the gluon-gluon interaction forces the colour field lines

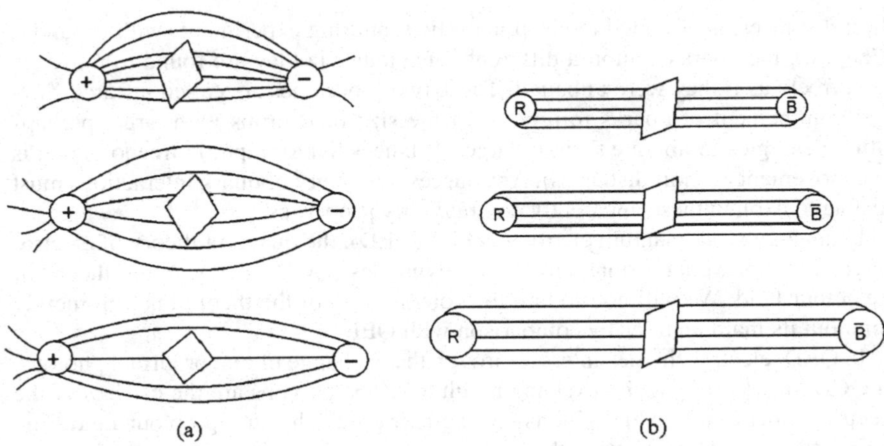

Fig. 5.27 Panels (**a**) and (**b**) show the difference between the electric field of two charges and the colour field of two quarks, respectively

into a narrow tube. The force (again represented as the density of field lines crossing a surface) remains roughly constant as the separation is increased. As we try to separate to large distance, the work will eventually exceed the production threshold for creation of a $q\overline{q}$ pair resulting in the formation of a meson. Thus putting energy into a nucleus is an attempt to liberate a quark goes to create new mesons, as is observed.

5.11 Hadron Decays and Zweig's Rule

Consider ω and ϕ mesons. Both of these mesons have the same quantum numbers, namely $J^{PC} = 1^{--}$, both are unstable, decaying principally into pseudoscalar mesons via strong interactions. Now, ω has mass of 782.7 MeV and ϕ has a mass of 1019.7 MeV. From phase space considerations we expect the ϕ to have larger width because it is more massive. However, it is found that ϕ has a width of 4.1 MeV and ω has a width of 10.0 MeV. This puzzle becomes more acute when we examine the partial decay width of the decay of ω and ϕ into three pions. The partial decay width is more than an order of magnitude larger for the ω than for ϕ.

The dynamical mechanism that inhibits the decay of the ϕ into three pions is not fully understood. But one can be guided by what is known as Zweig's rule [15], which applies to other processes as well.

Zweig's rule is most readily explained in terms of quark line diagrams. A quark line diagram of a hadron is a composite picture of the hadron in terms of its quark constituents, illustrated as lines in Fig. 5.28.

Quark line diagrams illustrating the decays $\Delta^{++} \rightarrow p + \pi^{+}$ and $\rho^{+} \rightarrow \pi^{+} + \pi^{0}$ are shown in Figs. 5.29 and 5.30, respectively.

Quark line diagram illustrating the reaction $\pi^{-} + p \rightarrow K^{0} + \Lambda$ is shown in Fig. 5.31.

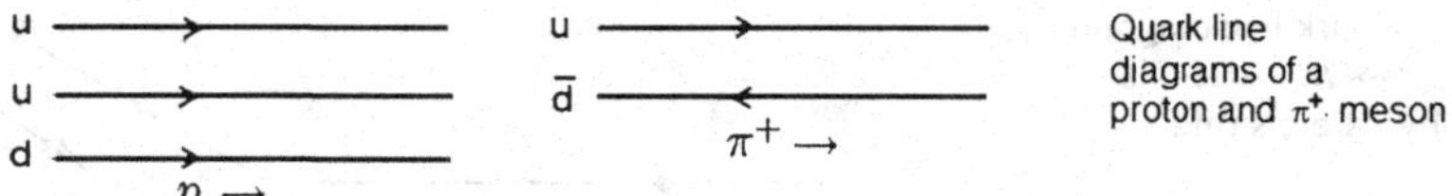

Fig. 5.28 A quark line diagram of a hadron

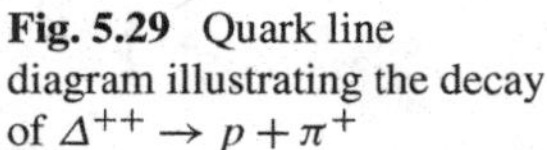

Fig. 5.29 Quark line diagram illustrating the decay of $\Delta^{++} \to p + \pi^{+}$

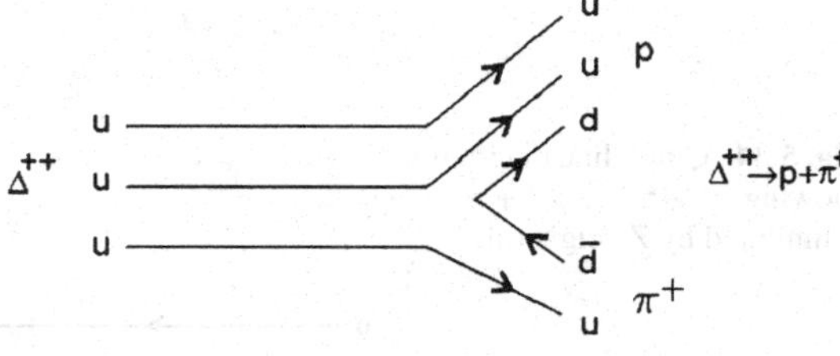

Fig. 5.30 Quark line diagram illustrating the decay of $\rho^{+} \to \pi^{+} + \pi^{0}$

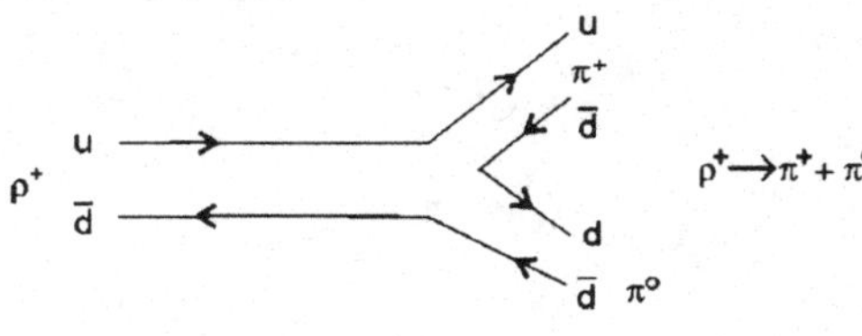

Fig. 5.31 Quark line diagram illustrating the reaction $\pi^{-} + p \to K^{0} + \Lambda$

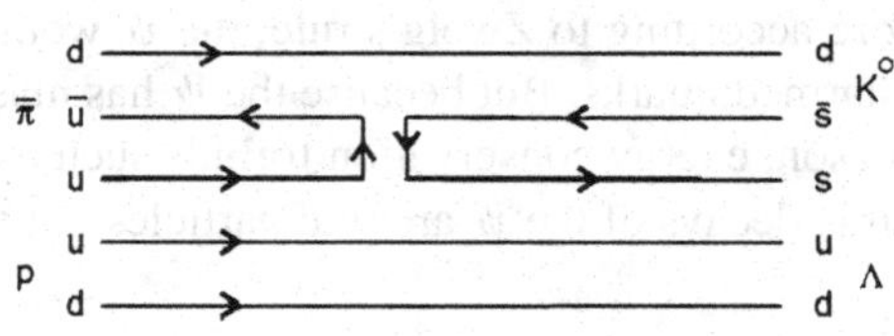

Fig. 5.32 Quark line diagram showing $\omega \to \pi^{+} + \pi^{0} + \pi^{-}$ is allowed by Zweig's rule

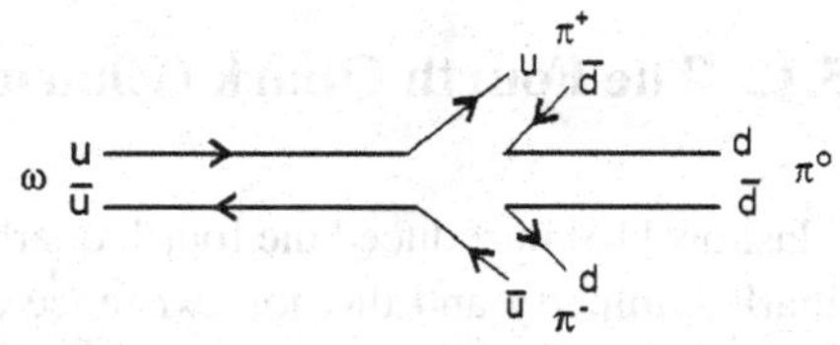

Quark line diagram showing $\omega \to \pi^{+} + \pi^{0} + \pi^{-}$ and $\phi \to K^{+} + K^{-}$ are allowed by Zweig's rule in Figs. 5.32 and 5.33, respectively, but $\phi \to \pi^{+} + \pi^{0} + \pi^{-}$ is inhibited as in Fig. 5.34.

A simplified statement of Zweig's rule is that processes which correspond to disconnected quark line diagrams are inhibited. A diagram is disconnected if one or more hadrons can be isolated by a line which does not cut any quark lines. Okubo postulated an essentially equivalent rule in terms of $SU(3)$ algebraic properties before the quark model was invented.

Zweig's rule also explains the very narrow width of the ψ meson. The ψ is considered as a composite of a charmed quark-antiquark pair (Sect. 5.12), and there-

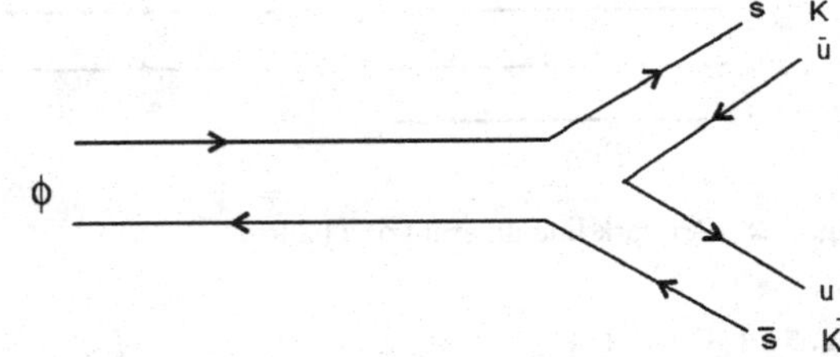

Fig. 5.33 Quark line diagram showing $\varphi \to K^+ + K^-$ is allowed by Zweig's rule

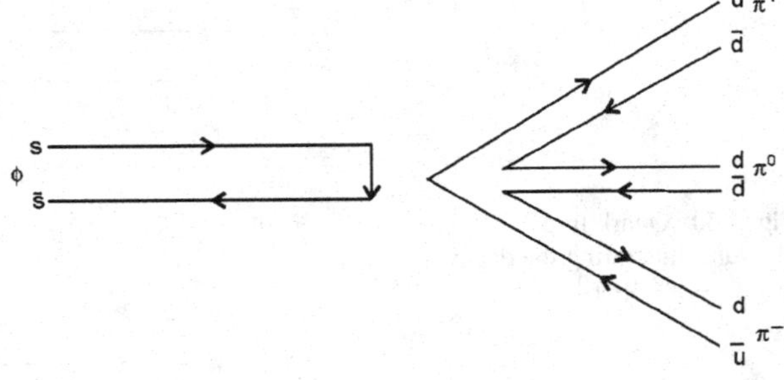

Fig. 5.34 Quark line diagram showing $\varphi \to \pi^+ + \pi^0 + \pi^-$ is inhibited by Zweig's rule

fore according to Zweig's rule, the ψ would tend to decay into mesons containing charmed quarks. But because the ψ has mass less than twice the mass of a charmed meson, energy conservation forbids such a decay. Thus, the only energetically possible decays of the ψ are into particles not containing charmed quarks.

5.12 The Fourth Quark (Charmed Quark)

Glashow [16] introduced the fourth quark known as the charmed quark from lepton-quark symmetry and the non-existence of strangeness-changing neutral weak currents (Chap. 6). Among the lighter quarks u- and d-quarks are assumed to form an isospin doublet. Similarly, the charmed (c) quark and the s-quark are the isospin doublet, although the mass difference is much larger, ~1200 MeV. Thus corresponding to four leptons (e^-, ν_e, μ^-, ν_μ) there are four flavours of quarks (d, u, s, c). The charmed quark was assigned a charge $Q/e = +2/3$. Its mass was calculated (Gaillard and Lee) to be $\simeq$1.5 GeV/c. In November 1974 the c-quark was dramatically discovered by Burton Richter at Slac (Stanford) and Samuel Ting at Brookhamen National Laboratory (Long Island N.Y.) almost simultaneously. At Slac the total cross-section for e^+e^- interactions was studied in the e^+e^- storage ring, as a function of total energy which was increased in small steps. At total energy of about 3.1 GeV the cross-section showed a narrow peak rising above the background by

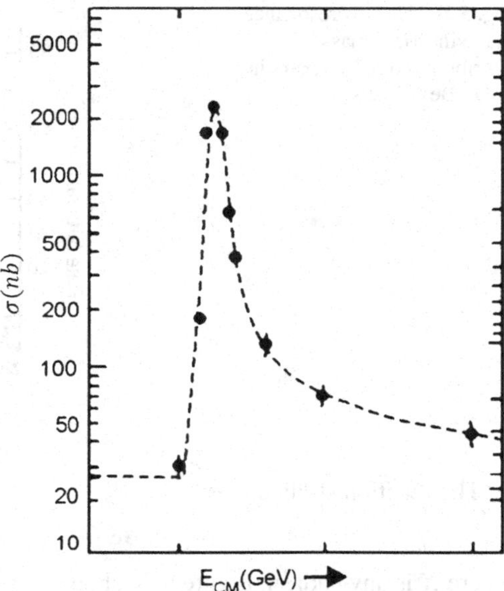

Fig. 5.35 Variation of cross-section for the multihadron production

more than two orders of magnitude. The peaks were also seen in the partial cross-sections for the processes.

$$\begin{aligned} e^+e^- &\rightarrow e^+e^- \\ &\rightarrow \mu^+\mu^- \\ &\rightarrow \text{hadrons} \end{aligned} \tag{5.75}$$

The variation of cross-section for the multihadron production [17] is shown in Fig. 5.35.

The experimental width of the peak was ~2 MeV; even this was due to spread in energy of $e^+$and e^- beams resulting from synchrotron radiation. The corrected intrinsic total width was found to be between 60 and 70 keV, an astonishingly small value for a particle with a mass of 3.1 GeV/c^2.

Richter used a multipurpose large solid angle magnetic detector, SLAC-LBL Mark I. The heart of the detector was a cylindrical magnetostrictive spark chamber inside a solenoidal magnet of 4.6 kG. This was surrounded by time-of-flight counters for particle velocity measurements, shower counters for photon detection and electron identification and proportional counters embedded in iron absorbers labs for muon identification.

Ting at Brookhaven National Laboratory used the alternating-gradient synchrotron to study the production of direct e^+e^- pairs from a Be target bombarded by protons of 28 GeV. The experiment used two magnetic spectrometers to measure separately the e^+ and e^-. Be target was used to minimize multiple Coulomb scattering. Electrons were identified by Cerenkov counters, time-of-flight method, and pulse height measurements.

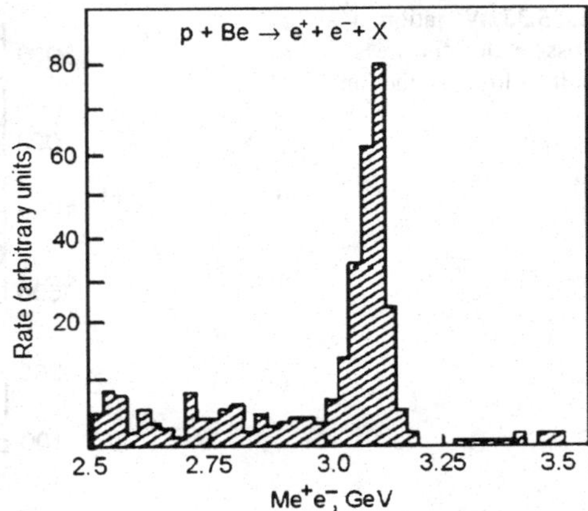

Fig. 5.36 Narrow resonance in the invariant mass distribution of e^+e^- pairs in the P–Be collisions

The reaction studied was

$$p + \mathrm{Be} \rightarrow e^+ + e^- + X \tag{5.76}$$

where X is any hadron. Figure 5.36 shows the results of Aubert et al. [18] indicating the narrow resonance in the invariant mass distribution of e^+e^- pairs in the P–Be collisions. This resonance was named as ψ by the SLAC group and J by the Brookhaven group. It is also known as J/ψ resonance.

Ten days after the discovery of ψ/J, a second narrow resonance was found, $\psi'(3685)$. Approximately 50 percent of the ψ's decay via the lower mass states in modes like $\psi' \rightarrow \psi\pi\pi$, $\psi\eta$. As in the case of ψ here also the width is narrow.

5.12.1 Properties of the J/ψ

The narrow width of the $J/\psi|(67$ keV) distinguishes it sharply from other strongly decaying vector mesons such as the $\rho(776$ MeV) with $\Gamma = 100$ MeV and $\omega(784$ MeV) with $\Gamma = 11$ MeV. The extreme narrowness of the ψ and ψ' states indicated that they could not be understood in terms of light quarks u, d and s and the antiquarks $\overline{u}$, $\overline{d}$ and $\overline{s}$. It had to be a new quark called charmed (c) quark postulated by Glashow four years earlier. A new quantum number c (for charm) was associated with the charmed quark, which like strangeness would be conserved in strong and electromagnetic interactions. The large masses of the ψ, ψ' mesons imply that the charmed quark itself must be massive ($\sim$1.5 GeV/c^2).

Charmonium It is postulated that ψ, ψ' consist of the vector combination of $c\overline{c}$ called charmonium just like positronium (e^+e^-), ϕ-meson ($s\overline{s}$) or ρ^0 meson ($u\overline{u}$ and $d\overline{d}$). The fact that its partial decay width for the pure electromagnetic decay $\Gamma(\psi \rightarrow e^+e^-) = 4$ keV is comparable with that of the other vector mesons, $T(\omega \rightarrow e^+e^-) = 0.8$ keV or $T(\phi \rightarrow e^+e^-) = 1.6$ keV, thereby supporting its vector nature, $J^P = 1^-$.

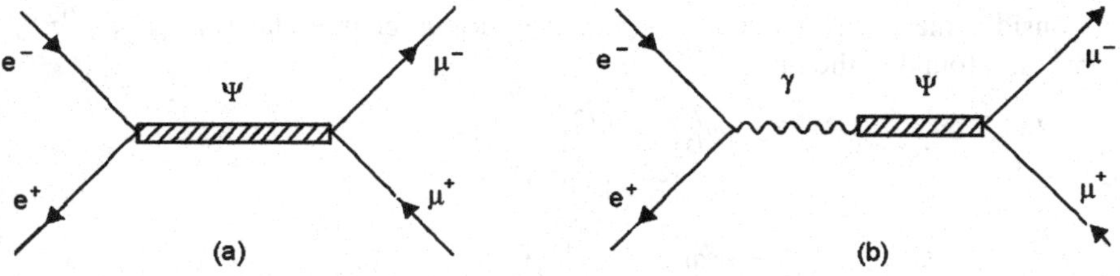

Fig. 5.37 The two-fold production mechanism of ψ via **(a)** direct channel **(b)** an intermediate virtual photon

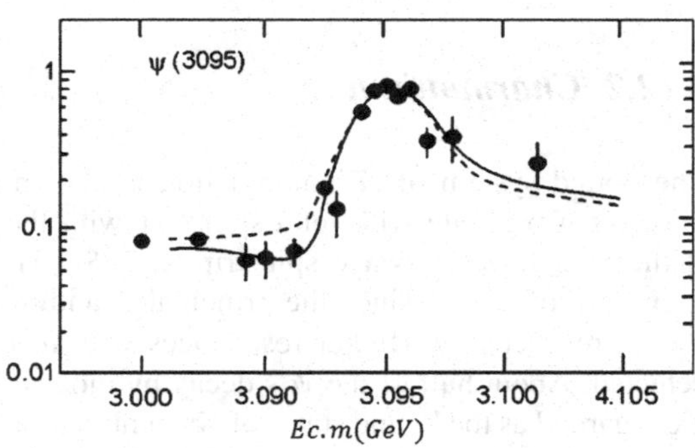

Fig. 5.38 The cross-section for $\mu^+\mu^-$ production in e^+e^- interactions for the two interfering amplitudes

The assumption $J^P = 1^-$, that is the vector nature of the ψ particle is corroborated by the shape of the resonance curve. The production mechanism is two-fold (a) direct channel (b) ψ production via an intermediate virtual photon, Fig. 5.37(a) and (b) respectively.

The cross-section for $\mu^+\mu^-$ production in e^+e^- interactions for the two interfering amplitudes is fitted well in Fig. 5.38 (SLAC-LBL collaboration). The solid line refers to the expected interference and dotted line to no interference. Similar results are obtained for ψ'.

ψ (3097) decays predominantly mostly into $(2n+1)\pi$ with a branching ratio of 86 per cent, via e^+e^- mode with 7 per cent and $\mu^+\mu^-$ mode with 7 per cent.

The intrinsic parity of charmonium is

$$P = (-1)^{L+1} \tag{5.77}$$

as for the positronium. With $L = 0$, $P = -1$, the C-parity

$$C = (-1)^{L+S} \tag{5.78}$$

as for the positronium. With $S = 1$, $C = -1$.

The G-parity is -1 since the charmonium mostly decays into $(2n+1)\pi$.

Also

$$G = (-1)^{I+J} \tag{5.79}$$

Putting $G = -1$ and $J = 1$, we get the result that the isospin $I =$ even, actually zero which can be proved from the Clebsch-Gordon coefficients for combining two states with $I = 1$.

Consider the decay mode $\psi \to \rho\pi$, the various charge states $\rho^+\pi^-$, $\rho^0\pi^0$, $\rho^-\pi^+$, are found in the ratio

$$|\phi^I_{I_3}|^2 = |\phi^2_0|^2 = 1:4:1$$
$$|\phi^1_0|^2 = 1:0:1$$
$$|\phi^0_0|^2 = 1:1:1$$

The experimentally observed ratio, $1:1:1$, is consistent with $I = 0$. The ψ is thus an isosinglet $^3S_1$ state.

5.12.2 Charmonium

The bound system of $c\bar{c}$ combination is known as charmonium. While ψ is the lowest s-wave state with total spin $= 1$ with the spectroscopic notation $1\,^3S_1$, ψ' is the next lowest s-wave spin triplet, $2\,^3S_1$. The latter may be called the radial excitation of the ψ, since the principal quantum numbers assigned are $n = 1$ and $n = 2$, respectively. Higher resonances with masses up to 4.4 GeV/c^2 have been detected. About half of the ψ's decay by modes, $\psi' \to \psi\pi\pi$, $\psi\eta$. Both ψ and ψ' are regarded as the bound states of $c\bar{c}$ combination. In analogy with the positronium, the charmonium was expected to possess the whole lot of other states. This was the beginning of a new chapter of spectroscopy. It was expected that between the s-wave states of ψ and ψ' there would be a set of p-wave states, $3p$ would have total angular momentum $J = 2, 1$ or 0, while the spin-singlet states would have $J = 1$. The $P_{2,1,0}$ states would have $J^{PC} = 2^{++}, 1^{++}, 0^{++}$, while the $^1P_1$ state would have $J^{PC} = 1^{+-}$. The ψ' was expected to decay radiatively to the c-even states which were denoted by $\chi\,(\psi' \to \gamma\chi)$. The existence of three such χ states was established by detecting the photons in radiative transitions. The s-wave spin-singlet state $1\,^1S_0$ and $2\,^1S_0$ (denoted by η_c and η_c') are located just below the corresponding spin-triplet states $1\,^3S_1$ and $2\,^3S_1$.

These states which have C-parity $+1$ and $J = 0$, can not be produced directly by e^+e^- annihilation through a virtual photon. They are formed through radiative decays of the ψ and ψ' in the manner of χ states. The photon energy resulting from the electromagnetic transitions ranges from 100 to 700 MeV. The electric dipole transitions obey the selection rules $\Delta L = 1$ and $\Delta S = 0$.

Intermediate states with total angular momentum 0, 1 or 2 and positive parity must therefore be created in such decays. The transitions are suppressed by kinematical and dynamical factors. They were identified only after a big effort. The masses are $m_{\psi 1} = 3.686$ GeV/c^2, $m_{\chi 2} = 3.556$ GeV/c^2, $m_{\chi 1} = 3.510$ GeV/c^2 and $m_{\chi 0} = 3.415$ GeV/c^2. The splitting of energy levels is due to magnetic interaction, Fig. 5.39.

The spin 0 charmonium states ($n\,^1S_0$), which are called η_c, and cannot be produced in e^+e^- collisions, are only produced in magnetic dipole transitions from J/ψ or $\psi(23S_1)$. These obey the selection rules $\Delta L = 0$ and $\Delta S = 1$ and thus

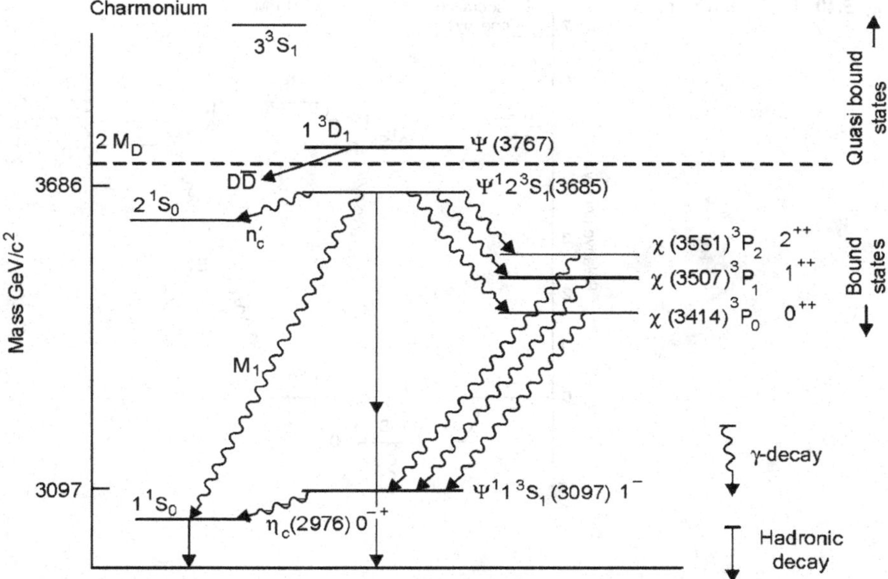

Fig. 5.39 Charmonium states

connect states with the same parity. They correspond to a spin flip of one of the c-quarks. Magnetic dipole transitions are weaker than electric dipole transitions.

5.12.3 The Positronium and Charmonium

It is instructive to compare the spectra of charmonium and positronium, Figs. 5.39 and 5.40. We find that the states with $n = 1$ and $n = 2$ are similarly arranged when the positronium scale is increased by a factor of about 10^8.

However, the higher charmonium states do not display the $1/n^2$ behaviour seen in the positronium.

For the positronium the energy and orbital radius are

$$E_n = -\frac{\alpha^2 mc^2}{2n^2}, \qquad r_n = \frac{\hbar c}{\alpha \cdot mc^2} \tag{5.80}$$

where $\alpha = 1/137$ is the fine structure constant, $m = \frac{m_e}{2} = 0.255$ MeV/c^2, is the reduced mass and the principal quantum number $n = N + l + 1$, where N is the number of nodes in the radial wave function and l is the orbital angular momentum in units of $\hbar$. The spin-orbit interaction (fine structure) and the spin-spin interaction (hyperfine structure) split the degeneracy of the principal energy levels. The orbital angular momentum $l = 0, 1, 2, 3, \ldots$ correspond to $s, p, d, f, \ldots$. The spectroscopic notation used is $n^{2S+1}L_J$. Thus $n = 2$, $S = L = J = 1$ is denoted by $2\,^3P_1$.

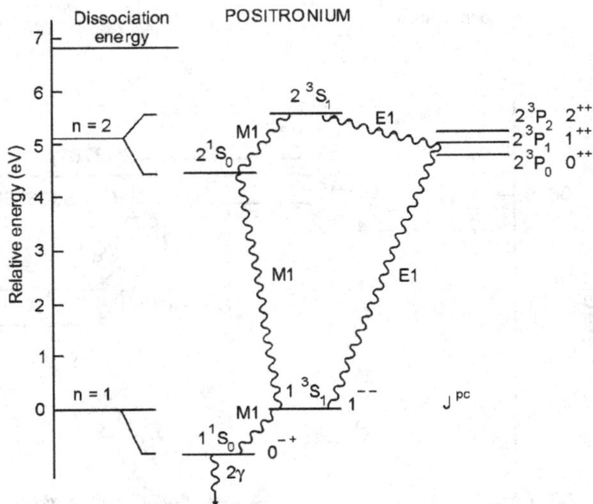

Fig. 5.40 Positronium states

In the case of positronium the energy levels are determined by Coulomb's potential ($V_c \propto 1/r$) apart from the reduced mass and the principal quantum number. For charmonium the actual potential is unknown. At short distance we expect a Coulomb potential, $V \sim 1/r$. On the other hand, at large distances we have to account for quark confinement, the potential must increase without limit. The simplest choice is a linear potential, $V \sim r$, corresponding to a constant force. The quark-antiquark potential is assumed to be of the form

$$V(r) = -\frac{4}{3}\alpha_s \frac{\hbar c}{r} + kr$$
$$V(r \to 0)\alpha \frac{1}{r} \quad \text{and} \quad V(r \to \infty) \to \infty \tag{5.81}$$

The quark-antiquark potential V, (5.81) as a function of distance of separation is displayed in Fig. 5.41.

The factor 4/3 is a theoretical consequence of quarks coming in three different colours. The strong coupling constant α_s is not a true constant, its value depends on the interquark distance of separation. In fact it is called running constant. For the charmonium system typical values are $\alpha_s \simeq 0.15$–0.25, $k \simeq 1$ GeV/fm and the c-quark mass $m_c \simeq 1.5$ GeV/c^2, m_c being the constituent mass (also called effective mass) of the c-quark. The α_s in the charmonium system is ∼20–30 times larger than the electromagnetic coupling $\alpha = 1/137$.

The $J/\psi(1\,^3S_1)$ has a radius of ≃0.4 fm (distance between the quark and antiquark) which is five orders of magnitude smaller than that of positronium. For the higher energy levels, other terms must be incorporated in the potential, and the splitting of energy levels described by the spin-orbit interaction.

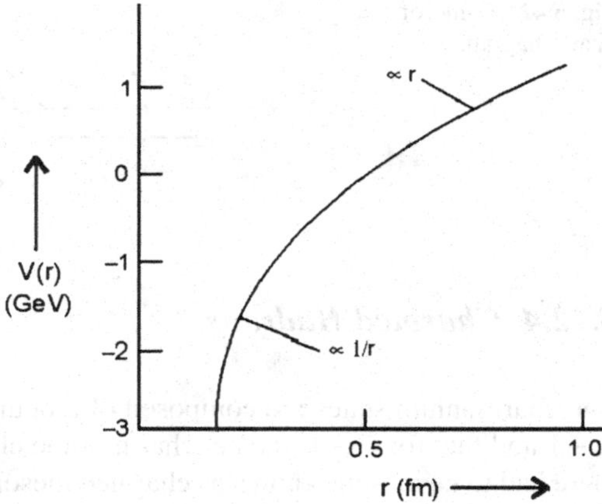

Fig. 5.41 Quark-antiquark potential

Table 5.10 Charmonium energy levels

State	Actual (MeV)	Predicted (MeV)
4*s*	4415	4460
3*s*	4030	4110
2*s*	3686	(3686)
1*s*	3097	(3097)

Since the *c*-quarks are relatively heavy, the non-relativistic Schrodinger equation may be used as a first approximation. Assuming *s*-states the radial equation for the present problem is

$$-\frac{h^2}{2m}\frac{d^2u}{dr^2} + V(r)u(r) - Eu(r) = 0 \tag{5.82}$$

where m is the reduced mass $m_q/2$ and $V(r)$ given by (5.81) has the form, $V(r) = Kr - \frac{b}{r}$.

Calculations have been performed by Eichten et al. [19]. The results for the expected spectrum of the *s*-states and the observations are shown in Table 5.10.

Values in the parenthesis are inputs for the model and are used to evaluate the parameters K and b. There is a reasonably good agreement between the predicted and observed energy levels.

The masses of the quarks themselves do not appear in any QCD calculations. In fact if the quarks are permanently confined, it makes little sense to discuss the rest mass of a free quark. Instead we can find the effective mass of a quark which is found in a hadron, this is usually called the constituent quark mass. We can make a rough estimate of the *c*-quark mass.

$$m_c c^2 \simeq \frac{1}{2} m_\psi c^2 \simeq 1.5 \text{ GeV} \tag{5.83}$$

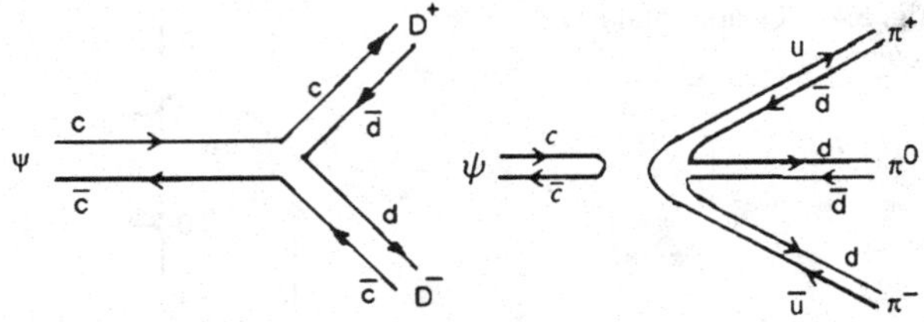

Fig. 5.42 D-meson decay quark diagrams

5.12.4 Charmed Hadrons

The charmonium states are composed of a bound $(c\overline{c})$ pair. Since the charm for C is $+1$ and that for $\overline{c}$, -1, the net charm of the charmonium states is zero. However, there had to be mesons known as charmed mesons with the composition $c\overline{u}$, $c\overline{d}$ etc., with non-zero charm. The D-mesons represent coupling of c-quark with u- and d-quark.

$$\begin{aligned} D^+(1869\ \text{MeV}) &= (c\overline{d}), \qquad D^- = (d\overline{c}) \\ D^0(1865\ \text{MeV}) &= (c\overline{u}), \qquad \overline{D} = (u\overline{c}) \end{aligned} \tag{5.84}$$

D^- is the antiparticle of D^+, $\overline{D^0}$ is the antiparticle of D^0. Similarly, the excited states (D^{*+}, D^{*0}) and $(D^{*-}, \overline{D}^{*0})$ form an analogous set.

The $\psi(3100)$ or $\psi'(3700)$ can not decay into a pair of D-mesons because of energy conservation. The sum of the masses $2M_D \sim 3740$ MeV, which is greater than $M_\psi(3097\ \text{MeV})$ or $M_{\psi'}(3686\ \text{MeV})$ forbids the decay, although such a decay is favoured by the quark line diagram, Fig. 5.42(a) according to the Zweig or OZI rule. OZI stands for Okubo, Zweig and Iizuka.

The decay corresponding to Fig. 5.42(b) is suppressed because of disconnected quark lines. The D-mesons can be produced from the decay of still more massive $c\overline{c}$ states above the ψ' state: $\psi(3770)$, $\psi(4030)$, $\psi(4160)$, $\psi(4415)$. All of them are 1^- states formed in more energetic e^+e^- collisions. They have much greater width (25–80 MeV) in comparison with ψ or ψ'. This is attributed to decay through strong interaction.

For higher excitations, decays into mesons with s-quarks are also possible.

$$c\overline{c} \rightarrow c\overline{s} + \overline{c}s \quad \left(D_s^+ \text{ and } D_s^-\right) \tag{5.85}$$

Previously D_s^+ was called F^+ and D_s^- was called F^-. To summarise, charmonium states with $n = 1$ and $n = 2$ are relatively long-lived because of the OZI rule which suppresses their strong decays. For $n \geq 3$ the charmonium masses lie above the threshold for (OZI-allowed) production of two charmed D-mesons (D^0, $\overline{D^0}$ at a mass of 1865 MeV/c^2 or $D^\pm$ at 1869 MeV/c^2). Their lifetimes are therefore much shorter; they are called "quasi-bound states." These quasi-bound states of charmonium have been observed up to at least as high as $n = 4$.

The D's share some of the properties of kaons. They are produced in pairs indicating the conservation of charm in strong and electromagnetic interactions. In addition the decays violate parity. Also the decay of D^{*+} (vector) into D^0

(pseudo scalar), $D^{*+} \rightarrow D^0\pi^+$ is the analogue of $K^* \rightarrow K\pi$. Further, the decay $D_S^* \rightarrow D_S\gamma$, has been identified.

The decays of D-mesons change both c and s by one unit. The observed decay width is less than 2 MeV which after correction becomes ~70 keV, the corresponding lifetime being 10^{-12} sec.

The charmed pseudoscalar mesons like $D^+(= c\overline{d})$, $D^0(= c\overline{u})$ and their antiparticles decay by weak interactions, with $\Delta c = 1$, into noncharmed states, with the decay into kaons favoured by the Cabibbo suppression factor (Sect. 7.12).

The weak decay of a particle containing a c-quark would yield an s-quark. Thus the decay of a $D^+(= c\overline{d})$ could produce a $K^-(= s\overline{u})$ but not a $K^+(= \overline{s}u)$. Similarly, D^0 can decay into a K^-

$$D^+ \rightarrow K^-\pi^+\pi^+ \tag{5.86}$$

$$D^0 \rightarrow K^-\pi^+ \tag{5.87}$$

$$\rightarrow K^-\pi^+\pi^-\pi^+ \tag{5.88}$$

The mode (5.86) was seen but $D^+ \rightarrow K^+\pi^-\pi^+$ was not.

The D_s meson is a charmed strange pseudo-scalar meson of mass 1969 MeV and lifetime 4.5×10^{-13}. It decays as

$$D_s \rightarrow \phi\pi \tag{5.89}$$

with $\Delta s = \Delta c = 1$. Other decays also have been seen.

The quark model requires charmed baryons, in addition to charmed mesons. The lowest mass charmed baryon has the composition udc and is denoted by Λ_c^+(2282 MeV/c^2). The spin parity assignment is $J^P = \frac{1^+}{2}$. Its strangeness is -1 and is identified in the decays

$$\begin{aligned} \Lambda^+ &\rightarrow \Lambda\pi^+\pi^+\pi^- \\ &\rightarrow \Lambda\pi^+ \\ &\rightarrow pK_s^0 \\ &\rightarrow pK^-\pi^+ \end{aligned} \tag{5.90}$$

Here also the c-quark changes to the s-quark.

$$\Lambda_c^+ \rightarrow \Lambda \tag{5.91}$$

Charmed Σ-hyperion etc., have also been discovered.

5.12.5 *SU*(4) Symmetry

We recall that the $SU(2)$ symmetry is the I-spin symmetry of hadron multiplets which is slightly broken by the electromagnetic interactions. The $SU(3)$ is the symmetry of I and Y in a super-multiplet which is badly broken by strong interactions.

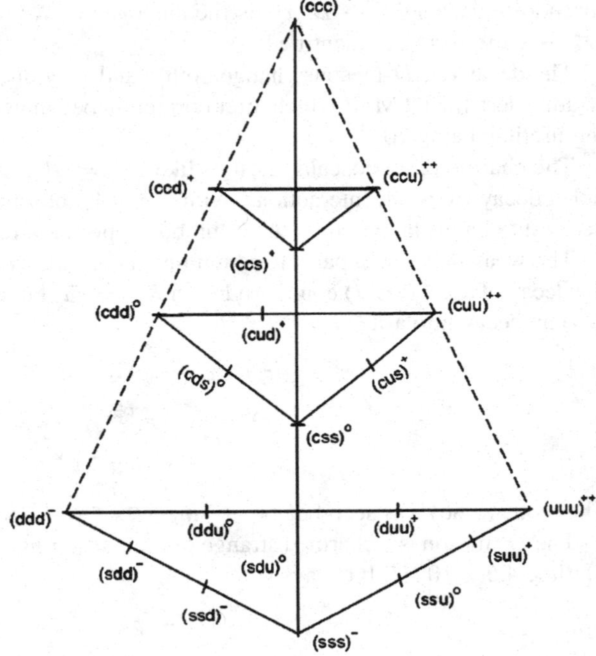

Fig. 5.43 The members of the baryon decuplet ($J^P = 3^+/2$) including the charmed states

By adding the fourth flavour of charm (c) we are concerned with the $SU(4)$ symmetry in C, I and Y. The $SU(4)$ is still more badly broken. Gell-Mann's formula is modified as

$$\frac{Q}{e} = I_3 + \frac{1}{2}(B + S + C) \tag{5.92}$$

and the previous weight diagrams of $SU(3)$ become pyramids with C along the z-axis. Because of the inclusion of charmed particles many more hadron resonances are accommodated in the enlarged diagrams.

Thus each of the meson nonets becomes a 16 plet through

$$4 \otimes \overline{4} = 1 \oplus 15 \tag{5.93}$$

implying the existence of seven new (charmed) states. For the pseudoscalar mesons the new states are

$$D^0(= c\overline{u}), \quad D^+(= c\overline{d}), \quad D_s^+(= c\overline{s}), \quad \overline{D^0}(= u\overline{c}),$$
$$D^-(= d\overline{c}), \quad D_s^-(= s\overline{c}) \quad \text{and} \quad \eta_c(= c\overline{c})$$

Similarly the vector meson nonet is enlarged by the inclusion of D^*, D_s^* and ψ states.

For the baryons the 27 plet now becomes 64 plet through

$$4 \otimes 4 \otimes 4 = 4 \oplus 20 \oplus 20 \oplus 20 \tag{5.94}$$

Table 5.11 Quark charges

Quark	d	u	s	c	b	t
Q/e	$-\frac{1}{3}$	$+\frac{2}{3}$	$-\frac{1}{3}$	$+\frac{2}{3}$	$-\frac{1}{3}$	$+\frac{2}{3}$

The $SU(3)$ octets are enhanced by 12 each and the decuplet by ten charmed states. In Fig. 5.43, the members of the baryon decuplet ($J^P = \frac{3^+}{2}$) including the charmed states are displayed.

5.13 The Fifth Quark (Bottom Quark)

The discovery of the J/ψ and the charmed quark C as well as the τ particle and its associated neutrino ν_τ, required two heavier quarks, b (for bottom or beauty) and t (for top or truth) for reasons of symmetry.

$$\begin{pmatrix} u \\ d \\ e \\ \nu_e \end{pmatrix} \begin{pmatrix} c \\ s \\ \mu \\ \nu_\mu \end{pmatrix} \begin{pmatrix} t \\ b \\ \tau \\ \nu_\tau \end{pmatrix} \tag{5.95}$$

The electric charge assigned to the six quarks are given in Table 5.11.

Both e^+e^- annihilation and hadron production were studied in the collisions of 400 GeV protons with target nuclei by Leon Lederman and his collaborators [20]. The technique employed was to measure the invariant mass of $\mu^+\mu^-$ pair by a double-arm spectrometer, at energy above 5 GeV. The hadrons were eliminated by the use of long beryllium filters in each arm. A clear peak for the $\mu^+\mu^-$ mass was revealed at around 9.5 GeV—a resonance which was named Υ (upsilon).

A more detailed analysis showed better agreement with two peaks with $M_r = 9.46 \pm 0.01$ GeV and $M_{\Upsilon'} = 10.17$ GeV. The partial width $\Gamma_{e^+e^-}(\Upsilon)$ for the decay $\Upsilon \to e^+e^-$ was found to be 1.3 ± 0.4 keV. The Υ'–Υ splitting is nearly the same as for ψ'–ψ. In 1980s Υ'' and Υ''' were also observed at 10.355 GeV and 10.577 GeV. The states Υ, Υ' and Υ'' are narrow while Υ''' is broader. They correspond to $1\,{}^3s_1, 2\,{}^3s_1, 3\,{}^3s_1$, and $4\,{}^3s_1$, states.

The B mesons contain a $\overline{b}$ quark and u- or d-quark. Thus,

$$B^+ = \overline{b}u, \qquad B^0 = \overline{b}d, \qquad B^- = b\overline{u}, \qquad \overline{B} = b\overline{d} \tag{5.96}$$

The decay modes of B mesons are

$$\begin{aligned} &B^- \to D^0\pi^- \\ &\qquad\quad \hookrightarrow K^-\pi^+ \\ &\overline{B}^0 \to D^{*+} + \pi^- \end{aligned} \tag{5.97}$$

The existence of a series of s-wave bound states required that there be p-wave states as well. These are observed through radiative transitions from the s-wave states, $\Upsilon' \to \chi_b\gamma$, where χ_b represents a $C = +1,\ P = +1$ p-wave state. Evidence was

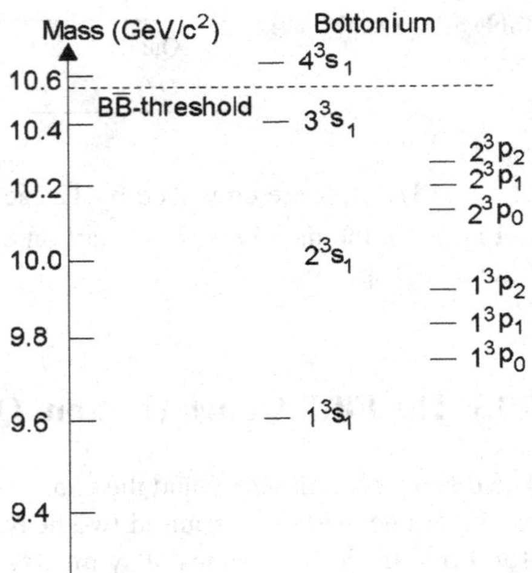

Fig. 5.44 Electromagnetic transitions between various bottomium states

obtained from the photon spectrum $\Upsilon' \to \gamma +$anything, and from the cascade $\Upsilon' \to \chi_b\gamma$, $\Gamma \to \ell^+\ell^-$, where ℓ represents e or μ.

Electromagnetic transitions between various bottomium states are also observed. The spectrum (Fig. 5.44) of these states closely parallels that of charmonium (Fig. 5.39). This indicates that the quark-antiquark potential is independent of quark flavour. The b-quark mass is about 3 times as large as that of c-quark. The radius of quarkonium in the ground state (Bohr's formula) is inversely proportional both to the quark mass and to strong coupling constant α_s. The $1s$ $b\overline{b}$ state thus has a radius ~0.2 fm, that is about half of that of equivalent $c\overline{c}$ state. The equal mass difference between $1s$ and $2s$ states is, however, astounding. A purely Coulombic potential would cause the levels to be proportional to the reduced mass of the system (Bohr's formula). It is clear that the long distance part of potential kr cancels the mass dependence of the energy levels.

The Υ (Upsilon) resonances are analogous to ψ resonance. The width of the lowest resonance is exceedingly small (44 keV), and the width remains small through the excited spin-1 states, analogous to ψ decays, until the state at 10575 MeV, with $\Gamma = 14$ MeV. The interpretation of these narrow width (exactly analogous to ψ) is that Υ is a bound state of yet another new quark called b (for bottom or beauty) and its antiquark $\overline{b}$. The decays are Zweig-suppressed until the energy threshold for production of $B^\pm$ mesons ($b\overline{u}$ and $\overline{b}u$) with a rest energy of 5271 MeV is reached. There are also B^0 and $\overline{B^0}$ mesons ($b\overline{d}$ and $\overline{b}d$); thus $B^\pm$, B^0, $\overline{B}$ form a set similar to D and K. As yet no mesons with quark combinations $b\overline{s}$ or $b\overline{c}$ have been discovered, but there is evidence for a baryon $\Lambda_b^0(udb)$ at 5425 MeV.

The excited spectrum of $b\overline{b}$ states bears a remarkable similarity to the $c\overline{c}$ states as shown in the Fig. 5.44 for the spin-1 states corresponding to parallel alignment of the two quarks which are in an orbital state $\ell = 0$. There are of course other spin states in both the schemes.

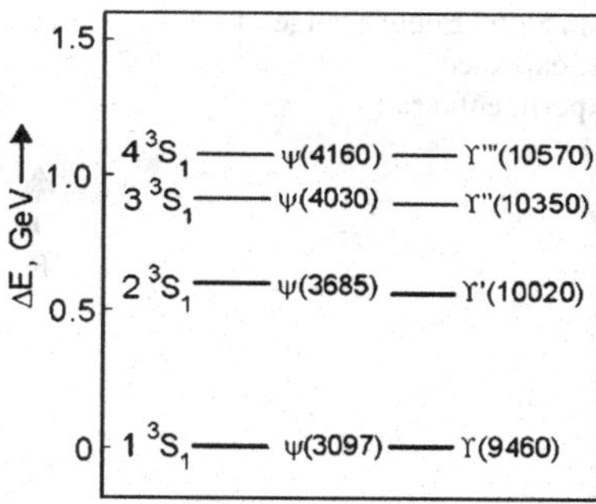

Fig. 5.45 Comparison of Charmonium and Bottonium states

Figure 5.45 shows the triplet s-states of the charmonium and bottomium. Observe the near equality of the $2\,^3s_1$–$1\,^3s_1$, level separation in the two systems.

Further the excitation energies are not strongly dependent on the quark mass.

We expect the leptonic width Γ_{ee} to be proportional to the squared sum of the quark charges $|\sum a_i Q_i|^2$, in the meson's wave function. Table 5.12 gives the Γ_{ee}, $|\sum a_i Q_i|^2$ and the ratio $\Gamma_{ee}/|\sum a_i Q_i|^2$ for the vector mesons ρ, ω, ϕ, and ψ along with their quark description. The value for Υ is obtained by assuming the charge $-e/3$ for the b-quark.

Another indication of the quark charge comes from the ratio R defined by,

$$\frac{\sigma(e^+e^- \to \text{hadrons})}{\sigma(e^+e^- \to \mu^+\mu^-)} = R = \frac{\sum_i Q^2_{qi}}{Q^2_\mu} \tag{5.98}$$

With three flavours (u, d, s) in three colours, the expected ratio is 2. Above the $c\bar{c}$ threshold, the sum should be carried out over four flavours (u, d, s, c) and three colours, and the expected value of ratio R above 4 GeV should be

$$3 \times \left[\left(\frac{2}{3}\right)^2 + \left(-\frac{2}{3}\right)^2 + \left(-\frac{1}{3}\right)^2 + \left(\frac{1}{3}\right)^2\right] = \frac{10}{3}$$

Figure 5.46 shows that in energy range 5–7 GeV, the experimental ratio R is in excellent agreement with the expected value. Beyond the threshold for $b\bar{b}$ production it is necessary to include five flavors and assigning $Q = -\frac{1}{3}e$ to the b quark. We expect $R \to \frac{11}{3}$. This is borne out by experimental data, Fig. 5.47.

Table 5.12 The ratio $\Gamma_{ee}/|\sum a_i Q_i|^2$ for the vector mesons ρ, ω, ϕ, ψ and Υ

Particle	$\rho(=\frac{u\bar{u}-d\bar{d}}{\sqrt{2}})$	$\omega(=\frac{u\bar{u}+d\bar{d}}{\sqrt{2}})$	$\phi(=s\bar{s})$	$\psi(=c\bar{c})$	$\Upsilon(=b\bar{b})$
Υ_{ee}(keV)	6.7 ± 0.8	0.76 ± 0.17	1.31 ± 0.10	4.8 ± 0.6	1.30 ± 0.05
$\lvert\sum a_i Q_i\rvert^2$	$\frac{1}{2}$	$\frac{1}{18}$	$\frac{1}{9}$	$\frac{4}{9}$	$\frac{1}{9}$
$\Upsilon_{ee}/\lvert\sum a_i Q_i\rvert^2$	13.4 ± 1.6	13.7 ± 3.1	11.8 ± 0.9	10.8 ± 1.1	11.7 ± 0.5

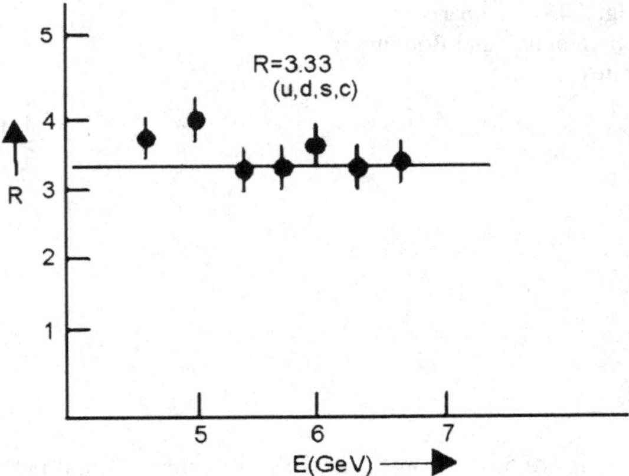

Fig. 5.46 Energy range of the expected value of the experimental ratio R

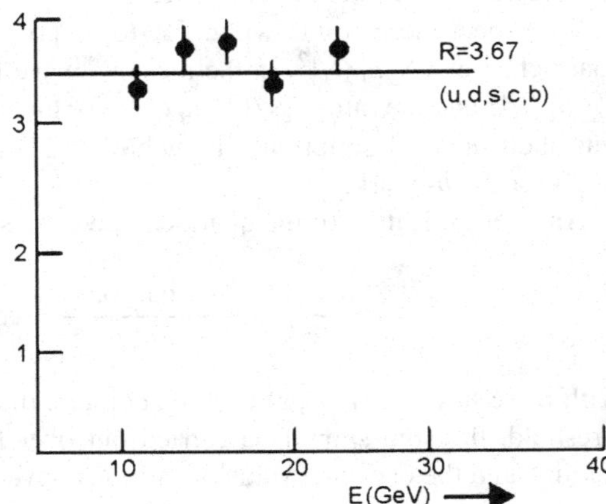

Fig. 5.47 Value of R borne out by experimental data

5.14 The Sixth Quark (Top Quark)

As quarks are seen to come in pairs it was expected that there is a partner to the bottom quark called t (for top or truth). This is the weak isospin partner of the bottom quark with a charge of $+\frac{2}{3}e$. Searches were made for evidence of the top quark through an increase of the hadronic cross section in the ratio R.

In the spring of 1995 at the Fermi Lab evidence for the production of top-antitop quark pairs was obtained in the collisions of p^{-}–p at 1.8 TeV CM energy by Wimpeny and Brian. The predicted dominant modes for the $t\bar{t}$ pair production are, $q\bar{q} \rightarrow t\bar{t}$ and $gg \rightarrow t\bar{t}$ (g for gluon). The lowest-order Feynman diagrams for the $t\bar{t}$ production are shown in Fig. 5.48.

The mass $M_t = 174\ \text{GeV}/c^2$ and the production cross-section $\sigma_{t\bar{t}}$ is found to be 5.9 pico barns.

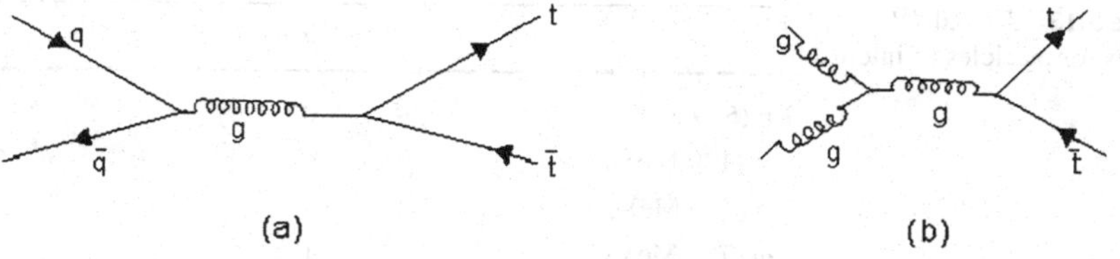

Fig. 5.48 The lowest-order Feynman diagrams for the $t\bar{t}$ production

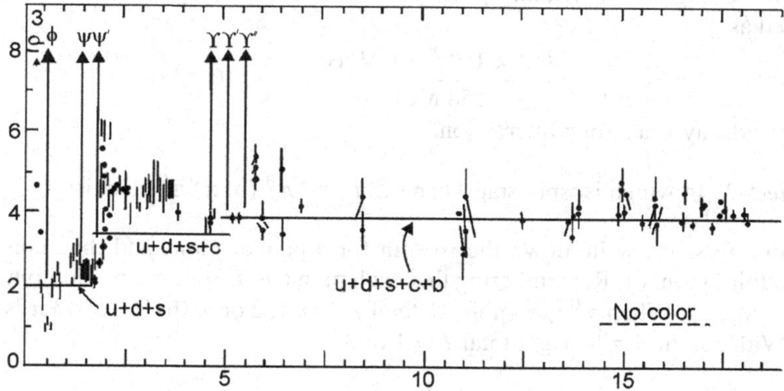

Fig. 5.49 R is plotted against electron energy (in GeV) [21]. Reprinted by permission of John Wiley & Sons, Inc.

With the inclusion of the top quark, Gell-Mann formula is modified as

$$\frac{Q}{e} = I_3 + \frac{1}{2}(B + s + c + b + t) \tag{5.99}$$

Above 175 GeV CMS energy the value of R levels off to 5.0 as it should for the six quarks (not shown in Fig. 5.49).

The t-quark due to its large mass has only a very short lifetime. Thus no pronounced $t\bar{t}$ states (toponium) are expected. Preferred decay for mesons containing a top quark will be by cascade, such as

$$\begin{aligned} t &\to b e^+ \nu_e \\ &\quad \hookrightarrow c\bar{\nu}_e \\ &\qquad \hookrightarrow s e^+ \nu_e \\ &\qquad\quad \hookrightarrow u e^- \bar{\nu}_e \end{aligned} \tag{5.100}$$

With the discovery of the top-quark the lepton-quark symmetry, as envisaged by the standard model, is complete.

Example 5.1 Given that the ρ-meson has a width of 158 MeV/c^2 in its mass, how would you classify the interaction from its decay?

Table 5.13 J^P and I^G values for particles of interest

	J^p	I^G
η (549 MeV)	0^-	0^+
π (139 MeV)	0^-	1^-
ρ (770 MeV)	1^-	1^+
ω (783 MeV)	1^-	0^-

Solution Using the uncertainty principle for time and energy, mean life time is estimated as

$$\tau \sim \frac{\hbar}{\Delta E} = \frac{6.2 \times 10^{-20}\ \text{MeV sec}}{158\ \text{MeV}} \simeq 4 \times 10^{-24}\ \text{sec}$$

Hence the decay is a strong interaction.

Example 5.2 In which isospin states can (a) $\pi^+\pi^-\pi^0$ (b) $\pi^0\pi^0\pi^0$ exist?

Solution First we write down the isospin for a pair of pions and then combine with the third pion. (a) Remembering that each pion has $T = 1$, $\pi^+\pi^-$ combination gives $I = 0, 1, 2$. When π^0 is combined, total $I = 0, 1, 2$ or 3. (b) Here two π^0s give $I = 2$. With the third π^0 we get total $I = 1$ or 3.

Example 5.3 Show that (a) $\rho \to \eta + \pi$ is forbidden as a strong decay (b) $\omega \to \eta + \pi$ is forbidden as an electromagnetic or strong decay.

Solution We tabulate in Table 5.13 J^p and I^G values for particles of interest. (a) G-parity is violated. (b) Isospin is violated.

Example 5.4 The mean life time of the ρ^0 is 10^{-23} sec where as that of the K^0 is 0.89×10^{-10} sec although both decay predominantly to $\pi^+ + \pi^-$. Explain.

Solution The decays are

$$\rho^0 \to \pi^+ + \pi^- \tau = 4 \times 10^{-24}\ \text{sec}$$
$$K^0 \to \pi^+ + \pi^- \tau = 8.9 \times 10^{-11}\ \text{sec}$$

The ρ decay conserves all quantum number and is caused by the strong interactions. However, K^0 decay does not conserve strangeness and is caused by weak interaction and has a mean life time greater by a factor of order 10^{13} than that for ρ^0.

Example 5.5 State which of the following decay modes of the ρ-meson ($J^p = 1^-$, $I = 1$) are allowed by the strong or electromagnetic interactions.

$$\rho^0 \to \pi^+\pi^- \quad (1)$$
$$\to \pi^0\pi^0 \quad (2)$$
$$\to \eta^0\pi^0 \quad (3)$$
$$\to \pi^0\gamma \quad (4)$$

Solution

1. Yes, for strong reactions.
2. No, for Bose symmetry.
3. No, because of C symmetry, $cp = -1$ since $\rho^0 \to e^+e^-$. $C_\eta = C_\pi = +1$ since both decay to two gamma rays.
4. Yes, for electromagnetic interaction.

Example 5.6 Which of the following particles may undergo two-pion decay? Give your reasoning for each of the three cases:

$$f^0(J^P, I) = (2^+, 0)$$

$$\omega^0(J^P, I) = (1^-, 0)$$

$$\eta^0(J^P, I) = (0^-, 0)$$

Solution For pions, $I = 1$. Two pions can exist in $I = 0, 1, 2$ state. Hence, none of the decays is forbidden by isospin conservation alone.

Now pions obey Bose statistics, so that ψ_{total} is symmetrical since $J = 0$ for pions and spatial part of pion system must be symmetrical. Hence allowed states correspond to $L = 0, 2, 4$. Now, the intrinsic parity of each pion being negative does not contribute to the parity of the state. The parity of the system is mainly determined by the L value.

$$p = (-1)^L$$

$\therefore$ Particles with quantum numbers (0^+), (2^+), can decay to two pions. We, therefore conclude that only the decay $f^0 \to \pi^+ + \pi^-$ is possible (the other two particles ω^0 and η^0 actually decay into three pions).

Example 5.7 The scattering of pions by protons shows evidence of a resonance at a centre of mass system momentum of 230 MeV/c. At this momentum, the cross-section for scattering of positive pions reaches a peak cross-section of 190 mb while that for negative pions is only 70 mb. Show that the amplitude $T = 1/2$ is negligible and that the resonance is to be assigned $T = 3/2$.

Solution

$$|a_{3/2}|^2 \propto 190$$

$$\frac{|a_{3/2}|^2}{\{\frac{1}{\sqrt{3}}|a_{3/2} + 2a_{1/2}|\}^2} = \frac{190}{70}$$

Solving, $a_{1/2} = 0.0256 a_{3/2}$. Thus the amplitude $a_{1/2}$ is negligible. The resonance is therefore characterised by $T = 3/2$.

Example 5.8 The cross-section for $K^- + p$ shows a resonance at $p_k \simeq 400$ MeV/c. This resonance appears in the reactions

$$\begin{aligned} K^- + p &\to \Sigma + \pi \\ &\to \Lambda + \pi + \pi \end{aligned}$$

but not in the reaction

$$K^- + p \to \Lambda + \pi^0$$

What conclusion can you draw on the isospin value of the resonance?

Solution The initial state can have $I = 1$ or 0. The resonance must have the same I values. In the first two reactions the final state has $I = 2$, 1 or 0. The value $I = 2$ is excluded as the initial state does not have this value. Now in the third reaction $I = 1$ only. But the resonance does not go through. Hence the conclusion is that resonance has isospin $I = 0$.

Example 5.9 Assuming that the decay $B^+ \to \omega^0 + \pi^+$ occurs in the s-state by strong interaction, deduce the spin, parity and the isospin of B^+ meson (for charge triplet pion $J^P = 0^-$, and for charge singlet ω^0, $J^P = 1^-$). How would your conclusions be altered under the assumption of a weak decay?

Solution

$$\begin{array}{lcc} B^+ \to & \omega^0 + & \pi^+ \\ J^P & 1^- & 0^- \\ I & 0 & 1 \end{array}$$

Assuming strong interaction

$$\begin{aligned} P_B &= P_\omega P_\pi (-1)^l = (-1)(-1)(-1)^0 = +1 \\ J_\omega &= 1 + 0 = 1 \quad (\because \ell = 0) \\ I &= 0 + 1 = 1 \end{aligned}$$

$\therefore$ For B^+ meson, $J^P = 1^+$ and $I = 1$.

In case of weak decay, the spin would still be 1 but it would not be meaningful to talk about I or P.

Example 5.10 Consider the formation of the resonance $\Delta(1236)$ due to the incidence of π^+ and π^- on p. Assuming that at the resonance energy the $I = \frac{1}{2}$ contribution to the $\pi^- + p$ interaction is negligible show that at the resonance peak.

$$\frac{\sigma(\pi^+ + p \to \Delta)}{\sigma(\pi^- + p \to \Delta)} = 3$$

Solution

$$\text{LHS} = \frac{|a_{3/2}|^2}{\{\frac{1}{\sqrt{3}}|a_{3/2} + 2a_{1/2}|\}^2}$$

Put $a_{1/2} = 0$ to get the required result.

Example 5.11 The Coulomb self-energy of a hadron with charge $+e$ or $-e$ is about 1 MeV. The quark content and rest energies (in MeV) of some hadrons are:

$$n(udd)940, \quad p(uud)938$$

$$\Sigma^-(dds)1197, \quad \Sigma^0(uds)1192, \quad \Sigma^+(uus)1189$$

$$K^0(d\bar{s})498, \quad K^+(u\bar{s})494$$

The u and d quarks make different contributions to the rest energy. Estimate this difference.

Solution Subtract 1 MeV from the rest energies of the charged particles.

$$(udd)940, \quad (uud)937; \quad 3\text{ MeV for } m_d - m_u$$

$$(dds)1196, \quad (uds)1192, \quad (uus)1188; \quad 4\text{ MeV for } m_d - m_u$$

$$(d\bar{s})498, \quad (u\bar{s})493; \quad 5\text{ MeV for } m_d - m_u$$

Interchanging a d quark for a u quark always increases the rest energy; in this sample by an average of 4 MeV.

Example 5.12

a. What are the quark constituents of Δ states?
b. Assuming the quarks are in states of zero angular momentum, what fundamental difficulty appears to be associated with the Δ states, which have $j = \frac{3}{2}$, and how is it resolved?
c. How do you explain the occurrence of excited states of the nucleon and Δ with higher values of j?
d. What approximate sequence in the parity of these higher states would your simple model predict?

Solution

a. $\Delta^-(ddd)$, $\Delta^0(ddu)$, $\Delta^+(duu)$, $\Delta^{++}(uuu)$.
b. The three identical quarks in Δ^- and Δ^{++} have a total wave function which is symmetric. Consequently, the Pauli exclusion principle is violated. This difficulty is overcome by the introduction of a new quantum number called colour.
c. Higher orbital angular momentum between the quarks gives rise to excited states.
d. The expected parity is determined by the factor $(-1)^l$ so that the expected parity is the sequence $+$ (ground state), $-, +, -, \ldots$. The higher value ℓ also increases the mass.

Example 5.13 Both Δ^0 baryon and Λ hyperon decay to proton and negative pion. But the lifetime of Δ^0 is $\sim 10^{-23}$ sec while that of Λ is 2.6×10^{-10} sec. Explain.

Solution The decay $\Delta^0 \to \pi^- + p$ is caused by the strong interaction as it conserves all the quantum numbers. The Q-value is large (~150 MeV) and so its mean life time is of the order of 10^{-23} sec. In the decay $\Lambda \to \pi^- p$, there is a strangeness change of +1 and this implies a change in quark flavour. This is attributed to the weak interaction. As the weak interaction has an effective strength 10^{-14} that of the strong interaction so that the mean life times is of the order of 10^{-8} sec.

Example 5.14 Use the additive quark model to prove the relation

$$\sigma(\Lambda p) = \sigma(pp) + \sigma(\overline{K}^0 p) - \sigma(\pi^+ p)$$

(Quark composition: $p = uud, \pi^+ = u\overline{d}, \Lambda = uds, \overline{K}^0 = \overline{d}s$)

Also deduce $\sigma(\Xi^- p)$, given $\Xi^- = ssd$.

Solution

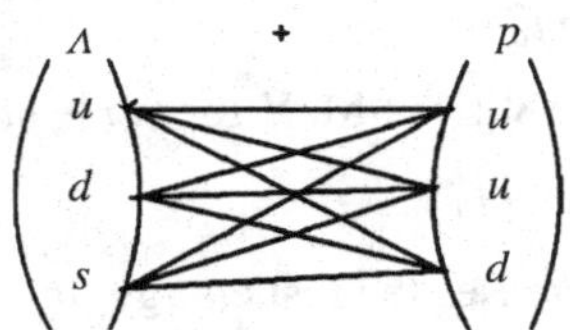

$$\sigma(\Lambda p) = \sigma \begin{bmatrix} uu & + & uu & + & ud \\ +du & + & du & + & dd \\ +su & + & su & + & sd \end{bmatrix}$$

$$= 2\sigma(su) + \sigma(sd) + 2\sigma(uu) + \sigma(ud) + 2\sigma(du) + \sigma(dd)$$

$$= 2\sigma(su) + \sigma(sd) + 2\sigma(uu) + 3\sigma(ud) + \sigma(dd)$$

since $\sigma(ud) = \sigma(d\overline{u})$.

$$\sigma(pp) = \sigma(uu) + \sigma(uu) + \sigma(ud) + \sigma(uu) + \sigma(uu) + \sigma(ud) + \sigma(du) + \sigma(du) + \sigma(dd)$$

$$= 4\sigma(uu) + 4\sigma(ud) + \sigma(dd)$$

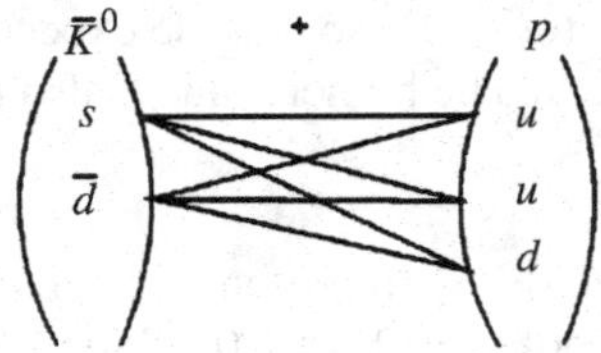

$$\begin{aligned}\sigma\left(\overline{K}^0 p\right) &= \sigma(su) + \sigma(su) + \sigma(sd) + \sigma(\overline{d}u) + \sigma(\overline{d}u) + \sigma(\overline{d}d)\\ &= 2\sigma(su) + \sigma(sd) + 2\sigma(\overline{d}u) + \sigma(\overline{d}d)\\ &= 2\sigma(su) + \sigma(sd) + 2\sigma(ud) + \sigma(dd)\end{aligned}$$

where it is assumed $\sigma(\overline{d}u) = \sigma(du) = \sigma(ud)$ and $\sigma(\overline{d}d) = \sigma(dd)$.

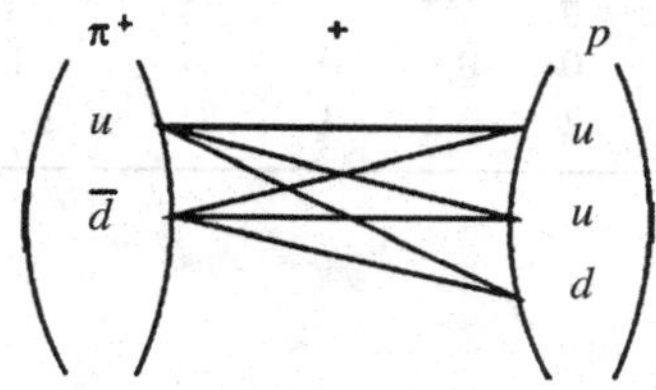

$$\begin{aligned}\sigma\left(\pi^+ p\right) &= \sigma(uu) + \sigma(uu) + \sigma(ud) + \sigma(\overline{d}u) + \sigma(\overline{d}u) + \sigma(\overline{d}d)\\ &= 2\sigma(uu) + 3\sigma(ud) + \sigma(dd)\end{aligned}$$

$$\begin{aligned}\therefore\quad \sigma(pp) + \sigma\left(\overline{K}^0 p\right) - \sigma\left(\pi^+ p\right) &= 2\sigma(uu) + 3\sigma(ud) + \sigma(dd)\\ &\quad + 2\sigma(su) + \sigma(sd)\\ &= \sigma(\Lambda p)\end{aligned}$$

$$\begin{aligned}\sigma\left(\Xi^- p\right) &= \sigma(du) + \sigma(du) + \sigma(dd) + \sigma(su) + \sigma(su) + \sigma(sd)\\ &\quad + \sigma(su) + \sigma(su) + \sigma(sd)\\ &= 2\sigma(ud) + \sigma(dd) + 4\sigma(su) + 2\sigma(sd)\end{aligned}$$

But

$$\begin{aligned}&\sigma\left(\overline{K}^0 p\right) = 2\sigma(su) + \sigma(sd) + 2\sigma(ud) + \sigma(dd)\\ &\sigma(pp) = 4\sigma(uu) + 4\sigma(ud) + \sigma(dd)\\ &\sigma\left(\pi^+ p\right) = 2\sigma(uu) + 3\sigma(ud) + \sigma(dd)\\ &\therefore\quad 2\sigma\left(\overline{K}^0 p\right) + \sigma(pp) - 2\sigma(\pi^+ p) = \sigma\left(\Xi^- p\right)\end{aligned}$$

Example 5.15 Use the quark model to determine the quark composition of

(a) Σ^+, Σ^-, n and p
(b) K^+, K^-, π^+, π^- mesons

Table 5.14 Quantum numbers assumed for quarks

	u	$\overline{u}$	d	$\overline{d}$	s	$\overline{s}$
I, isospin	$\frac{1}{2}$	$\frac{1}{2}$	$\frac{1}{2}$	$\frac{1}{2}$	0	0
I_3, z-component of I	$\frac{1}{2}$	$-\frac{1}{2}$	$-\frac{1}{2}$	$\frac{1}{2}$	0	0
Y, hypercharge	$\frac{1}{3}$	$-\frac{1}{3}$	$\frac{1}{3}$	$-\frac{1}{3}$	$-\frac{2}{3}$	$\frac{2}{3}$
Q/e, charge	$\frac{2}{3}$	$-\frac{2}{3}$	$-\frac{1}{3}$	$\frac{1}{3}$	$-\frac{1}{3}$	$\frac{1}{3}$
S, strangeness	0	0	0	0	-1	$+1$
B, Baryon number	$\frac{1}{3}$	$-\frac{1}{3}$	$\frac{1}{3}$	$-\frac{1}{3}$	$\frac{1}{3}$	$-\frac{1}{3}$

Solution

(a) $\Sigma^+ = uus$, since

$$\begin{aligned} S &= 0 + 0 + (-1) = -1 \\ B &= \frac{1}{3} + \frac{1}{3} + \frac{1}{3} = 1 \\ Q/e &= \frac{2}{3} + \frac{2}{3} - \frac{1}{3} = 1 \\ I_3 &= \frac{1}{2} + \frac{1}{2} + 0 = 1 \\ I &= \frac{1}{2} + \frac{1}{2} + 0 = 1 \end{aligned}$$

Similarly,

$$\begin{aligned} \Sigma^- &= dds \\ n &= udd \\ p &= uud \end{aligned}$$

(b)

$$\begin{aligned} K^+ &= u\overline{s} \\ K^- &= \overline{u}s \\ \pi^+ &= u\overline{d} \\ \pi^- &= \overline{u}d \end{aligned}$$

Example 5.16 Conventionally nucleon is given positive parity. What does one say about deuteron's parity and the intrinsic parities of u- and d-quarks?

Solution Parity of deuteron $= \pi_p \pi_n (-1)^0 = +1$.

Since the quarks have zero orbital angular momentum in the nucleons, the quark's intrinsic parities must be positive.

Example 5.17 A 12 GeV electron collides with proton and emerges from the collision with a 10° deflection and energy of 8 GeV. Calculate the rest mass W of the recoiling hadron state.

Solution Neglect the electron mass. Let W, E_w, p_w be the mass, energy and momentum of the recoiling hadronic state, M the nucleon mass, E_0 and E the energy of the incident electron and scattered electron, θ the angle of deflection of electron, $\vec{p}_0$ and $\vec{p}$ the momentum of incident electron and scattered electron.

$$\vec{p}_w = \vec{p}_0 - \vec{p}$$

$$P_w^2 = p_0^2 + p^2 - 2p_0 p \cos\theta \qquad (1)$$

$$E_w = E_0 + M - E \qquad (2)$$

$$E_w^2 - P_w^2 = M_w^2$$

$$= M^2 + E_0^2 - p_0^2 + E^2 - p^2 + 2M(E_0 - E) + 2p_0 p \cos\theta - 2E_0 E$$

$$\because \quad p_0 \simeq E_0 \text{ and } p \simeq E$$

Further, $E_0^2 - p_0^2 = m_e^2 \cong 0$

$$M_w^2 = M^2 + 2M(E_0 - E) - 2E_0 E(1 - \cos\theta)$$

Put $E_0 = 12$ GeV, $E = 8$ GeV, $M = 0.94$ GeV, $\theta = 10°$

$$M_w = 2.5 \text{ GeV}/c^2$$

Example 5.18 Why is the decay $\eta \to 4\pi$ not observed?

Solution The mass of η-meson is 549 MeV/c^2 while the mass of four pions is 558 MeV/c^2 if all the four pions are charged and 540 MeV/c^2 if the pions are uncharged. In the first case the decay cannot occur because of energy nonconservation, and in the second case the decay will be suppressed because of small available phase space. Apart from this the only possible pionic decay is the 3-body mode, the strong decay into two pions or four pions being forbidden by parity conservation.

Example 5.19 The $\Delta(1232)$ is a resonance with $I = \frac{3}{2}$. What is the predicted branching ratio for $(\Delta^0 \to p\pi^-)/(\Delta^0 \to n\pi^0)$? What would be this ratio for a resonance with $I = \frac{1}{2}$?

Solution The nucleon has $I = \frac{1}{2}$ and pion $I = 1$. The Δ^0 has $I_3 = -\frac{1}{3}$. Using Clebsch-Gordan coefficients for $1 \times \frac{1}{2}$, we have

$$\underset{\Delta^0}{\left|\frac{3}{2}, -\frac{1}{2}\right\rangle} = \sqrt{\frac{2}{3}}\underset{\pi^0}{|1, 0\rangle} \underset{n}{\left|\frac{1}{2}, -\frac{1}{2}\right\rangle} + \sqrt{\frac{1}{3}}\underset{\pi^-}{|1, -1\rangle} \underset{p}{\left|\frac{1}{2}, \frac{1}{2}\right\rangle}$$

The ratio of the amplitudes for the decays $\Delta^0 \to \pi^- p$ and $\Delta^0 \to \pi^0 n$ is just the ratio of the corresponding Clebsh-Gordan Coefficients. Hence the branching ratio

$$\Delta^0 \to p\pi^-/\Delta^0 \to n\pi_0 \quad \text{is} \quad \left(\frac{1}{\sqrt{3}}\right)^2 \Big/ \left(\sqrt{\frac{2}{3}}\right)^2 = \frac{1}{2}$$

If the Δ had $I = \frac{1}{2}$, we would have

$$\underset{\Delta^0}{\left|\frac{1}{2}, -\frac{1}{2}\right\rangle} = \sqrt{\frac{1}{3}}\underset{\pi^0}{|1,0\rangle}\underset{n}{\left|\frac{1}{2}, -\frac{1}{2}\right\rangle} + \sqrt{\frac{2}{3}}\underset{\pi^-}{|1,-1\rangle}\underset{p}{\left|\frac{1}{2}, \frac{1}{2}\right\rangle}$$

In this case the branching ratio would be

$$\left(\sqrt{\frac{2}{3}}\right)^2 \Big/ \left(\sqrt{\frac{1}{3}}\right)^2 = 2$$

5.15 Questions

5.1 Give the quark structure of (a) Λ; (b) Ξ^-; (c) p^-; (d) K^0.
[Ans. (a) uds; (b) dss; (c) $\overline{u}\,\overline{u}\overline{d}$; (d) $d\overline{s}$]

5.2 What are the quark constituents of (a) Δ^-; (b) Δ^0; (c) Δ^+; (d) Δ^{++}?
[Ans. (a) ddd; (b) ddu; (c) duu; (d) uuu]

5.3 What are the three particles described by taking three identical quarks?
[Ans. Δ^{++}, Δ^-, Ω^-]

5.4 Give the quark structure of (a) π^0; (b) $\overline{n}$; (c) $\overline{\Lambda}$.
[Ans. (a) $(u\overline{u} - d\overline{d})/\sqrt{2}$; (b) $\overline{u}\overline{d}\,\overline{d}$; (c) $\overline{s}\,\overline{u}\overline{d}$]

5.5 If a resonance has $I = \frac{3}{2}$, $B = 1$ and $S = 0$, what are its charge states?
[Ans. Δ^-, Δ^0, Δ^+, Δ^{++}]

5.6 Use the Gell-Mann-Nishijima formula to verify the charges of u, d, s quarks and their antiquarks.

5.7 What values of electric charge are possible for (a) a baryon and (b) a meson in the quark model?
[Ans. (a) $++, +, 0$ and $-$; (b) $+, 0$ and $-$]

5.8 Determine the quark content of the K^+, $K^0(S = +1)$ and K^-, $\overline{K}^0(S = -1)$ mesons.
[Ans. $K^+ = \overline{s}u$ and $K^0 = \overline{s}d$; $K^- = s\overline{u}$ and $\overline{K}^0 = s\overline{d}$]

5.9 The Σ^0 hyperon decays to $\Lambda + \gamma$ with a mean life time of 7.4×10^{-20} sec. Estimate its width.
[$\Delta E \simeq \hbar/(7.4 \times 10^{-20})$ sec $= 1.05 \times 10^{-21}/(7.4 \times 10^{-20}) = 0.014$ MeV $= 14$ keV typical of EM decay]

5.10 The Λ is an isosinglet baryon with strangeness $S = -1$. What is its quark content?
[Ans. uds]

5.11 What are the quark content of Ξ hyperon?
[Ans. $\Xi^0 = ssu$; $\Xi^- = ssd$]

5.12 Using quark model explain whether the decay $K^- \rightarrow \pi^- + \pi^0$, is strong, electromagnetic or weak?
[Ans. As an s quark changes to a d quark, it is weak]

5.13 The charmed mesons $D^{+,0}$, have $J^\pi = 0^-$ and their excited states $D^{*+0}(2010)$ have $J^P = 1^-$. Determine their quark content and give the values of total quark spin and orbital angular momentum.
[Ans. $D^+(c\overline{d})$, $D^0(c\overline{u})$ have $S = 0$, $L = 0$, $D^{*+}(c\overline{d})$, $D^{*0}(c\overline{u})$ have $S = 1$, $L = 0$]

5.14 Indicate at the quark level whether the following reactions or decays can proceed through the strong, the electromagnetic, or the weak interaction.

(a) $\pi^- + p \rightarrow K^0 + \Lambda$
(b) $D^+ \rightarrow K^- + \pi^+ + \pi^+$
(c) $\Xi^- \rightarrow \Lambda + \pi^-$
(d) $n + p \rightarrow d + \gamma$
(e) $p + p \rightarrow d + \pi^+$
(f) $K^- + p \rightarrow K^0 + n$

Solution

(a) All quarks flavours are conserved. The associated production implies that an $s\overline{s}$ quark pair has to be produced.
(b) There is a quark flavour change $c \rightarrow s$. The decay therefore proceeds through weak interaction.
(c) There is a quark flavour change, $s \rightarrow u$, and so it must be a weak process.
(d) As photon is produced, it is an electromagnetic process.
(e) This involves hadrons alone and there is no change in the quark flavour. It is therefore a strong interaction process.
(f) This reaction implies $\Delta s = +2$ (where $S =$ strangeness) and requires a second order weak interaction, the cross-section being unobservably small.

5.15 Explain what is meant by 'colour' in the context of quarks. What property of the Ω^- baryon led to the concept of colour?

5.16 How does the property of colour make Quantum Chromodynamics much less easy to use than Quantum electrodynamics?

5.17 What are the quark contents of the K^+ and K^- mesons?

5.18 Plot the S-I_3 diagram for the pseudo-scalar mesons.

5.19 Explain the term four-momentum transfer q.

5.20 The ρ^0 meson has an intrinsic spin of h, and pions have zero spin. If parity is conserved in the decay $\rho^0 \to \pi^+\pi^-$, what must the intrinsic parity of the ρ^0 be?

5.21 Give the Y-I_3 diagram for the baryon decuplet (spin-parity $\frac{3}{2}^+$).

5.22 The following charmed baryons have been discovered, Σ_c^{++}, Σ_c^+, Σ_c^0 with $S = 0$ and masses $\simeq$2450 MeV/c^2, Σ_c^+ with $S = -1$ and mass $\simeq$2450 MeV/c^2. What is their likely quark content?
[Ans. uuc, udc, ddc and usc respectively]

5.23 Explain why the strong force is limited in range, despite the fact that gluons are massless.

5.24 Define the term 'Deep inelastic scattering'. How is this technique used to explore the structure of proton?

5.25 What are the quantum numbers of the b quark? Give its spin, electric charge, isospin and beauty.

5.16 Problems

5.1 At certain energy $\sigma(\pi^+ n) = \sigma(\pi^- n)$, while $\sigma(K^- p) = 2.5\sigma(K^+ n)$. Explain.

5.2 Show that the magnetic moment $\vec{\mu}_\Lambda$, of the Λ is simply the magnetic moment $\vec{\mu}_s$ of the S-quark and that $\mu_\Lambda = -\frac{m}{3m_s}\mu_p$, where μ_p is the proton magnetic moment and m is the u, d quark constituent mass.

5.3 An energetic ρ^0 decays into two charged pions of energy 1.38 GeV each, separated at an angle of 39°. Calculate the mass of ρ^0 (mass of charged pion is 0.14 GeV/c^2).
[Ans. 760 MeV/c^2]

5.4 A D^0 at rest decays according to $D^0 \to K^-\pi^+$. The momentum of the pion is found to be 862 MeV/c. Calculate the mass of the D^0 given $m_k = 494$ MeV/c^2 and $m_\pi = 140$ MeV/c^2.
[Ans. 1867 MeV/c^2]

5.5 The wave functions of three gluon states may be written as

$$\sqrt{\frac{1}{2}}(r\bar{r} + g\bar{g}), \qquad \sqrt{\frac{1}{3}}(r\bar{r} + g\bar{g} + b\bar{b}) \quad \text{and} \quad \sqrt{\frac{1}{6}}(r\bar{r} + g\bar{g} - 2b\bar{b}).$$

Why does $b\overline{b}$ appear differently to the other two combinations? Determine, explaining your reasoning, which of these belong to the octet and which is the singlet.

5.6 Give the I_3-Y diagram for the light baryons octet $(\frac{1}{2}^+)$.

5.7 A particle ω decays at rest via $\omega \rightarrow \pi^+ + \pi^-$, each pion having a kinetic energy of 251 MeV. Calculate the mass of the particle ω. An alternative decay mode is $\omega \rightarrow \pi^0 + \gamma$, determine the momenta, in MeV/c, of the decay products.
[Ans. 403 MeV/c]

5.8 At an energy of 2 GeV, the cross-section for $e^+e^- \rightarrow \mu^+\mu^-$ is 20 nb. At this energy u, d, and s quarks can be produced. What is the cross-section for hadron production?
[Ans. 40 nb]

5.9 Assuming that $\sigma(qq) = \sigma(q\overline{q})$, prove the relation, $\sigma(\Lambda p) = \sigma(pp) + \sigma(k^0 p) - \sigma(\pi^+ p)^-$.

5.10 Explain why the decay $\rho \rightarrow \eta + \pi$ is forbidden.

5.11 (a) Show how the decay $\eta \rightarrow 2\pi$ is forbidden by parity conservation. (b) The decay $\eta \rightarrow 3\pi^0$ by isospin conservation.

5.12 Verify that isospin invariance excludes the decay $\omega \rightarrow 3\pi^0$.

5.13 Show that for an $I = \frac{3}{2}$ resonance in πp scattering the $\Delta(1232)$ produced yields a $1 + 3\cos^2\theta$ angular distribution.

5.14 Write down Okuba's mass formula and explain how omega minus was predicted.

5.15 What are quark lines? What is the dominant decay modes of D-mesons.

References

1. Anderson, Fermi, Long, Nagle (1952)
2. M.H. Alston, M. Ferio-Luzzi, Rev. Mod. Phys. **33**, 416 (1961)
3. Shafer et al. (1963)
4. D.H. Perkins, High Energy Phys.
5. Pjerrou et al. (1962)
6. Bertanza et al. (1962)
7. Maglic et al. (1961)
8. Pevsner et al. (1961)
9. Alston et al. (1961)
10. Fermi, Yang (1949)

11. Sakata (1956)
12. Gell-Mann, Zweig (1964)
13. Barnes et al. (1964)
14. Greenberg (1964)
15. Zweig (1964)
16. Glashow (1970)
17. Augustin et al. (1974)
18. Aubert et al. (1974)
19. Eichten et al., Phys. Rev. D **21**, 203 (1980)
20. L. Lederman et al. (1977)
21. F. Halzen, A.D. Martin, *Quarks and Leptons* (Wiley, New York, 1984), p. 229

Chapter 6
Electromagnetic Interactions

Electromagnetic forces are familiar in atomic and molecular physics, and the quantum theory concerned with these forces predicts results to a high degree of precision which have been verified by experiments. The words force and interaction are interchangeably used. The theory is mainly concerned with the interactions of charged particles and photon. According to Dirac's relativistic theory an electron can have positive energy as well as negative energy states. All the negative energy states are assumed to be filled. If an energetic photon ($E_\gamma \geq 2mc^2$) hits an electron, then it can be lifted from the sea of negative energy states into a positive energy state, leaving a hole behind. The hole is emitted as a positron (e^+) along with the electron. This is known as the electron-positron pair production process.

The incident photon below the pair production threshold (1.02 MeV) may be absorbed by bound electron resulting in its ejection (provided the photon energy is above the threshold for this process to occur). This type of interaction is known as photoelectric effect.

Alternatively, the incident photon may undergo an elastic scattering with a nearby free electron and continue its motion with a deviation with reduced energy, the electron recoiling at some angle, satisfying the energy and momentum conservation. The above processes were described in detail in AAK 1,2. Apart from these, there are other processes involving electrons and positrons which will be considered later.

6.1 Feynman Diagrams

In 1940s a pictorial technique was invented by Feynman for the analysis of elementary particle interactions such as the scattering of a particle off the other, decay of a particle etc. Originally, these diagrams were introduced as a shorthand for writing down individual terms in the calculation of transition matrix elements in electromagnetic process within the framework of quantum electrodynamics (QED).

A. Kamal, *Particle Physics*, Graduate Texts in Physics,
DOI 10.1007/978-3-642-38661-9_6, © Springer-Verlag Berlin Heidelberg 2014

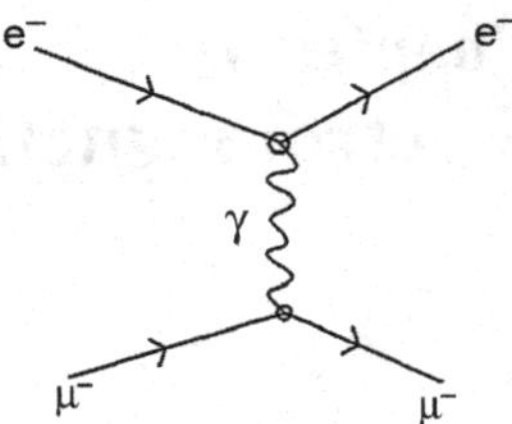

Fig. 6.1 A first-order Feynman diagram for e^{-}–μ^{-} scattering

The general mathematical theory of Feynman diagrams is in the preview of full advanced quantum mechanics texts and is well beyond the scope of this book. However, even qualitative use of these diagrams provides considerable insight into the mechanism of interactions in question.

Consider the diagram shown in Fig. 6.1. The solid straight lines are fermion lines, an electron for the top line and a muon for the bottom line. We use the convention that the time runs from left to right on the horizontal axis. (Some authors take time axis vertically up.) The dots corresponding to vertices are points where interactions occur. The wriggly line is a photon line. The given diagram represents an electron emitting one photon which is then absorbed by the muon (or vice versa), i.e. $e^{-} + \mu^{-} \rightarrow e^{-} + \mu^{-}$. This corresponds to the lowest order of perturbation theory, and is known as the first order Feynman diagram or leading Feynman diagram. The Feynman methodology includes spins, relativity, Fermi, Dirac and Bose statistics, strength of interaction, and is equivalent to a full field theory. Every allowed diagram is a term in the perturbation series. Each part of the figure has a mathematical meaning. The detailed correspondence will not be used in the text, nonetheless some features presented in the diagrams will be invaluable in understanding the interactions.

In the picture, the solid lines have arrows indicating direction. An anti-particle is viewed in these diagrams as a particle going backward in time. In Dirac theory anti-particles come from negative-energy solutions of the Dirac equation. Feynman viewed these as positive-energy particles moving backward in time with the charge reversed from the positive-energy solution. In the case of photon, its antiparticle is photon itself, that is why the photon line does not need an arrow.

The internal lines (those which begin and end within the diagram) represent particles that are not observed (virtual particles). Only the external lines (those which enter or leave the diagram) represent "real" (observable) particles. The external lines then tell us what physical process is occurring, the internal lines describe the mechanism involved.

Feynman diagrams are purely symbolic, they do not represent particle trajectories. The horizontal dimension on the time axis and the spacing between vertices do not correspond to reality. The divergence of lines also does not suggest attraction or repulsion between particles. The Feynman diagram can be twisted in any position so long the rules of the game are obeyed.

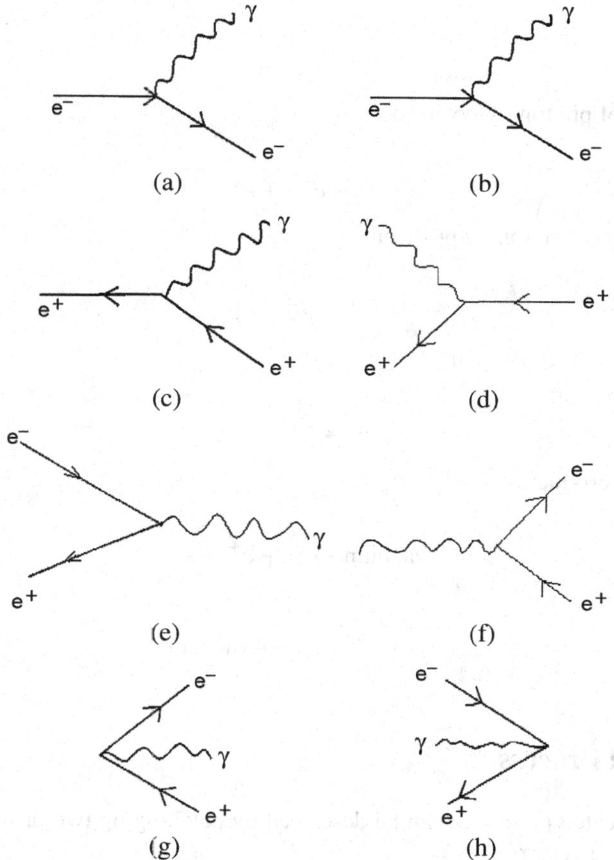

Fig. 6.2 Feynman diagrams for eight basic electromagnetic processes

6.2 Basic Electromagnetic Processes

The eight basic electromagnetic processes may be represented by the Feynman diagrams (Fig. 6.2).

a.

$$e^- \rightarrow e^- + \gamma \tag{6.1}$$

Emission of photon by an electron which corresponds to transition between positive energy states of the electron

b.

$$\gamma + e^- \rightarrow e^- \tag{6.2}$$

Absorption of photon by electron which also corresponds to transition between positive energy states of the electron

c.

$$e^+ \rightarrow e^+ + \gamma \tag{6.3}$$

Emission of photon by positron

d.

$$\gamma + e^+ \rightarrow e^+ \tag{6.4}$$

Absorption of photon by positron

e.

$$e^+ + e^- \rightarrow \gamma \tag{6.5}$$

Annihilation of e^+e^- pair

f.

$$\gamma \rightarrow e^+ + e^- \tag{6.6}$$

e^+e^- pair production

g.

$$\text{Vacuum} \rightarrow \gamma + e^+ + e^- \tag{6.7}$$

h.

$$\gamma + e^+ + e^- \rightarrow \text{vacuum} \tag{6.8}$$

6.3 Actual Process

More complicated processes can be described by patching up two or more replicas of the primitive vertex.

Figure 6.3 shows that two electrons enter, a photon is emitted by one and absorbed by the other, and the two exit. In the classical theory the process is called the repulsion of the like charges. In QED it is called Moller scattering, an interaction mediated by the exchange of a virtual photon of momentum q. The direction of arrows on the fermion lines is such that the charge is conserved at each vertex. Further, four-momentum (Sect. 6.18) is also conserved at all the vertices. The virtual particles (here photon) are present neither in the initial state nor in the final state and exist briefly during the interaction.

In each of the diagrams (a)–(h) in Fig. 6.2 energy is violated for all the basic processes which are virtual and cannot occur in isolation in free space. For the real process, two or more virtual processes must be combined in such a way that energy conservation is only violated for a short period of time τ compatible with energy-time uncertainty principle

$$\tau \cdot \Delta E \sim \hbar \tag{6.9}$$

Because of the uncertainty principle, virtual particles need not satisfy the energy-momentum relation $E^2 = c^2p^2 + m^2c^4$. A virtual particle can have any mass. It does

Fig. 6.3 Moller scattering

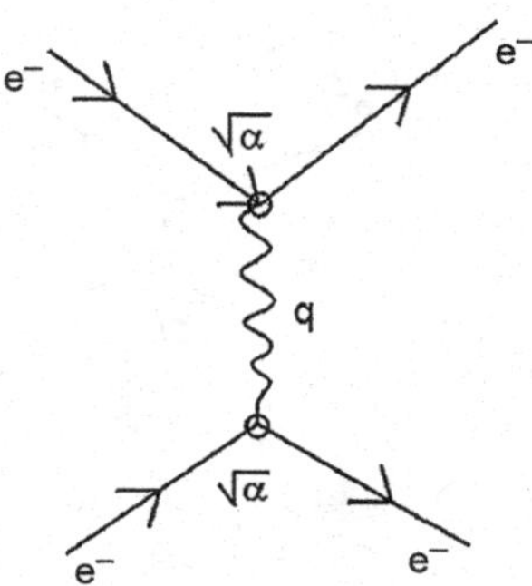

not carry the same mass as that of the corresponding free particle. One says that virtual particles do not lie on their mass shell. External lines, by contrast, represent real particles, and these do carry the correct mass.

In Fig. 6.3 the exchanged photon couples to the charge of one electron at the top vertex and the second one at the bottom vertex. For each vertex the transition amplitude carries a factor which is proportional to e, that is $\sqrt{\alpha}$ (square root of fine structure constant). The transition matrix will be proportional to $\sqrt{\alpha}\sqrt{\alpha}$ or α. The exchanged particle also introduces a propagation term in the matrix element, the general form being

$$\frac{1}{Q^2 + M^2C^2} \tag{6.10}$$

where Q^2 is the square of the four-momentum transfer for the interaction and M is the mass of the exchange particle. The 4-momentum transfer $Q = (p, iE)$, where $Q^2 = p^2 - E^2$, in units of $\hbar = c = 1$. In the case of the virtual photon, (6.10) reduces to $1/Q^2$ for the amplitude and $1/Q^4$ for the cross-section. Thus, in the Rutherford scattering, the momentum transfer is $2p \sin\frac{\theta}{2}$, the differential cross-section is proportional to $\frac{1}{p^4 \sin^4 \theta/2}$ or $\frac{1}{E^2 \sin^4 \theta/2}$, as it should.

In general, a number of Feynman diagrams can be drawn for a given event for the same end result. The transition matrix element includes the superposition of amplitudes of all such diagrams resulting in the same final state.

6.3.1 Moller Scattering

Consider photon exchange again in the Moller scattering (e^-e^- scattering).

Figure 6.4(a) represents a process where by an electron emits a photon which is subsequently absorbed by a second electron. Although energy conservation is violated at the first vertex, this can be compensated by a similar violation at the second vertex to give an overall energy conservation.

Figure 6.4(a) represents the contribution to the physical scattering process

$$e^- + e^- \to e^- + e^-$$

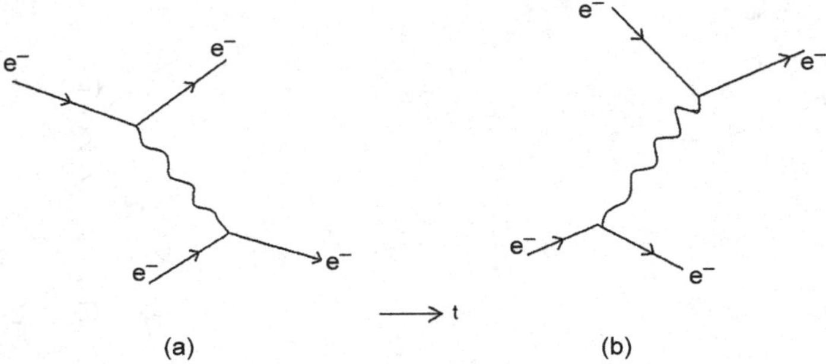

Fig. 6.4 Moller scattering. Two ordered diagrams

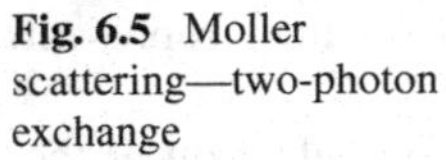
Fig. 6.5 Moller scattering—two-photon exchange

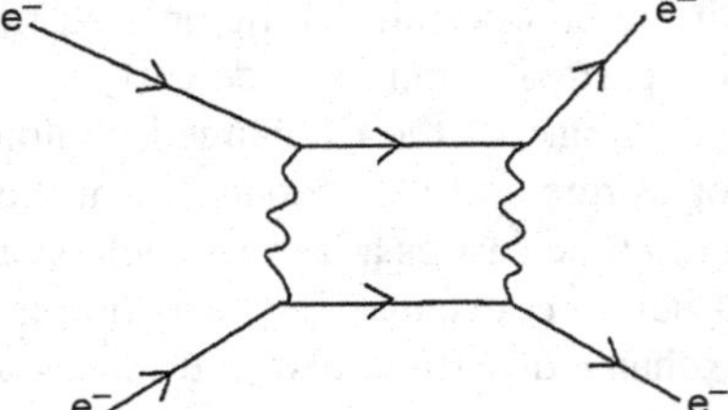

from single-photon exchange. There is also a second contribution, represented by Fig. 6.4(b) in which the other electron emits the photon which is exchanged. Both processes contribute to the observed scattering. Scattering can also occur via multi-photon exchange. As an example a diagram corresponding to two-photon exchange is shown in Fig. 6.5. Contributions of such diagrams are, however, by far smaller than the one-photon exchange contribution. This is explained by considering the number of vertices in each diagram called its order. Each vertex represents a basic process whose probability of occurrence is proportional to $\alpha \cong 1/137 \ll 1$ and any diagram of order n gives a contribution of order α^n. Thus for a single-photon exchange contribution is of order α^2, for two-photon exchange it is of order α^4 and more generally for n-photon exchange it is of order α^{2n}. In practice one photon-exchange gives sufficiently good accuracy.

In what follows we shall consider some other familiar electromagnetic processes depicted by Feynman diagrams.

6.3.2 e^+–e^- *Annihilation*

Figures 6.6(a) and 6.6(b) are the leading diagrams for the two-photon annihilation. The two diagrams are related by 'time ordering'. Figure 6.6(a) is obtained by combining Figs. 6.2(a) and (e), and Fig. 6.6(b) by combining Figs. 6.2(c) and (e).

The three-photon annihilation $e^+ + e^- \to \gamma + \gamma + \gamma$ is represented in Fig. 6.7 as a possible diagram. It is obtained by combining (a), (c) and (e) of Fig. 6.2. Since

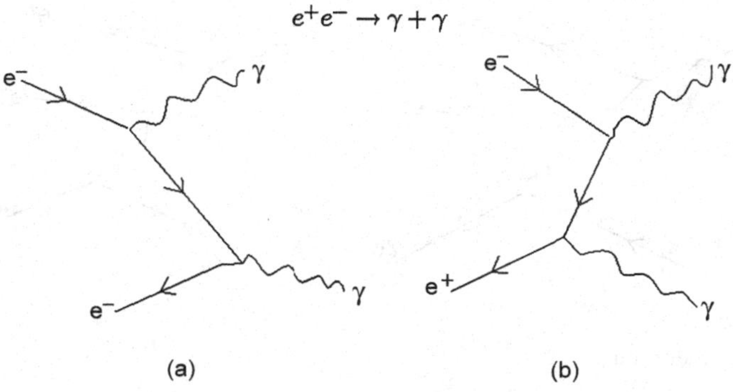

Fig. 6.6 Two-photon annihilation

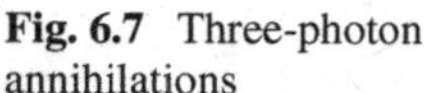

Fig. 6.7 Three-photon annihilations

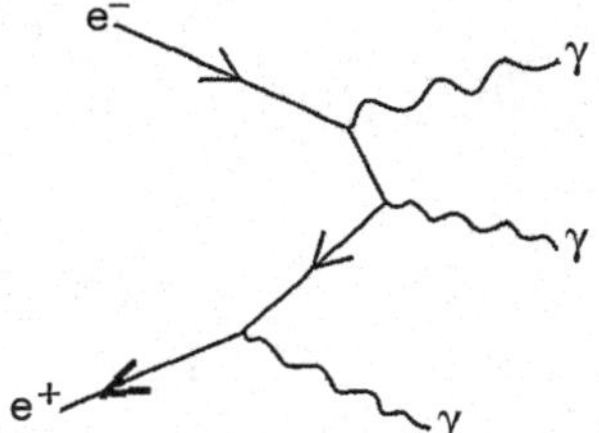

there are three vertices, there are $3! = 6$ different ways of ordering them in time. Other five related time-ordered diagrams are not shown but are implied.

From the previous discussion we expect

$$R = \frac{\text{Rate}(e^+e^- \to 3\gamma)}{\text{Rate}(e^+e^- \to 2\gamma)} = O(\alpha) \tag{6.11}$$

Here O is for order. This prediction can be tested by the observed 2γ and 3γ decay rates of positronium. The experimental value of $R = 0.9 \times 10^{-3}$ is some what smaller than $\alpha = 0.7 \times 10^{-2}$.

Note that one-photon annihilation is not possible as it would violate the conservation of momentum.

6.3.3 Pair Production $\gamma \to e^+ + e^-$

This basic process cannot conserve both energy and momentum simultaneously, but can proceed in the presence of a nucleus.

$$\gamma + (Z, A) \to e^+ + e^- + (Z, A)$$

The Pair production process in the lowest order is shown by the two diagrams in Fig. 6.8. The two diagrams which represent distinct contributions to the process are not related by time ordering. The expected rate is of order $z^2\alpha^3$ which is experimentally confirmed.

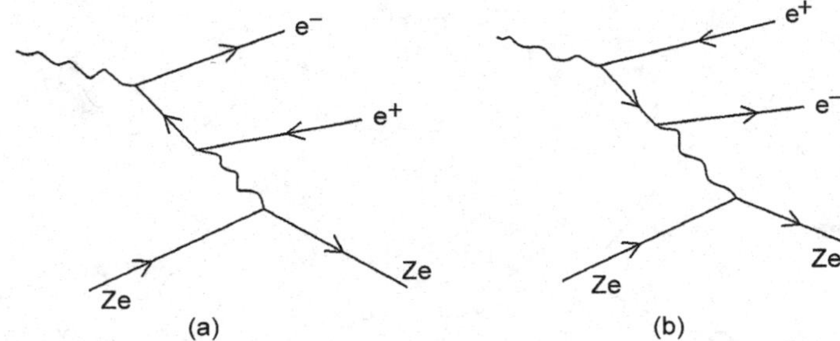

Fig. 6.8 Pair production

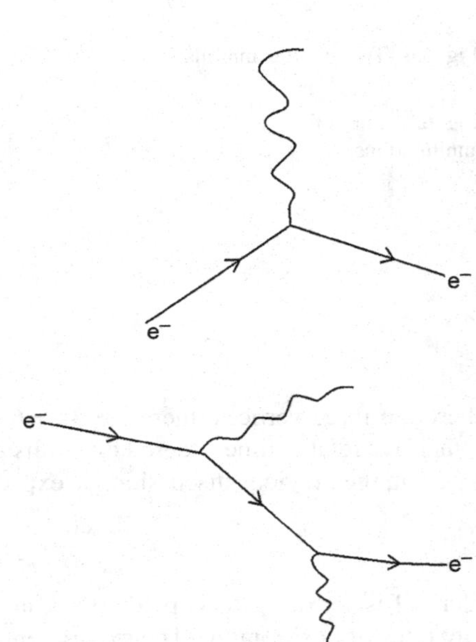

Fig. 6.9 Photoelectric effect

Fig. 6.10 Bremsstrahlung

6.3.4 Photoelectric Effect and Bremsstrahlung

The diagram in Fig. 6.9 is for photoelectric effect which is identical with the basic diagram of Fig. 6.2(b). The diagram in Fig. 6.10 corresponds to Bremsstrahlung.

6.4 Second-Order Diagrams

Sometimes diagrams can have lines that form closed loops, as in Fig. 6.11(b). In this methodology the internal four-momentum is integrated over all closed loops.

Figure 6.11(a) depicts the annihilation of e^+–e^- pair into a μ^-–μ^+ pair; via the production of photon in the intermediate state. Figure 6.11(b) is a slightly com-

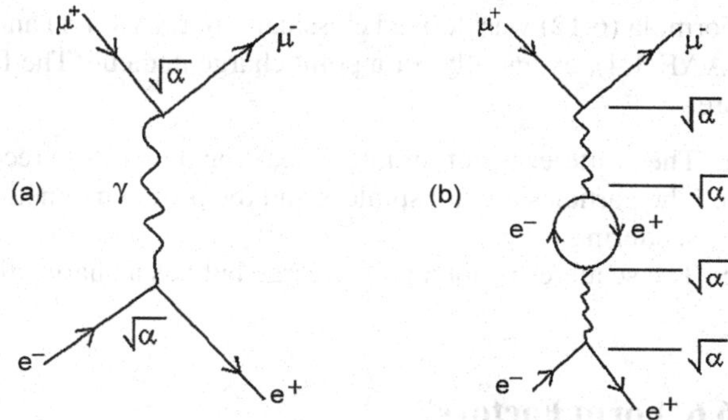

Fig. 6.11 Annihilation of e^+–e^- into $\mu^+\mu^-$ pair

plicated version of the same process. Here the photon is briefly transformed into an intermediate state consisting of a e^+–e^- pair. This and additional more complicated diagrams contributing to the same process are known as higher-order diagrams, while a diagram as in Fig. 6.9 with a single vertex is known as the lowest order or leading diagram.

The field theory used to compute the cross-sections for such electromagnetic process is called quantum electrodynamics (QED). One important property of QED is that of renormalizability. In processes as in Figs. 6.8, 6.9 and 6.10 divergent terms will be present in the QED calculations, the theoretically calculated "bare" mass m_0 or charge e_0 becoming infinite. Fortunately, it is found possible to dump all the divergences into m_0 or e_0 and then redefine the mass and charge, replacing them by their physical values e and m. This process in known as renormalization. When this is done in terms of the physical charge e and mass m, the results give finite and highly accurate value for cross-sections, decay rates etc.

6.5 Scattering of Electrons

For the purpose of exploring the structure of a nucleus or nucleon, a beam of protons is not suitable as bombarding particles because of the complications arising from strong interaction. In contradistinction e–p and e–n scattering results can be interpreted directly.

The e–p scattering may be visualized as an interaction due to exchange of single virtual photon as a good approximation.

6.5.1 Elastic Scattering of Spinless Electrons by Nuclei

If $E \gg mc^2$, the elastic scattering of spinless electron from a static point charge ze is given by

$$\frac{d\sigma}{d\Omega} = \frac{z^2}{4}\left(\frac{e^2}{mv^2}\right)^2 \frac{1}{\sin^4\frac{\theta}{2}} \quad \text{(Rutherford formula)} \tag{6.12}$$

Formula (6.12) was derived classically in (AAK 1,1) and by Born approximation in (AAK 1,4), essentially for a point charge nucleus. The factors which modify (6.12) are

a. The scatterer is not infinitely heavy and therefore recoils.
b. The particles are not spinless and the proton magnetic moment contributes to the scattering.
c. The scatterer is not a point charge but has a charge distribution $e\rho(x)$.

6.6 Form Factors

In general, the cross-section for scattering by an extended target nucleus can be obtained from that for a point charge with the inclusion of a form factor $F(q^2)$

$$\frac{d\sigma}{d\Omega} = \left(\frac{d\sigma}{d\Omega}\right)_{point} |F(q^2)|^2 \tag{6.13}$$

where the momentum transfer is from the projectile to the target.

$$q = p_i - p_f \tag{6.14}$$

and p_i and p_f are the initial and final momenta, respectively. The form factor can be shown to be the Fourier transform of the charge distribution (AAK 1,4)

$$F(q) = \int \rho(x) e^{iq\cdot x} d^3x \tag{6.15}$$

In (AAK 1,4), it was shown that the nuclear form factor for the spherical charge distribution is

$$F(q^2) = \int \rho(r) \frac{\sin qr}{qr} 4\pi r^2 dr \tag{6.16}$$

Using the units $\hbar = c = 1$, $\mu_0 = \varepsilon_0 = 1$, and assuming that the scattered electrons are extremely relativistic and the recoiling nucleus is nonrelativistic and the nuclear recoil momentum $p' = q \ll p_i$ so that $p_i = p_f$,

$$q = 2p_i \sin \frac{1}{2}\theta$$

or

$$q^2 = 4p_i^2 \sin^2 \frac{1}{2}\theta \tag{6.17a}$$

$$= 2p_i^2(1 - \cos\theta) \tag{6.17b}$$

where θ is the scattering angle, Fig. 6.12. All the quantities refer to lab system.

Formula (6.12) can then be written as

$$\frac{d\sigma}{d\Omega} = \frac{Z^2\alpha^2[F(q^2)]^2}{4p_i^2 \sin^4 \frac{1}{2}\theta} \tag{6.18}$$

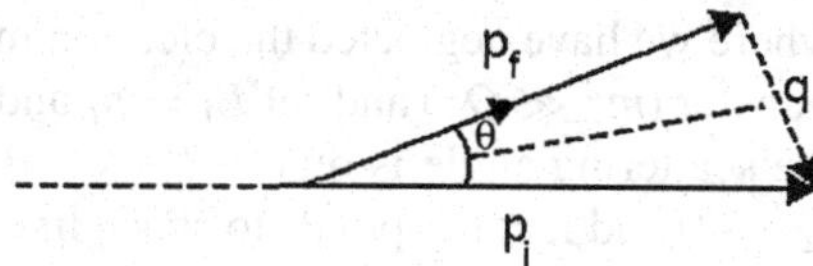

Fig. 6.12 Kinematics of elastic collision

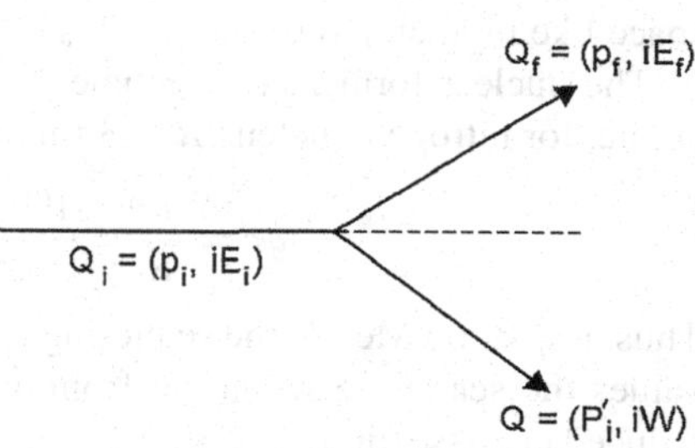

Fig. 6.13 Four-momentum transfer of a particle

where $\alpha = \frac{e^2}{4\pi}$ is the fine-structure constant.

It is useful to express (6.18) in terms of momentum transfer

$$d\Omega = 2\pi d(\cos\theta) = \pi \frac{dq^2}{p_i^2} \tag{6.19}$$

where we have used (6.17b)

$$\frac{d\sigma}{dq^2} = \frac{4\pi\alpha^2 Z^2 [F(q^2)]^2}{q^4} \tag{6.20}$$

For low values of q, which is appropriate for a point charge nucleus, $F(q^2) \to 1$ (6.18) and (6.20) reduce to the familiar Rutherford scattering formulas.

6.7 Four-Momentum Transfer

Equation (6.20) expresses the scattering in terms of the square of 3-momentum transfer $q^2 = |q|^2$ between incident and target particle in the lab system. However, if the results are to be expressed in a covariant form, that is the form is the same in any inertial system, say lab system or CM-system, then the 4-momentum transfer must be introduced (Fig. 6.13). The 4-momentum of the particle is defined by three space and one time component.

The square of the 4-momentum vector (of a real particle) is given by

$$Q^2 = p^2 + (iE)^2 = p^2 - E^2 = -m^2$$

and is clearly invariant. Likewise the 4-momentum transfer squared, between the incident and scattered electron is invariant with a value

$$\begin{aligned} Q^2 &= (p_i - p_f)^2 - (E_i - E_f)^2 = -2m^2 - 2p_i p_f \cos\theta + 2E_i E_f \\ &= 2p_i p_f (1 - \cos\theta) = 4p_i p_f \sin^2 \frac{1}{2}\theta \end{aligned} \tag{6.21}$$

where we have neglected the electron mass in comparison with the four-momentum transfer ($m^2 \ll Q^2$) and set $E_i = p_i$ and $E_f = p_f$. The quantity Q^2 is positive since the scattering angle is real ($-1 < \cos\theta < 1$). For the exchange of a virtual particle $Q^2 > 0$ and it corresponds to space like momentum transfer, while for the exchange of real particle $Q^2 < 0$ and is termed as time-like. All scattering processes have space like momentum transfer.

The nuclear form factor becomes important for $qR \gg 1$ (AAK 1,4). As an example, for nitrogen nucleus $R \simeq 3$ fm and $qR = 1$ implies that

$$qc = \frac{hc}{R} = \frac{197\ \text{MeV fm}}{3\ \text{fm}} \simeq 66\ \text{MeV}$$

Thus, if $q \ll 66$ MeV/c the scattering is from a point charge nucleus while for large values the scattering would be from an extended charge nucleus, resulting in the reduced cross-section.

For the scattering of a spinless electron by a spinless pointlike proton the differential cross-section (6.20) reduces to

$$\frac{d\sigma}{dq^2} = \frac{4\pi\alpha^2}{q^4} \tag{6.22}$$

The form of (6.22) can be explained by invoking for the Feynman diagram in which the matrix element is given by the product of two vertex factors and a photon propagator term

$$M = \sqrt{\alpha} \cdot \sqrt{\alpha} \cdot \frac{1}{q^2}$$

The phase-space factor has a value of πdq^2 and

$$\frac{d\sigma}{dq^2} \simeq |M|^2 \simeq \frac{\alpha^2}{q^4}$$

6.8 Elastic Scattering of Electrons with Spin by Spinless Nuclei (Mott Scattering)

The scattering is given by the Mott formula

$$\left(\frac{d\sigma}{d\Omega}\right)_{Mott} = \frac{Z^2\alpha^2\cos^2\frac{1}{2}\theta}{4p_i^2\sin^4\frac{1}{2}\theta[1 + (2p_i/M)\sin^2\frac{1}{2}\theta]} \tag{6.23}$$

The effect of the electron spin on the scattering by a spinless nucleus is to introduce a factor $\cos^2(\theta/2)$ in the cross-section. The term in the square brackets takes into account the recoil of the nucleus. As before for an extended nucleus the above expression must be multiplied by the square of the form factor at large momentum transfer q^2.

6.8.1 Elastic Scattering of an Electron by a Point-Like Dirac Particle of Mass M

Here the spin of the target particle as well as the incident electron is taken into account. Consider the scattering of electrons by point like protons. Apart from the electrical (Coulomb) interaction, there will be magnetic interaction which produces spin flip of the particles. Since the magnetic field due to the proton moment varies as r^{-3} compared with r^{-2} for the electric field, close collisions and large q^2 will be more important for the magnetic interaction. The scattering cross-section for the point like Dirac particles is

$$\left(\frac{d\sigma}{d\Omega}\right)_{Dirac} = \left(\frac{d\sigma}{d\Omega}\right)_{Rutherford}\left(\cos^2\frac{1}{2}\theta + \frac{q^2\sin^2\frac{\theta}{2}}{2M^2}\right) \tag{6.24}$$

where

$$\left(\frac{d\sigma}{d\Omega}\right)_{Rutherford} = \frac{\alpha^2}{(4p_i^2\sin^4\frac{1}{2}\theta)[1+(2p_i/M)\sin^2\frac{1}{2}\theta]}$$

The cos term in (6.24) represents the Mott scattering while the sin term results from the magnetic spin-flip scattering.

6.9 Rosenbluth Formula

Protons and neutrons are not point like and their magnetic moments are anomalous in that they differ from the predicted values for point like Dirac particles. While the predicted value for proton is $\mu_p = e\hbar/2Mc = 1$ nm, the observed value is $+2.79$ nm, for neutron the predicted and observed values are zero and -1.91 nm respectively. The structure of protons and neutrons is described by two form factors, one electric and the other magnetic, in analogy with the nuclear form factors. The cross-section then takes the form

$$\frac{d\sigma}{d\Omega} = \left(\frac{d\sigma}{d\Omega}\right)_{Mott}\left\{\left(\frac{G_E^2 + (\frac{Q^2}{4M^2})G_M^2}{1+(\frac{Q^2}{4M^2})}\right) + \frac{Q^2}{4M^2}\cdot 2G_M^2\tan^2\frac{\theta}{2}\right\} \tag{6.25}$$

where $G_E = G_E(a^2)$, $G_M = G_M(Q^2)$ with the normalization $G_E^p(0) = 1$, $G_E^N(0) = 0$, $G_M^p(0) = 2.79$. This is known as the Rosenbluth formula, it was first derived in 1950. Formula (6.25) can be expressed as

$$\left(\frac{d\sigma}{d\Omega}\right)\Big/\left(\frac{d\sigma}{d\Omega}\right)_{Mott} = A(Q^2) + B(Q^2)\tan^2\frac{\theta}{2} \tag{6.26}$$

so that the plots of the cross-section for different incident momenta and different scattering angle, with Q^2 remaining fixed must have a linear dependence on $\tan^2(\theta/2)$. The observed linearity verifies the prediction, Fig. 6.14.

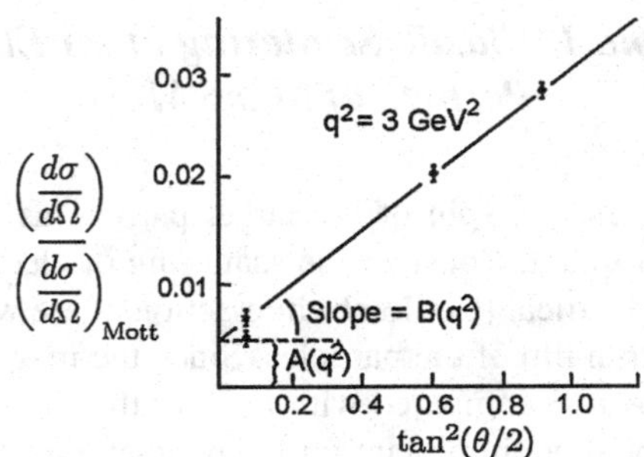

Fig. 6.14 The electron-scattering cross-section plotted for fixed q^2 and different scattering angles θ

Figure 6.14 is the $d\sigma/d\Omega$ versus $\tan^2\frac{\theta}{2}$ plot, known as Rosenbluth plot, for fixed q^2 [1].

Formula (6.25) is derived on the assumption that Born's approximation is valid and that the interaction is mediated by the single photon exchange.

Proton form factor has been determined from the observation of high energy electron beams (400 MeV–16 GeV) elastically scattered from hydrogen target and employing magnetic spectrometers for precise measurement of the momentum and the angle of scattering. Elastic events were selected from the kinematical relation

$$\frac{p_f}{p_i} = \frac{1}{1 + \frac{p_i}{M}(1 - \cos\theta)} \tag{6.27}$$

where M is the proton mass. For the neutron form factor scattering is observed with deuterium targets and a subtraction procedure is adopted which corrects for the fact that neutron is bound in deuteron.

6.10 The Scaling Law

An important result from high energy electron scattering experiments is that the form factors obey the simple *scaling law*.

$$G_E^p(q^2) = \frac{G_M^p(q^2)}{|\mu_p|} = \frac{G_M^n(q^2)}{|\mu_n|} = G(q^2), \quad G_E^n(q^2) = 0 \tag{6.28}$$

and the empirical dipole formula

$$G(Q^2) = \left(1 + \frac{Q^2}{a^2}\right)^{-2}, \quad \text{with } a^2 = (0.84\ \text{GeV}/c)^2 \tag{6.29}$$

The dipole formula (6.29) fits well the experimental range of Q^2 from 2 (GeV/c)^2 to 25 (GeV/c)^2, Fig. 6.15. Over a range of momentum transfer square, $q^2 = 0$ to 25 (GeV/c)^2, the form factor square falls by a factor of 10^6.

From the inverse Fourier transform of (6.16) we can extract the charge-magnetic-moment distribution for the dipole formula (6.29) for exponential distribution of the form

$$\rho(r) = \rho_0 e^{-ar} \tag{6.30}$$

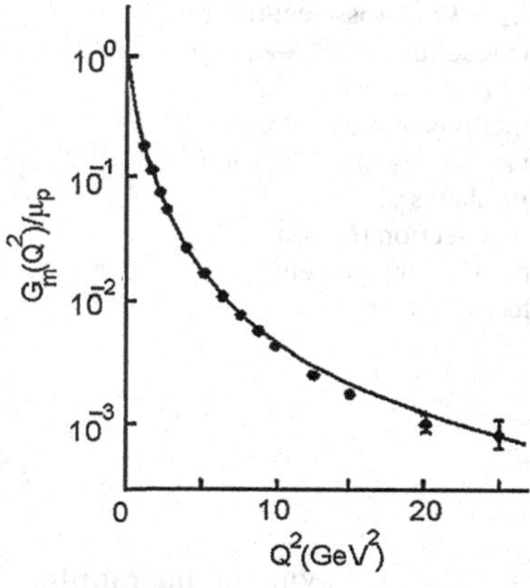

Fig. 6.15 Comparison between the measured values of the proton magnetic form factor and the 'dipole fit' [2]

with a root-mean square radius

$$r_{rms} = \frac{\sqrt{12}}{a} = 0.80 \text{ fm} \tag{6.31}$$

6.11 Muon Pair Production in e^+e^- Annihilation

The differential cross-section for the process $e^+e^- \to \mu^+\mu^-$, is given by

$$\frac{d\sigma}{d\Omega} = \frac{\alpha^2}{2s}\left(1 + \cos^2\theta\right) \tag{6.32}$$

where θ is the angle of emission of the muons with respect to the incident beam direction in the CMS, $S = 4E_1E_2$ is the square of the CMS energy, E_1 and E_2 being the energies of the electron and positron colliding head-on, Fig. 6.16(a).

The mechanism is identical with that of two-jet events discussed in Chap. 5.

All lepton masses have been neglected in comparison with the CMS energy. Figure 6.16(b) is the Feynman diagram for the process $e^+e^- \to \mu^+\mu^-$ which is identi-

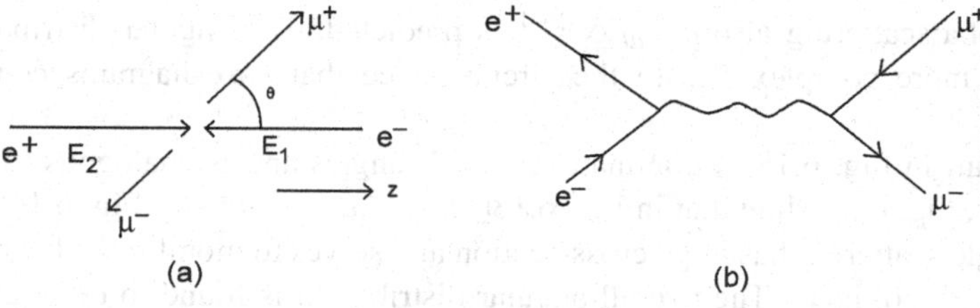

Fig. 6.16 (**a**) $e^+e^- \to \mu^+\mu^-$ process. (**b**) Feynman diagram for the process in (**a**)

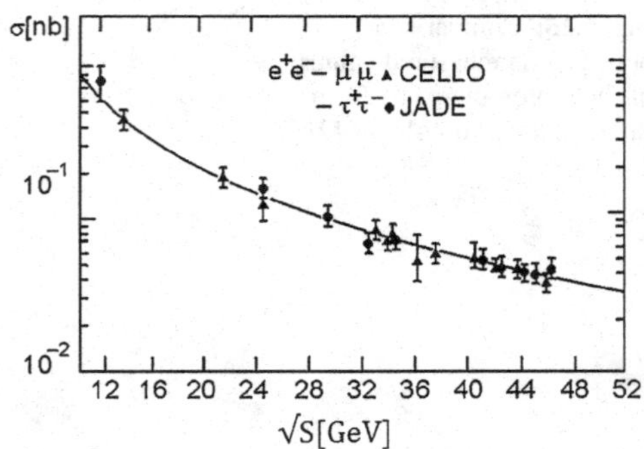

Fig. 6.17 Cross-sections of the reactions $e^+e^- \to \mu^+\mu^-$ and $e^+e^- \to \tau^+\tau^-$ as functions of the centre of mass energy $\sqrt{s}$. The *solid line* shows the cross-section (6.33a) predicted by quantum electrodynamics

cal with the Feynman diagram for electron-muon scattering (Fig. 6.1) turned on its side. The total cross-section is given by integrating (6.32) over $\theta = 0$ to π.

$$\sigma_{total} = \frac{4\pi\alpha^2}{3s} \tag{6.33}$$

Numerically,

$$\sigma\left(e^+e^- \to \mu^+\mu^-\right) = 21.7 \text{ nbarn}/\left(E^2/\text{GeV}^2\right) \tag{6.33a}$$

This is also the first-order Born approximation cross-section for e^+e^- annihilation into any point spin $1/2$ fermion particle-antiparticle pair with unit electric charge. For point quarks one inserts the square of the electric charge. This formula was already used when the evidence for colour was discussed in Chap. 5.

Note that the cross-section falls as $1/s$. The prediction is borne out by the results of cello collaboration and Jade collaboration, Fig. 6.17 (Behrand et al. and Bartel et al.). Since (6.33a) describes the experimental cross-section so well, the form factors of the μ and τ are unity, which means that they are point like particles.

In view of the universality of weak interaction, muon and tauon may be regarded as massive version of electron.

6.12 Bhabha Scattering $e + e^- \to e + e^-$

For Bhabha scattering also $\sigma_{total} \propto s^{-1}$ is predicted. The angular distribution is, however, more complex due to the circumstance that two diagrams contribute, Fig. 6.18.

Diagram in Fig. 6.18(a) dominates at small angles and is analogous to Moller scattering, Fig. 6.3, while that in Fig. 6.18(b) to $e^+e^- \to \mu^+\mu^-$, Fig. 6.16(b). The small angle scattering has large cross-section and serves to monitor the beam intensity in e^+e^- colliders. The overall angular distribution is found to be in excellent agreement with the QED predictions.

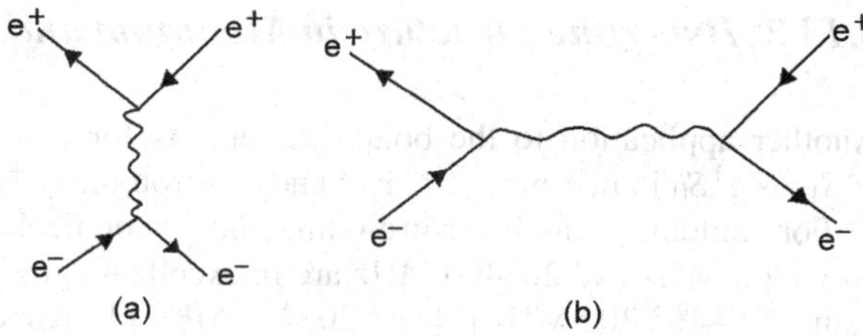

Fig. 6.18 Diagrams contributing to Bhabha scattering $e^+e^- \to e^+e^-$

6.13 Tests of QED

The predictions of QED have been checked with a high degree of accuracy. One of the early successes of QED for the bound states was the explanation of the famous Lamb shift.

6.13.1 The Lamb Shift

In the Dirac theory, the fine structure splitting is given by

$$\Delta E_{fs} = -\alpha^4 mc^2 \frac{1}{4n^2}\left(\frac{2n}{(j+\frac{1}{2})} - \frac{3}{2}\right) \tag{6.34}$$

A striking feature of the fine structure formula (6.34) is that it depends only on the principle quantum number n and the total angular momentum number j and not on the orbital angular momentum number ℓ. In general, two different values of ℓ share the same energy. For example, the $2S_{1/2}$ ($n = 2$, $\ell = 0$, $j = 1/2$) and $2P_{1/2}$ ($n = 2$, $\ell = 1$, $j = 1/2$) should remain perfectly degenerate. In 1947 Lamb and Rutherford demonstrated that the energy levels depend also on the orbital angular momentum ℓ so that the $2p_{1/2}$ and $2S_{1/2}$ electronic levels are not degenerate. The Lamb shift was explained by Bethe, Feynman, Schwinger and Tomonaga. It is due to the quantization of the electromagnetic field itself. The Lamb shift is an example of a radiative correction which has no analogue in classical theory. The nth Bhor level (fine line) splits into n sublevels characterized by $j = 1/2, 3/2, \ldots, (n - 1/2)$. Each $\ell = 0$ level is divided into two, the singlet (when the electron and proton spins are oppositely aligned) and the triplet (when the spins are parallel). The singlet is pushed down and the triplet pushed up.

For $n = 1$, the energy gap is

$$\varepsilon = E_{triplet} - E_{singlet} = \frac{32\gamma_p E_1^2}{3m_p c^2} \tag{6.35}$$

with $E_1 = 13.6$ eV and $\gamma_p = 2.7928$.

The resulting photon has the wavelength

$$\lambda = \frac{2\pi\hbar c}{\varepsilon} = 21.1 \text{ cm}$$

This is the transition that gives rise to the famous “21 centimeter line” in microwave astronomy.

6.13.2 Hyperfine Structure in Muonium and Positronium

Another application to the bound systems is for the hyperfine structure interval $1^3S_1 \to 1^1S_0$ in muonium (μ^+e^-) and positronium (e^+e^-).

For muonium and positronium, the theoretical transition frequencies of 4463.304 MHz and 203400 MHz are in excellent agreement with the experimental values of 4463.302 MHz [3] and 203387 MHz [4], respectively.

6.13.3 The g-Factors

In the Dirac theory, an electron or muon is pointlike and possesses a magnetic moment equal to the Bohr magneton

$$\mu_B = \frac{e\hbar}{2mc} \tag{6.36}$$

where m is the lepton mass. The magnetic moment μ is related to the spin vector s by

$$\mu = g\mu_B S \tag{6.37}$$

where g is known as the Lande g-factor, and $g\mu_B = \mu/s$ is the gyromagnetic ratio which is the ratio of magnetic to mechanical moment. Thus for spin $-1/2$ particles the Dirac theory predicts.

$$g = 2$$

The experimental g-values which have been determined with great precision, however, differ by a small amount (0.2 %) from the value 2. This implies that Dirac's assumption that these leptons are structureless is not entirely correct.

The magnetic moment of a charged particle is determined by the e/m ratio. Deviations from the g-values of 2 for a spin $-1/2$ particle are attributed to distortion of charge and mass distributions. It is assumed that the pointlike or "bare" object is surrounded by a cloud of one or more virtual photons, which are continually being emitted and reabsorbed. Thus, part of the mass energy is carried by the photon cloud, leading to a small increase in the e/m ratio under action of external magnetic field B. The amount of correction will be proportional to the emission of probability of $1, 2, \ldots$ virtual photons which in turn is given by $\alpha, \alpha^2, \ldots$. Figure 6.19 gives some typical diagrams for various virtual processes involving radiative corrections to the magnetic moments of leptons.

Figure 6.19(a) represents interaction with "bare" electron, Fig. 6.19(b) to emission of one photon and therefore of order α, Fig. 6.19(c) one of two-photon emission and one of pair formation, and of order α^2. The leading diagram predicts the anomaly for electron

$$\left(\frac{g-2}{2}\right)_e = \frac{\alpha}{2\pi} = 0.001623041$$

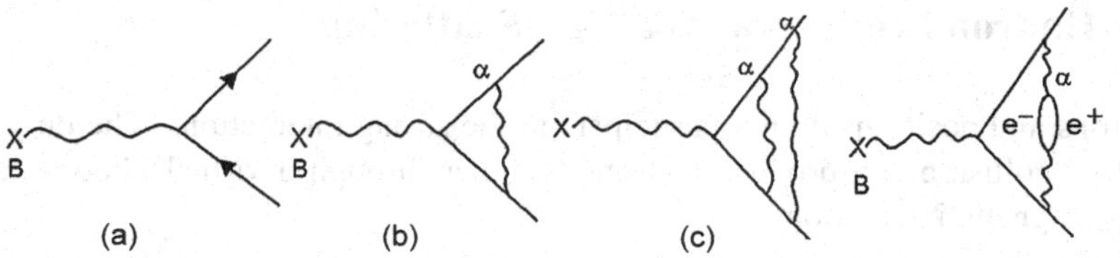

Fig. 6.19 Feynman diagrams indicating some of the radiative corrections to the magnetic moments of leptons

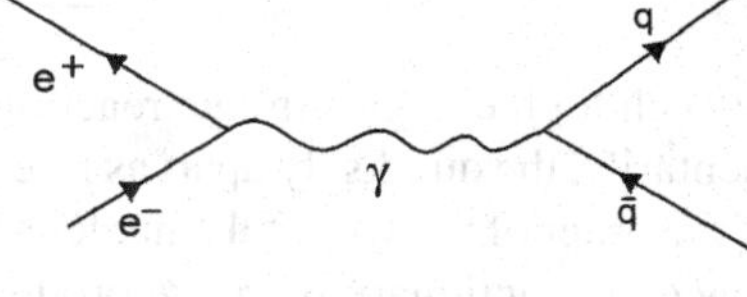

Fig. 6.20 Production of hadrons in electron-positron annihilation

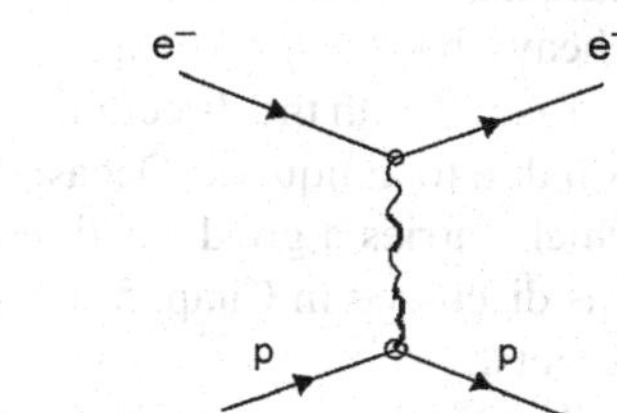

Fig. 6.21 High-energy electron-proton scattering

When higher order terms are added, the predicted value becomes $0.0011596524 \pm (4)$ which is in exact agreement with the experimental vale of $0.0011596524 \pm (2)$.

For muon a virtual photon can transform not only into a lepton pair but also into hadrons. When this extra contribution is taken into account, the predicted anomaly is $0.001165918 \pm (10)$ is to be compared with the experimental value of $0.001165924(\pm 9)$. The numbers in brackets are errors in the last place of decimal.

6.14 Electrodynamics of Quarks and Hadrons

The electrodynamics applied to electrons and muons works equally well to quarks by the use of appropriate charge $2/3e$ or $-1/3e$. From the point of view of experiments the quarks themselves are unobserved and their presence must be inferred by the hadrons into which they are fragmented. Here we shall consider two examples (1) the production of hadrons in electron-positron annihilation for which the relevant diagram is Fig. 6.20 and (2) high-energy electron-proton scattering "deep inelastic scattering" for which the basic diagram is Fig. 6.21.

6.15 Hadron Production in e^+–e^- Scattering

Electrons and positrons do not participate in the strong interactions. The only way an e^+e^- collision can produce hadrons is either through a virtual photon via an electromagnetic interaction.

$$e^+ + e^- \rightarrow \gamma \rightarrow q + \overline{q} \rightarrow \text{hadrons} \tag{6.38}$$

or by Z^0 production at high energies (above the threshold for Z^0 production) via weak interaction

$$e^+ + e^- \rightarrow Z^0 \rightarrow q + \overline{q} \rightarrow \text{hadrons} \tag{6.39}$$

At such high energies in fact reaction (6.39) would be a dominant mechanism. Momentarily, the quarks fly apart as free particles, but when the distance of separation is of the order of the size of the nucleon ($\sim$1 fm), their strong interaction is so great that new quark-antiquark pairs are produced mainly from gluons. Countless quarks and antiquarks hadronize into baryons and mesons which are actually recorded. Against a heavy background of tracks can be seen two back-to-back collimated "jets", one correlated with the direction of the original quark and the other in the opposite direction due to antiquark. Occasionally a three-jet event is observed attributed to gluon which carries a good fraction of total energy. The chromodynamics of such events was discussed in Chap. 5. Here we are mainly concerned with the electrodynamics aspects.

The total cross-section for the process (6.38) is given by

$$\sigma = \frac{\pi q^2}{3}\left(\frac{\hbar c\alpha}{E}\right)^2 \sqrt{1-(Mc^2/E)^2}\left[1+\frac{1}{2}\left(\frac{Mc^2}{E}\right)^2\right]\left[1+\frac{1}{2}\left(\frac{mc^2}{E}\right)^2\right] \tag{6.40}$$

where m is the mass of the electron and M is that of the quark. For $E < Mc^2$ (below threshold for quark-antiquark production), the cross-section becomes imaginary, meaning the process is kinematically forbidden when there is not enough energy to create the $q\overline{q}$ pair. At energies much above threshold ($E > Mc^2 \gg mc^2$), (6.40) simplifies considerably.

$$\sigma = \frac{\pi}{3}\left(\frac{\hbar q c\alpha}{E}\right)^2 \tag{6.41}$$

As we break up the beam energy in various intervals, a succession of such thresholds are encountered—first the light quarks, followed by the charm quark (at above 1500 MeV), the tau at 1784 MeV, the bottom quark (4700 MeV) and finally the top quark at 175 GeV.

The structure of the graph is revealed by plotting the ratio of the hadron production to that for muon pairs

$$R = \frac{\sigma(e^+e^- \rightarrow \text{hadrons})}{\sigma(e^+e^- \rightarrow \mu^+\mu^-)}$$

where

$$\sigma\left(e^+e^- \to \mu^+\mu^-\right) = \frac{4\pi\alpha^2}{3s} \tag{6.42}$$

with $s = E^2_{CMS}$.

In Sect. 5.14 it was shown that the ratio R is easily calculated in the quark model. The behaviour of R vide Fig. 5.49 was the compelling reason for accepting 'colour' (3 in number) for each quark flavour. The Rutherford cross-section for electron-quark scattering will be proportional to $\sum e_i^2$, the square of the quark charge summed over all contributing quark flavour. At energies devoid of the resonance region associated with ρ, ω, ϕ, ψ and γ-series of levels, we expect a "stair case" graph for $R(E)$, Fig. 5.49, which ascends by one step at each new quark threshold, with the height of the rise determined by the quark's charge.

At low energy ($E_{cm} = 1.0$ to 3.0 GeV) only the u, d and s quarks contribute, we expect

$$R = 3\left[\left(-\frac{1}{3}\right)^2 + \left(\frac{2}{3}\right)^2 + \left(-\frac{1}{3}\right)^2\right] = 2$$

Between the c threshold and b threshold ($E_{cm} = 5$ to 8 GeV) we ought to have

$$R = 2 + 3\left(\frac{2}{3}\right)^2 = \frac{10}{3}$$

Between the b threshold and t threshold ($E = 10$ to 175 GeV)

$$R = \frac{10}{3} + 3\left(-\frac{1}{3}\right)^2 = \frac{11}{3}$$

Above the t threshold ($E > 180$ GeV), and well past the Z^0 resonance

$$R = \frac{11}{3} + 3\left(\frac{2}{3}\right)^2 = 5$$

The agreement between theory and experiment is reasonably good. Further, the angular distribution of the jets obeys the $(1 + \cos^2\theta)$ law, appropriate for spin $-1/2$ structureless particles. Thus, in the process $e^+e^- \to$ hadrons, the constancy of R at various steps up the "staircase" is a proof for the pointlike constituents of hadrons. These constituents were named as Partons by Feynman which were later identified as quarks. The value of R is what is expected if the partons have fractional charge and come in three colours.

6.16 Natural Units

It is convenient to adopt natural units in Elementary particle physics, with the choice of fundamental constants.

$$\hbar = 1, \qquad c = 1 \tag{6.43}$$

The constants $\hbar$ and c figure quite frequently in particle physics due to the use of quantum mechanics and relativity. With the choice (6.43) the quantities $\hbar$ and c are omitted resulting in considerable simplification of equations. As an example, the relativistic energy relation

$$E^2 = p^2c^2 + m^2c^4$$

becomes $E^2 = p^2 + m^2$ while the Fermi constant

$$\frac{G_F}{(\hbar c)^3} = 1.166 \times 10^{-5}\ \text{GeV}^{-2}$$

used in weak interactions becomes

$$G_F = 1.166 \times 10^{-5}\ \text{GeV}^{-2} \tag{6.44}$$

In natural units all quantities have the dimension of a power of energy since they can all be expressed in terms of $\hbar$, c and energy. In particular, mass, length and time can be expressed in the form

$$M = E/c^2, \qquad L = \hbar c/E, \qquad T = \hbar/E$$

so that a quantity with MKS dimensions $M^p L^q T^r$ has the dimension E^{p-q-r} in natural units. Since $\hbar$ and c are suppressed in natural units, this is the only dimension which is relevant. Further estimates and dimensional checks are converted into 'practical' units in which the experimental results are usually expressed. This is done by restoring the $\hbar$, c factors and using conversion factors. Note that in natural units many different quantities have the same dimension, e.g. mass, energy and momentum.

6.16.1 Conversion Factors

Length: We start with the equation

$$\hbar c = 197.3\ \text{MeV fm} \tag{6.45}$$

Set $\hbar c = 1$. Then

$$\begin{aligned} 1 = 197.3\ \text{MeV fm} \quad \text{or} \quad 1\ \text{fm} = \frac{1}{197.3}\ \text{MeV}^{-1} \\ \therefore \quad 1\ \text{fm} = \frac{10^3}{197.3}\ \text{GeV}^{-1} = 5.068\ \text{GeV}^{-1} \end{aligned} \tag{6.46}$$

Cross-section:

$$\begin{aligned} 1\ \text{mb} &= 10^{-31}\ \text{m}^2 = 10^{-1}\ \text{fm}^2 = 10^{-1} \times (5.068)^2\ \text{GeV}^{-2} \\ &= 2.568\ \text{GeV}^{-2} \end{aligned} \tag{6.47}$$

Time:

$$\begin{aligned} c &= 3 \times 10^8\ \text{m/sec} \quad \text{and if } c = 1 \\ 1\ \text{sec} &= 3 \times 10^8\ \text{m} = 3 \times 10^8 \times 10^{15}\ \text{fm} = 3 \times 10^{23} \times 5\ \text{GeV}^{-1} \\ 1\ \text{sec} &= 1.5 \times 10^{24}\ \text{GeV}^{-1} \end{aligned} \tag{6.48}$$

Fine structure constant (α): In natural units apart from $h = c = 1$, we also have $\varepsilon_0 = \mu_0 = 1$.

$$\begin{aligned} \alpha &= \frac{e^2}{4\pi\varepsilon\hbar c} = \frac{1}{137} \\ \alpha &= \frac{e^2}{4\pi} = \frac{1}{137} \end{aligned} \tag{6.49}$$

6.16.2 Restoration of Practical Units

We shall illustrate how the practical units may be restored from natural units by an example. *V–A* theory gives the formula for the width (Γ_μ) of the muon decay.

$$\Gamma_\mu = \frac{\hbar}{\tau} = \frac{G_F^2 m_\mu^5}{192\pi^3} \quad \text{(natural units)} \tag{6.50}$$

Let

$$\Gamma_\mu = \frac{G_F^2 m_\mu^5 \hbar^a c^b}{192\pi^3}$$

Using the dimensions, $[\Gamma_\mu] = [\mathrm{ML^2T^{-2}}]$, $[m_\mu] = [\mathrm{M}]$, $[\hbar] = [\mathrm{ML^2T^{-1}}]$, $[c] = [\mathrm{LT^{-1}}]$, $[G_F] = [\hbar c]^3[\mathrm{ML^2T^{-2}}]^{-2}$, we find $a = -6$, $b = 4$, so that

$$\Gamma_\mu = \frac{G_F^2 (m_\mu c^2)^5}{192\pi^3 (\hbar c)^6} \quad \text{(practical units)} \tag{6.51}$$

6.17 Deep Inelastic Lepton-Hadron Scattering

The internal structure of hadrons (essentially nucleons) is conveniently explored by the scattering of high energy leptons, that is electrons, muons and neutrinos. This provides a dynamical method (1968) for establishing the parton model of hadrons as opposed to the static study described in Chap. 5. Partons are pointlike constituents named by Feynman and later identified with quarks. The leptons are particularly suitable for such investigations as the electromagnetic and weak forces between quarks which constitute the hadrons and leptons are well understood. This is in contrast with hadron-hadron collisions which complicate the processes due to strong interaction. The electromagnetic and weak interactions are feeble in the sense that the first Born approximation provides an accurate description of the collision.

The internal structure of hadrons is revealed most clearly in "deep inelastic scattering" i.e. scattering in which the momentum transfer (q) is much larger than the relative momenta of the hadron's constituents, and a large amount of energy (ν) is transferred to the hadron (nucleon). Here large means that $qc \gg Mc^2$ and $\nu \gg Mc^2$, where M is the nucleon mass and inelastic means that the final state is not a single nucleon but consists of at least one baryon and one or more mesons.

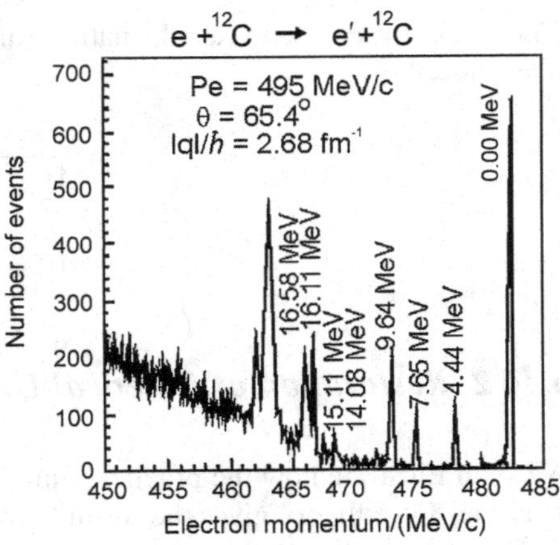

Fig. 6.22 Spectrum of electron scattering of ^{12}C. The sharp peaks correspond to elastic scattering and to the excitation of discrete energy levels in the ^{12}C nucleus by inelastic scattering. The excitation energy of the nucleus is given for each peak. The 495 MeV electrons were accelerated with the linear accelerator MAMI-B in Mainz and were detected using a high-resolution magnetic spectrometer at a scattering angle of 65.4° (Walcher, Rasner and Mainz)

First, we shall consider electron scattering from atomic nuclei for electron energy E less than ε, the lowest excitation energy of the nucleus. The scattering will be elastic, the nucleus remaining in the ground state and the electron does not suffer any energy loss in the CMS. The angular distribution (or equivalently the dependence on the momentum transfer q), gives information about the charge distribution in the nuclear ground state, which is expressed by the form factor $F(q)$ (AAK 1,4).

If the incident energy E is higher than the lowest excitation energy and the momentum transfer is not considerably larger than the internal momenta of the nucleons in the nucleus, then it is inelastic scattering. In the CMS the scattered electron energy E' is then distributed over many values. In the CMS the scattering cross section for a given ε, as a function of ε', exhibits peaks whenever $E - \varepsilon$ is equal to an excitation energy of the nucleus. Elastic scattering also occurs, but becomes less prominent as more states can be excited. When $E - \varepsilon$ is larger than the energy necessary to remove a nucleon from the nucleus, we find a continuum instead of peaks. The dependence of the cross section on the momentum transfer is complicated, since it depends on the properties of the initial and the final states of the nucleus.

When the momentum transfer is much larger than the average internal momentum of the nucleons in the ground state, the process is called "deep inelastic" scattering. Under these conditions the interaction between the electron and the nucleons is so sudden, and involves such a large change of momentum that one can neglect the binding forces between the constituents during the collision. In a first approximation, the constituents behave like free particles with a momentum distribution given by the momentum-space wave function of the ground state. This is called the impulse approximation: the electron can be considered as being scattered by one of the "free" nucleons. The total scattering cross section is then the sum of the cross section for each of the constituent "free" nucleons s the scattering is assumed to be incoherent (quasi-elastic scattering).

We apply this analysis to the scattering by an isolated nucleon. If the lepton energy is low ($\leq$300 MeV), the first excited state (Δ) cannot be reached, and only elastic scattering takes place (Fig. 6.22). Such measurements are used to determine

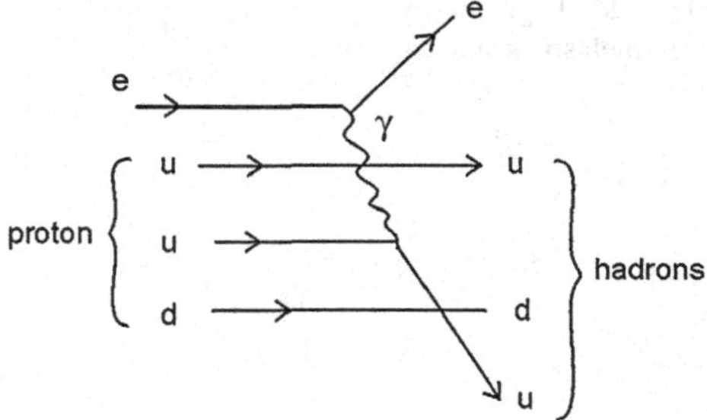

Fig. 6.23 Quark model for deep e–p inelastic scattering

the form factor of the nucleon, which leads to the radius $R \approx 10^{-13}$ cm. The average internal quark momentum therefore is of the order $R^{-1} \approx 200$ MeV/c. The cross section will have peaks when the energy loss equals one of the excitation energies of the nucleon. For lepton energies of several GeV, the momentum and energy transfer can be considerably larger than 200 MeV/c. We then have deep inelastic scattering of leptons by nucleons.

Let us exploit the analogy with the scattering of electron by nuclei. The role of the electrons is played by one of the leptons (electron, muon or neutrino), that of the nucleus is played by the nucleon, and that of nucleons within the nucleus by quarks. But there is one important difference: Quarks cannot be ejected from the nucleon (since we believe that no free quarks exist). However, this neither plays a significant role in the evaluation of the scattering cross section, nor in the momentum distribution of the scattered leptons.

Since quark confinement does not, in any appreciable way influence the scattering of high-energy leptons by quarks, we may use the same description of deep inelastic scattering by hadrons as used in the case of lepton-nucleus scattering. We neglect the quark binding and treat the nucleon as an assembly of free quarks with a momentum distribution given by the nucleon ground state. The analysis is for the scattering of ultra relativistic leptons by free pointlike fermions. The net scattering is given by the sum of incoherent scattering from various quarks.

When the highly energetic quark attempts to leave the hadron, quark-antiquark pairs are produced and these eventually materialize into a jet of mesons moving in the general direction of the struck quark. Figure 6.23 shows the quark model in which a virtual photon is exchanged between the incident electron and a proton in the deep inelastic scattering. The electron loses energy ν and the momentum $\vec{q}$, which are transferred by means of a virtual photon to one quark in the nucleon which is effectively $e^- q$ elastic scattering. The struck quark parts company from the spectator quarks and initiates fragmentation into hadrons.

In electron-nucleus scattering as q^2 increases from low values, the coherent nuclear scattering fades out and the incoherent elastic scattering from individual nucleons becomes progressively more important. Increasing q^2 is equivalent to decreasing the wavelength of the probe (electron) and so also the resolution, and at sufficiently large q^2, the scattering occurs from the individual nucleons rather than the entire nucleus. In e–p scattering if q^2 is increased to still larger value then the electron would see the constituents of proton, and the scattering would be essen-

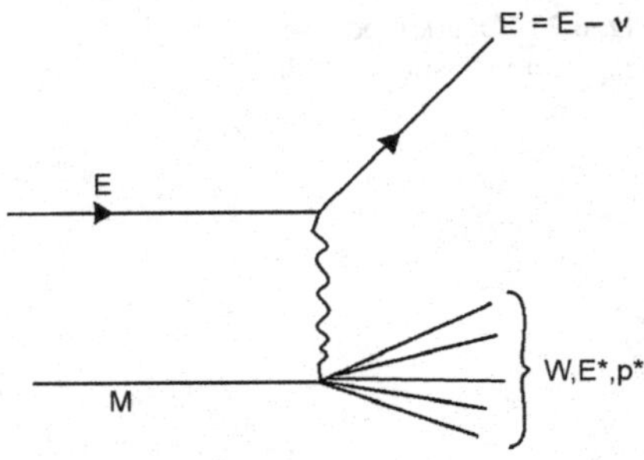

Fig. 6.24 High energy *e-p* deep inelastic scattering

tially inelastic as hadrons are produced in such collisions (Fig. 6.24). Denoting the 3-momentum, energy, and invariant mass of the final hadron state by p^*, E^*, and W, we get

$$Q^2 = (p^* - 0)^2 - (E^* - M)^2 = p^{*2} - (E^* - M)^2$$
$$\nu = E^* - M$$
$$W^2 = E^{*2} - p^{*2}$$

so that

$$Q^2 = M^2 + 2\nu M - W^2 \tag{6.52}$$

The energy $\nu = (E^* - M)$ acquired by the hadronic system is equal to the energy transfered by the virtual photon.

Note that in elastic scattering $W = M$ and $Q^2 = 2\nu M$ so that Q and are directly related. However, for inelastic scattering this is not so because of the production of particles of varying mass and excitation.

In analogy with (6.24), the cross-section can be written in terms of two variables q^2 and ν.

$$\frac{d^2\sigma}{dq^2 d\nu} = \frac{4\pi\alpha^2}{q^4}\frac{E'}{EM}\left[W_2(q^2,\nu)\cos^2\frac{\theta}{2} + 2W_1(q^2,\nu)\sin^2\frac{\theta}{2}\right] \tag{6.53}$$

where E and E' are the energies of the incident and scattered electron (E, $E' \gg mc^2$), $W_2(q^2, \nu)$ and $W_1(q^2, \nu)$ are arbitrary electric and magnetic structure functions. The validity of (6.53) is established from the linearity of Rosenbluth plot similar to the one for elastic scattering.

Such a plot affords the determination of the ratio W_1/W_2. It is convenient to choose the scaling variable

$$x = \frac{Q^2}{2M\nu} \tag{6.54}$$

This dimensionless quantity x is a measure of inelasticity, and has the range $0 \le x \le 1$. For elastic scattering, $Q^2 = 2M\nu$, and $x = 1$, and for totally inelastic scattering $x = 0$. Figure 6.25 shows the excitation curve of inelastic *ep* from the experiments at the DESY electron accelerator [5] for $E = 4.879$ GeV at $\theta = 10.0°$. The elastic peak occurs at $x = 1$, followed by successive peaks due to nucleon resonance $\Delta(1232)$, $N(1450)$, $\Delta(1688)$, ... at $x = Q^2/(Q^2 + W^2 - M^2)$.

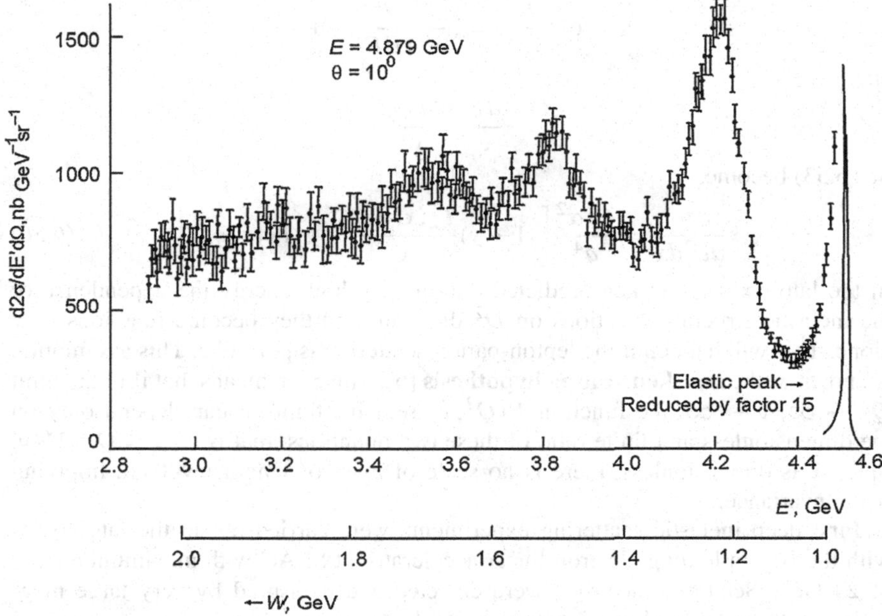

Fig. 6.25 Excitation curve of inelastic ep scattering, obtained at the DESY electron accelerator [5]. E and E' are the energies of the incident and the scattered electron, and W is the mass of the recoiling hadronic state. The peaks due to the pion-nucleon resonances of masses 1.24, 1.51, and 1.69 GeV are clearly visible

6.18 Bjorken Scaling and Parton Model

Rewriting (6.53) in terms of related functions and variables.

$$F_1(Q^2, \nu) = W_1(Q^2, \nu),$$

$$F_2(Q^2, \nu) = \frac{\nu W_2(Q^2, \nu)}{M}$$

$$y = \frac{\nu}{E},$$

$$\frac{E'}{E} = 1 - y$$

$$Q^2 = 2MExy,$$

$$\frac{d^2\sigma}{dQ^2 d\nu} = \frac{4\pi\alpha^2}{Q^4}\frac{E'}{E\nu}\left[F_2(Q^2, \nu)\cos^2\frac{\theta}{2} + \frac{2\nu}{M}F_1(Q^2, \nu)\sin^2\frac{\theta}{2}\right] \quad (6.55)$$

Now from (6.21)

$$\sin^2\frac{\theta}{2} = \frac{Q^2}{4EE'} = \frac{Mxy}{2E'}$$

$$\cos^2\frac{\theta}{2} = 1 - \frac{Q^2}{4EE'} \simeq 1$$

and

$$\frac{d\nu}{\nu} = \frac{dx}{x}$$

So (6.53) becomes

$$\frac{d^2\sigma}{dq^2dx} = \frac{4\pi\alpha^2}{q^4}\left[(1-y)\frac{F_2(x,q^2)}{x} + y^2F_1(x,q^2)\right] \tag{6.56}$$

In the late sixties, Bjorken predicted that at very high energy the dependence of the inelastic structure functions on Q^2 dies out, and they become functions of x alone. This will happen if the lepton-parton scattering is pointlike. This assumption is known as the **Bjorken scaling hypothesis** [6]. This then means that if in the limit $Q^2 \to \infty$, $\nu \to \infty$, the function $F(Q^2, \nu)$ remains finite, it can depend only on the dimensionless and finite ratio of these two quantities, that is on $x = Q^2/2M\nu$. Since x is dimensionless, there is no scale of mass or length involved, implying scale invariance.

First deep inelastic scattering experiments were carried out in the late sixties with the two-mile long electron linear accelerator at SLAC with maximum energy of 25 GeV. Scattered electrons were detected and measured by very large magnetic spectrometer. Inelastic interactions of the type, $ep \to ep\pi\pi$, $ep \to en\pi\pi, \ldots$ dominated. In inelastic scattering excitation energy of proton adds one more degree of freedom. Hence the structure functions and cross-sections are functions of the two independent parameters, for example (E', θ) or (Q^2, ν). From the direction and energy E' of the scattered electron 4-momentum transfer can be calculated and $d^2\sigma/d\Omega dE'$ determined as a function of E' and Q^2. Figure 6.26 shows the excitation spectra off hydrogen obtained at a fixed scattering angle $\theta = 4°$. In the graphs $d^2\sigma/d\Omega dE'$ is plotted as a function of W. Each spectrum covers a different range of Q^2 from $0.06 < Q^2 < 0.09$ (GeV/c)^2 to $1.45 < Q^2 < 1.84$ (GeV/c)^2. The cross-sections drop off rapidly with Q^2 in the range of nucleon resonances. For increasing W, the fall-off is, however less pronounced.

Figure 6.27 shows the variation of the ratio $\frac{d^2\sigma}{d\Omega dE'}/(\frac{d\sigma}{d\Omega})_{Mott}$ measured in these experiments as a function of Q^2 at different values of W. It is seen that the ratio only depends weakly for $W > 2$ GeV/c^2, in contrast to the rapid drop with $|G(\text{dipole})|^2 \simeq \frac{1}{Q^8}$ for elastic scattering. Thus in deep inelastic scattering the structure functions W_1 and W_2 are nearly independent of Q^2 for fixed values of the invariant mass W.

If one extracts $F_1(x, Q^2)$ and $F_2(x, Q^2)$ from the cross-sections, one observes at fixed values of x they depend only weakly, or not at all on Q^2. This is indicated in Fig. 6.28 where $F_2(x, Q^2)$ is displayed as a function of x, for data covering a range of Q^2 between 2 (GeV/c)^2 and 18 (GeV/c)^2.

Experimentally, Bjorken scaling sets in for $Q^2 \gtrsim 1$ (GeV/c)^2 and $M\upsilon \gtrsim$ 3.5 (GeV/c)^2.

The fact that x is dimensionless and has no associated scale affords the description of the phenomenon as scale invariance. Figure 6.29 gives at plot of νW_2 against Q^2, showing the weak dependence.

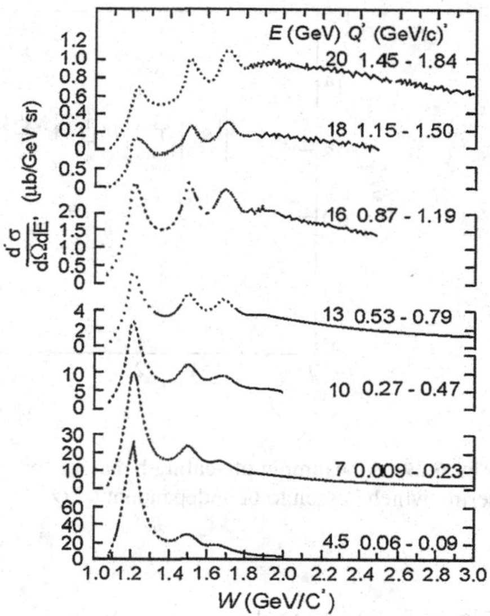

Fig. 6.26 Electron-proton scattering: excitation spectra measured in deep inelastic electron-nucleon scattering as functions of the invariant mass W. Note the different scales of the y-axis. The measurements were taken at a fixed scattering angle, $\theta = 4°$. The average Q^2-range of the data increases with increasing beam energy E. The resonances, in particular the first one ($W = 1.232$ GeV/c^2), become less and less pronounced, but the continuum ($W \geq 2.5$ GeV/c^2) decreases only slightly

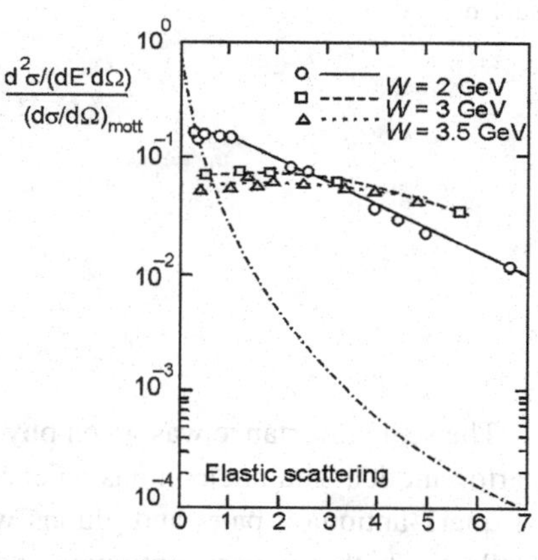

Fig. 6.27 Electron-proton scattering measured cross-sections normalized to the Mott cross-section as functions of Q^2 at different values of the invariant mass W (Breidenbach et al.)

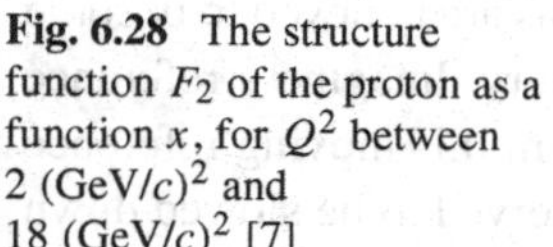

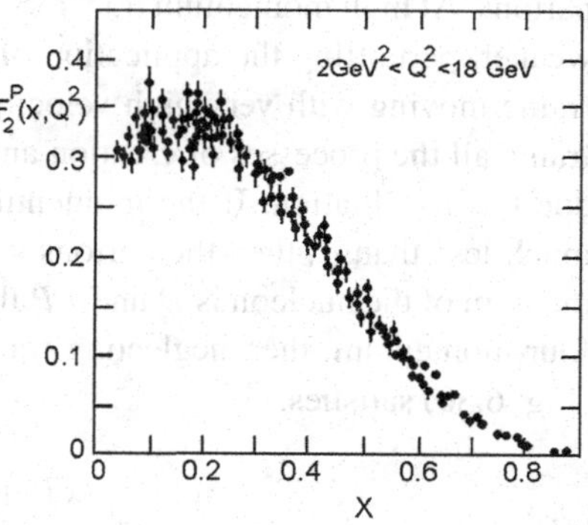

Fig. 6.28 The structure function F_2 of the proton as a function x, for Q^2 between 2 (GeV/c)^2 and 18 (GeV/c)^2 [7]

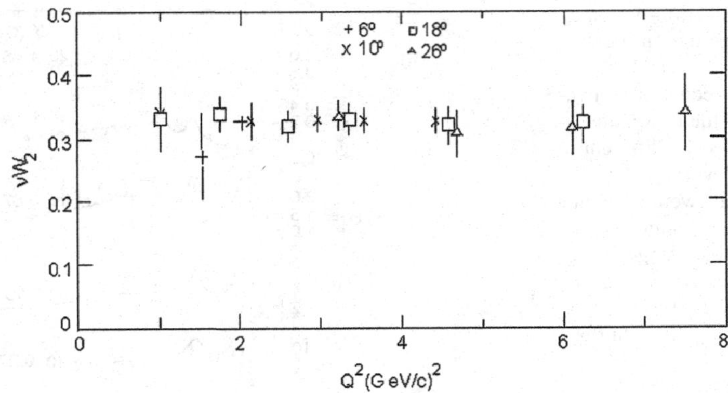

Fig. 6.29 An example of scaling behavior for the structure function W_2 for electron-proton scattering which is seen to be independent of Q^2

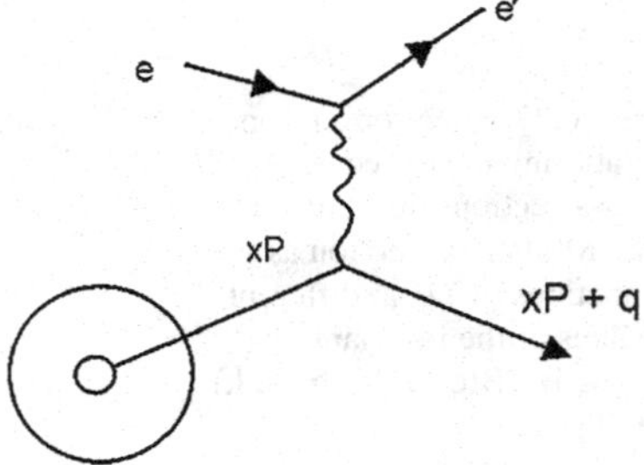

Fig. 6.30 Scattering of an electron by a parton in a nucleon

The scale invariance was given physical interpretation by Feynman [8] using the parton model. The nucleon has valence quarks as well as a sea of transient clouds of quark-antiquark pairs and gluons which are continually created and destroyed. Collectively these quark, antiquark and gluon constituents of the nucleon are called partons. At high momentum transfers, the strong interactions are believed to become weaker, permitting the application of the perturbation theory. Imagine a reference frame moving with very high velocity past a nucleon. From this moving reference frame all the processes of creation and destruction are observed to be slowed down due to time dilation. If the momentum transfers are quite high so that binding is much less than q, then the partons would appear almost free. If the total four momentum of the nucleon is P and xP that carried by the proton, and q is the absorbed four-momentum, then neglecting transverse-momentum the final-state momentum (Fig. 6.30) satisfies.

$$(xP + q)^2 = m^2 \simeq 0 \tag{6.57}$$

where m is the parton mass which is negligible in comparison with P or q. Essentially, $P = M$. If $|x^2 P^2| = x^2 M^2 \ll q^2$

$$x = \frac{-q^2}{2P \cdot q} = \frac{q^2}{2M\nu} \tag{6.58}$$

In (6.58) x represents the fractional 3 momentum of the parton in the moving frame of reference.

If the previous argument is applied to a free quark, outside the nucleon same calculations would apply except that $x = 1$ and $M \to m$. Hence for a free parton.

$$Q^2 = -q^2 = 2m\nu \tag{6.59}$$

Comparing (6.58) with (6.59) it is concluded that

$$x = \frac{m}{M} \tag{6.60}$$

6.19 Parton Spin and Callan-Gross Relation

Using the relations

$$d\Omega = 2\pi d(\cos\theta) = \frac{\pi dq^2}{p_c^2} \tag{6.61}$$

the Dirac equation (6.24) for scattering of point like particles of charge ze and spin 1/2 can be written as

$$\left(\frac{d\sigma}{dq^2}\right)_{Dirac} = \frac{4\pi\alpha^2 z^2}{q^4}\frac{E'}{E}\left(\cos^2\frac{\theta}{2} + \frac{q^2}{2m^2}\sin^2\frac{\theta}{2}\right)\frac{1}{x} \tag{6.62}$$

Compare (6.62) with the expression

$$\left(\frac{d\sigma^2}{dq^2 dx}\right)_{inelas} = \frac{4\pi\alpha^2}{q^4}\frac{E'}{E}\left(F_2(x)\cos^2\frac{\theta}{2} + \frac{q^2}{2M^2x^2}2xF_1(x)\sin^2\frac{\theta}{2}\right)\frac{1}{x} \tag{6.63}$$

We find

$$\frac{2xF_1(x)}{F_2(x)}\frac{m^2}{M^2x^2} = 1 \tag{6.64}$$

Using (6.60) in (6.64) we get

$$\frac{2xF_1(x)}{F_2(x)} = 1 \tag{6.65}$$

Equation (6.65) is known as **Callan-Gross relation** (1968).

Figure 6.31 shows that the observed ratios $2xF_1/F_2$ are consistent with the spin 1/2 for partons. For zero-spin partons the ratios would have been zero. It is natural to identify partons with quarks. The significance of Bjorken scaling and the Callan-

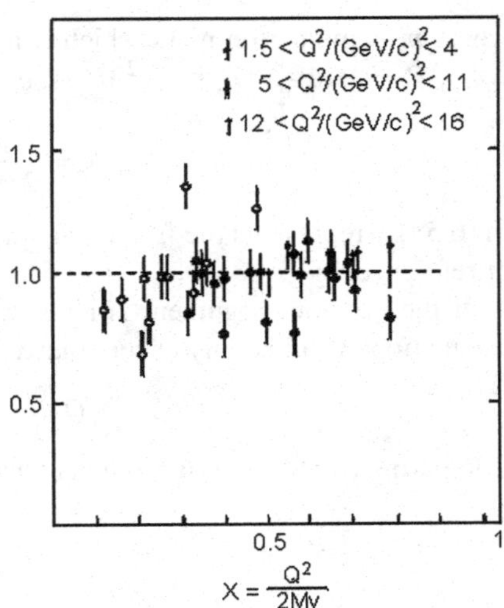

Fig. 6.31 The ratio $2xF_1/F_2$ at fixed x

Gross formula is that they provide a direct test of the quark-parton model in the deep inelastic scattering regime. Furthermore, instead of describing the deep inelastic scattering with two unknown functions of two variables [$W_1(q^2, x)$ and $W_2(q^2, x)$], it is sufficient to deal with only one unknown function say $F_1(x)$. With this stunning simplification deep inelastic scattering becomes a powerful tool for providing the structure of nucleon.

6.19.1 Parton Charges

Since partons are identified with quarks, the charges are established as $Q_u = +\frac{2}{3}$, $Q_d = -1/3$, etc. (Chap. 5). An independent confirmation also comes from the comparison of deep inelastic scattering data on $F_2(x)$ for electron-nucleon scattering and neutrino-nucleon scattering. Theory gives

$$F_2^{\nu N}(x) = \frac{18}{5} F_2^{eN}(x) \tag{6.66}$$

where the number results directly from the fractional quark-charge assignment and the small (few percent) contribution of $s, c, \ldots$ quarks is ignored. The measured $F_2^{\nu N}$ in neutrino-nucleon scattering in the heavy-liquid bubble chamber in a PS neutrino beam at CERN and the F_2^{eN} from electron-nucleon scattering, in the same region of q^2 obtained at SLAC, were shown to be in excellent agreement with (6.66), Fig. 6.32. The total area under the curve represents the total momentum fraction in the nucleon carried by quarks. The measured area is about 0.5. The remaining mass is attributed to gluon constituents which are the carriers of the interquark colour field.

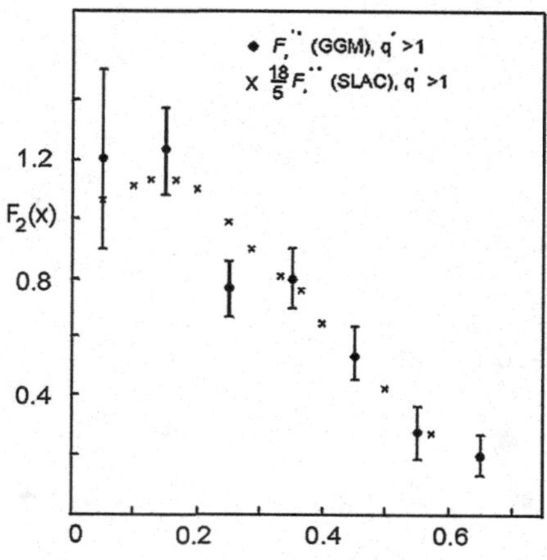

Fig. 6.32 The comparison of $F_2^{\nu N}(x)$ and $18/5F_2^{\nu e}(x)$ from the data at CERN and SLAC, respectively [9]

References

1. Weber (1967)
2. W. Panofsky, in *International Conference on High Energy Physics*, Vienna (1968)
3. Bodwin, Yennie (1978)
4. Berko, Pendleton (1980)
5. Bartel et al. (1968)
6. Bjorken (1967)
7. W.B. Atwood, Prog. Phys. **4** (1982)
8. Feynman (1969)
9. D.H. Perkins, *Introduction to High Energy Physics* (Addison-Wesley, Reading, 1986)

Chapter 7
Weak Interactions

7.1 Characteristics

The life times of elementary particles fall into three categories. First one corresponds to life times of the order of 10^{-23} sec, resonances like $\Delta 1232$ being a typical example. Decays of this kind are strong interaction processes in which all conservation laws like conservation of electric charge, baryon number, isospin, strangeness, and parity, are satisfied. In the second category there are few processes like $\pi^0 \to 2\gamma\, \Sigma^0 \to \Lambda + \gamma$ which are electromagnetic in origin and are characterised by lifetimes of the order of 10^{-18} sec–10^{-20} sec. In the third one there are many decays with lifetimes greater than about 10^{-13} sec. The weak decays which include nuclear β-decay are attributed to the weak interaction. The characteristic features of such interactions are:

(a) Long lifetimes ($\sim 10^{-10}$ sec) for weak decay process.
(b) Small interaction cross sections ($\sim 10^{-39}$ cm^2) via weak processes.
(c) Leptons experience only weak interaction and if charged, the electromagnetic interaction as well.
(d) Weak interactions do not conserve isospin. The rule is $|\Delta I| = 1/2$.
(e) Weak interactions of strange particles do not conserve strangeness (Chap. 4). The rule is $|\Delta S| = 1$, for example $\Lambda \to p + \pi^-$; $\Xi^- \to \Lambda + \pi^-$.
(f) The weak processes do not conserve parity.

The weak decays can be divided into three classes depending on the extent to which leptons are involved. These classes are referred to as (i) leptonic (ii) semileptonic (iii) non-leptonic decay.

7.2 Leptonic Decays

The decay products consist of leptons only and the lepton number L_e, L_μ and L_τ are conserved. Examples of leptonic decays are

A. Kamal, *Particle Physics*, Graduate Texts in Physics,
DOI 10.1007/978-3-642-38661-9_7, © Springer-Verlag Berlin Heidelberg 2014

$$\begin{aligned}
\mu^- &\to \nu_\mu + e^- + \overline{\nu}_e \\
\mu^+ &\to \overline{\nu}_\mu + e^+ + \nu_e \\
\tau^- &\to \mu^- + \overline{\nu}_\mu + \nu_\tau \\
&\to e^- + \overline{\nu}_e + \nu_\tau \\
\tau^+ &\to \mu^+ + \nu_\mu + \overline{\nu}_\tau \\
&\to e^+ + \nu_e + \overline{\nu}_\tau
\end{aligned}$$

Recall that there are three distinct types of neutrinos, ν_e, ν_μ, ν_τ each associated with the member of the family of charged leptons, e^-, μ^-, τ^-, respectively. Further, each of the neutrinos has its own antiparticle, $\overline{\nu}_e$, $\overline{\nu}_\mu$, $\overline{\nu}_\tau$ just as the charged leptons have their own antiparticles, e^+, μ^+, τ^+ and that each one of the three lepton numbers L_e, L_μ and L_τ assigned to the leptons is separately conserved. Thus the interaction $e^+ + e^- \to \nu_\mu + \overline{\nu}_\mu$ is allowed while $e^+ + e^- \to \nu_e + \nu_\mu$ is forbidden. Experiments show that the upper limits for any violation of these conservation laws in electromagnetic or weak processes are very small for example the ratio.

$$\frac{T(\mu^\pm \to e^\pm \gamma)}{T(\mu^\pm \to \text{all channels})} < 5 \times 10^{-11}$$

not withstanding the fact that the phase-space for the decay, $\mu^\pm \to e^\pm \gamma$, is very large. It also proves that muon is not an excited state of electron, but is a distinct particle.

Note that these decays have a form similar to processes involving β-decay.

$$\begin{aligned}
n &\to p + e^- + \overline{\nu}_e \\
p &\to n + e^+ + \nu_e
\end{aligned}$$

7.3 Semileptonic Decays

Here the decays involve both hadrons and leptons. Typical examples for the cases $\Delta S = 0$ and $\Delta S = \pm 1$ are

(a) for $\Delta S = 0$

$$\begin{aligned}
n &\to p + e^- + \overline{\nu}_e \\
\pi^- &\to \mu^- + \overline{\nu}_\mu \\
&\to e^- + \nu_e
\end{aligned}$$

(b) $\Delta S = \pm 1$

$$\begin{aligned}
K^0 &\to \pi^- + e^+ + \nu_e \\
K^+ &\to \pi^0 + e^+ + \nu_e \\
K^+ &\to \mu^+ + \nu_\mu \\
\Lambda &\to p + e^- + \overline{\nu}_e
\end{aligned}$$

7.4 Non-leptonic Decay

Here the decays do not involve any of the leptons. The concerned processes do not conserve parity, isospin or strangeness and obey the selection rules, $\Delta S = \pm 1$, $\Delta I = \pm 1/2$. Typical examples are

$$\begin{aligned} \Lambda &\rightarrow p + \pi^- \\ &\rightarrow n + \pi^0 \\ \Sigma^+ &\rightarrow p + \pi^0 \\ &\rightarrow n + \pi^+ \\ \Xi^- &\rightarrow \Lambda + \pi^- \\ \Omega^- &\rightarrow \Lambda + K^- \\ &\rightarrow \Xi^0 + \pi^- \\ &\rightarrow \Xi^- \pi^0 \\ K^+ &\rightarrow \pi^+ + \pi^0 \\ &\rightarrow \pi^+ + \pi^+ + \pi^- \end{aligned}$$

7.5 Weak Interaction and Quarks

The semi-leptonic decays can be easily interpreted by invoking for the quark structure of hadrons. For example, consider the decay of neutron,

$$n \rightarrow p + e^- + \overline{\nu}_e$$

In terms of the quark structure, the above decay would be

$$udd \rightarrow udu + e^- + \overline{\nu}_e$$

This involves the transformation of a d-quark into a u-quark that is

$$d \rightarrow u + e^- + \overline{\nu}_e$$

We can also explain the decay of π^- by noting that the creation of a particle is equivalent to annihilation of the antiparticle. The last process can be rewritten as

$$d + \overline{u} \rightarrow e^- + \overline{\nu}_e$$

That is

$$\pi^- \rightarrow e^- + \overline{\nu}_e$$

for the $\Delta S = 0$ in the semileptonic process.

Similarly, we can account for the $\Delta S = \pm 1$ semileptonic process by introducing the $\mu^- \overline{\nu}_\mu$ pair.

The $\Delta S = \pm 1$ semileptonic process for Λ can be accounted for by the decay of s quark.

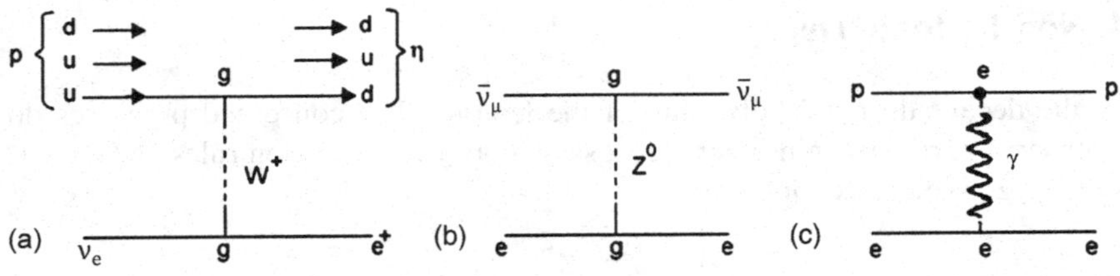

Fig. 7.1 Different types of interactions

$$\Lambda \rightarrow p + e^- + \overline{\nu}_e$$

$$sud \rightarrow uud + e^- + \overline{\nu}_e$$

that is,

$$s \rightarrow u + e^- + \overline{\nu}_e$$

We thus find that in weak decays the flavour of the quark changes in contrast with strong or electromagnetic decays where the flavour is conserved.

The weak interactions take place between all the quark and lepton constituents. However, the interaction is so feeble that it is usually swamped by the much stronger electromagnetic and strong interactions, unless they are forbidden by conservation laws. The observed weak interactions therefore either involve neutrinos which are neutral weakly interacting particles or quarks with a flavour change ($\Delta S = 1$, $\Delta C = 1$ etc., forbidden for electromagnetic or strong interactions).

7.6 Charge Current and Neutral Current Weak Interactions

The weak interactions are mediated by massive bosons $W^{\pm}$ and Z^0 analogous to photon exchange in electromagnetic interactions. The masses of $W^{\pm}$ and Z^0 are, respectively, 81 GeV and 94 GeV. $W^{\pm}$ exchange results in change of charge of the lepton as in the antineutrino absorption

$$\overline{\nu}_e + p \rightarrow n + e^+$$

Also, the neutron decay $n \rightarrow p + e^- + \overline{\nu}_e$, has the inverse process

$$\nu_e + n \rightarrow p + e^-$$

in which the lepton changes its charge. Reactions in which the lepton changes its charge and hadrons participate are called "charge current" reactions. On the other hand, if the lepton does not change its charge then it is termed as "Neutral Current" reaction. Figures 7.1(a), (b) and (c) indicate, respectively the charged-current weak interaction ($\overline{\nu}_e + p \rightarrow n + e^+$), the neutral-current weak interaction ($\overline{\nu}_\mu + e^- \rightarrow \overline{\nu}_\mu + e^-$) and the electro magnetic interaction ($\overline{e} + p \rightarrow \overline{e} + p$).

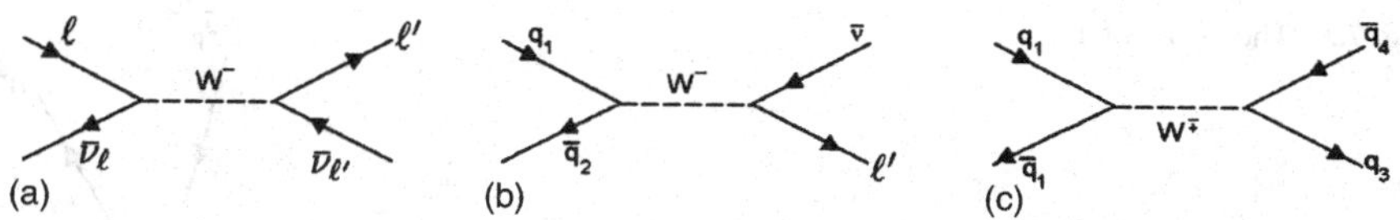

Fig. 7.2 Charge current for (**a**) leptonic (**b**) semi leptonic (**c**) non-leptonic processes

The lifetime for weak decays depends on phase-space factor as well as the weak coupling constant G but are long ($>10^{-13}$ sec) compared with typical lifetimes for electromagnetic decays ($\simeq 10^{-19}$ sec) or strong decays ($\simeq 10^{-23}$ sec); the cross-sections being correspondingly small. Thus, for example the cross section for $\nu_\mu + N \to N + \pi + \mu^-$ at 1 GeV, σ is 10^{-38} cm^2, a value smaller by a factor 10^{12} than for $\pi + N \to N + \pi + \mu^-$ at the same energy.

It was pointed out that the weak interaction can transform a charged lepton into its family's neutrino and that it can produce a charged lepton (antilepton) and its antineutrino (neutrino). In the same manner quarks of one flavour can be transformed into quarks with another flavour in weak interaction, for example in the β-decay of a neutron a d-quark is transformed into a u-quark.

$$n \to p + e^- + \overline{\nu}_e$$

$$udd \to uud + e^- + \overline{\nu}_e$$

In all such reactions the identity of the quarks and leptons involved changes and simultaneously, the charge changes by $+1e$ or $-1e$. Such reactions are described by Charged Current. They are mediated by charged bosons the W^+ and W^-. Originally only such reactions were known to exist. In 1973 the neutral currents were discovered which were predicted by Weinberg, Salam and Glashow. In the weak reactions caused by Neutral currents, the interactions proceed via the exchange of an electrically neutral heavy boson, Z^0. In this case the quarks and leptons are not changed. An example is the neutrino-electron scattering.

$$\nu_\mu + e^- \to \nu_\mu + e^-$$

The charged currents may be divided into three categories, Fig. 7.2(a) for the leptonic processes, (b) for the semi leptonic processes and (c) for the non-leptonic processes (ℓ denotes a lepton).

Leptonic Processes: If the W boson only couples to leptons then we are concerned with a leptonic process. The underlying reaction is

$$l + \overline{\nu}_l \leftrightarrow l' + \overline{\nu}_{l'}$$

Example of this is the leptonic decay of the τ-lepton (Fig. 7.3).

$$\tau^- \to \mu^- + \overline{\nu}_\mu + \nu_\tau$$

$$\tau^- \to e^- + \overline{\nu}_e + \nu_\tau$$

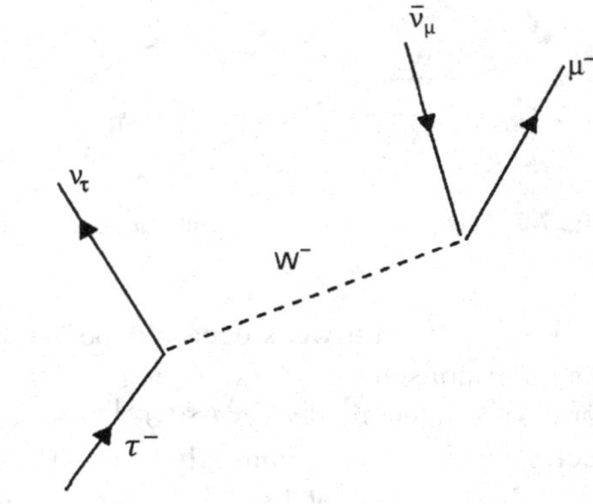

Fig. 7.3 The decay of τ^-

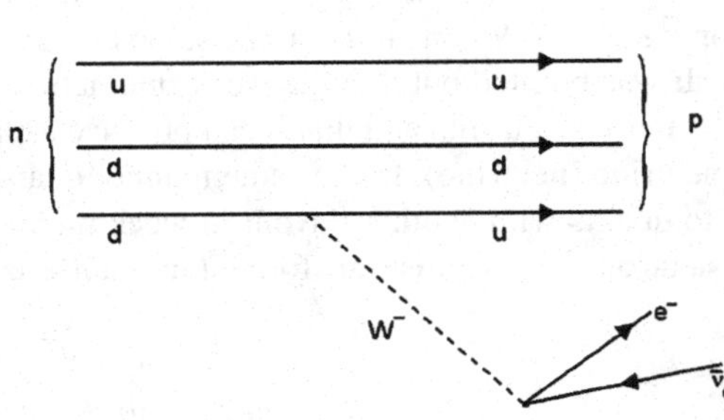

Fig. 7.4 Spectator quarks are shown as $u \to u$ and $d \to d$

And the scattering process

$$\nu_\mu + e^- \to \mu^- + \nu_e$$

Semileptonic Processes: Semileptonic processes are those where the exchanged W boson couples to both leptons and quarks. Here the fundamental process is

$$q_1 + \overline{q}_2 \leftrightarrow \ell + \overline{\nu}_l$$

Examples of this are charged pion decay, the decay of the K^- or the β-decay of the neutron.

Decay	**Quark description**
$\pi^- \to \mu^- + \overline{\nu}_\mu$	$d + \overline{u} \to \mu^- + \overline{\nu}_\mu$
$K^- \to \mu^- + \overline{\nu}_\mu$	$s + \overline{u} \to \mu^- + \overline{\nu}_\nu$
$n \to p + e^- + \overline{\nu}_e$	$d \to u + e^- + \overline{\nu}_e$

The β-decay of a neutron may be reduced to the decay of a d-quark in which the two other quarks are not involved. The latter are known as spectator quarks, Fig. 7.4.

Non-leptonic Processes: Non-leptonic processes do not involve leptons at all. The basic process is

$$q_1 + \overline{q_2} \leftrightarrow q_3 + \overline{q_4}$$

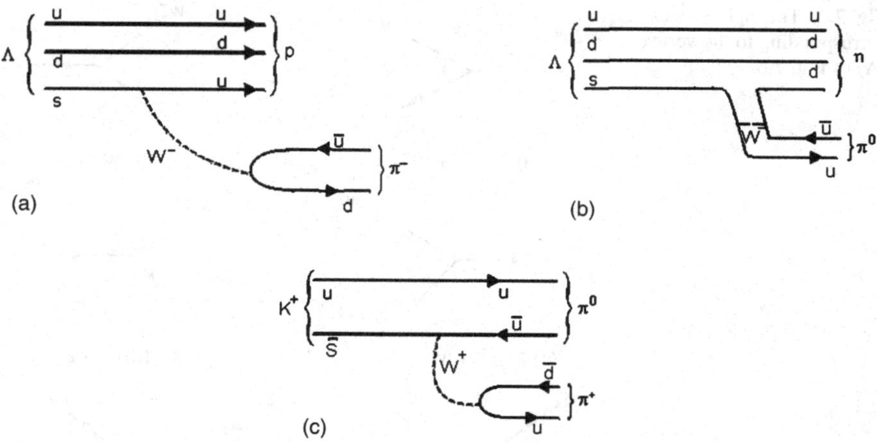

Fig. 7.5 Quark description of non-leptonic process (**a**) $\Lambda \to p + \pi^-$; (**b**) $\Lambda \to n + \pi^0$; (**c**) $K^+ \to \pi^+ + \pi^0$

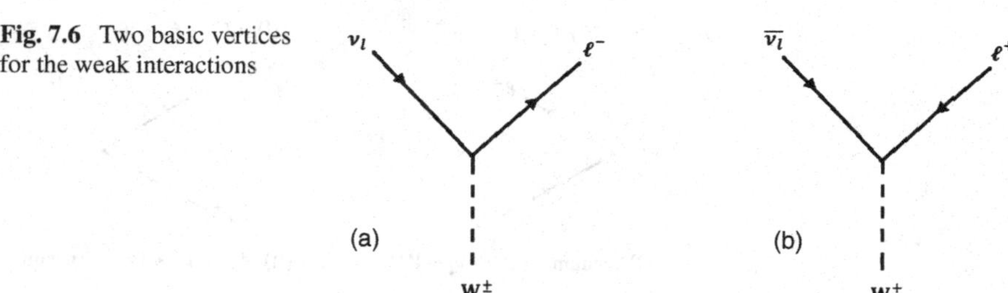

Fig. 7.6 Two basic vertices for the weak interactions

Charge conservation requires that the only allowed quark combinations have a total charge $\pm 1e$ i.e. examples are the hadronic decays of strange baryons and strange mesons, such as the decay of the Λ-hyperon (uds) into a nucleon and a pion or that of $K^+(u\bar{s})$ into two pions, Fig. 7.5.

7.7 Basic Vertices for $W^{\pm}$-Lepton Interactions

It was pointed out that all the electromagnetic interactions of electrons and positrons could be built from eight basic interactions in which a photon from a lepton is either emitted or absorbed (Fig. 7.2). In the same way weak interactions processes can also be built from limited number of basic reactions. For charged current reactions for each lepton, $\ell = e, \mu, \tau$, there are 16 such basic reactions corresponding to the two vertices of Fig. 7.6.

The eight processes corresponding to the vertex (B) are not shown explicitly as they can be obtained from the set in Fig. 7.7(a)–(h) by replacing all the particles by their antiparticles. Thus in the set (A), Fig. 7.7(e) is transformed as in Fig. 7.8.

Fig. 7.7 The eight processes corresponding to the vertex (A) of Fig. 7.6

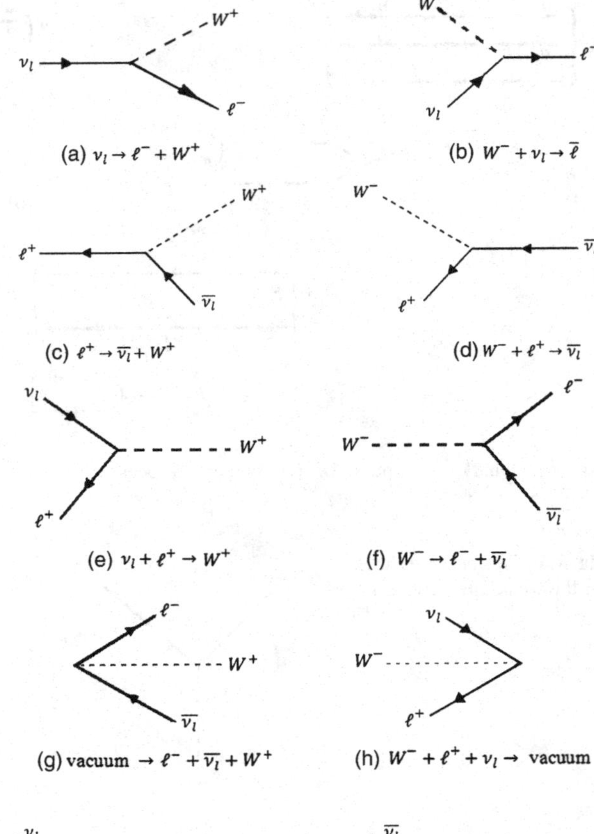

Fig. 7.8 A typical weak process for the vertex (B) of Fig. 7.6 analogous to Fig. 7.7(e)

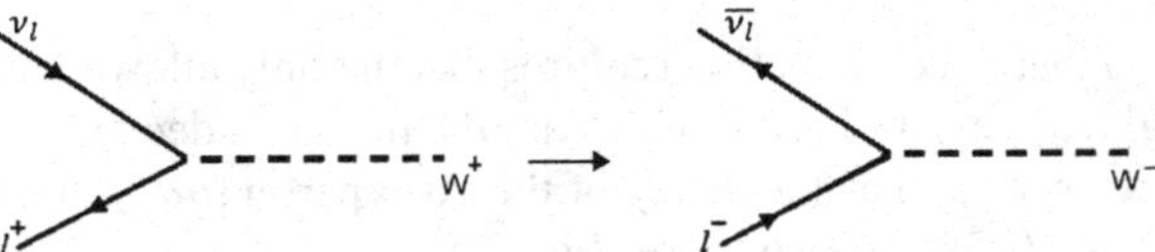

In Figs. 7.7(a)–(h) all the processes conserve lepton numbers. However, they do not occur as isolated reactions in free space in analogy with the e.m interactions. However, two or more such diagrams can be combined in such a way that energy is conserved on the whole.

7.8 Coupling Strength of the Charged Current

For charged current a weak charge g is assumed to play the same role in weak interactions as the electric charge e in electromagnetic interactions. As for Mott scattering or e^+e^- annihilation, the transition matrix element in the case of weak

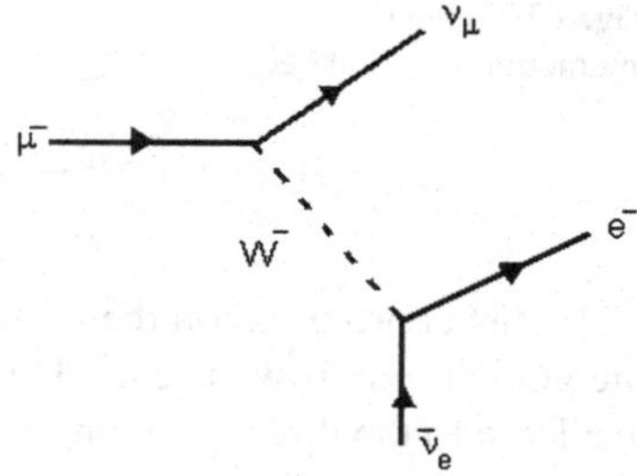

Fig. 7.9 Graph for muon decay

interactions is proportional to the square of the weak charge g to which the W boson is coupled and to the propagator $1/(q^2c^2 + M_w^2c^4)$ of a massive spin 1 particle.

$$M_{fi} \propto g \frac{g}{Q^2c^2 + M_W^2c^4} \xrightarrow{Q^2 \to 0} \frac{g^2}{M_W^2c^4} \tag{7.1}$$

It is seen that the propagator is nearly constant for small momenta transfer $Q^2 \ll M_W^2c^2$. It turns out that the weak charge g and the electric charge e are of a similar size. The large mass of the exchanged boson implies that at small Q^2 the weak interaction appears to be much weaker then the electromagnetic interaction for which the propagator is $(Qc)^{-2}$. It also follows that its range. $\hbar/M_W c \simeq 2.5 \times 10^{-3}$ fm is very limited.

In the domain of small momentum transfer the interaction may be described as a point like interaction of the four particles involved. This was in fact the original description of the weak interaction before the concept of W boson was entertained. The coupling strength of this interaction is described by the Fermi constant G_F, which is proportional to the square of the weak charge g, similar to the electromagnetic coupling constant $\alpha = e^2/(4\pi\varepsilon_0\hbar c)$ which is proportional to the square of the electric charge e. It is so defined that $G_F/(\hbar c)^3$ has dimensions of $[1/\text{energy}]^2$ and related to g by

$$\frac{G_F}{\sqrt{2}} = \frac{\pi\alpha}{2} \cdot \frac{g^2}{e^2} \cdot \frac{(\hbar c)^3}{M_W^2c^4} \tag{7.2}$$

7.8.1 The Decay of the Muon

The most exact value for the Fermi constant is obtained from the muon decay.

Recall the muon decay, $\mu^- \to e^- + \overline{\nu}_e + \nu_\mu$. The dominant Feynman diagram for muon decay is shown in Fig. 7.9.

Since the muon mass is small compared to that of the W boson it is reasonable to treat this interaction as point like and describe the coupling via the Fermi constant, Fig. 7.10.

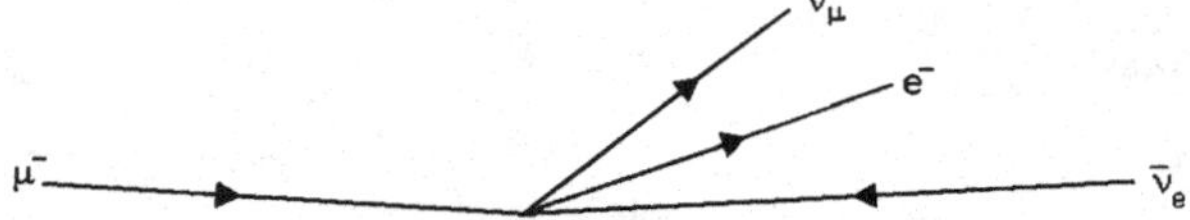

Fig. 7.10 Point like interaction in muon decay

In this approximation the lifetime of the muon may be calculated with the aid of the golden rule, if we use the Dirac equation and take into account the phase space available to the three out going leptons. One finds that the decay width is

$$T_{\mu} = \frac{\hbar}{\tau_{\mu}} = \frac{G_F^2 \cdot (m_{\mu}c^2)^5 \cdot (1+\varepsilon)}{192\pi^3(\hbar c)^6} \tag{7.3}$$

The correction term ε, which reflects higher order radiation corrections and phase space effects resulting from the finite electron mass, is small.

Note that the transition rate is proportional to the fifth power of the energy and hence the mass of the decaying muon. The dependence on the mass to the fifth power follows the Sargent rule for the three body decay.

We can use (7.3) to estimate the mean lifetime τ_2 in a three body decay of particle 2 with branching ratio B_2, knowing τ_1 and B_1 for particle 1, from formula

$$\frac{\tau_2}{\tau_1} = \frac{B_2}{B_1}\left(\frac{m_1}{m_2}\right)^5 \tag{7.4}$$

In this way we can estimate the mean life τ_{τ} for tauon with $B_{\tau} = 0.117$ for $\tau^- \to e^-\overline{\nu}_e\nu_{\tau}$, $\tau_{\mu} = 2.197 \times 10^{-6}$ sec, $m_{\tau} = 1784$ MeV, $m_{\mu} = 105.659$ MeV. We find $\tau_{\tau} = 2.824 \times 10^{-13}$ sec; this is in good agreement with the experimental value 2.8×10^{-13} sec. The pure lepton decay modes of tauon are $\tau^- \to e^- + \overline{\nu}_e + \nu_{\tau}$ and $\tau^- \to \mu^- + \overline{\nu}_{\mu} + \nu_{\tau}$. The e–μ universality implies that the branching ratios for these two decay modes are equal—a feature that has been verified by experiment.

The muon mass and lifetime have been measured to a high precision.

$$m_{\mu} = (105.658389 \pm 0.000034)\ \text{MeV}/c^2$$
$$\tau_{\mu} = (2.197035 \pm 0.000040) \times 10^{-6}\ \text{sec.}$$

This yields a value for the Fermi constant

$$\frac{G_F}{(\hbar c)^3} = (1.16639 \pm 0.00001) \times 10^{-5}\ \text{GeV}^{-2} \tag{7.5}$$

7.9 Neutrino-Electron Scattering

This is inverse of muon decay

$$\nu_{\mu} + e^- \to \mu^- + \nu_e$$

Figure 7.11 shows the two time-ordered Feynman diagrams in which the muon-neutrinos scatter off electrons and ν_{μ} is transformed into a μ^- via W^- exchange.

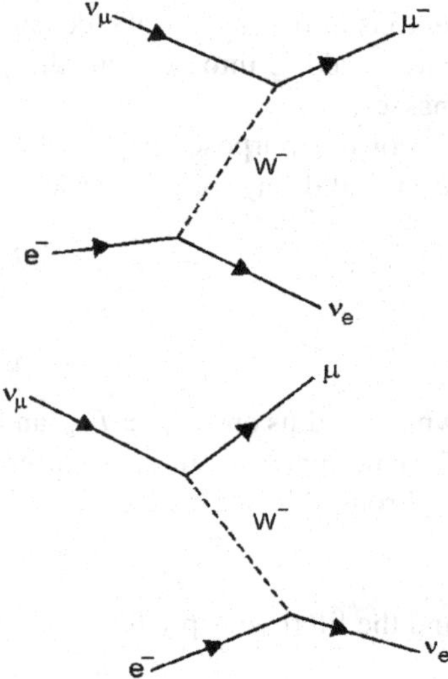

Fig. 7.11 Time ordered diagrams for inverses muon decay

As before, time flows from left to right. By convention, one of these diagrams is usually drawn, leaving the existence of the other implied.

For small four-momenta the total cross-section for neutrino-electron scattering is given by

$$\sigma = \frac{G_F^2 \cdot s}{\pi(\hbar c)^4} \tag{7.6}$$

where s is the square of the centre of mass energy. In a fixed target experiment with m satisfying $mc^2 \ll E$, s is related to E by

$$s \simeq 2mc^2 \cdot E \tag{7.7}$$

$$\sigma_{lab} = 1.7 \times 10^{-41}\ \text{cm}^2 \cdot E_\nu/\text{GeV} \tag{7.8}$$

At very high energies the simple formula (7.6) is no longer valid. For, at large four-momentum transfer, $Q^2 \gg M_W^2 c^2$ the propagator term primarily determines the energy dependence of the cross-section and the point like approximation is untenable. In that case the total cross-section is given by

$$\sigma = \frac{G_F^2}{\pi(\hbar c)^4} \cdot \frac{M_W^2 c^4 \cdot s}{s + M_W^2 c^4} \tag{7.9}$$

The cross-section does not increase linearly with s, as the point-like approximation implies, rather it asymptotically approaches a constant value.

7.10 Universality of the Weak Interaction

If we assume that the weak charge g is the same for all quarks and leptons then (7.3) must hold for all possible charged decays of the fundamental fermions into lighter leptons or quarks. All the decay channels then contribute equally to the total decay width, taking into account the phase space correction coming from the different masses.

For this purpose consider the decay of the τ-lepton. There are essentially three modes of decay for this particle.

$$\tau^- \to \nu_\tau + \overline{\nu}_e + e^-$$
$$\tau^- \to \nu_\tau + \overline{\nu}_\mu + \mu^-$$
$$\tau^- \to \nu_\tau + \overline{u} + d$$

whose widths are $\Gamma_{\tau e} \simeq \Gamma_{\tau\mu}$ and $\Gamma_{d\overline{u}} \simeq 3\Gamma_{\tau\mu}$. The factor of 3 is introduced for the $\overline{u}d$-pair appearing in three different colour combinations ($r\overline{r}$, $b\overline{b}$, $g\overline{g}$).

From (7.3) we have

$$T_{\tau e} = (m_\tau/m_\mu)^5 \cdot T_{\mu e} \tag{7.10}$$

and the lifetime is predicted to be

$$\tau_\tau = \frac{\hbar}{\tau_{\tau e} + \tau_{\tau\mu} + \tau_{\tau d\overline{u}}} \simeq \frac{\tau_\mu}{5 \times (m_\tau/m_\mu)^5} \simeq 3.1 \times 10^{-13}\ \text{sec}$$

The predicted lifetime is in good agreement with the experimental value

$$\tau_\tau(\text{expt}) = (2.900 \pm 0.012) \times 10^{-13}\ \text{sec}$$

and confirms that quarks occur in three different colours and strongly supports the assumption that the weak charges for leptons and quarks are identical.

7.11 Lepton-Quark Symmetry

The result that the coupling constant is identical for both leptons and quarks is based on the sum of the decay widths for the leptonic and hadronic processes in the decay at τ^--particle, rather than separately. Hence the proof for universality of the coupling constant was subject to further scrutiny.

The coupling to quarks can be better determined from semi-leptonic hadron decays. This gives a smaller value for the coupling than that obtained from muon data. As an example, in β-decay of the neutron, a d-quark is transformed into a u-quark, the coupling constant is found to be 4 % smaller. In the Λ-decay which is a purely hadronic decay an s-quark is transformed into a u-quark and the coupling constant appears to be 20 times smaller. The purely hadronic decays are not so well understood as the leptonic ones as the final-state particles interact strongly with each other resulting in effects which are difficult to calculate. In what follows we shall therefore be concerned with the semi-leptonic interactions and restrict ourselves to the first two generation of quarks.

$$\begin{pmatrix} u \\ d \end{pmatrix} \quad \text{and} \quad \begin{pmatrix} c \\ s \end{pmatrix} \tag{7.11}$$

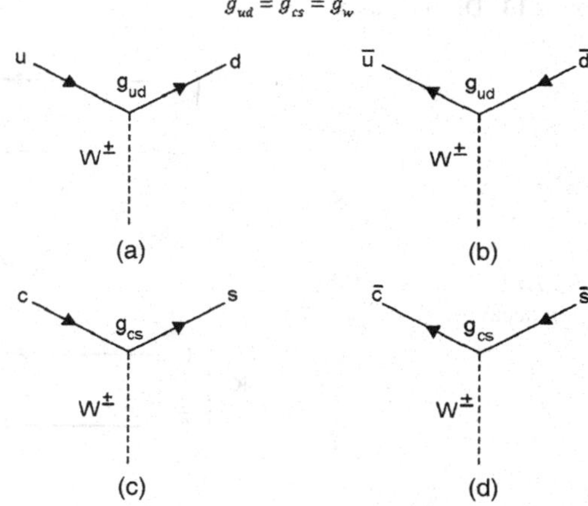

Fig. 7.12 Basic $W^{\pm}$-quark vertices

of which numerous known hadrons are composed.

The weak interactions of quarks are best understood in terms of two ideas, lepton-quark symmetry and quark mixing. First we shall focus on $W^{\pm}$ boson-quark interaction and deduce the results for Z^0-boson-quark interaction and show that the observed weak interactions of hadrons are explained in both cases.

The lepton-quark symmetry asserts that the two generations of quarks (7.11) and the two generations of leptons

$$\begin{pmatrix} \nu_e \\ e^- \end{pmatrix} \quad \text{and} \quad \begin{pmatrix} \nu_\mu \\ \mu^- \end{pmatrix} \tag{7.12}$$

have identical weak interactions. In other words we obtain the basic $W^{\pm}$ quark vertices by the replacements $\nu_e \to u$, $e^- \to d$, $\nu_\mu \to c$ and $\mu^- \to s$ in the basic $W^{\pm}$ lepton vertices (A) and (B), Fig. 7.7, leaving the coupling constant g_W unchanged.

In this way we can obtain the vertices of Fig. 7.12 with

$$g_{ud} = g_{cs} = g_W$$

This works well for many reactions, like pion decay

$$\pi^+ \to \mu^+ + \nu_\mu$$

which corresponds to

$$u\bar{d} \to \mu^+ + \nu_\mu$$

for which the Feynman diagram is shown in Fig. 7.13.

However, many other decays which are experimentally observed are forbidden in this scheme. An example is the kaon decay

$$K^+ \to \mu^+ + \nu_\mu$$

which corresponds to

$$u\bar{s} \to \mu^+ + \nu_\mu$$

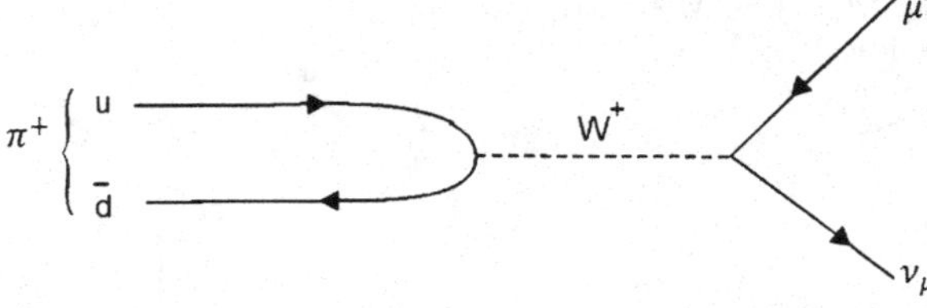

Fig. 7.13 Diagram for the pion decay

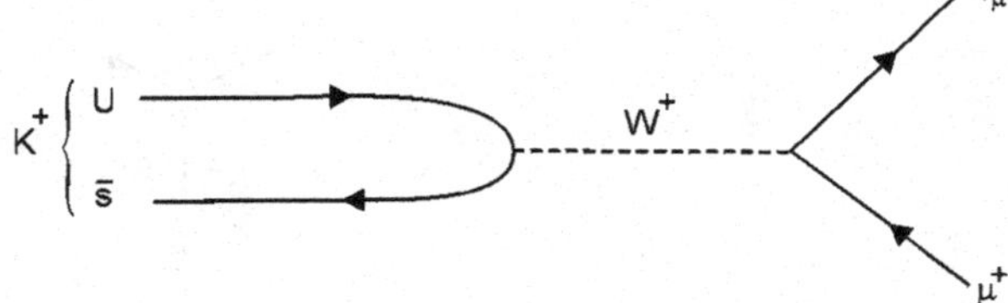

Fig. 7.14 Diagram for the kaon decay

at the quark level which is shown by the Feynman diagram of Fig. 7.14.

It includes a 'usW' vertex which is not included in the vertices of Fig. 7.12. It is, however, incorporated in Cabibbo's theory.

7.12 The Cabibbo Theory

An explanation of these findings was proposed by Cabibbo as early as 1963, at a time at which quarks had not been introduced. In weak interactions with charged currents, the leptons in the same doublet can be transformed into each other, $e^- \leftrightarrow \nu_e$ and $\mu^- \leftrightarrow \nu_\mu$ for example. Quark transitions are observed not only within a family but to a lesser degree, from one family to another. For charged currents, the partner of the flavour eigen state $|u\rangle$ is therefore not the flavour state $|d\rangle$ but a linear combination of $|d\rangle$ and $|s\rangle$. We call this linear combination $|d'\rangle$. Similarly the partner of the c-quark is a linear combination of $|s\rangle$ and $|d\rangle$, called $|s'\rangle$ which is orthogonal to $|d'\rangle$.

According to Cabibbo d and s should be mixed by a rotation parameter, known as the Cabibbo angle θ_c. The quark eigen states $|d'\rangle$ and $|s'\rangle$ of W exchange are related to the eigen states $|d\rangle$ and $|s\rangle$ of the strong interaction by a rotation through θ_c.

$$\begin{aligned} |d'\rangle &= |d\rangle \cos\theta_c + |s\rangle \sin\theta_c \\ |s'\rangle &= -|d\rangle \sin\theta_c + |s\rangle \cos\theta_c \end{aligned} \tag{7.13}$$

In matrix form

$$\begin{pmatrix} |d'\rangle \\ |s'\rangle \end{pmatrix} = \begin{pmatrix} \cos\theta_c & \sin\theta_c \\ -\sin\theta_c & \cos\theta_c \end{pmatrix} \begin{pmatrix} |d\rangle \\ |s\rangle \end{pmatrix} \tag{7.14}$$

This means that the lepton-quark symmetry now applies to the doublets

$$\begin{pmatrix} u \\ d' \end{pmatrix} \quad \text{and} \quad \begin{pmatrix} c \\ s' \end{pmatrix}$$

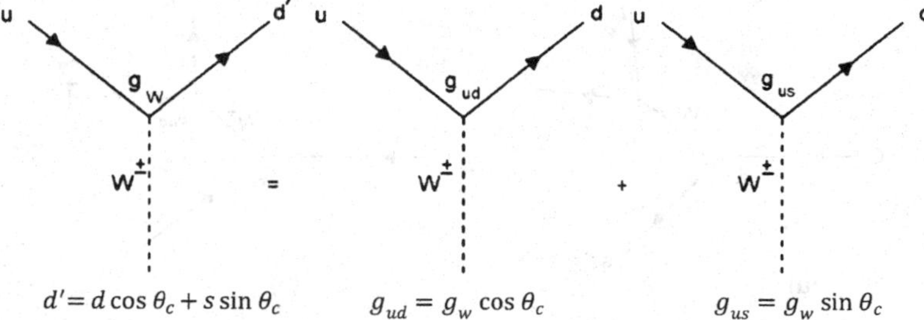

Fig. 7.15 The lepton-quark symmetry applied to the doublets

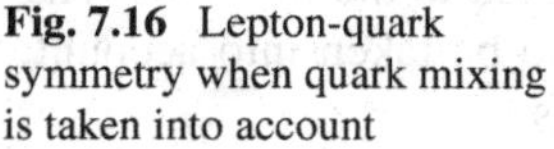

Fig. 7.16 Lepton-quark symmetry when quark mixing is taken into account

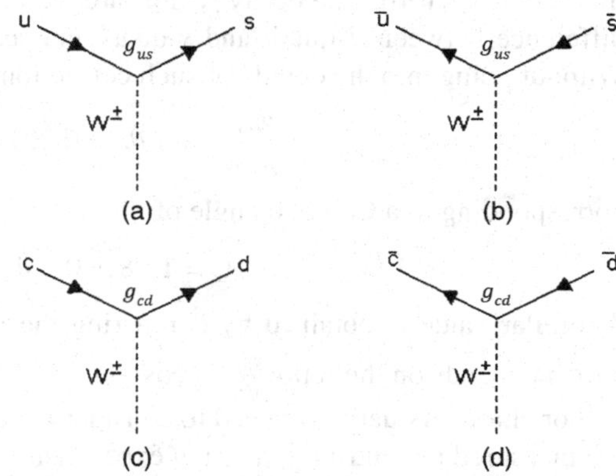

Figure 7.15 illustrates the first of these.

Thus the previously allowed udW vertex in Fig. 7.12(a) is suppressed by a factor $\cos\theta_c$ compared to $g_{ud} = g_W$ and the previously forbidden usW vertex is now allowed with a coupling $g_W \sin\theta_c$. Same reasoning is applied to the other three vertices in Fig. 7.12 so that in addition to the four vertices with the couplings.

$$g_{ud} = g_{cs} = g_w \cos\theta_c \tag{7.15a}$$

we have the vertices of Fig. 7.16 with the couplings

$$g_{us} = -g_{cd} = g_w \sin\theta_c \tag{7.15b}$$

because of lepton-quark symmetry when quark mixing is taken into account.

The value of Cabibbo angle θ_c can be obtained from the ratio of decay width of these hadrons which depend on the ratio g_{us}/g_{ud}. Compare the decay rates of K^+ and π^+. The decays differ in that a $\bar{d}$ quark has been replaced by an $\bar{s}$ quark in the initial state, and a coupling g_{ud} replaced by g_{us} so that the ratio of their rate

$$\frac{\Gamma(K^+ \to \mu^+\nu_\mu)}{\Gamma(\pi^+ \to \mu^+\nu_\mu)} \propto \frac{g_{us}^2}{g_{ud}^2} = \tan^2\theta_c \tag{7.16}$$

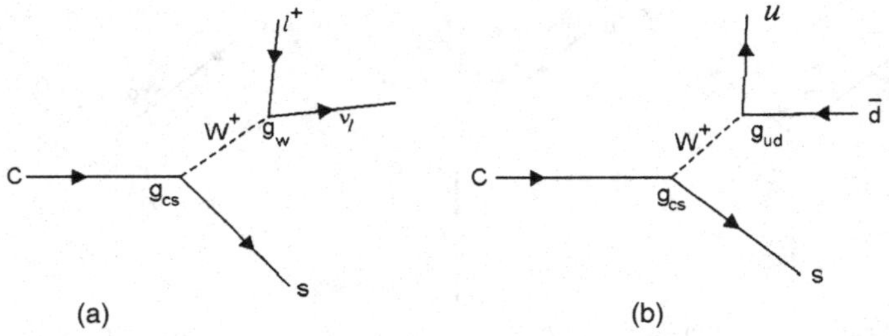

Fig. 7.17 Production s-s quark due to the decay of a charmed quark to a lighter quark and leptons

by (7.15a), (7.15b). The decay widths are reciprocally related to the lifetimes. The difference between d-quark and s-quark masses must also be taken into account. Without going into the details of such corrections one finds

$$\frac{g_{us}}{g_{ud}} = \tan\theta_c = 0.227 \pm 0.004 \tag{7.17}$$

corresponding to a Cabibbo angle of

$$\theta_c = 12.8 \pm 0.2 \text{ degrees} \tag{7.18}$$

A similar value is obtained by comparing the rates for neutron and muon decay, which depends on the ratio $\frac{g_{ud}^2}{g_w^2} = \cos^2\theta_c$.

For charmed quarks we need to consider the quark couplings g_{cd} and g_{cs}. These are measured in neutrino scattering experiments and yield a value

$$\theta_c = 12 \pm 1 \text{ degrees}$$

in agreement with (7.18).

The most striking result is for charmed particles decays, which almost always decay into strange particles. This can be explained by noting that decays which involve the couplings (7.15b) are called Cabibbo-suppressed because their rates are typically reduced by a factor of order

$$\frac{g_{us}^2}{g_{ud}^2} = \frac{g_{cd}^2}{g_{cs}^2} = \tan^2\theta_c = \frac{1}{20} \tag{7.19}$$

compared with similar Cabibbo-allowed decays which involve the couplings (7.15a).

The Cabibbo allowed decays

$$c \to s + \ell^+ + +\nu_l \quad (\ell = e, \mu) \tag{7.20a}$$

and

$$c \to s + u + \overline{d} \tag{7.20b}$$

of a charmed quark to a lighter quark and leptons are shown in Fig. 7.17 and they necessarily produce an s-squark in the final state in agreement with the experiments.

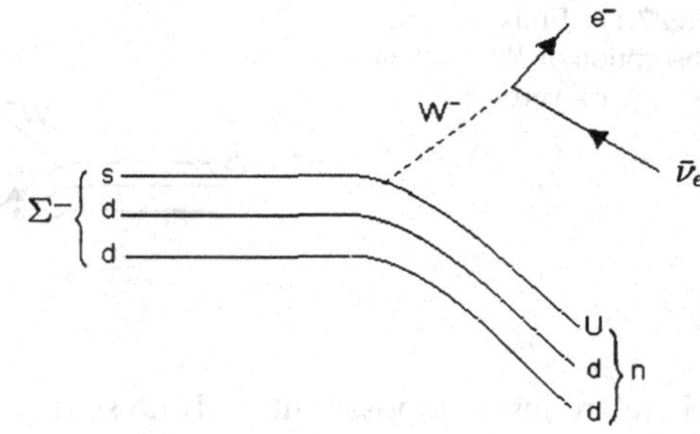

Fig. 7.18 Feynman diagram of quark decay by weak interaction

7.13 Selection Rules in Weak Decays

Many of the weak decays or their forbiddingness can be explained by certain simple selection rules, based on the $W^{\pm}$-quark exchange. For example, consider the decays

$$\Sigma^- \rightarrow n + e^- + \overline{\nu}_e \tag{7.21}$$

and

$$\Sigma^+ \rightarrow n + e^+ + \nu_e \tag{7.22}$$

which appear to be similar where

$$\Sigma^+(1189) = uus \quad \text{and} \quad \Sigma^-(1197) = dds$$

However while reaction (7.21) is observed, (7.22) is not. Experimentally, the ratio of the transition rates of decay

$$\frac{\Gamma(\Sigma^+ \rightarrow n + e^+ + \nu_e)}{\Gamma(\Sigma^- \rightarrow n + e^- + \overline{\nu}_e)} < 5 \times 10^{-3}$$

The reason for this is that (7.21) can go via mechanism of Fig. 7.18 whereas no diagram can be drawn for (7.22) which at the quark level is

$$uss \rightarrow udd + e^+ + \nu_e$$

and would require a mechanism involving the emission and absorption of two W bosons, but this being of higher order will have negligible contribution.

This and many other forbidden reactions can be explained (despite the fact that they satisfy all the appropriate conservation laws) using a number of selection rules for single W^+ exchange processes which can be deduced from the vertices of Figs. 7.12 & 7.16.

7.13.1 Semi Leptonic Decays

First consider the semi leptonic decays like those of Figs. 7.12 and 7.16. Since these involve a single $W^{\pm}$-quark vertex, the changes in the strangeness and electric charge of the hadrons are given by the possible changes in S and Q at this vertex.

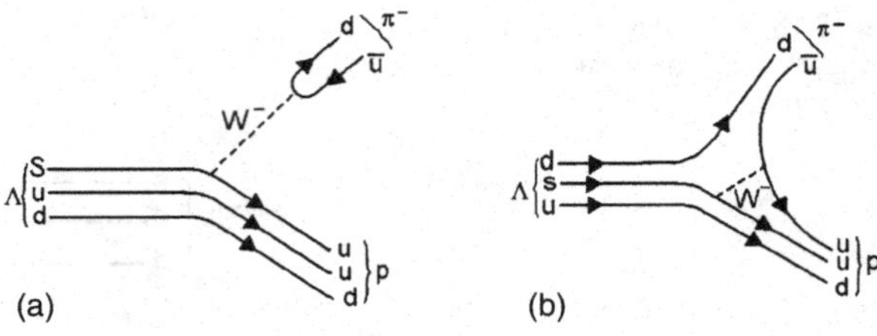

Fig. 7.19 Emission and absorption of W boson at $W^{\pm}$-quark vertices

There are just two possibilities. If no strange quarks are involved at the vertex as in Fig. 7.12(a), (b) and Fig. 7.16(c) and (d) there is obviously no change in strangeness, while the quark charge changes by ± 1 depending on the charge of the w boson. Hence the changes ΔS and ΔQ in the strangeness and the electric charge of the hadrons satisfy

$$\Delta S = 0, \qquad \Delta Q = \pm 1 \tag{7.23}$$

On the other hand, those vertices like Figs. 7.12(c) and (d) and Figs. 7.16(a) and (b) which do involve a strange quark give rise to processes like

$$u \rightarrow s + W^{+} \quad \text{or} \quad W^{-} \rightarrow s + \overline{c}$$

in which the total quark charge and strangeness both decrease, giving $\Delta S = \Delta Q = -1$; or processes like

$$s \rightarrow u + W^{-} \quad \text{or} \quad W^{+} \rightarrow \overline{s} + c$$

in which the total quark charge and strangeness both increase, gives $\Delta S = \Delta Q = 1$. Thus the allowed semi leptonic decays are characterised by the selection rules (7.23) and

$$\Delta S = \Delta Q = \pm 1 \tag{7.24}$$

where ΔQ is the change in the charge of the hadrons only. The latter is called the $\Delta S = \Delta Q$ rule for strangeness changing decays, and decays with

$$\Delta S = -\Delta Q = \pm 1 \tag{7.25}$$

are forbidden. Reaction (7.22) is a typical example of a forbidden $\Delta S = -\Delta Q$ reaction requiring changes (7.25) since the Σ^{+} has strangeness $S = -1$ and $Q = +1$ while the neutron has $S = 0$ and $Q = 0$.

7.13.2 Pure Hadronic Decays

In such decays, the exchanged W boson must be both emitted and absorbed at $W^{\pm}$-quark vertices as illustrated by the dominant diagrams for Λ-decay, $(\Lambda \rightarrow p + \pi^{-})$, Fig. 7.19.

Applying the selection rules (7.23) and (7.24) to each individual vertex, subject to the constraint that the change in the hadron charge must now be $\Delta Q = 0$ overall since no leptons are involved and the total charge must of course be conserved. If two vertices satisfying (7.23) are involved strangeness is conserved and $\Delta S = 0$, while if one satisfies (7.23) and the other (7.24) $\Delta S = \pm 1$. Finally if two vertices satisfying (7.23) are involved, $\Delta S = 0$ overall because of $\Delta Q = 0$ condition. The selection rule is

$$\Delta S = 0, \pm 1 \tag{7.26}$$

for hadronic weak decays.

Applying (7.26) to the cascade hyperons

$$\Xi^0[1315] = ssu, \qquad \Xi^-(1321) = ssd \quad (S = -2)$$
$$\Omega^-(1672) = sss \quad (S = -3)$$

because of baryon conservation these hyperons ultimately decay into a proton or neutron. But direct decay to a nucleon is forbidden because of (7.26). Thus, the decay, $\Xi^- \to n + \pi^-$ is forbidden. However, they can decay in steps so that at each stage (7.26) is obeyed

$$\Xi^- \to \Lambda + \pi^-$$
$$\hookrightarrow$$

$$\Omega^- \to \Xi^0 + \pi^-$$
$$\hookrightarrow \Lambda + \pi^0$$
$$\hookrightarrow p + \pi^-$$

7.13.3 Decay of Charmed Particles

Cabibbo-allowed decays involve the CSW vertex of Fig. 7.16(a) giving rise to the selection rule $\Delta C = \Delta S = \Delta Q = \pm 1$. Cabibbo suppressed decays involve the CdW vertex of Fig. 7.16(c), giving rise to the selection rule $\Delta C = \Delta Q = \pm 1$, $\Delta S = 0$.

7.14 Leptonic Decays of Vector Mesons

We now consider the partial width of the leptonic decays of the vector mesons which is given by the Van Royen-Weisskopf [1] formula

$$\Gamma(V \to l^+ l^-) = \frac{16\pi\alpha^2 Q^2}{M_v^2} |\psi(0)|^2 \tag{7.27}$$

where the particle V decays into $l = e$, μ, and the decay is assumed to proceed via exchange of a single virtual photon, $Q^2 = |\Sigma a_i, \ Q_i|^2$ is the squared sum of the charges of the quarks in the meson, $\psi(0)$ is the amplitude of the $q\overline{q}$ wave functions at the origin, and M_v is the meson mass.

We consider specifically the mesons ρ^0 (765 MeV), ω^0 (785 MeV) and ϕ^0 (1020 MeV) which have similar masses. In that case we expect $|\psi(0)|^2/M_v^2$ to be approximately constant and $\Gamma \propto Q^2$. Recalling

$$\rho^0 = \frac{1}{\sqrt{2}}(u\overline{u} - d\overline{d})$$
$$\omega^0 = \frac{1}{\sqrt{2}}(u\overline{u} + d\overline{d})$$
$$\phi^0 = s\overline{s}$$

and inserting the quark charges,

$$\rho^0: \quad \left[\frac{1}{\sqrt{2}}\left(\frac{2}{3} - \left(-\frac{1}{3}\right)\right)\right]^2 e^2 = \frac{1}{2}e^2$$
$$\omega^0: \quad \left[\frac{1}{\sqrt{2}}\left(\frac{2}{3} - \frac{1}{3}\right)\right]^2 e^2 = \frac{1}{18}e^2$$
$$\phi^0: \quad \left(\frac{1}{3}\right)^2 e^2 = \frac{1}{9}e^2$$

The predicted ratios of the leptonic widths

$$\Gamma(\rho^0) : \Gamma(\omega^0) : \Gamma(\phi^0) = 9 : 1 : 2$$

which are in excellent agreement with the observed ratios, 8.8 : 1 : 1.7, thereby confirming the correct assignment of vector mesons and of the quark charges.

7.15 Non Conservation of Parity

The τ–θ Puzzle Of various types of K-mesons, two turned out to be puzzling, one called θ-meson (now $K\pi_2$) decaying into two pions and the other called τ-meson (now $K\pi_3$) decaying into three pions.

$$\theta^{\pm} \rightarrow \pi^{\pm} + \pi^0 \qquad \tau^{\pm} \rightarrow \pi^{\pm} + \pi^{+} + \pi^{-}$$

In the period 1949–1955 detailed studies revealed that their masses, mean life times and interaction cross-sections appeared to be identical. It was therefore logical to conclude that the θ-meson and τ-meson were one and the same particle and that the two-pion and three-pion decay modes were merely the competitive decay modes. In the decay of θ-meson, the two pions will emerge with zero angular momentum ($\ell = 0$) as the energy available is small and the overall parity for two-pion system will be $(-1)^2(-1)^l = (-1)^2(-1)^0 = +1$. If parity is conserved in the decay the θ-meson will have $j^P = 0^+$. Similarly the three-pion system has an odd parity. Thus

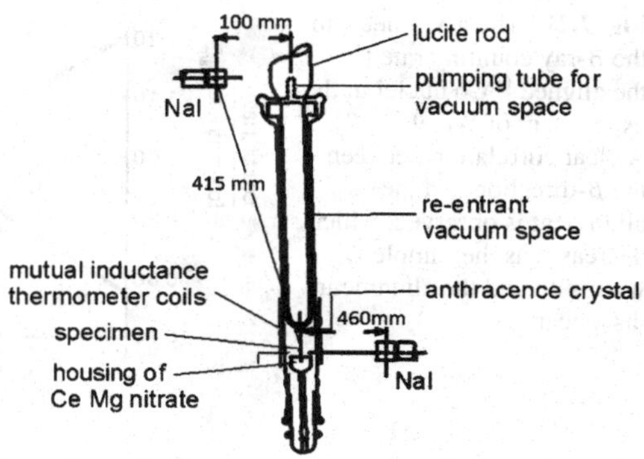

Fig. 7.20 The cryostat and counters used by Wu et al. [2] in the detection of parity violation in ^{60}Co decay

τ-and θ-mesons had to be assigned opposite parities in view of the conservation of parity. In that case τ and θ had to be different particles. This dilemma is known as the τ–θ puzzle which remained unresolved for quite some time. The scenario was so desperate that it seems Wheeler called for a wild idea and Feynman responded "parity may be violated". It was left to T.D. Lee and C.N. Yang to conclude after detailed analysis that while there existed strong evidence to suggest the conservation of parity in strong and electromagnetic interactions, there was none for or against the weak interactions. They also proposed a number of experiments to verify their suggestion. This was followed by the famous experiment of Wu et al. concerned with the beta decay of ^{60}Co nuclei in which it was unequivocally revealed that parity is indeed violated in weak interactions.

7.16 Experiment of Wu and Her Collaborators (1957)

The critical experiment which conclusively proved that parity was not conserved in weak nuclear β-decay was undertaken by Wu, Ambler, Hayward, Hoppes and Hudson in 1957. In this experiment the angular distribution of β-particles emitted by aligned radioactive nuclei was investigated. The source used was ^{60}Co, which is a β-emitter with maximum electron energy 0.312 MeV, Fig. 7.20.

The β-emission is followed by γ-rays of energy 1.19 and 1.32 MeV in cascade to yield ^{60}Ni. The ^{60}Co spins were aligned by depositing the source in a crystal of cerium magnesium nitrate which provides a strong internal magnetic field. The crystal was cooled to 0.01 K by adiabatic demagnetisation in order to minimise the misalignment due to thermal agitation. The degree of alignment was measured by observing the anisotropy of the emitted γ-rays, which were detected by scintillation counters using NaI crystals. After the cooling was complete, the magnetic field coils were removed, the emitted electrons were counted by anthracene scintillator located above the sample. The counting rate was measured as a function of time for nuclei with their spins aligned upward as well as downward.

The results displayed in Fig. 7.21 clearly show an anisotropy in the angular distribution of the emitted β particles. When the sample warms up the spin alignment

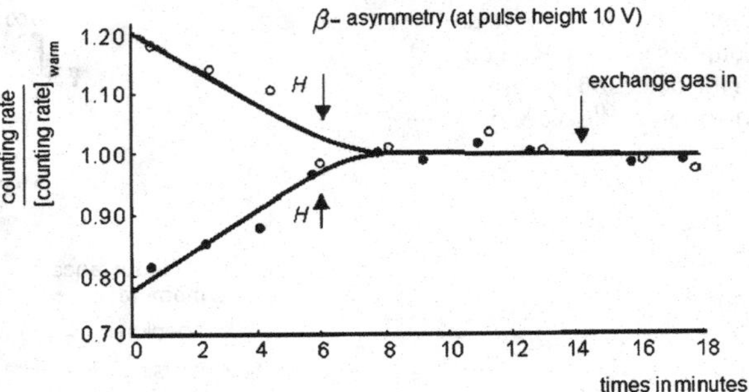

Fig. 7.21 The asymmetry in the β-ray counting rate for the aligned ^{60}Co nuclei in the experiment of Wu et al. [2]. A clear correlation between the β-direction and the alignment is observed, which decreases as the sample warms up and the alignment disappears

disappears, so does the anisotropy. The observations demonstrated the parity violation in β-decay. Several other examples may be concerned in which the Hamiltonian describing the interaction is not invariant under parity transformation leading to parity violation. Any pseudo-scalar term would produce parity non-conservation. A pseudo-scalar quantity can be constructed from the scalar product of an axial vector like the angular momentum and a polar vector like the linear momentum, that is the quantity $\langle L\rangle \cdot p$ is a pseudo-scalar quantity, where $\langle L\rangle$ is the expectation value of the spin vector for the aligned nuclei. If party is not conserved then the intensity of electrons will be a function of this quantity resulting in the angular distribution of the form

$$1 + A\cos\theta$$

where $\cos\theta = L \cdot P/|L \cdot P|$ and A is a constant.

7.16.1 Parity Non-conservation in β-Decay

A consequence of non-conservation of parity in β-decay accompanied by neutrinos is that relativistic electrons emitted even from non-aligned nuclei are polarised with their spins oriented along the direction of motion. This implies that $\cos\theta = +v/c \simeq 1$, where θ is the angle between the spin and the direction of motion. The cosine of the angle is defined as the helicity.

Similar observations in the decay of pion and muon and any other weak decay have established that parity is not conserved.

7.16.2 Parity Non-conservation in Λ-Decay

Assume that the Λ-spin is $1/2$, then in its decay, $\Lambda \rightarrow p + \pi^-$, the relative angular momentum of p and π^- can be 0 or 1, with the angular momentum state with positive or negative parity, respectively. If parity is conserved in the decay then only one of the parities would characterize the final state. But if parity is not conserved both the states will be present, resulting in interference between them. If Λ particles are polarised preferentially along or against the direction of motion as would be the case in many of the production processes, then the effect of interference will be revealed in the angular distribution of the decay products in the CMS. The angular distribution will exhibit a forward-backward anisotropy which can be represented by the form

$$1 + \alpha \cos\theta$$

where α is a constant. In case the Λ-particles are not fully polarized then the anisotropy is decreased and the angular distribution has the form

$$1 + \alpha P \cos\theta \tag{7.28}$$

where P is the polarisation of the Λ's given by

$$P = \frac{N^+ - N^-}{N^+ + N^-} \tag{7.29}$$

where N^+ and N^- are the number of Λ's with spins along and against the direction of motion. A number of experiments have demonstrated the asymmetry in the Λ-decay.

7.17 The Dirac Equation

In order to discuss the weak interactions it is necessary to summarise the results of Dirac's relativistic equation for the spin-$1/2$ particles. Dirac desired to write a relativistic equation to describe the electron including its spin and to satisfy various constraints. The equation had to be linear in the time derivative like Schrodinger equation to guarantee conservation of probability and linear in the space derivative to satisfy relativistic invariance.

We use natural units ($\hbar = c = 1$). The time-dependent Schrodinger wave equation expressed in terms of a Hamiltonian operator is

$$i\frac{\partial}{\partial l}\psi(r,t) = H\psi(r,t) \tag{7.30}$$

The classical relativistic Hamiltonian for a free particle is the positive square root of the right side of

$$E^2 = p^2 + m^2 \tag{7.31}$$

However, if this is substituted into (7.30) and p is replaced by $-i\,\mathrm{grad}$, the resulting wave equation is unsymmetrical with respect to space and time derivatives, and hence not relativistic. Dirac therefore modified the Hamiltonian in such a way as to make it linear in space derivatives.

7.17.1 Free-Particle Equation

The simplest Hamiltonian that is linear in the momentum and mass terms is

$$H = \alpha \cdot \boldsymbol{p} + \beta m \tag{7.32}$$

Substitution in (7.30) results in the wave equation

$$(E - \alpha \cdot \boldsymbol{p} - \beta m)\psi = 0 \tag{7.33}$$

or $(i\frac{\partial}{\partial t} + i\alpha \cdot \text{grad} - \beta m)\psi = 0$.

Consider the four quantities α_1, α_2, α_3 and β. If (7.33) is to describe a free particle, the Hamiltonian cannot depend on the space coordinator or time. Also, only P and E may carry space and time derivations but not α and β since (7.33) ought to be linear in all these derivatives. Hence α and β are independent of r, t, p and E and commute with all of them. It does not follow that α and β are numbers since they need not commute with each other.

Multiply (7.33) on the left by $(E - \alpha \cdot p - \beta m)$ to obtain

$$\begin{aligned}\{E^2 - [&\alpha_1^2 p_1^2 + \alpha_2^2 p_2^2 + \alpha_3^2 p_3^2 + (\alpha_1\alpha_2 + \alpha_2\alpha_1)p_1 p_2 \\ &+ (\alpha_2\alpha_3 + \alpha_3\alpha_2)p_2 p_3 + (\alpha_3\alpha_1 + \alpha_1\alpha_3)p_3 p_1] - m^2\beta^2 \\ &- m[(\alpha_1\beta + \beta\alpha_1)p_1 + (\alpha_2\beta + \beta\alpha_2)p_2 + (\alpha_3\beta + \beta\alpha_3)p_3]\}\psi = 0\end{aligned} \tag{7.34}$$

where the substitutions, $E \to \frac{i\partial}{\partial t}$, $p \to -i\text{grad}$ for E and p are implied. Equation (7.34) agrees with (7.31) if α, β, satisfy the relations

$$\begin{aligned}&\alpha_1^2 = \alpha_2^2 = \alpha_3^2 = \beta^2 = 1 \\ &\alpha_1\alpha_2 + \alpha_2\alpha_1 = \alpha_2\alpha_3 + \alpha_3\alpha_2 = \alpha_3\alpha_1 + \alpha_1\alpha_3 = 0 \\ &\alpha_1\beta + \beta\alpha_1 = \alpha_2\beta + \beta\alpha_2 = \alpha_3\beta + \beta\alpha_3 = 0\end{aligned} \tag{7.35}$$

The four quantities are said to anticommute in pairs, and their squares are unity.

Since α, β anticommute rather than commute with each other, they cannot be numbers. Such quantities can be represented by matrices. Further, since the H given by (7.32) is Hermitian, each of the four matrices must be Hermitian and square. If one of these matrices is diagonal then the others cannot be diagonal since they do not commute with this one. The lowest possible rank matrices will be 4×4 matrices. Choose β as a diagonal matrix of the form

$$\beta = \begin{pmatrix} 1 & 0 \\ 0 & -1 \end{pmatrix} \quad \text{and} \quad \alpha = \begin{pmatrix} 0 & \boldsymbol{\sigma} \\ \boldsymbol{\sigma} & 0 \end{pmatrix} \tag{7.36}$$

where the 4×4 matrices are written in blocks of 2×2 matrices. In (7.36) 0 stands for the zero 2×2 matrix, 1 for the unit 2×2 matrix and $\sigma = (\sigma_1, \sigma_2, \sigma_3)$ where

$$\sigma_1 = \begin{pmatrix} 0 & 1 \\ 1 & 0 \end{pmatrix}, \quad \sigma_2 = \begin{pmatrix} 0 & -i \\ i & 0 \end{pmatrix}, \quad \sigma_3 = \begin{pmatrix} 1 & 0 \\ 0 & -1 \end{pmatrix} \tag{7.37}$$

are the usual Pauli spin matrices. In (7.37) the elements are numbers. Thus for example

$$\alpha_1 = \begin{pmatrix} 0 & \sigma_1 \\ \sigma_1 & 0 \end{pmatrix} = \begin{pmatrix} 0 & 0 & 0 & 1 \\ 0 & 0 & 1 & 0 \\ 0 & 1 & 0 & 0 \\ 1 & 0 & 0 & 0 \end{pmatrix}$$

although this is not a unique choice. Note that

$$\sigma_1^2 = \sigma_2^2 = \sigma_3^2 = \begin{pmatrix} 1 & 0 \\ 0 & 1 \end{pmatrix} \tag{7.38}$$

In non-relativistic theory a spin $1/2$ wave function has two components corresponding to spin up and spin down. In the Dirac theory the wave function has four components corresponding to spin up, spin down and particle, antiparticle. This is because negative energy and negative momentum are allowed by (7.31).

The smallest matrices that can work are 4×4 matrices. These are known as γ-matrices which are related to α matrices by

$$\gamma_j = -i\beta\alpha_j \quad (j = 1, 2, 3), \qquad \gamma_4 = \beta \tag{7.39}$$

They obey the relations

$$\gamma_\nu\gamma_\mu + \gamma_\mu\gamma_\nu = 2\delta_{\mu\nu}$$

where

$$\delta_{\mu\nu} = \begin{cases} 1, & \nu = \mu \\ 0, & \nu \neq \mu \end{cases} \tag{7.40}$$

A representation of the γ-matrices obeying these commutation relations written in full is

$$\gamma_1 = \begin{pmatrix} 0 & 0 & 0 & -i \\ 0 & 0 & -i & 0 \\ 0 & i & 0 & 0 \\ i & 0 & 0 & 0 \end{pmatrix} \qquad \gamma_2 = \begin{pmatrix} 0 & 0 & 0 & -1 \\ 0 & 0 & 1 & 0 \\ 0 & 1 & 0 & 0 \\ -1 & 0 & 0 & 0 \end{pmatrix}$$
$$\gamma_3 = \begin{pmatrix} 0 & 0 & -i & 0 \\ 0 & 0 & 0 & i \\ i & 0 & 0 & 0 \\ 0 & -i & 0 & 0 \end{pmatrix} \qquad \gamma_4 = \begin{pmatrix} 1 & 0 & 0 & 0 \\ 0 & 1 & 0 & 0 \\ 0 & 0 & -1 & 0 \\ 0 & 0 & 0 & -1 \end{pmatrix} \tag{7.41}$$

The product matrix is defined as

$$\gamma_5 = \gamma_1\gamma_2\gamma_3\gamma_4 = \begin{pmatrix} 0 & 0 & -1 & 0 \\ 0 & 0 & 0 & -1 \\ -1 & 0 & 0 & 0 \\ 0 & -1 & 0 & 0 \end{pmatrix} \tag{7.42}$$

with

$$\gamma_5^2 = 1, \qquad \gamma_5\gamma_\mu + \gamma_\mu\gamma_5 = 0, \qquad \mu = 1 \ldots 4 \tag{7.43}$$

7.18 Free-Particle Solutions

Now that α and β are represented by matrices, (7.33) has no meaning unless the wave function ψ is itself a matrix with four rows and one column:

$$\psi(r,t) = \begin{pmatrix} \psi_1(r,t) \\ \psi_2(r,t) \\ \psi_3(r,t) \\ \psi_4(r,t) \end{pmatrix} \tag{7.44}$$

Then (7.33) is equivalent to four simultaneous first-order partial differential equations that are linear and homogeneous in the four ψ's. The four-element column matrix ψ is called 'Dirac Spinor'. Although it has four components, this object is not a four-vector as it is transformed differently.

Plane wave solutions are of the form

$$\begin{aligned} \psi &= u_r e^{ip_\mu x_\mu} \\ \psi &= v_r e^{-ip_\mu x_\mu} \end{aligned} \tag{7.45}$$

where $r = 1, 2$ and the u_r and v_r are four-component spinors, $u_{1,2}$ correspond to a particle of momentum p and energy E, and $v_{1,2}$ to a state of momentum-p and energy-E, that is to a negative energy state. Substituting (7.45) in the Dirac equation

$$\left(\gamma_\mu \frac{\partial}{\partial x_\mu} + m\right)\psi = 0 \tag{7.46}$$

we get

$$(i\gamma_\mu p_\mu + m)u = 0 \tag{7.47}$$

$$(-i\gamma_\mu p_\mu + m)v = 0 \tag{7.48}$$

Using (7.41) in (7.47), four simultaneous equations are obtained which can be solved up to a constant factor N to yield.

$$u_1 = \begin{bmatrix} 1 \\ 0 \\ p_3/(E+m) \\ (p_i + ip_2)/(E+m) \end{bmatrix} N, \qquad u_2 = \begin{bmatrix} 0 \\ 1 \\ (p_1 - ip_2)/(E+m) \\ -p_3/(E+m) \end{bmatrix} N \tag{7.49}$$

$$v_1 = \begin{bmatrix} p_3/(|E|+m) \\ (p_1 + ip_2)/(|E|+m) \\ 1 \\ 0 \end{bmatrix} N, \qquad v_2 = \begin{bmatrix} (p_1 - ip_2)/(|E|+m) \\ -p_3/(|E|+m) \\ 0 \\ 1 \end{bmatrix} N \tag{7.50}$$

Note that u and v correspond to particle and antiparticle respectively and r is the spin index, $r = 1$ for spin up and $r = 2$ for spin down.

The normalization factor

$$N = \sqrt{(|E| + m)/2m} \tag{7.51}$$

7.19 Parity of Particle and Antiparticle

Let $\psi(r,t)$ satisfy the Dirac equation

$$\left(\gamma_\mu \frac{\partial}{\partial x_\mu} + m\right)\psi = 0$$

Then the space inversion, $\psi(-r,t)$ does not satisfy Dirac's equation. But $\gamma_4\psi(-r,t)$ or $\overline{\psi}(-r,t)\gamma_4$ does

$$\left(\gamma_4 \frac{\partial}{\partial x_4} - \gamma_k \frac{\partial}{\partial x_k} + m\right)\psi(-\boldsymbol{r},t) = 0$$

Multiplying through from left by γ_4, and using the relation $\gamma_4\gamma_k + \gamma_k\gamma_4 = 0$

$$\left(\gamma_\mu \frac{\partial}{\partial x_\mu} + m\right)\gamma_4\psi(-r,t) = 0$$

Thus $\gamma_4\psi(-r,t)$ is the inversion of the original state $\psi(r,t)$, and γ_4 is the parity operator for the Dirac function. Now consider a spin-up positive energy state in the rest frame of the particle ($p = 0$), (7.49) and (7.50) become

$$u_1 = \begin{bmatrix} 1 \\ 0 \\ 0 \\ 0 \end{bmatrix}, \quad \gamma_4 u_1 = \begin{bmatrix} 1 \\ 0 \\ 0 \\ 0 \end{bmatrix}$$

For a spin-up negative-energy state

$$v_1 = \begin{bmatrix} 0 \\ 0 \\ 1 \\ 0 \end{bmatrix}, \quad \gamma_4 v_1 = -\begin{bmatrix} 0 \\ 0 \\ 1 \\ 0 \end{bmatrix}$$

Thus, if the positive-energy state, (particle) is assigned a positive (even) intrinsic parity, then the negative-energy state (anti particle) has negative (odd) parity. Thus conventionally for fermions the intrinsic parity is assumed to be positive and for antifermions negative. However, for bosons no such distinction is made.

7.20 Left-Handed and Right-Handed Fermions

Since the γ-matrices are written in 2×2 form, we assume

$$\psi = \begin{pmatrix} \psi_R \\ \psi_L \end{pmatrix} \tag{7.52}$$

where ψ_L and ψ_R are a pair of two-component spinors. At the moment the labels L and R just serve to identify the two solutions, but they will have additional interpretation. Here, it is convenient to use a new notation

$$\gamma_5 = \begin{pmatrix} +1 & 0 \\ 0 & -1 \end{pmatrix} \tag{7.53}$$

Consider the operators

$$P_L = \frac{1-\gamma_5}{2} = \begin{pmatrix} 0 & 0 \\ 0 & 1 \end{pmatrix} \tag{7.54}$$

$$P_R = \frac{1+\gamma_5}{2} = \begin{pmatrix} 1 & 0 \\ 0 & 0 \end{pmatrix} \tag{7.55}$$

They are projected operators i.e.

$$\left.\begin{array}{l} P_L^2 = P_L \\ P_R^2 = P_R \\ P_L + P_R = 1 \\ P_L P_R = 0 \end{array}\right\} \tag{7.56}$$

For any massive fermion described by a solution u of the Dirac-equation, we can define a left-handed projection

$$u_L = P_L u \tag{7.57}$$

Explicitly

$$u_L = \begin{pmatrix} 0 & 0 \\ 0 & 1 \end{pmatrix} \begin{pmatrix} u_R \\ u_L \end{pmatrix} = \begin{pmatrix} 0 \\ u_L \end{pmatrix} \tag{7.58}$$

So we can think of U_L as the lower components of ψ.

Similarly

$$u_R = P_R u \tag{7.59}$$

and

$$u_R = \begin{pmatrix} 1 & 0 \\ 0 & 0 \end{pmatrix} \begin{pmatrix} u_R \\ u_L \end{pmatrix} = \begin{pmatrix} u_R \\ 0 \end{pmatrix} \tag{7.60}$$

In the case of electrons, both solutions exist, i.e. there are both left-handed and right-handed electrons. But they interact differently: e_L can interact directly with a neutrino, but e_R cannot. This is the subtle means nature uses to violate parity invariance. The separation ((7.57) to (7.60)) into f_L and f_R for any fermion f is one of the most important technical points in the structure of the standard Model.

7.21 Useful Relations

The current has the form

$$\begin{aligned} \overline{\psi}\gamma_\mu\psi &= \overline{\psi}(P_L + P_R)\gamma_\mu(P_L + P_R)\psi \\ &= \overline{\psi}P_L\gamma_\mu P_L\psi + \overline{\psi}P_R\gamma_\mu P_L\psi + \overline{\psi}P_L\gamma_\mu P_R\psi + \overline{\psi}P_R\gamma_\mu P_R\psi \end{aligned} \tag{7.61}$$

Using the commutation relation, $P_L\gamma_\mu = \gamma_\mu P_R$ and $P_R\gamma_\mu = \gamma_\mu P_L$, the first and fourth terms vanish.

Noting that

$$\overline{\psi_L} = (P_L\psi)^{\dagger}\gamma_4 = \psi_L^{\dagger}P_L\gamma_4 = \overline{\psi}P_R \tag{7.62}$$

and

$$\overline{\psi_R} = \overline{\psi}P_L \tag{7.63}$$

we get an important relation

$$\overline{\psi}\gamma_\mu\psi = \overline{\psi_L}\gamma_\mu\psi_L + \overline{\psi_R}\gamma_\mu\psi_R \tag{7.64}$$

This says that the helicity is preserved whenever the interaction is of the form $\overline{\psi}\gamma_\mu\psi$.

On the other hand a mass term has the form $m\overline{\psi}\psi$. To express $\overline{\psi}\psi$ in terms of left-handed and right handed states, we have

$$\begin{aligned}\overline{\psi}\psi &= \overline{\psi}(P_L^2 + P_R^2)\psi \\ &= (\overline{\psi}P_L P_L\psi + \overline{\psi}P_R P_R\psi) \\ &= (\overline{\psi_R}\psi_L + \overline{\psi_L}\psi_R)\end{aligned} \tag{7.65}$$

Thus a mass term is equivalent to a helicity flip and conversely.

The parity-conserving electromagnetic interaction of (7.64) has both LL and RR terms, i.e. they are equally probable.

If for some reason only the $\overline{\psi_L}\gamma_\mu\psi_L$ term could occur, we would have

$$\begin{aligned}\overline{\psi_L}\gamma_\mu\psi_L &= \frac{1}{4}\overline{\psi}(1+\gamma_5)\gamma_\mu(1-\gamma_5)\psi \\ &= \frac{1}{2}\overline{\psi}\gamma_\mu(1-\gamma_5^2)\psi = 0\end{aligned} \tag{7.66}$$

which has two pieces, one transforming as a normal four-vector; this is known as a V-A interaction, or a left-handed interaction or current. If only $\psi_R\gamma_\mu\psi_R$ were to occur, it would be a right handed current $\frac{1}{2}\overline{\psi}\gamma_\mu(1+\gamma_5)\psi$. Because of (7.62) and the lines above it only the bottom component of ψ' in any current $\overline{\psi'}\gamma_\mu\psi_L$ can interact. Even if the top component of ψ' corresponds to a physical state, it will not undergo any interaction with the state represented by ψ_L.

7.22 Currents

The Dirac equation is

$$i\gamma_4\frac{\partial\psi}{\partial t} + i\gamma_k\frac{\partial\psi}{\partial x_k} - m\psi = 0 \tag{7.67}$$

where k is summed for $k = 1, 2, 3$. The Hermitian conjugate is

$$-i\frac{\partial\psi^\dagger}{\partial t}\gamma_4 - i\frac{\partial\psi^\dagger}{\partial x_k}\gamma_k^\dagger - m\psi^\dagger = 0 \tag{7.68}$$

Using

$$\alpha_i = \begin{pmatrix} 0 & \sigma_i \\ \sigma_i & 0 \end{pmatrix}, \qquad \beta = \begin{pmatrix} 1 & 0 \\ 0 & -1 \end{pmatrix}$$

$$\gamma_i = \beta\alpha_i \quad \text{and} \quad \gamma_4 = \beta$$

and knowing that the Pauli matrices are Hermitian,

$$\left.\begin{aligned} &\gamma_4^\dagger = \gamma_4 \\ &\gamma_k^\dagger = (\beta\alpha_k)^\dagger = \alpha_k^\dagger\beta - \alpha_k\beta = -\beta\alpha_k = -\gamma_4 \\ &(\gamma_4)^2 = \gamma_4\gamma_4 = 1 \\ &\gamma_k\gamma_k = \beta\alpha_k\beta\alpha_k = -1 \end{aligned}\right\} \tag{7.69}$$

In the last of these equations there is no sum over k. Multiply (7.68) by $\overline{\psi}$ from the right by γ_4, recall that $\overline{\psi} = \psi^\dagger\gamma_4$, and use $\gamma_\mu\gamma_\nu + \gamma_\nu\gamma_\mu = 2\delta_{\mu\nu}$, to interchange γ_k and γ_4.

Then (7.68) gives

$$i\frac{\partial\overline{\psi}}{\partial t}\gamma_4 + \frac{i\partial\overline{\psi}}{\partial x_k}\gamma_4 + m\overline{\psi} = 0 \tag{7.70}$$

or

$$i\left(\frac{\partial}{\partial x_4}\overline{\psi}\right)\gamma_\mu + m\overline{\psi} = 0 \tag{7.71}$$

Next multiply (7.67) by $\overline{\psi}$ from the left in order to have a scalar in spin space, and multiply (7.71) by ψ from the right, and add. The $m\overline{\psi}\psi$ terms drop out and we get

$$\overline{\psi}\gamma_\mu\frac{\partial\psi}{\partial x_\mu} + \left(\frac{\partial}{x_\mu}\overline{\psi}\right)\gamma_\mu\psi = 0$$

which can be written as

$$\frac{\partial}{\partial x_\mu}(\overline{\psi}\gamma_\mu\psi) = 0$$

we define the current

$$j_\mu = \overline{\psi}\gamma_\mu\psi$$

which is conserved, i.e.

$$\frac{\partial}{\partial x_\mu}j_\mu = 0$$

For electric current

$$j_\mu(\text{electric}) = -e\overline{\psi}\gamma_\mu\psi$$

But the current for any of the kinds of charges is denoted by the term $\overline{\psi}\gamma_\mu\psi$.

7.23 Application of the Dirac Theory to Beta Decay

In the four-fermion β-decay, in all there are 16 products of the form $\psi_i^* \psi_j$ (taking one component from ψ^* and one from ψ), since i and j run from 1 to 4. These 16 products which are Lorentz invariant can be added together in various linear combinations to construct quantities with distinct transformation behavior as follows:

(1) $\overline{\psi}\psi = \psi^\dagger \gamma_4 \psi$	scalar	S (one component)
(2) $\overline{\psi}\gamma_\mu \psi$	4-vector	V (four components)
(3) $i\overline{\psi}\gamma_\mu \gamma_\nu \psi$	anti symmetric tensor of second order	T (six components)
(4) $i\overline{\psi}\gamma_5 \gamma_\mu \psi$	axial 4-vector	A (four components)
(5) $\overline{\psi}\gamma_5 \psi$	pseudoscalar	P (one component)

We can verify, for example, that $\overline{\psi}\psi$ is indeed a scalar. To do this we investigate the behaviour of this quantity under space inversion. As previously pointed out $\psi(-r, t)$ does not satisfy the Dirac equation but $\gamma_4 \psi(-r, t)$ or $\psi(-r, t)\gamma_4$ does. Thus under inversion

$$\psi\overline{\psi} \to \psi\gamma_4\gamma_4\overline{\psi} = \psi\gamma_4^2\overline{\psi} = \psi\overline{\psi}$$

and is therefore a scalar.

On the other hand,

$$\overline{\psi}\gamma_5\psi \to \overline{\psi}\gamma_4\gamma_5\gamma_4\psi = \overline{\psi}(-\gamma_5\gamma_4)\gamma_4\psi = -\overline{\psi}\gamma_5\psi$$

changes sign and is pseudoscalar.

7.24 Helicity

The helicity operation is defined as

$$H = \frac{\sigma \cdot \boldsymbol{p}}{|\boldsymbol{p}|} \tag{7.72}$$

Take the particular case,

$$\begin{aligned} &p = p_3, \qquad p_1 = p_2 = 0 \\ &Hu_1 = \frac{\sigma \cdot \boldsymbol{p}}{|\boldsymbol{p}|} u_1 = u_1 \\ &Hu_2 = -u_2 \\ &Hv_1 = v_1 \\ &Hv_2 = -v_2 \end{aligned} \tag{7.73}$$

The states u_1 and u_2 (or v_1 and v_2) are said to be states of helicity $H = +1$ and -1 respectively. Thus the fermion spin is aligned with the direction of motion.

7.25 Fermi's Theory of β-Decay-Revisited

In AAK1,3 Fermi's theory of β-decay (1934) was described elaborately. It was shown that many of the properties of the β-decay are explained successfully. The theory is concerned with the interaction of four particles at a point

$$n \to p + e^- + \overline{\nu}$$

which can be replaced by the more symmetrical form

$$\nu_e + n \to p + e^-$$

These are known as contact interactions.

Fermi developed the theory in analogy with the emission of electromagnetic radiation where the amplitude is proportional to electric current four-vector. For β-decay we have simultaneous transformation of $n \to p$ and $\nu_e \to e^-$, so that the matrix element is written as proportional to the product of two currents.

$$M = C j_{lepton} \cdot j_{baryon}$$

where C is the coupling constant for the overall interaction strength.

For the spin $1/2$ particles involved in β-decay, the appropriate wave functions are the four component spinors satisfying the Dirac equation. The matrix element has dimensions of an energy density. The lepton current has the form

$$j_{lepton} \propto \overline{\psi} O \psi_\nu$$

where O is a matrix operator transforming $\nu \to e$. Relativistic invariance places certain constraints on the form of the matrix element. Five covariant forms associated with the Dirac equation satisfy the requirements and in general

$$M = \sum_i C_i (\overline{\psi}_p O_i \psi_n)(\overline{\psi}_e O_i \psi_\nu)$$

The O_i may have scalar (S) vector (V) tensor (T), axial-vector (A) or pseudoscalar (P). It may be recalled that in pure Fermi transition $\Delta J = 0$, $\Delta \pi = 0$ and in Gamow-Teller transition $\Delta J = 1$, $\Delta \pi = 0$. The S and V interaction can only be associated with Fermi transition. The T and A interactions can produce spin changes and are thus possible forms to account for Gamow-Teller transitions. The P interaction results in a factor of v/c in the matrix elements where the nucleon velocity v in transitions is always much smaller than c so that the interaction is certainly unimportant. Thus the complete matrix element can be written as

$$M = \underbrace{\sum_{i=sV} C_i (\overline{\psi}_p O_i \psi_n)(\overline{\psi}_e O_i \psi_V)}_{\text{Fermi}} + \underbrace{\sum_{j=T,A} C_j (\overline{\psi}_p O_j \psi_n)(\overline{\psi}_e O_j \psi_\nu)}_{\text{Gamow-Teller}} \qquad (7.74)$$

The C's here are coupling constants for different interactions. The effective interactions can be found out by appealing to experimental evidences.

7.25.1 Electron Energy Spectrum

Integrating the expression for the transition probability over the angles we find for electrons or positrons

$$\frac{dn}{dE_e} = \frac{p_s E_e}{2\pi^3}(E_0 - E_e)^2 \times \left[|M_F|^2(C_s^2 + C_V^2) + |M_{GT}|^2(C_T^2 + C_A^2) + \frac{2m_e}{E_e}\left(|M_F|^2 C_s C_V + |M_{GT}|^2 C_T C_A\right) \right] \tag{7.75}$$

where the M_F and M_{GT} are the Fermi and Gamow-Teller matrix elements and E_0 is the maximum possible electron energy.

Since pure Fermi and pure Gamow-Teller transitions are observed we cannot have

$$C_S = C_V = 0 \quad \text{or} \quad C_A = C_T = 0$$

An analysis of the detailed spectrum shapes in pure Fermi and pure Gamow-Teller transitions allows determinations of the ratios

$$\frac{C_S C_V}{C_S^2 + C_V^2} = 0.00 \pm 0.15, \qquad \frac{C_T C_A}{C_T^2 + C_A^2} = 0.00 \pm 0.02$$

From these data Michel [3] concluded that either C_S or C_V and either C_A or C_T equal zero, i.e. there are only two effective couplings. We shall see that these are C_V and C_A.

7.25.2 Electron-Neutrino Correlations

The correlations between electron and neutrino momenta P_e and P_ν are given by

$$\underset{\text{Fermi}}{-(C_S^2 - C_V^2)\frac{P_e \cdot P_\nu}{E_e E_\nu}} \quad \text{and} \quad \underset{\text{Gamow-Teller}}{\frac{1}{3}(C_T^2 - C_A^2)\frac{P_e \cdot P_\nu}{E_e E_\nu}}$$

(the factor $\frac{1}{3}$ is due to the three possible orientations for the $J = 1$ total angular momentum of the leptons thus reducing the correlations).

Writing $\frac{P_e \cdot P_\nu}{E_e E_\nu} = a_e \cos\theta_e$ the transition probability for a pure Fermi transition is proportional to

$$C_S^2(1 - a_e \cos\theta_e) \quad \text{if } C_V = 0$$

$$C_V^2(1 + a_e \cos\theta_e) \quad \text{if } C_S = 0$$

and for a pure Gamow-Teller transition

$$C_T^2\left(1 + \frac{1}{3}a_e \cos\theta_e\right) \quad \text{if } C_A = 0$$

$$C_A^2\left(1-\frac{1}{3}a_e\cos\theta_e\right) \quad \text{if } C_T=0$$

The experiments measure the electron and nuclear recoil directions from which the neutrino direction can be inferred. The results provide unambiguous evidence for pure V, A couplings.

7.25.3 'ft Values' and Coupling Constants

Putting $C_S=C_T=0$ in (7.75) we get

$$\frac{dn}{dE_e}=\frac{p_eE_e}{2\pi^3}(E_0-E)^2\left[C_V^2|M_F|^2+C_A^2|M_{GT}|^2\right]$$

The mean lifetime τ for the decay is obtained by integrating over energy

$$\frac{1}{\tau}=\frac{1}{2\pi^3}\left[C_V^2|M_F|^2+C_A^2|M_{GT}|^2\right]\int_{me}^{E_0}p_eE_e(E_e-E)^2dE_e$$

The integral is conventionally denoted by m_e^5f so that f is dimensionless.

The ft value is then given by

$$(ft)^{-1}=\frac{m_e^5}{2\pi^3}\left[C_V^2|M_F|^2+C_A^2|M_{GT}|^2\right]$$

and the value of the coupling constant may be obtained from measurement of (f_T). It is conventional to use the half life $T_{1/2}$ rather than the mean life τ. Typical ft values for Fermi transitions are of the order of 3000 sec, which yield a value of

$$C_V=1.4\times10^{-49}\ \text{erg cm}^3=\frac{10^{-5}}{M_p^2} \qquad (\hbar=c=1)$$

For allowed transitions the nuclear matrix elements

$$|M_F|^2=|M_{GT}|^2=1$$

The decay of free neutron is a mixed transition. Since there are three possible spin orientations in the Gamow-Teller transition there is a weighing factor of 3 associated with the G–T term and

$$\frac{1}{ft}=\left(C_V^2+3C_A^2\right)\frac{m_e^5}{2\pi^3\ell n2}=(1080\pm16)^{-1}\ \text{sec}$$

We can obtain the ratio of C_V/C_A by comparison with a pure vector or axial-vector decay, the conventional example being the (Fermi) decay of $^{14}O\rightarrow{}^{14}N$. In this case the decay can come from two protons outside the ^{12}C core, thus requiring inclusion of an additional factor 2

$$\frac{(ft)^{14}O}{(ft)n}=\frac{3100\pm20}{1080\pm16}=\frac{C_V^2+3C_A^2}{2C_V^2}$$

yielding

$$\left|\frac{C_A}{C_V}\right| = 1.25 \pm 0.02$$

7.26 Parity Non-conservation in β-Decay

In view of non-conservation of parity in weak interactions, the matrix element of (7.71) which is a scalar would be modified by addition of a pseudo-scalar so that C_V is replaced by

$$(C_i + C_i'\gamma_5)\frac{1}{\sqrt{2}}$$

The factor of $1/\sqrt{2}$ is introduced to retain the numerical value of C_V given above. γ_5 in this expression is a matrix operator which ensures that the second term $C_i'\overline{\psi_e}O_i\gamma_5\psi_\nu$ is pseudo-scalar. The matrix element thus becomes

$$M = \sum_{V,A}(\overline{\psi}_p O_i\psi_n)[\psi_e O_i(C_i + C_i'\gamma_5)\psi_\nu]$$

We may write the lepton bracket as

$$C_i\overline{\psi}_e O_i\psi_\nu + C_1'\overline{\psi}_e O_i\gamma_5\psi_\nu = \frac{(C_i + C_i')}{2}\overline{\psi}_e O_i(1+\gamma_5)\psi_\nu + \frac{(C_i - C_i')}{2}\overline{\psi}_e O_i(1-\gamma_5)\psi_\nu \tag{7.76}$$

$C_i' = C_i$ will be relevant for S, P, T interactions. $C_i' = -C_i$ will be relevant for V, A interactions.

In the muon decay or any other purely leptonic weak interaction $C_A = -C_V$, that is the axial-vector and vector couplings are equal in magnitude but opposite in sign, we obtain the V-A interaction. But this is not true for decays involving hadrons. The axial coupling C_A but not the vector coupling C_V is affected by strong interaction effects, for example for nucleons $C_A = -1.25C_V$.

It follows from the Dirac equation applied to massless particles that the operator $(1+\gamma_5)$ projects out positive helicity states for $\overline{\nu}$ and negative helicity states for ν while the operator $(1-\gamma_5)$ has the opposite effect

	$\overline{\nu}$	ν
$1+\gamma_5$:	right-handed positive helicity	left-handed negative helicity
$1-\gamma_5$:	left-handed negative helicity	right-handed positive helicity

In other words, the $(1\pm\gamma_5)\psi$ are the positive and negative eigen states of the helicity operator.

The relative sign of V and A coupling was determined in a study of the decay of polarized neutrons. It turns out that the amplitude of $(C_A + C_V)$ corresponds to electrons polarised parallel to, the neutron polarisation direction. In the experiment of Burgy et al. [4] neutrons were polarized by reflection from magnetized Cobalt mirrors, the polarization direction being reversible by switching the direction of magnetization. The link between polarization and momentum (helicity) for the leptons then leads to different ratios of electrons along and opposite to the neutron spin for the $C_A + C_V$ and $C_A - C_V$ amplitudes. In the experiment both electron and proton were detected so that the momenta of all the particles were determined. The result gives

$$\frac{C_A}{C_V} = -1.26 \pm 0.02$$

consistent in magnitude with the number quoted earlier on the basis of ft-values and in addition demonstrating that the couplings are of opposite sign interaction u is of the V-A form.

The properties of muon decay

$$\mu \to e\nu\overline{\nu}$$

muon capture

$$\mu^- + p \to n + \nu$$

and pion decays

$$\pi \to \mu\nu$$

$$\pi \to e\nu$$

$$\pi^- \to \pi^0 e^- \overline{\nu}$$

may also be compared with the predictions for the V-A interaction. The experimental results are found to be in close agreement with the V-A expectations. For instance, the ratio $(\pi \to e\nu)/(\pi \to \mu\nu)$ for purely axial coupling is found to be $(1.267 \pm 0.023) \times 10^{-4}$, ratio $\pi^\pm \to \pi^0 e^\pm \nu/\pi^\pm \to \mu^\pm \nu = (1.02 \pm 0.07) \times 10^{-8}$ (the last result also involves the additional, conserved vector current hypothesis).

7.27 β-Decay Theory After Parity Non-conservation

Fermi assumed the conservation of parity, but his calculations involved only scalar quantities. The results of his theory are largely valid in spite of the fundamental changes produced by the discovery of the non-conservation of parity. The theory successfully explains the form of beta spectra, the relation between the maximum energy of the decay and the mean life (Sargent's law) the classification of beta transitions and the formulation of selection rules. The fundamental ideas of Fermi's theory are general enough to be adopted to changes of interaction like non-conservation of parity, still preserving the validity of many of the results.

Fermi developed the theory of β-decay in analogy with electromagnetic interactions. The matrix operator O for the weak interaction was assumed to be a vector as for electromagnetic interaction.

Prior to the discovery of parity violation in 1956 the vector interaction was satisfactory in describing Fermi transitions. However, it could not account for Gamow-Teller transitions, since it could not produce a flipover of the nucleon spin. There are five independent forms for the operator O which meet the requirements of relativistic invariance. They are scalar (S), vector (V), tensor (T), axial vector (A) and pseudoscalar (P). Their names are connected with the transformation properties of the weak currents under space inversion. The S- and V-interaction produce Fermi transitions, while T and A produce Gamow-Teller transitions. The pseudo-scalar interaction P is unimportant in β-decay since it introduces a factor of $v^2/c^2 \sim 10^{-6}$ in M, where v is the nucleon velocity.

The V- and A-interactions result in lepton and antilepton of opposite helicities while S, T- and P-interaction produce lepton and antilepton with the same helicity. Experiments show that V- and A-interactions are involved and not S, T or P.

Note that Fermi's theory was successful in accounting for the β-decay phenomenon without the use of the propagator which involves the heavy intermediate boson. This is because of the largeness of the boson mass the momentum transfer barely affects the propagator term. The β-interaction, therefore could be taken as a point interaction.

Other features like the helicity of neutrino and the angular distribution of electrons from the polarized nuclei are explained by the V-A theory.

7.28 Helicity Conservation in Vector Interactions

Any fermion spinor can be written as a superposition of helicity states

$$u = u_L + u_R \tag{7.77}$$

A vector operator γ_μ acting on an extreme relativistic fermion current produces

$$\begin{aligned} \overline{u}\gamma_\mu u &= (\overline{u}_L + \overline{u}_R)\gamma_\mu(u_L + u_R) \\ &= \overline{u}_L\gamma_\mu u_L + \overline{u}_L\gamma_\mu u_R + \overline{u}_R\gamma_\mu u_L + \overline{u}_R\gamma_\mu u_R \end{aligned} \tag{7.78}$$

The second and third terms cancel as

$$\overline{u}_L = \frac{1}{2}\overline{u}(1-\gamma_5)$$

$$u_R = \frac{1}{2}(1-\gamma_5)u$$

so that for a V-interaction

$$\overline{u}_L\gamma_\mu u_R = \frac{\overline{u}}{4}(1-\gamma_5)\gamma_\mu(1-\gamma_5)u = \frac{\overline{u}}{4}\gamma_\mu(1-\gamma_5^2)u = 0 \tag{7.79}$$

and similarly $\overline{u}_R\gamma_\mu u_L = 0$. This is also true for the A-operator.

For a scalar operator S,

$$\overline{u}_L u_R = \frac{\overline{u}}{4}(1-\gamma_5)(1-\gamma_5)u = \frac{\overline{u}}{2}(1+\gamma_5)u \neq 0 \tag{7.80}$$

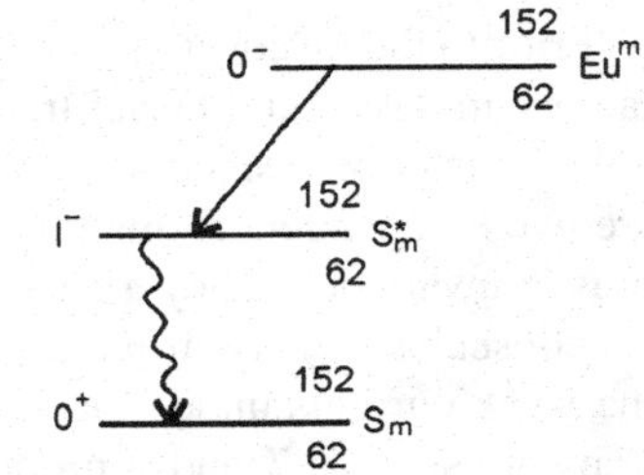

Fig. 7.22 The process of electron K-capture in a $0^- \to 1^-$ transition

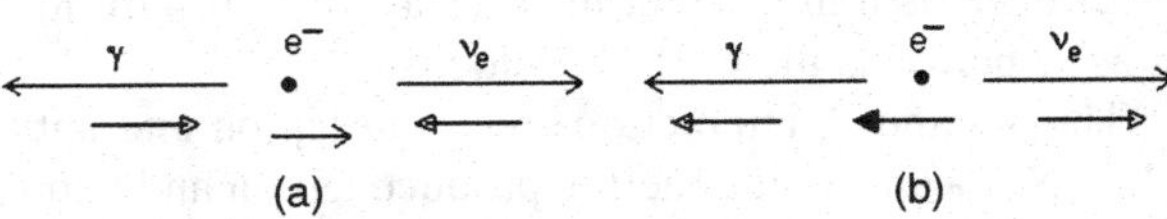

Fig. 7.23 The three collinear final-state particles and the emergence of photon in opposite direction

where as

$$\overline{u}_L u_L = \frac{\overline{u}}{4}(1-\gamma_5)(1+\gamma_5)u = 0 \tag{7.81}$$

and similarly for P- and T-interactions. As a result of (7.78) helicity of a fermion is unchanged by a V- or A-interaction.

7.29 Helicity of Electron Neutrino

The helicity of ν_e has been determined by Goldhaber, Grodzins and Sunyar [5] by an elegant method which used the process of electron K-capture in a $0^- \to 1^-$ transition.

An isomer state of ${}^{152}_{62}\mathrm{Eu}^m\,(J=0)$ nucleus can decay into a $J=1$ state of ${}^{152}_{62}\mathrm{Sm}^*$ via K capture. The latter has an excitation energy of 960 keV. This then emits a photon in going to the ground state ($J=0$), Fig. 7.22. This decay is a pure G-T transition.

$$e^- + {}^{151}\mathrm{Eu}^m \to {}^{152}\mathrm{Sm}^* + \nu$$
$${}^{152}\mathrm{Sm}^* \to {}^{152}\mathrm{Sm} + \gamma$$

Since the electron is captured from the K-shell, with zero initial momentum, the neutrino and ${}^{152}\mathrm{Sm}^*$ nucleus recoil in opposite direction. The experiment selected events in which the photon was emitted in the direction of motion of the decaying ${}^{152}\mathrm{Sm}^*$ nucleus so that overall the observed reaction was

$$e^- + \mathrm{Eu}(J=0) \to {}^{152}\mathrm{Sm}(J=0) + \nu_e + \gamma$$

where the three final-state particles were collinear and the photon emerged in opposite direction Fig. 7.23.

Apply the angular momentum conservation about the event axis to the overall reaction. In doing so no orbital angular momentum is involved because the initial electron is captured from the K-shell and the final-state particles all move along the event axis. Hence the spin components of the neutrino and photon, which can

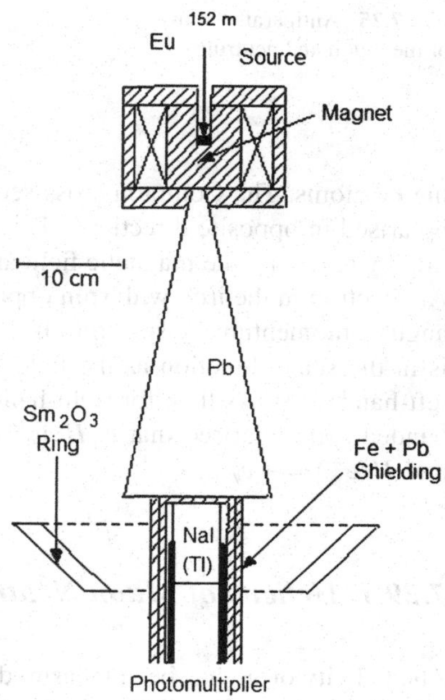

Fig. 7.24 Set up of the Goldhaber experiment. Photons from the $^{152}Eu^m$ source are scattered in the Sm_2O_3 ring and detected in a NaI (TI) scintillation detector

be $\pm 1/2$ and ± 1, respectively must add to give the spin-component of the initial electron, which can be $\pm 1/2$. This gives two spin configurations as shown in Figs. 7.23(a) and (b), and in each case the helicities of the photon and neutrino are the same.

The fact that only those photons were accepted which moved in the direction of the recoiling ^{152}Sm atoms, was ensured by resonant scattering from a second samarium target. It relies on the fact that these γ-rays traveling in the opposite direction to the neutrino have slightly more energy than those emitted in other direction, and only the former have enough energy to excite the resonance level. If the photon is emitted in the recoil direction then its helicity will be equal to that of the neutrino. Thus to determine the neutrino's helicity one must measure the helicity of the photon, which corresponds to circular polarisation, and at the same time make sure that only those photons are considered which are emitted in the direction of the recoiling nucleus (in a direction opposite to that of neutrino). The experimental apparatus is shown in Fig. 7.24.

The photons can only reach the scintillation detector if they are resonantly scattered in a ring of Sm_2O_3. Resonant absorption is normally impossible in Nuclear Physics since the states are narrower than the shift due to the recoil (AAK1,2). The photons are emitted by $^{152}Sm^*$ nuclei that are already moving. If a nucleus is moving towards the Sm_2O_3 absorber before the γ emission, then the photon has a small amount of extra energy, which is sufficient to allow resonant absorption. In this way one can fix the recoil direction of the ^{152}Sm nucleus and hence that of neutrino. The ^{152}Eu source is inside a Fe magnet which the photons must cross to reach the ring of Sm_2O_3. Some of the photons undergo Compton scattering off the electrons in

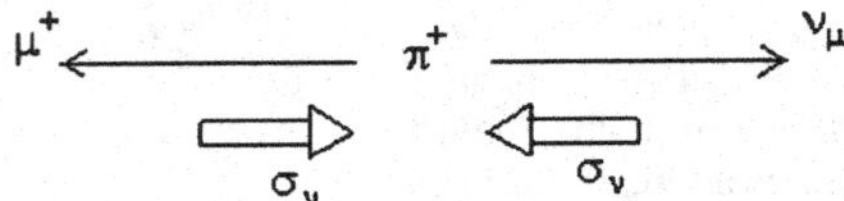

Fig. 7.25 Antiparallel spins of the muon and neutrino

the Fe atoms. The Compton cross-section is larger if the electrons and photons are polarised in opposite directions. This permits us to determine the photon polarisation by reversing the magnetic field and comparing the old and new counting rates. An electron in the iron with spin opposite to that of the photon can absorb a unit of angular momentum by spin-flip, if the spin is parallel it cannot. If the γ-ray beam is in the same direction as the field B, the transmission of the iron is greater for left-handed γ-rays than for right-handed. It is concluded that the neutrinos are left handed spin polarised that is $H(\nu_e) = -1$. Note that $H(\overline{\nu}_e) = +1$, $H(\overline{e}) = -v/c$ and $H(e^+) = +v/c$.

7.29.1 Helicity of Muon Neutrino

The helicity of ν_μ has been measured through a determination of the muon helicity in $\pi^- \rightarrow \mu^- \overline{\nu}_\mu$ decay. This is accomplished by Mott scattering of the polarised muon. It is found that the helicity of $\overline{\nu}_\mu$ is $+1$.

7.30 Pion and Muon Decay

The decay schemes for pion and muon are

$$\pi^+ \rightarrow \mu^+ + \nu_\mu$$
$$\mu^+ \rightarrow e^+ + \nu_e + \overline{\nu}_\mu$$

In view of the fact that the pion has spin zero, the muon and neutrino must have spins antiparallel, Fig. 7.25.

If the neutrino has negative helicity ($H = -1$) as in β-decay, the μ^+ too must have $H = -1$.

In the subsequent decay, the positron spectrum is peaked in the neighborhood of maximum energy. In the rest frame of the muon the momentum of positron is maximum if the momentum of the neutrinos are parallel to each other, and antiparallel to the momentum of the positron. From Fig. 7.26 it is obvious that the spin of the emitted positron must be in the same direction as that of the muon since the spins of the $(\nu_e, \overline{\nu}_\mu)$ pair cancel.

Experimentally it is observed that electrons from polarised muon (μ^-) decays are preferentially emitted with their spins opposite their momentum, that is they are left handed. On the other hand in the decay of μ^+, the positrons are right-handed. This left-right symmetry is a manifestation of parity violation.

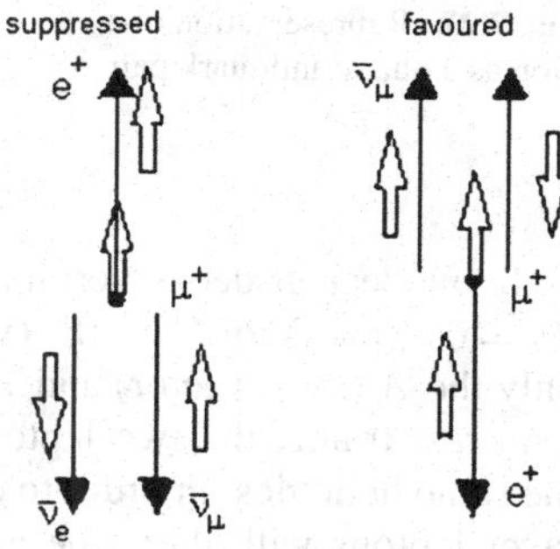

Fig. 7.26 The spin of the emitted positron and the muon are in the same direction

In the experiments, positive pions decayed in flight, and those decay muons are projected in the forward direction, and therefore with negative helicity, were selected. These μ^+ are arrested in a carbon absorber, and the e^+ angular distribution relative to the original muon momentum P_μ is observed. In the carbon material there will be no depolarisation of muon spin. Therefore the muon spin σ should be opposite to P_μ. The angular distribution is found to be of the form

$$\frac{dN}{d\Omega} = 1 - \frac{\alpha}{3} \cos\theta \tag{7.82}$$

where θ is the angle between P_μ, the initial momentum and P_e, the electron momentum, and $\alpha = 1$. The same value of α is found both for μ^+ and μ^-. This is precisely the form predicted by the V–A theory. The helicity of the electrons (positrons) is measured to be $+v/c$ $(-v/c)$.

The muon lifetime from the V–A theory is given by

$$\tau^{-1} = \frac{G^2 m_\mu^5}{192\pi^3} \tag{7.83}$$

where the dependence on the mass to the fifth power follows from the Sargent rule for the three-body decay. From the measured values $m_\mu c^2 = 105.6593$ MeV, $\tau = 2.19709$ μs, the value of the Fermi constant is

$$G = 1.16637 \times 10^{-5}\ \mathrm{GeV}^{-2}$$

after taking into account 0.5 % radiative correction.

7.30.1 The $\pi \to \mu$ and $\pi \to e$ Branching Ratios

Pion decay provides an important test of the V–A theory, via the decay branching ratio

$$\frac{\pi \to e\nu}{\pi \to \mu\nu}$$

The pion decay is an example of the four-fermion interaction, if the pion is represented as a quark-antiquark pair, Fig. 7.27.

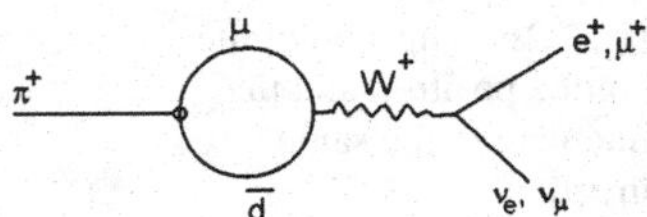

Fig. 7.27 Representation of a pion as a quark-antiquark pair

In nuclear β-decay terminology, the transition is from a hadronic state of $J^p = 0^-$ (pion) to $J^P - 0^+$ (vacuum), so of the five operators mentioned earlier, only the A (axial vector) and P (pseudoscalar) operators could be involved. In the pion rest frame, the two leptons must be emitted in opposite directions but with the same helicities, in order to conserve angular momentum. P- and A-interactions favor leptons with the same and opposite helicities, respectively. In any case the neutrino must have helicity -1. The probability for the A-interaction to produce a e^+ (or μ^+) of velocity v and unfavored helicity $(-v/c)$ will be proportional to $(1 - v/c)$; where as for the p coupling favoring the same helicity for e^+ and neutrino it will be proportional to $(1 + v/c)$.

The stated branching ratio follows if one includes the phase-space factor. This is

$$\frac{dN}{dE_0} = \text{const}\, p^2 \frac{dp}{dE_0}$$

Here p is the momentum of the charged lepton in the pion rest frame, v its velocity and m its rest mass. The neutrino momentum is then $-p$, Fig. 7.28.

In units $c = 1$, the total energy is

$$E_0 = m_\pi = p + \sqrt{p^2 + m^2}$$

Hence

$$p^2 \frac{dp}{dE_0} = \frac{(m_\pi^2 + m^2)(m_\pi^2 - m^2)^2}{4m_\pi^4}$$

$$1 + \frac{v}{c} = \frac{2m_\pi^2}{m_\pi^2 + m^2}$$

$$1 - \frac{v}{c} = \frac{2m^2}{m_\pi^2 + m^2}$$

For A-coupling the decay rate is given by

$$p^2 \frac{dp}{dE_0}\left(1 - \frac{v}{c}\right) = \frac{m^2}{2}\left(1 - \frac{m^2}{m_\pi^2}\right)^2$$

and for P-coupling it is

$$P^2 \frac{dp}{dE_0}\left(1 - \frac{v}{c}\right) = \frac{m_\pi^2}{2}\left(1 - \frac{m^2}{m_\pi^2}\right)^2$$

The predicted branching ratios become, with the approximation $m_e^2/m_\pi^2 \ll 1$,

$$A\text{-coupling} \quad R = \frac{\pi \to e\nu}{\pi \to \mu\nu} = \frac{m_e^2}{m_\mu^2} \frac{1}{(1 - m_\mu^2/m_\pi^2)^2} = 1.275 \times 10^{-4}$$

$$P\text{-coupling} \quad R = \frac{\pi \to e\nu}{\pi \to \mu\nu} = \frac{1}{(1 - m_\mu^2/m_\pi^2)^2} = 5.5$$

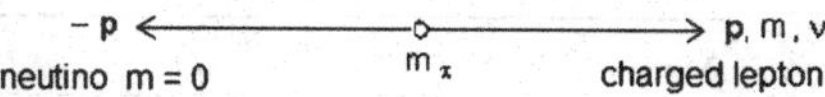

Fig. 7.28 Charged lepton in the pion rest frame with momentum p, velocity v and rest mass m and $-p$ is the neutrino momentum

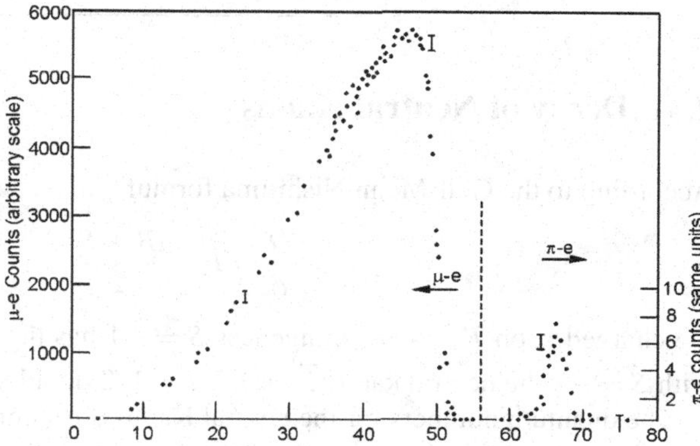

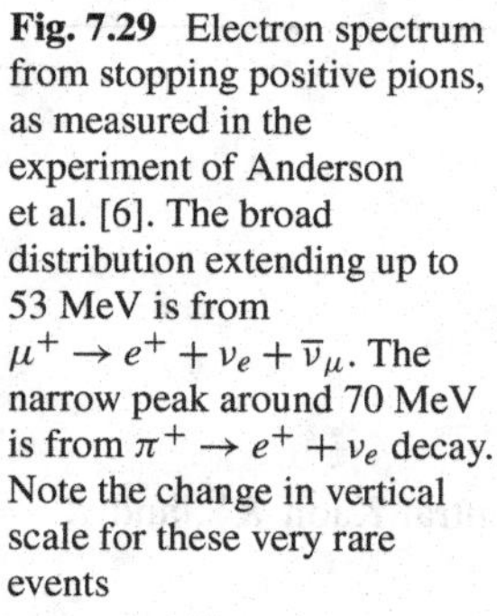

Fig. 7.29 Electron spectrum from stopping positive pions, as measured in the experiment of Anderson et al. [6]. The broad distribution extending up to 53 MeV is from $\mu^+ \to e^+ + \nu_e + \overline{\nu}_\mu$. The narrow peak around 70 MeV is from $\pi^+ \to e^+ + \nu_e$ decay. Note the change in vertical scale for these very rare events

The dramatic difference in the branching ratio for the two types of coupling is due to conservation of angular momentum which compels the electron or muon to have the wrong helicity for A-coupling. The phase-space factor for the electron is greater than for muon decay but the factor $(1 - v/c)$ strongly inhibits decay to the lighter lepton. The current measured value for the ratio is

$$R_{\text{exp}} = (1.267 \pm 0.023) \times 10^{-4}$$

The result was a major triumph for the V-A theory, and proves that the pseudo-scalar coupling is zero or extremely small.

Figure 7.29 shows typical positron spectrum observed from positive pions stopping in an absorber in the experiment of Anderson et al. [6]. The rare $\pi \to e\nu$ process yields positrons of unique energy of about 70 MeV. They are accompanied by the much more numerous positrons from the decay sequences $\pi \to \mu\nu$, $\mu \to e\nu\overline{\nu}$. The spectrum from muon decay extends to 53 MeV. Rejection of electrons from $\pi \to \mu \to e$ decays is based on momentum, timing (the mean life of the pion is 25 ns, that of the muon 2200 ns) as well as the absence of a muon pulse in the counters.

For Kaon decay,

$$R_K = \frac{k \to e\nu}{k \to \mu\nu} = 2.5 \times 10^{-5}$$

compared with the experimentally measured value of $(2.43 \pm 0.14) \times 10^{-5}$. Here the electron is even more relativistic and therefore this decay is more strongly suppressed than in pion decay.

In the above analysis we have assumed the same coupling G for $\pi \to \mu\nu$ and $\pi \to e\nu$, and the data support this assumption. This hypothesis of universality in the coupling of electrons, muons and τ leptons has to be modified for quarks.

Table 7.1 Quantum numbers for the neutral kaons

S	I_3	
	$+1/2$	$-1/2$
$+1$	K^+	K^0
-1	$\overline{K}^0$	K^-

7.31 Decay of Neutral Kaons

According to the Gell-Mann Nishijima formula

$$\frac{Q}{e} = I_3 + \frac{B+S}{2}$$

the charged kaon K^+ with strangeness $S=+1$ has the neutral Kaon K^0, and K^- with $S=-1$ the neutral kaon $\overline{K}^0$ as the $I=1/2$ doublets.

The quantum numbers for the neutral Kaons are summarised in Table 7.1.

K^0 and $\overline{K^0}$ which are particle and antiparticle have opposite strangeness. They are connected by the process of charge conjugation which implies a reversal of I_3 and a change of strangeness $\Delta S = 2$.

The K^0 can be produced in association with a hyperon in N-N or π-N collisions.

$$\begin{array}{lllllll} & p & + p \rightarrow & p & + \Sigma^+ & + K^0 & (K_{threshold} = 1.796\ \text{GeV}) \quad \text{(i)} \\ S & 0 & 0 & 0 & -1 & +1 & \end{array}$$

$$\begin{array}{lllllll} & p & + n \rightarrow & p & + \Sigma^0 & + K^0 & (K_{threshold} = 1.8\ \text{GeV}) \quad \text{(ii)} \\ S & 0 & 0 & 0 & -1 & +1 & \end{array}$$

$$\begin{array}{llllll} & \pi^- & + p \rightarrow & \Lambda & + K^0 & (K_{threshold} = 0.77\ \text{GeV}) \quad \text{(iii)} \\ S & 0 & 0 & -1 & +1 & \end{array}$$

$$\begin{array}{llllll} & \pi^+ & + n \rightarrow & \Sigma^+ & + K^0 & (K_{threshold} = 0.896\ \text{GeV}) \quad \text{(iv)} \\ S & 0 & 0 & -1 & +1 & \end{array}$$

However, a $\overline{K}^0$ can be produced in association with a kaon or antihyperon, of $S=+1$

$$\begin{array}{llllllll} & p & + p \rightarrow & p & + \ p & + K^0 & + \overline{K}^0 & (K_{threshold} = 2.52\ \text{GeV}) \quad \text{(v)} \\ S & 0 & 0 & 0 & 0 & +1 & -1 & \end{array}$$

$$\begin{array}{lllllll} & \pi^+ & + p \rightarrow & K^+ & + \overline{K}^0 & + \ p & (K_{threshold} = 1.5\ \text{GeV}) \quad \text{(vi)} \\ S & 0 & 0 & +1 & -1 & 0 & \end{array}$$

$$\begin{array}{llllllll} & \pi^- & + p \rightarrow & \overline{\Lambda} & + \overline{K}^0 & + \ n & + \ n & (K_{threshold} = 6.0\ \text{GeV}) \quad \text{(vii)} \\ S & 0 & 0 & +1 & -1 & 0 & 0 & \end{array}$$

It is seen that the threshold energy for the production of K^0 is lower than that for $\overline{K}^0$. These two conjugate mesons are distinguished from each other only by their opposite strangeness. However because of the CPT theorem they can not be distinguished by their decay modes. That they can be identified through their strong interaction process is illustrated by the following example

$$K^0 + p \to K^+ + n \quad \text{allowed}$$

but

$$\left.\begin{array}{l} K^0 + p^- \to \Sigma^+ + \pi^0 \\ K^0 + n \to K^- + p \end{array}\right\} \quad \text{forbidden}$$

because of strangeness conservation.

On the other hand

$$\left.\begin{array}{l} \overline{K}^0 + p \to \Sigma^+ + \pi^0 \\ \overline{K}^0 + n \to K^- + p \end{array}\right\} \quad \text{allowed}$$

but

$$\overline{K}^0 + p^- \to K^+ + n \quad \text{forbidden}$$

7.31.1 Behaviour of Neutral Particles Under the Joint Action of CP

In view of the C and P violation, it was thought that the joint action of CP is invariant and if T is separately invariant then the invariance for the overall action of TCP would be preserved for all the three types of interactions. We may consider three kinds of neutral particles.

a. Particles which transform into themselves or are self conjugate, for example π^0, Positronium (e^+e^-), Protonium ($p^- p$). They are characterised by $B = Q = S = 0$. They can be eigen state of C.
b. Particles which are transformed by C into distinct antiparticles, examples are $n, \Lambda, \Sigma^0, \Xi^0$.
c. Particles which are distinct from their antiparticles and which can be eigen states of CP as a particle-antiparticle mixture, for example neutral K-mesons. K^0 is distinguished from $\overline{K}^0$ through strangeness, K^0 can be transformed into $\overline{K}^0$ through virtual second-order weak interaction. This is the only example of particle and antiparticle which are distinct yet coupled, since this requires non-zero strangeness but zero charge, zero lepton numbers and zero baryon number.

The difference between (b) and (c) is that in (b) n and $\overline{n}$ can never mix because of conservation of B.

Consider the weak two pion decay mode, $K^0 \to \pi^+\pi^-$ with $|\Delta S| = 1$.

Then its charge conjugate decay, $K^0 \to \pi^-\pi^+$ gives the connection $K^0 \leftrightarrow 2\pi \leftrightarrow \overline{K}^0$. Similarly, for the three pion decay mode $K^0 \leftrightarrow 3\pi \leftrightarrow \overline{K}^0$. These transitions are $\Delta S = 2$ and thus second-order weak interaction. The mixing can occur via virtual intermediate pion states. We can easily see that, neither $|K^0\rangle$ nor $|\overline{K}^0\rangle$ is an eigen state of $|CP\rangle$. For, the operation of CP on the states $|K^0\rangle$ and $|\overline{K}^0\rangle$ gives

$$CP|K^0\rangle \to \eta|\overline{K}^0\rangle$$
$$CP|\overline{K}^0\rangle \to \eta'|K^0\rangle$$

where η, η' are arbitrary phase factors, which can be defined as $\eta = \eta' = 1$.

However, we can form the linear combinations

$$|K_1\rangle = \frac{1}{\sqrt{2}}(|K^0\rangle + |\overline{K}^0\rangle) \tag{7.84}$$

$$CP|K_1\rangle = \frac{1}{\sqrt{2}}(|\overline{K}^0\rangle + |K^0\rangle), \qquad CP|K_1\rangle = |K_1\rangle$$

$$CP = +1$$

$$|K_2\rangle = \frac{1}{\sqrt{2}}(|K^0\rangle - |\overline{K}^0\rangle) \tag{7.85}$$

$$CP|K_2\rangle = \frac{1}{\sqrt{2}}(|K^0\rangle - |\overline{K}^0\rangle) \tag{7.86}$$

$$CP|K_2\rangle = \frac{1}{\sqrt{2}}(|\overline{K}^0\rangle - |K^0\rangle), \qquad CP|K_2\rangle = -|K_2\rangle$$

$$CP = -1$$

Unlike K^0 and $\overline{K}^0$, distinguished by their mode of production, K_1 and K_2 are distinguished by their mode of decay.

First consider the 2π-decay modes $\pi^0\pi^0$, $\pi^+\pi^-$. In both the cases, Bose symmetry requires that the total wave function should be symmetric under interchange of the two particles. As spin is not involved, this is equivalent to the operation C followed by P, so that $CP = +1$.

Next consider the 3π-decay mode $\pi^+\pi^-\pi^0$. The three pions are expected to be in a relative S-state ($l = 0$) since Q-value for the decay is small (70 MeV). By the previous reasoning, the CP-parity of $\pi^+\pi^-$ is $+1$.

As π^0 decays into two γ's, $P = -1$ and therefore $CP = -1$. Combining the π^0 with the $\pi^+\pi^-$ system; we obtain $CP = -1$ overall. For $l > 0$ both positive and negative CP-eigen values are possible, but such decays are strongly suppressed by angular momentum barrier effects.

Finally, consider the 3π decay mode $\pi^0\pi^0\pi^0$. Bose symmetry requires that the orbital angular momentum l between any two pions must be even. It follows that the l-value of the third pion about the dipion system is also even, since $J_K = 0$. Hence the overall parity is the product of the intrinsic pion parities so that $P = -1$. As the neutral pion has $C = +1$, we conclude that $CP = -1$, irrespective of the l-values resulting in the decay.

To summarize, the 2π state must have $CP = +1$ where as the 3π-state can have $CP = +1$ or -1, with $CP = -1$ heavily favored from kinematical considerations. The two pion decay mode has larger Q-value than the three-pion decay mode, and

Table 7.2 Branching fractions for the decays of K_1 and K_2

Particle	τ (second)	Main decay modes	Branching fraction
K_1	0.9×10^{-10}	$\pi^+\pi^-$	69 %
		$\pi^0\pi^0$	31 %
K_2	0.5×10^{-7}	$\pi^0\pi^0\pi^0$	21 %
		$\pi^+\pi^-\pi^0$	12 %
		$\pi\mu\nu_\mu$	27 %
		$\pi e\nu_e$	39 %

so larger phase-space factor and much faster ($\tau_1 = 0.9 \times 10^{-10}$ sec) than the three-pion decay mode ($\tau_2 = 0.5 \times 10^{-7}$ sec). The other modes of decay for K_2^0 are $\pi\mu\nu_\mu$ and $\pi e\nu_e$. K_1 is identified with K_s (s for short lived) and K_2 with K_L (L for long lived). Table 7.2 gives the branching fractions for the decays of K_1 and K_2.

7.31.2 Strangeness Oscillations

The transitions of $K^0 \to \overline{K}^0$ through 2π or 3π modes correspond to $\Delta S = 2$ and therefore of second-order weak interactions. Although extremely weak, it implies that if we start with a pure beam of K^0 at $t = 0$, at any later time t there will be a superposition of both K^0 and $\overline{K}^0$.

The relative phase of K_1- and K_2-states of a given momentum will only be constant with time if these particles have identical masses. However K_1 and K_2 are not charge - conjugate states, as they have different decay modes and lifetimes. Just as the mass difference of neutron and proton is attributed to the difference in electromagnetic coupling, an extremely small $K_1 - K_2$ mass difference is to be expected because of their different weak couplings.

Let K^0 beam be formed through a strong interaction like $\pi^- p \to K^0\Lambda$. At $t = 0$, the wave function of the system will have the form

$$\psi(0) = |K^0\rangle = \frac{1}{\sqrt{2}}(|K_1\rangle + |K_2\rangle) \tag{7.87}$$

As time progresses K_1 and K_2 amplitudes decay with their characteristic life times. The intensity of K_1 or K_2 components can be obtained by squaring the appropriate coefficient in $\psi(t)$. The amplitudes therefore contain a factor $e^{-t/2\tau}$. They also contain a factor $e^{-iEt/\hbar}$ which describes the time dependence of an energy eigen function in quantum mechanics.

In the rest frame of the K^0 we can write the factor $e^{-iEt/\hbar}$ as $e^{-imc^2t/\hbar}$, where m is the mass. The complete wave function for the system can therefore be written as

$$\psi(t) = \frac{1}{\sqrt{2}}\left[|K_1\rangle e^{-t\left(\frac{1}{2\tau_1}+\frac{im_1c^2}{\hbar}\right)} + |K_2\rangle e^{-t\left(\frac{1}{2\tau_2}+\frac{im_2c^2}{\hbar}\right)}\right]$$

$$= \frac{e^{-i\frac{m_1c^2t}{\hbar}}}{\sqrt{2}}\left[|K_1\rangle e^{-\frac{t}{2\tau_1}} + |K_2\rangle e^{i\frac{\Delta mc^2t}{\hbar}}\right]$$

where $\Delta m = m_1 - m_2$ and we have neglected the factor $e^{-t/2\tau_2}$ which varies slowly ($\tau_2 \simeq 500\tau_1$).

Re-expressing $|K_1\rangle$ and $|K_2\rangle$ in terms of $|K^0\rangle$ and $|\overline{K}^0\rangle$

$$|K_1\rangle = \frac{1}{\sqrt{2}}\left(|K^0\rangle + |\overline{K}^0\rangle\right)$$

$$|K_2\rangle = \frac{1}{\sqrt{2}}\left(|K^0\rangle - |\overline{K}^0\rangle\right)$$

$$\psi(t) = \frac{1}{2}e^{-imc^2t/\hbar}\left[e^{-t/2\tau_1}\left(|K_0\rangle + |\overline{K}^0\rangle\right) + e^{i\Delta mc^2t/\hbar}\left(|K^0\rangle - |\overline{K}^0\rangle\right)\right]$$

$$= \frac{1}{2}e^{-im_1c^2t/\hbar}\left[|K^0\rangle\left(e^{-t/2\tau_1} + e^{i\Delta mc^2t/\hbar}\right) + |\overline{K}^0\rangle\left(e^{-t/2\tau_1} - e^{i\Delta mc^2t/\hbar}\right)\right]$$

The intensity of the component is obtained by taking the absolute square of the coefficient of $|K^0\rangle$.

$$I\left(|K^0\rangle\right) = \frac{1}{4}\left[e^{-t/\tau_1} + 1 + 2e^{-t/2\tau_1}\cos\left(\Delta mc^2t/\hbar\right)\right] \tag{7.88}$$

Similarly,

$$I\left(|\overline{K}^0\rangle\right) = \frac{1}{4}\left[e^{-\frac{t}{\tau_1}} + 1 - 2e^{-\frac{t}{2\tau_1}}\cos\left(\Delta mc^2t/\hbar\right)\right] \tag{7.89}$$

The time variation of the intensity of $\overline{K}^0$ component of a beam with time, initially pure $\overline{K}^0$, is plotted in Fig. 7.30 for three value of Δmc^2. The most interesting feature of the curves is the oscillating behavior caused by the interference of the $|K_1\rangle$ and $|K_2\rangle$ amplitudes with their slightly different time dependence. Note that this oscillation is on a macroscopic scale (~10 cm wavelength for typical K^0 beam).

At the time of production each neutral K-meson has a probability of 1/2 of showing up either a K_1 or as K_2. Initially K_1 is exhibited via 2π mode because of shorter mean life time. After considerable time, say 20 mean time of K_1, these mesons appear only under K_2 form, the K_1 mesons having disappeared through their 2π decay, that is far from the production region, the neutral K mesons appear as a pure K_2 beam Fig. 7.31.

K^0- and $\overline{K}^0$-intensities oscillate with the frequency $\Delta mc^2/\hbar$. If one measures the number of $\overline{K}^0$ interaction events (i.e. the hyperon yield) as a function of position from the K^0-source, one can deduce $|\Delta m|$. The mass difference is found to be very small

$$\Delta m = 3.52 \times 10^{-6} ev$$

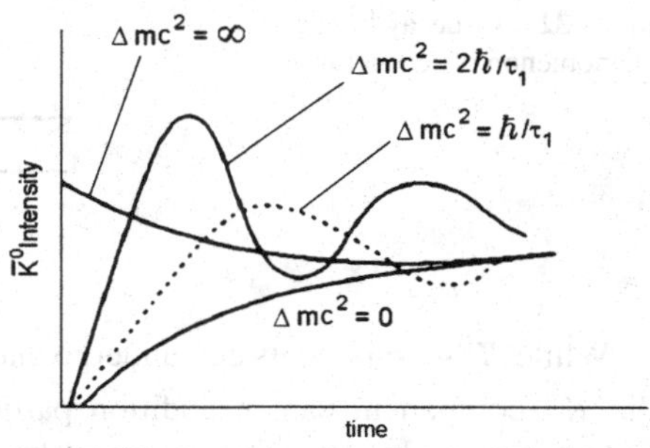

Fig. 7.30 The time variation of the intensity of $\overline{K}^0$ plotted for three value of Δmc^2

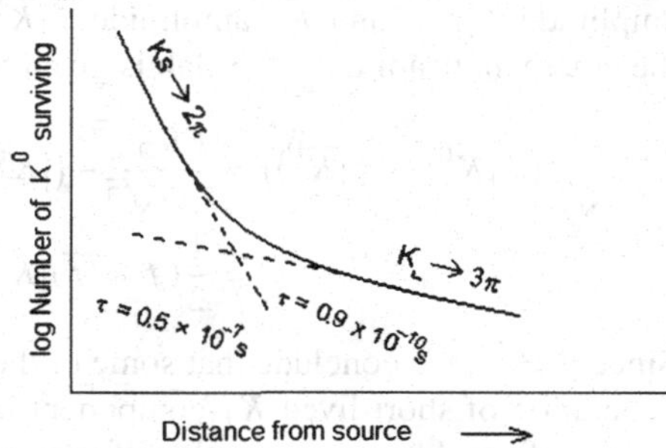

Fig. 7.31 Appearance of K mesons

where $\Delta m = m_1 - m_2$, and in separate regeneration Fig. 7.31 experiments, it is found that $m_2 > m_1$.

The fractional mass difference

$$\frac{\Delta m}{m} = 0.7 \times 10^{-14}$$

7.31.3 The K^0 Regeneration Phenomenon

Gell-Mann and Pais had predicted the K_2-decay which was observed by Pais and Piccioni [7] via the phenomenon of regeneration. Suppose we start with a pure beam which traverses in vacuum for a time of the order of $100K_1$ mean lives, so that all the K_1-component has decayed and we are left with K_2 only. Let the K_2-beam traverse a material say a carbon screen and interact—Fig. 7.32. The strong interactions would occur with $K^0(S = +1)$ and $\overline{K}^0(S = -1)$ components of the beam, i.e.,

$$|K_2\rangle = \frac{1}{\sqrt{2}}\left(|K^0\rangle - |\overline{K}^0\rangle\right) \tag{7.90}$$

Of the original K^0-beam intensity, 50 % has disappeared by K_1-decay. The remaining K_2-component consists of 50 % K^0 and 50 % $\overline{K}^0$. Upon traversing the material the existence of $\overline{K}^0$ with $S = -1$ is revealed by the production of hyperons in a typical reaction, $\overline{K}^0 + p \to \Lambda + \pi^+$.

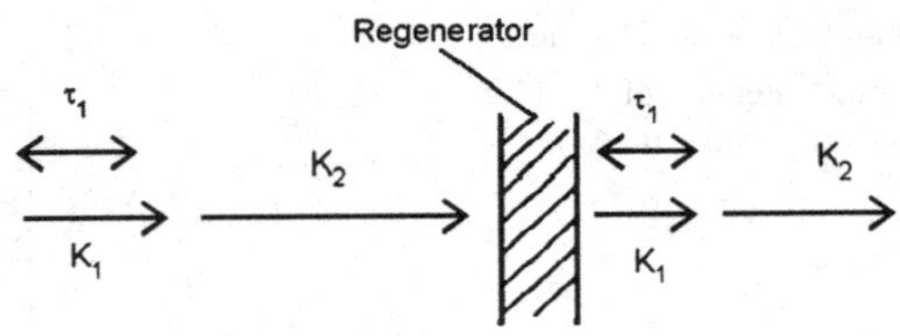

Fig. 7.32 K_2-decay by the phenomenon of regeneration

While K^0-components can undergo elastic and charge-exchange scattering only, the $\overline{K}^0$ component can in addition participate in absorption processes resulting in hyperon production. The emergent beam from the slab will then have the K^0-amplitude $f|K^0\rangle$ and $\overline{K}^0$ amplitude $\overline{f}|\overline{K}^0\rangle$ with $\overline{f} < f < 1$. The composition of the emergent beam from the slab is given by modifying (7.90)

$$\frac{1}{\sqrt{2}}(f|K^0\rangle - \overline{f}|\overline{K}^0\rangle) = \frac{(f+\overline{f})}{2\sqrt{2}}(|K^0\rangle - |\overline{K}^0\rangle) + \frac{(f-\overline{f})}{2\sqrt{2}}(|K^0\rangle + |\overline{K}^0\rangle)$$
$$= \frac{1}{2}(f+\overline{f})|K_2\rangle + \frac{1}{2}(f-\overline{f})|K_1\rangle \tag{7.91}$$

Since $f \neq \overline{f}$, we conclude that some of the K_1-state has been regenerating. The regeneration of short-lived K_1-component in a long-lived K_2-beam was experimentally confirmed from observation of two-pion decay mode (1956).

The phenomenon is similar to the change of polarization suffered by a beam of linearly polarized light propagating in a birefringent medium.

7.31.4 *CP* Violation in K^0-Decay

After the discovery of parity violation in weak decays in 1957, it was believed that the weak interactions are at least invariant under the CP-operation. This aspect was actually the basis for regarding the components K_1 and K_2 as the CP-eigenstates of neutral kaons. However, in 1964 the experiment by Christenson, Cronin, Fitch and Turlay first established that the long-lived state called K_2 could also decay to $\pi^+\pi^-$ with a branching ratio of order 10^{-3}. The experimental arrangement of Christenson et al. is shown in Fig. 7.33. The K^0-beam is incident from the left and consists of K_2 only, the K_s-component having disappeared, the charged products resulting from the decay occurring in a helium bag are analyzed by two spectrometers consisting of bending magnets and spark chambers triggered by scintillators.

The rare two-pion decays are distinguished from the common three-pion and leptonic decay from the invariant mass of the pair and the direction θ of their resultant momentum vector relative to the incident beam.

Figure 7.34 shows the distribution of events with $490 < M_{\pi\pi} < 510$ MeV. The distribution is that expected for three-body decays based on Monte Carlo calculations (dashed line), but with some 50 events (shaded) collinear with the beam due to $\pi^+\pi^-$ decay mode.

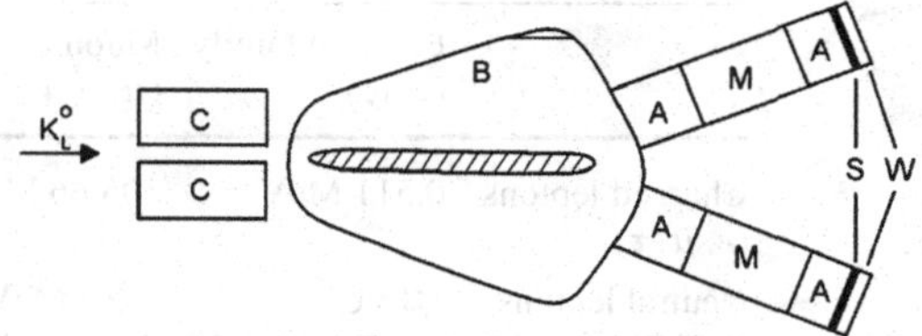

Fig. 7.33 Schematic diagram of the apparatus used in the discovery of CP-violation. The K_L^0 beam entered a helium filed bag B through a lead collimator C, and those CP-violating decays which occurred in the *shaded region* were detected by the symmetrically spaced spectrometers. Each of these contained a pair of spark chambers A separated by a magnet M, followed by a scintillation counter S and a water Cerenkov counter W [8]

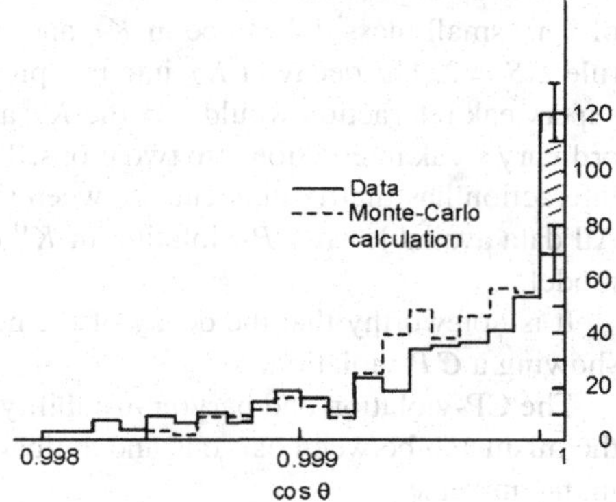

Fig. 7.34 The distribution of events expected for three-body decays

The degree of CP-violation is estimated by the amplitude ratio

$$|\eta| = \frac{\text{Ampl}(K_L \to \pi^+\pi^-)}{\text{Ampl}(K_S \to \pi^+\pi^-)} = (2.27 \pm 0.02) \times 10^{-3}$$

The state $|K_S\rangle$ consists principally of a $CP = +1$ amplitude with a small admixture of $CP = -1$, and $|K_L\rangle$ vice versa. Since both K_L and K_S decay to two pions, interference effects in the $\pi^+\pi^-$ signal are expected as a function of the time evolution of an initially pure K^0 beam. Such interference effects have indeed been observed. Similar CP-violating effects are found for the decay $K_L \to \pi^0$.

CP-invariance is also observed in the leptonic decay modes of K_L. These modes are

$$K_L \to e^+\nu_e\pi^- \tag{7.92}$$

$$K_L \to e^-\overline{\nu}_e\pi^+ \tag{7.93}$$

with similar ones for muons replacing electrons. The decays (7.92) and (7.93) transform one into the other under the CP-operation, and if CP-invariance is violated, a small charge asymmetry is expected. The asymmetry is

$$\Delta = \frac{\text{rate}(K_L \to e^+\nu_e\pi^-) - \text{rate}(K_L \to e^-\overline{\nu}_e\pi^+)}{\text{rate}(K_L \to e^+\nu_e\pi^-) + \text{rate}(K_L \to e^-\overline{\nu}_e\pi^+)} = (3.30 \pm 0.12) \times 10^{-3}$$

Table 7.3 Lepton masses

	Electron family (e, ν_e)	Muon family (μ, ν_μ)	Tau family (τ, ν_τ)
Charged leptons (e, μ, τ)	0.511 MeV	105.66 MeV	1784 ± 4 MeV
Neutral leptons	$\leq$15 eV	<0.5 MeV	<70 MeV

The source of CP-violation is a mystery. In principle, the CP-violation can occur in the strong, electromagnetic or weak interactions.

Of several hypotheses proposed to explain the CP violation the most plausible one is due to Wolfenstein [9]. This supposes the existence of a superweak interaction with a coupling constant of the order of 10^{-9} of that of the weak interaction because of very small mass difference in K_L and K_s component and obeying the selection rule $\Delta S = 2$. The decay of K_L into two pions would then be a two-step process. The superweak interaction would mix the K_L and K_s states and K_L would decay by the ordinary weak interaction into two pions. The mixing of K_L and K_s requires that the interaction has matrix elements between states differing by 2 units of strangeness. All data available on CP-violation in K^0 decay are consistent with the superweak model.

It is noteworthy that the decay of the neutral K is the only known phenomenon showing a CP violation.

The CP-violation and baryon instability have been postulated in order to explain the mismatch between baryons and antibaryons and the ratio of baryons to photons in the universe.

7.32 Neutrino Oscillations

The assumption that neutrinos are all massless is being seriously questioned. From the point of view of cosmology, 90 % of total gravitational mass of universe consists of invisible or "dark matter". It is thought that an important component of dark matter could be massive neutrinos. Such massive neutrinos would dramatically remove the mismatch between the motional energy of galaxies and their gravitational potential energy, the former being an order of magnitude larger than the latter. Further massless neutrinos might lead to difficulties in our understanding of expanding universe. The observed masses of the three charged leptons and the three neutral leptons are displayed in Table 7.3.

At present it is unlikely that the mass values of various types of neutrinos would improve.

The knowledge of accurate mass values of neutrinos is equally crucial in particle physics. Thus, in the standard model developed by Weinberg, Salam and Glashow the neutrino mass $m_\nu = 0$ and that it exists in only one helicity state. On the other hand, the grand unification theories predict a non-zero mass. If $m_\nu \neq 0$, then the standard model would have to be revised.

If the neutrino masses are exactly zero then ν and $\overline{\nu}$ can be distinguished from their helicity. Neutrinos are left-handed ($H = -1$), that is their spin points in the opposite direction of momentum; and antineutrinos are right-handed ($H = +1$). This is a relativistic invariant description analogous to the right and left circularly polarized light. However, for a neutrino with finite mass, this is no longer true. One could always find a reference frame traveling faster than the neutrino resulting in the reversal of sign of the helicity.

The quantum numbers L_e, L_μ, L_τ are introduced to distinguish the three types of leptons.

7.32.1 Solar Neutrino Problem

The recorded neutrino fluxes are $1/3$ to $1/2$ that of predicted fluxes from the standard solar model. This discrepancy is known as Solar neutrino problem.

7.32.2 Standard Solar Model (SSM)

The model is based on the assumption that the structure and temporal evolution of a star are unambiguously determined by its mass and chemical composition. The central temperature is estimated as 1.56×10^7 K and density 148 g/cm^3. The light atoms are completely ionized and form a plasma. The energy generated in the interior of the sun by fusion of light nuclei, escapes from the surface in the form of radiation, neutrinos, and particles. Source of energy in the sun is a series of nuclear reactions which convert hydrogen into helium. The predicted flux of neutrinos is $\simeq 10^{11}\ \mathrm{cm}^{-2}\,\mathrm{s}^{-1}$ at the earth. Sun obtains energy mainly from the *pp* cycle, while the CNO cycle plays an insignificant role.

***p–p* Cycle**

$$p + p \rightarrow {}^2\mathrm{H} + e^+ + \nu_e\left(E_\nu^{\max} = 0.42\ \mathrm{MeV}\right)$$

A less probable reaction is, $p + e^- + p \rightarrow {}^2\mathrm{H} + \nu_e (E_\nu = 1.44\ \mathrm{MeV})$

$$\begin{aligned} &p + {}^2\mathrm{H} \rightarrow {}^3\mathrm{He} + \gamma \\ &{}^3\mathrm{He} + {}^3\mathrm{He} \rightarrow {}^4\mathrm{He} + 2p \end{aligned}$$

Alternatively

$$\left.\begin{aligned} &{}^3\mathrm{He} + {}^4\mathrm{He} \rightarrow {}^7\mathrm{Be} + \gamma \\ &{}^7\mathrm{Be} + p \rightarrow {}^8\mathrm{B} + \gamma \\ &{}^8\mathrm{B} \rightarrow {}^8\mathrm{Be}^* + e^+ + \nu_e \\ &{}^8\mathrm{Be}^* \rightarrow {}^4_2\mathrm{He} \end{aligned}\right\}$$

Table 7.4 Sources of solar neutrinos

pp cycle		CNO cycle	
Source	ϕ_ν (10^{10} cm^{-2} s^{-1})	Source	ϕ_ν (10^{10} cm^{-2} s^{-1})
pp	6.0	^{13}N	0.06
pep	0.014	^{15}O	0.05
^{7}Be	0.47	^{17}F	5.2×10^{-4}
^{8}B	5.8×10^{-4}		

only one in a thousand ^{7}Be nuclei undergo this particular process, the rest are converted to ^{7}Li by electron capture

$$^7\text{Be} + e^- \to {}^7\text{Li}\left(\text{or } {}^7\text{Li}^*\right) + \nu_e(E_\nu = 0.862 \text{ or } 0.383 \text{ MeV})$$

followed by

$$^7\text{Li} + p \to {}^4_2\text{He}$$

CNO cycle

$$^{13}\text{N} \to {}^{13}\text{C} + e^+ + \nu_e(E_\nu \le 1.199 \text{ MeV})$$
$$^{15}\text{O} \to {}^{15}\text{N} + e^+ + \nu_e(E_\nu \le 1.732 \text{ MeV})$$
$$^{17}\text{F} \to {}^{17}\text{O} + e^+ + \nu_e(E_\nu \le 1.74 \text{ MeV})$$

7.32.3 SSM Predictions of Solar Neutrinos on Earth

The cross-sections for individual reactions in the sun are energy dependent so that the production rates are highly dependent on the interior temperature of the sun. For *pp* neutrinos $\sigma \alpha \tau^4$ and for ^{8}B neutrinos $\sigma \propto \tau^{19}$.

In Fig. 7.35 are displayed the energy spectrum for the *pp* and ^{8}B sources. Different energy ranges of the spectrum correspond to different neutrino sources.

There are three main sources of solar neutrinos.

a. *p–p* neutrinos. The spectrum is continuous with $E_{\text{max}} = 420$ keV.
b. ^{7}Be neutrinos which are monoenergetic at $E_\nu = 862$ keV (90 %) and 383 keV (10 %).
c. ^{8}B neutrinos which are most energetic with $E_{\text{max}} = 14$ MeV, but the fluxes are only 10^{-4} of those of *p–p* neutrinos.

There are also neutrinos arising from *p-e-p* reactions and to a lesser extent from ^{13}N, ^{15}O and ^{17}F decays in the CNO cycle of the sun. These contributions are much weaker than that from the *P–P* reaction and much less energetic than from ^{8}B source (Table 7.4).

Pioneering work was done in the ^{37}Cl experiment of Davis et al. during 1970–1988. The experiment has a threshold of 0.81 MeV and was therefore sensitive

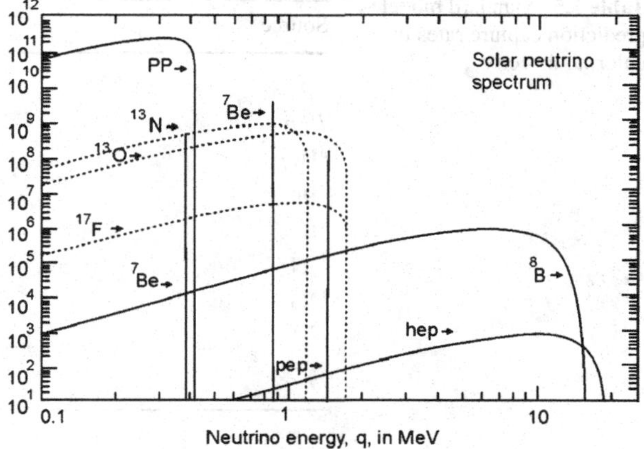

Fig. 7.35 Solar ν_e spectrum calculated by Bachall

mainly to ^{7}Be and ^{8}B neutrinos. In Davis' experiment radiochemical detectors were mainly employed, and the neutrinos were detected via the neutrino capture in ^{37}Cl with the formation of radioactive ^{37}Ar.

$$^{37}\mathrm{Cl} + \nu_e \rightarrow {}^{37}\mathrm{Ar} + e^-(E_{threshold} = 814\ \mathrm{keV})$$

^{37}Cl has favourable ft value ($\log ft = 5$). Obviously this detector does not respond to pp neutrinos.

$$e^- + {}^{37}\mathrm{Ar} \rightarrow {}^{37}\mathrm{Cl} + \nu \quad (T_{1/2} = 35D)$$

The reaction is detected by recording 2.82 keV Auger electrons following electron capture. The neutrino flux ϕ_ν was determined from the amount of Argon produced. The detector consisted of a large tank (38×10^4 liters) of C_2Cl_4 (perchloro-ethylene) and was located in the gold mine of South Dakota, at a depth of 1.4 km. The argon was rinsed out periodically and detected in a proportional counter. Few decays were recorded in the measured time, the background being a serious problem.

Since the expected rates for neutrino interactions are small, the solar neutrino unit (SNU) was introduced. Table 7.5 shows standard model prediction capture rates in chlorine detector.

Chlorine experiments are mainly sensitive to ^{8}B neutrinos which contribute to the extent of 77 % of the total event rate. In the period 1970–1988 the average value for all the data was found to be equal to $\Sigma\phi_i\sigma_i = 2.2 \pm 0.3$ SNU/day which is to be compared with the theoretical value of 5.35 SNU/day. The discrepancy may be due to one of the following factors.

1. Current models of the sun's interior may not be correct; they may be over-estimating the production of $^8_5$B. The flux of ^{8}B neutrinos sensitively depends on the central temperature of the sun. A fall of temperature from 15.6 to 14.6 million degrees, that is a mere 6 % would already account for the said discrepancy.
2. The standard model of the sun may be correct but the neutrinos have a small but non-zero mass, resulting in the oscillation between the three flavours,

Table 7.5 Standard model prediction capture rates in chlorine detector

Source	$\phi_i \sigma_i$ [SNU]
pp	0
Pep	0.2
^{3}Hep	0.03
^{7}Be	1.1
^{8}B	6.1
^{13}N	0.1
^{15}O	0.3
^{17}F	0.003
	$\sum_i = 7.9$

$\nu_e \leftrightarrow \nu_\mu \leftrightarrow \nu_\tau$. It is then possible that the original flux of ν_e is reduced by the time it is detected on account of its conversion into a different flavour.

7.32.4 Neutrino Oscillations

If lepton number is not absolutely conserved (and there are no compelling reasons for believing it should) and neutrinos have finite masses, the mixing may occur between different types of neutrinos (ν_e, ν_μ, ν_τ). This possibility was first investigated by Maki et al. [10], Pontecorvo [11]. The weak-interaction eigenstates, ν_e, ν_μ, ν_τ are expressed as combination of mass eigenstates say ν_1, ν_2, ν_3 which propagate with slightly different frequencies due to their mass differences. If one starts with a pure ν_e-beam, for example, oscillations would occur and at subsequent times one would have admixtures of ν_e with ν_μ and ν_τ. For simplicity, consider the case of two types of neutrino, ray ν_e and ν_μ. Each will be a linear combination of the two mass eigenstates, say ν_1 and ν_2 as given by the unitary transformation involving an arbitrary mixing angle θ.

$$\begin{pmatrix} \nu_\mu \\ \nu_e \end{pmatrix} = \begin{pmatrix} \cos\theta & \sin\theta \\ -\sin\theta & \cos\theta \end{pmatrix} \begin{pmatrix} \nu_1 \\ \nu_2 \end{pmatrix} \tag{7.94}$$

so that the wave functions

$$\begin{aligned} \nu_\mu &= \nu_1 \cos\theta + \nu_2 \sin\theta \\ \nu_e &= -\nu_1 \sin\theta + \nu_2 \cos\theta \end{aligned}$$

are orthonormal states. The states ν_μ and ν_e are those produced in a weak decay process, for example $\pi \to \mu + \nu_\mu$. However, propagation in space-time is determined by the characteristic frequencies of the mass eigenstates,

$$\begin{aligned} \nu_1(t) &= \nu_1(0) e^{-iE_1 t}, \\ \nu_2(t) &= \nu_2(0) e^{-iE_2 t} \end{aligned} \tag{7.95}$$

where we have set $\hbar = c = 1$. Since momentum is conserved we know that the states $\nu_1(t)$ and $\nu_2(t)$ must have the same momentum p. Then, if the mass $m_i \ll E_i$ $(i = 1, 2)$.

$$E_i = p + \frac{m_i^2}{2p} \tag{7.96}$$

Suppose we start at $t = 0$ with muon-type neutrinos, so $\nu_\mu(0) = 1$ and $\nu_e(0) = 0$. Then from (7.94) we find

$$\begin{aligned} \nu_2(0) &= \nu_\mu(0)\sin\theta \\ \nu_1(0) &= \nu_\mu(0)\cos\theta \end{aligned} \tag{7.97}$$

and

$$\nu_\mu(t) = \cos\theta\,\nu_1(t) + \sin\theta\,\nu_2(t)$$

Using (7.95) & (7.97)

$$\frac{\nu_\mu(t)}{\nu_\mu(0)} = \cos^2\theta e^{-iE_1t} + \sin^2\theta e^{-iE_2t}$$

and the intensity is

$$\begin{aligned} \frac{I_\mu(t)}{I_\mu(0)} = \left|\frac{\nu_\mu(t)}{\nu_\mu(0)}\right|^2 &= \cos^4\theta + \sin^4\theta + \sin^2\theta\cos^2\theta\left[e^{i(E_2-E_1)t} + e^{-i(E_2-E_1)t}\right] \\ &= 1 - \sin^2 2\theta \sin^2\left[\frac{(E_2 - E_1)t}{2}\right] \end{aligned}$$

Writing $\Delta m^2 = m_2^2 - m_1^2$, and with the help of (7.96) we find the probability of finding ν_μ or μ_e after time t:

$$P(\nu_\mu \to \nu_e) = 1 - P(\nu_\mu \to \nu_\mu) \tag{7.98}$$

$$P(\nu_\mu \to \nu_\mu) = 1 - \sin^2 2\theta \sin^2\left(\frac{1.27\Delta m^2 L}{E}\right) \tag{7.99}$$

Notice that for $\Delta m^2 = 0$ there is no mixing. The numerical constant 1.27 applies if Δm^2 is expressed in $(ev/c^2)^2$, while L, the distance from the source, is expressed in metres, and E, the mean energy is in MeV. Equations (7.98) and (7.99) show that the intensities of ν_μ and ν_e oscillate as a function of distance from the source. For example, at a reactor source (of antineutrinos) $E \simeq 1$ MeV, and for $\Delta m \simeq 1$ eV/c^2 the oscillation length will be a few meters.

The neutrino oscillation problem has become important in view of the so-called solar neutrino problem. The rate of detection of solar neutrinos in the reaction $\nu_e + {}^{37}\text{Cl} \to {}^{37}\text{Ar} + \overline{e}$ in found to be about a factor 3 smaller than expected [12]. However, there are uncertainties in the solar model calculation and the discrepancy is hardly an evidence for neutrino oscillations (although mixing μ_e with ν_μ and ν_τ would clearly give such a factor). At a reactor one measures the number of events of the type $\overline{\nu}_e + p \to n + e^+$ as a function of distance from the core and positron energy. It is sensitive to the disappearance of the $\overline{\nu}_e$ signal i.e. to transformations $\overline{\nu}_e \to \overline{\nu}_X$, where X is another type of lepton. Accelerator neutrino experiments are at higher energy (1–50 GeV) and the beams are 99.5 % pure ν_μ. They search for the appearance of anomalously large number of interactions of the type $\nu_e + N \to e^- + \cdots$, $\nu_\tau + N \to \tau^- + \cdots$, arising for oscillations $\nu_\mu \to \nu_e$, ν_τ. There is however at present no convincing evidence for oscillation phenomenon from laboratory using reactors or accelerators as the neutrino sources.

References

1. Van Royen, Weisskopf (1967)
2. Wu et al. (1957)
3. Michel (1957)
4. Burgy et al. (1958)
5. Goldhaber, Grodzins, Sunyar (1958)
6. Anderson et al. (1960)
7. Pais, Piccioni (1955)
8. J.H. Christenson et al., Phys. Rev. Lett. **13**, 138 (1964)
9. Wolfenstein (1964)
10. Maki et al. (1962)
11. Pontecorvo (1968)
12. Davis et al. (1968)

Chapter 8
Electroweak Interactions

8.1 Neutral Currents

Until 1973 all observed weak interactions were consistent with the hypothesis that they were mediated by the exchange of heavy charged bosons $W^{\pm}$ only. They are called charged current interactions because they change the electric charge of the lepton, for example, in the interaction, $\overline{\nu}_{\mu} + p \to \mu^{+} + n$, the charge of neutrino is changed when it is destroyed and a new particle (μ^{+}) is produced. Another example is $\nu_{\mu} + n \to \mu^{-} + p$. Although charged weak processes were recognized from the beginning as in beta decay, the possibility of neutral weak processes was not appreciated until 1958. Typical examples of neutral current events are

$$\overline{\nu}_{\mu} + e^{-} \to \overline{\nu}_{\mu} + e^{-}, \qquad \nu_{\mu} + e^{-} \to \nu_{\mu} + e^{-}, \qquad \nu_{\mu} + N \to \nu_{\mu} + \text{hadron},$$
$$\overline{\nu}_{\mu} + p \to \overline{\nu}_{\mu} + p, \qquad \nu_{\mu} + p \to \nu_{\mu} + p$$

They demonstrate the existence of neutral currents because the current connects only muonic lepton with muonic lepton. Both muonic leptons have the same charge (zero) and thus the current must be neutral. Same reasoning applies to electronic lepton and hadron.

Fermi's theory is equivalent to the exchange of only charged weak bosons, for example $\nu_{\mu} e^{-} \to \mu^{-} \nu_{e}$ is viewed as emission of a W^{+} by initial neutrino which turns into an electron neutrino, Fig. 8.1.

When the W is emitted or absorbed the charges of interacting particles are changed. The currents to which the W attaches, for example, $e^{-}\gamma_{\mu}(1-\gamma_{5})\nu$, are called charged currents.

The process $\nu_{\mu} e^{-} \to \nu_{\mu} e^{-}$ can not proceed in the Fermi theory because charge current can change ν_{μ} only to μ^{-}, not to e^{-}. In the above example no charge is transferred. This is possible if the interaction is mediated by Z^{0} boson. The ν_{μ} can emit a Z which is absorbed by the electron permitting the process $\nu_{\mu} \overline{e} \to \nu_{\mu} \overline{e}$, Fig. 8.2.

A. Kamal, *Particle Physics*, Graduate Texts in Physics,
DOI 10.1007/978-3-642-38661-9_8, © Springer-Verlag Berlin Heidelberg 2014

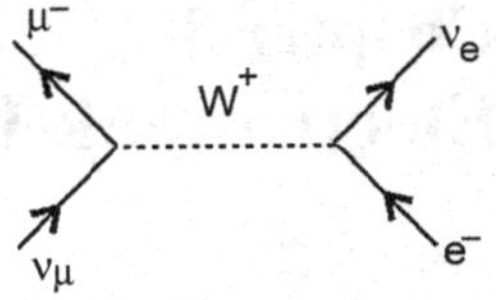

Fig. 8.1 Emission of a W^+ by initial neutrino which turns into an electron neutrino

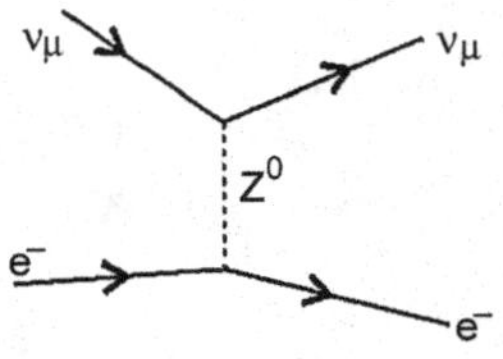

Fig. 8.2 Emission of Z by ν_μ

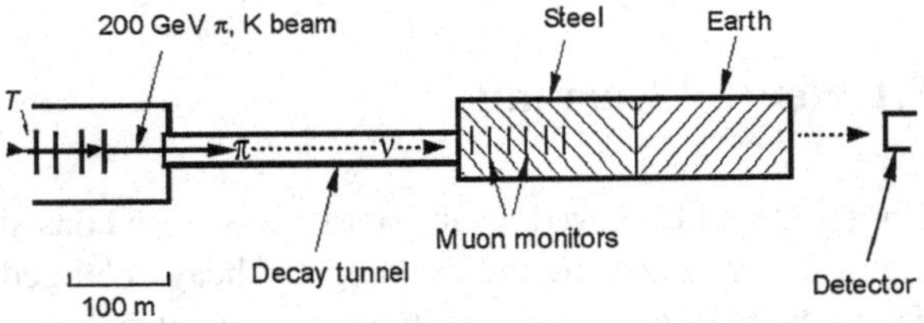

Fig. 8.3 Shown is a fraction of decay of pions to muons and neutrinos ($\pi^+ \to \mu^+\nu_\mu, \pi^- \to \mu^-\overline{\nu}_\mu$)

8.2 Layout of Neutrino Beam

The experiment consists of production of secondary pions and kaons from high-energy proton collisions in a target T. Dipole bending magnets, quadrupole focusing magnets and collimating slits are employed to direct high energy (~ 200 GeV) beams of pions and kaons into the decay tunnel, Fig. 8.3.

Muons are arrested by a thick iron shield, and the interactions of the neutrinos observed in the detector. Such experiments gave the first evidence in 1961 for the separate identity of electron and muon neutrinos. Note that in the actual experiment there is a kinematic relation between the energy and direction of neutrinos from monochromatic beams of pions or kaons, so that the energies of neutrino events are correlated with the distance, the more energetic events occurring closer to the beam axis. The neutral-current events were first discovered by A. Lagarrigue, P. Musset, D.H. Perkins, A. Rousset and co-workers using the Gargamelle bubble chamber at CERN in 1973. The candidates looked for were the charge lepton free events of the hadronic type $\nu_\mu N \to \nu_\mu X$, $\overline{\nu}_\mu N \to \overline{\nu}_\mu X$, where X is any hadronic state.

The freon (CF_3Br) bubble chamber exposed to antineutrino beam revealed three such events in 1.4×10^6 pictures each with $10^9 \overline{\nu}_\mu$ per pulse. In these events the identity of the neutrino or antineutrino is unchanged. The neutrino events were evenly distributed in the chamber as ν's have practically no attenuation. Otherwise neutron events would have shown exponential distribution.

The reason for the enormous delay in the observation of neutral-current is two fold (1) nobody was looking for them and (2) they are masked by much stronger electromagnetic effects. For example, Z can be exchanged between two electrons, but so can the photon, Fig. 8.4.

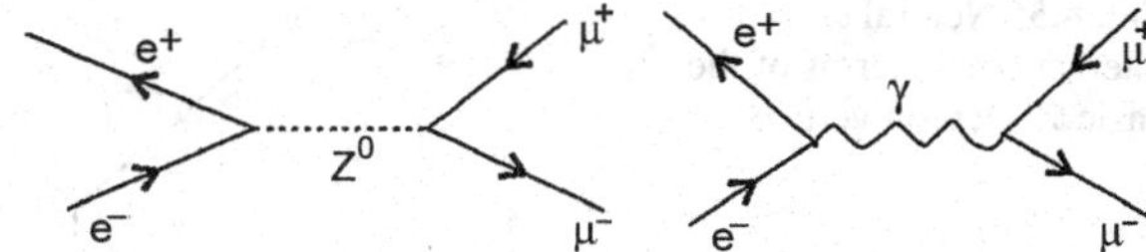

Fig. 8.4 Exchange of both Z and photon between two electrons

By comparing the couplings of the Z to that of the W, it is possible to derive a relation for the ratio of neutral-current events to charged-current events in deep inelastic neutrino scattering, NC/CC, using the parton model. The predictions are

$$R_\nu = \left(\frac{NC}{CC}\right)_\nu = \frac{1}{2} - \sin^2\theta_W + \frac{20}{27}\sin^4\theta_W \tag{8.1}$$

$$R_{\overline{\nu}} = \left(\frac{NC}{CC}\right)_{\overline{\nu}} = \frac{1}{2} - \sin^2\theta_W + \frac{20}{9}\sin^4\theta_W \tag{8.2}$$

Using the accepted value of $\sin^2\theta_w$ (θ_w is the Weinberg angle) the predicted ratios $R_\nu = 0.3$ and $R_{\overline{\nu}} = 0.38$ are in excellent agreement with the observations. Historically the experimental values of R_ν and $R_{\overline{\nu}}$ were used to determine Weinberg's angle.

8.3 The Basic Vertices

In Sect. 7.7 it was shown that all the charged current interactions could be explained with the aid of the basic $W^\pm$-lepton vertices as shown in Fig. 7.7. Similarly all known neutral current interactions can be understood in terms of the basic Z^0-lepton vertices of Fig. 8.5. Similar to $W^\pm$-lepton vertices these conserve the lepton numbers. The corresponding quark vertices can be obtained from the lepton numbers L_e, L_μ and L_τ, apart from the electric charge Q. The corresponding quark vertices can be obtained from the lepton vertices by using lepton-quark symmetry and quark mixing in the same way that the $W^\pm$-quark vertices were obtained from $W^\pm$-lepton vertices. Confining to the first two generations, the quark vertices are obtained from the lepton vertices, Fig. 8.5 by noting the substitutions

$$\nu_e \to u, \qquad \nu_\mu \to e, \qquad e^- \to d', \qquad \mu^- \to s' \tag{8.3}$$

where d' and s' are the mixtures (7.13). For $l = e$ or μ the lepton vertices of Fig. 8.5 can be obtained

$$\nu_e\nu_e Z^0, \qquad \nu_\mu\nu_\mu Z^0, \qquad e^-e^-Z^0, \qquad \mu^-\mu^-Z^0 \tag{8.4}$$

The corresponding quark vertices are uuZ^0, ccZ^0, $d'd'Z^0$, $s's'Z^0$.

In Fig. 7.12 we had a vertex

$$ud'W = udW\cos\theta_c + usW\sin\theta_c$$

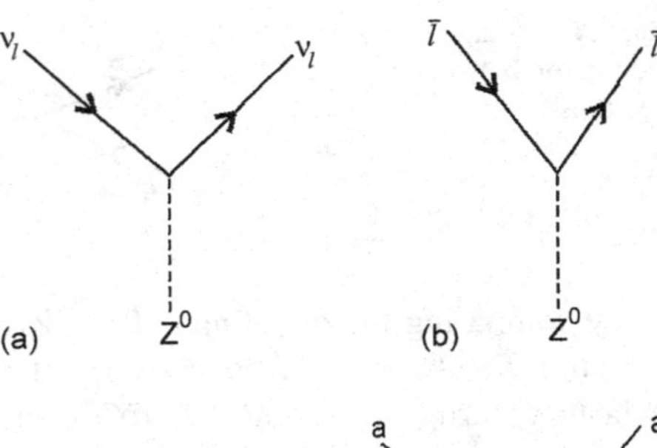

Fig. 8.5 Neutral current interactions in terms of the basic Z^0-lepton vertices

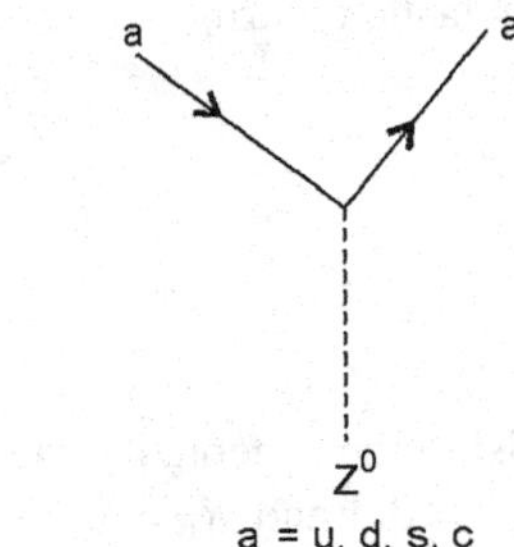

Fig. 8.6 The basic Z^0-quark vertices

which was interpreted as udW and usW vertices with relative strength $\cos\theta_c$ and $\sin\theta_c$ as in Fig. 7.15. In the same way we have

$$\begin{aligned} d'd'Z^0 &= (d\cos\theta_c + s\sin\theta_c)(d\cos\theta_c + s\sin\theta_c)Z^0 \\ &= ddZ^0\cos^2\theta_c + ssZ^0\sin^2\theta_c + \left(dsZ^0 + sdZ^0\right)\sin\theta_c\cos\theta_c \end{aligned} \tag{8.5}$$

and

$$\begin{aligned} s's'Z^0 &= (-d\sin\theta_c + s\cos\theta_c)(-d\sin\theta_c + s\cos\theta_c)Z^0 \\ &= ddZ^0\sin^2\theta_c + ssZ^0\cos^2\theta_c - \left(dsZ^0 + sdZ^0\right)\sin\theta_c\cos\theta_c \end{aligned} \tag{8.6}$$

so that the primed vertices $d'd'Z^0$ and $s's'Z^0$ both contribute to the vertices ddZ^0, ssZ^0, sdZ^0 and dsZ^0 for the real particles d and s. However, when both the sets of contributions (8.5) and (8.6) are combined

$$d'd'Z^0 + s's'Z^0 = ddZ^0 + ssZ^0 \tag{8.7}$$

so that the four vertices (8.4) can be replaced by the equivalent vertices (the basic Z^0-quark vertices)

$$uuZ^0, \qquad ccZ^0, \qquad ddZ^0, \qquad ssZ^0 \tag{8.8}$$

as in Fig. 8.6. The four vertices of Fig. 8.6 conserve strangeness and charm, where as the 'flavour changing' vertices ucZ^0 and dsZ^0 do not occur.

The conclusion is, neutral current interaction conserve strangeness, Fig. 8.6 and charm, in contrast to the charged current interactions which do not in general conserve these quantum numbers.

This prediction is confirmed by experiments e.g. strangeness-changing decay

$$K^+ \to \pi^0 + \mu^+ + \nu_\mu \tag{8.9}$$

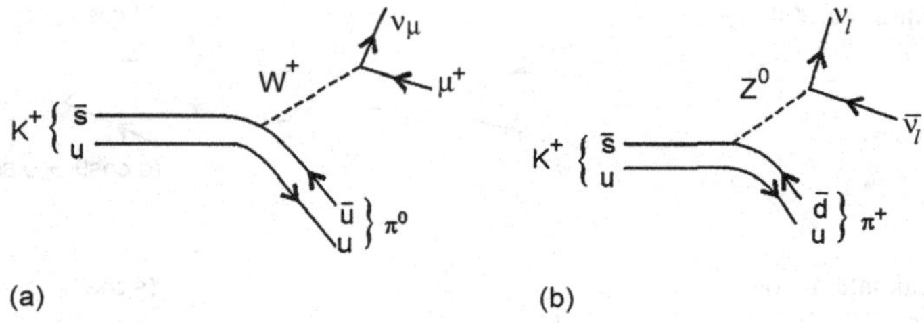

Fig. 8.7 (**a**) Strangeness-changing decay. (**b**) Forbidden neutral vertex dsZ^0

and

$$K^+ \to \pi^+ + \nu_l + \overline{\nu} \tag{8.10}$$

shown is Fig. 8.7(a) where as the second involves the forbidden neutral vertex dsZ^0. Experimentally (8.10) is not observed

$$\frac{\sum_l \Gamma(K^+ \to \pi^+ + \nu_l + \overline{\nu}_l)}{\Gamma(K^+ \to \pi^0 + \mu^+ + \nu_\mu)} < 10^{-6}$$

8.4 The GIM Model and Charm

In 1970 Glashow, Iliopoulos and Maiani (GIM) showed that charmed quarks were the simplest way to explain the absence of neutral strangeness changing weak currents, characterized by $\Delta S = 0$.

Until 1973 only weak currents that change charge had been observed. For example in μ-decay μ turns into ν_μ, and its charge changes by one unit. Neutral weak current, for example in the interaction $\nu p \to \nu p$ does not change strangeness. If strangeness could be changed by a neutral current, the decays $K^0 \to \mu^+\mu^-$ and $K^+ \to \pi^+e^+e^-$ would be possible. The GIM model shows that if in addition to the charged weak current changing an s-quark into a u-quark, there was another changing an s-quark into a c-quark, there would be cancellation of the second order terms.

Consider the decay of K^+ for which the ratio of neutral to charged-current rates is known to be extremely small.

$$\frac{K^+ \to \pi^+\nu\overline{\nu}}{K^+ \to \pi^0\mu^+\nu_\mu} < 10^{-5} \quad (\Delta S = 1)$$

The neutral-current coupling, Fig. 8.8 will be of the form

$$\underbrace{u\overline{u} + \left(d\overline{d}\cos^2\theta_c + s\overline{s}\sin^2\theta_c\right)}_{\Delta S=0} + \underbrace{(s\overline{d} + \overline{s}d)\sin\theta_c\cos\theta_c}_{\Delta S=1}$$

so that $\Delta S = 1$ neutral current should be possible since $\sin\theta_c \neq 0$. In 1970 the GIM model introduced a new quark of charge $+2/3$ and flavor C for 'charm'. This

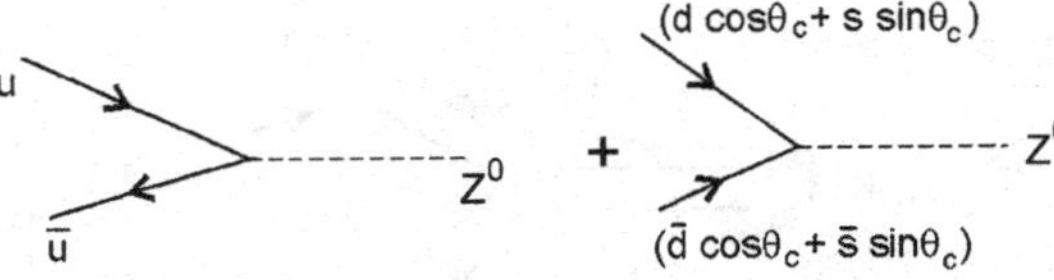

Fig. 8.8 Neutral-current coupling

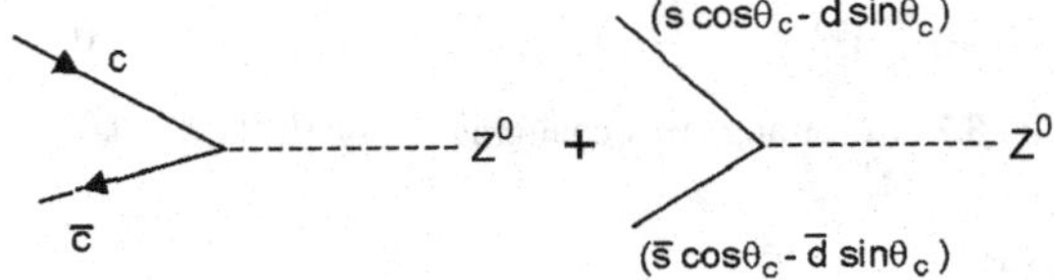

Fig. 8.9 Weak interaction neutral current

led to the introduction of the second doublet consisting of a C-quark and the s, d combination orthogonal to d in Cabibbo's theory. Thus the two doublets are

$$\begin{pmatrix} u \\ d_c \end{pmatrix} = \begin{pmatrix} u \\ d\cos\theta_c + s\sin\theta_c \end{pmatrix}, \quad \begin{pmatrix} c \\ s_c \end{pmatrix} = \begin{pmatrix} c \\ s\cos\theta_c - d\sin\theta_c \end{pmatrix} \tag{8.11}$$

This then means that extra terms are to be added to Fig. 8.8, as shown in Fig. 8.9 to obtain the weak interaction neutral current matrix element

$$\underbrace{u\overline{u} + c\overline{c} + (d\overline{d} + s\overline{s})\cos^2\theta_c + (s\overline{s} + d\overline{d})\sin^2\theta_c}_{\Delta S=0} + \underbrace{(s\overline{d} + \overline{s}d - \overline{s}d - s\overline{d})\sin\theta_c\cos\theta_c}_{\Delta S=1}$$

The Cabibbo angle would be simply a rotation mixing the d and s quarks. Thus with the inclusion of a new quark in the scheme, the unwanted $\Delta S = 1$ neutral currents have been canceled. It was a singular triumph for the theory when heavy quark states ($\psi = C\overline{C}$, $D = C\overline{U}$ etc.) were discovered in 1974–1976 as described in Chap. 4. Further according to (8.11) in the decays of charmed mesons to noncharmed mesons ($\Delta C = 1$), $C \rightarrow S$ transitions (coupling $\cos^2\theta_c$) would dominate $C \rightarrow d$ transitions (coupling $\sin^2\theta_c$). Thus D^0 meson (1893) decays predominantly in the mode $K^- + \pi'S$, while $D^0 \rightarrow \pi'S$ is strongly suppressed. Also, the mechanism that cancels $\Delta S = 1$ neutral current is also responsible for inhibiting $\Delta C = 1$ neutral current. In fact the ratio

$$\frac{\Delta C = 1}{\Delta C = 0} < 3\ \%$$

8.5 The Cabibbo-Kolayashi-Maskawa (CKM) Matrix

When the τ-lepton was discovered the quark-lepton symmetry was destroyed. However, with the discovery of the bottom quark and the anticipated top quark, the symmetry was restored.

Recall that the absence of strangeness-changing neutral-current process led to a picture in which the mass eigenstates d, s are mixed and related to the weak-interaction eigen states d', s' by

$$\begin{pmatrix} d' \\ s' \end{pmatrix} = \begin{pmatrix} \cos\theta_c & \sin\theta_c \\ -\sin\theta_c & \cos\theta_c \end{pmatrix} \begin{pmatrix} d \\ s \end{pmatrix}$$

θ_c is the Cabibbo angle.

Adding the third generation of quarks the 2×2 matrix is replaced by a 3×3 U-matrix

$$\begin{pmatrix} d' \\ s' \\ b' \end{pmatrix} = \begin{pmatrix} U_{ud} & U_{us} & U_{ub} \\ U_{cd} & U_{cs} & U_{cb} \\ U_{td} & U_{ts} & U_{tb} \end{pmatrix} \begin{pmatrix} d \\ s \\ b \end{pmatrix} \tag{8.12}$$

The probability for a transition from a quark q to a quark q' is proportional to $|U_{qq'}|^2$, the square of the magnitude of the matrix element.

With six quark flavors, the weak currents will be described by unitary transformations among three quark doublets, characterized by three Euler angles θ_1, θ_2 and θ_3 and six phases. Of the latter, five are arbitrary and unobservable, leaving one non-trivial phase δ. The matrix elements are by now well known. They are correlated since the matrix is unitary. The total number of independent parameters is four, three real angles, and an imaginary phase. The phase affects weak processes of higher order via the interference terms. The CP violation is attributed to the existence of this imaginary phase. We can have some idea about the approximate relative values of the matrix elements.

$|U_{ud}|$ (analogous to $\cos\theta_o$ in 2×2 matrix) from β-decay $= 0.973$

$|U_{us}|$ (analogous to $\sin\theta_o$ in 2×2 matrix) from kaon and hyperon semileptonic decays $= 0.23$

$|U_{cs}|$ from D-meson decay, $D^+ \to \overline{K^0} e^+ \nu_e \sim 0.97$

$|U_{ub}|$ from B-meson decays directly to non-strange particles ($b \to u e^- \overline{\nu_e}$) is small and consistent with zero ~ 0

$|U_{cb}|$ from B-meson decays to charmed particles is small ~ 0.05

$$(|U_{ij}|) = \begin{pmatrix} 0.973 & 0.23 & 0 \\ 0.23 & 0.97 & 0.05 \\ 0 & 0.05 & 1 \end{pmatrix} \tag{8.13}$$

The diagonal elements of this matrix describe transitions within a family; they deviate from unity by only a few percent. The values of the matrix elements U_{cb} and U_{ts} are nearly one order of magnitude smaller than those of U_{us} and U_{cd}. Accordingly, transitions from the third to the second generation ($t \to s$, $b \to c$) are suppressed by nearly two orders of magnitude compared to transitions from the second to the first generation ($s \to u$, $c \to d$). This applies to an even higher degree for transitions from the third to the first generation. The direct transition $b \to u$ was detected in the semi-leptonic decay of B-mesons into non-charmed mesons. Thus the model makes predictions about the weak decay sequence of mesons containing b-quarks. The decay chain

$$b \to c \to s \to u$$

should be favored over

$$b \rightarrow u$$

and hence in the decay of such bottom states, strange particles (kaons) should be frequent among the hadron secondaries. This is indeed the case.

Preferred decay for mesons containing a top quark would be

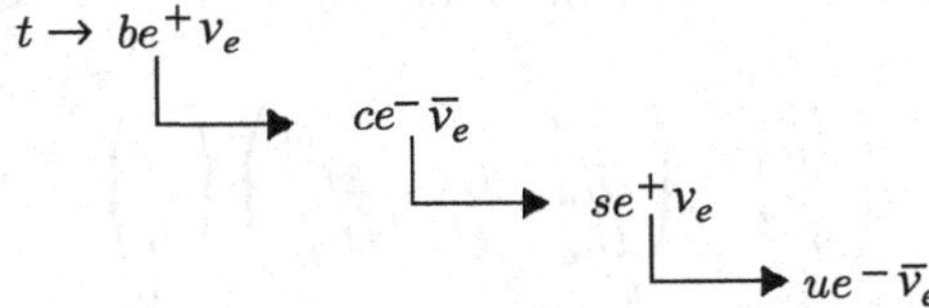

8.6 The Electroweak Unification

A high level of understanding is reached in physics whenever two different theories are unified, two apparently unrelated classes of facts are discovered to be intimately connected, forming two sub-classes of a larger set of phenomena described by a larger unitary theory. A notable example is the unification of electricity and magnetism in the form of electromagnetism by Maxwell and Faraday in the nineteenth century. A further unification of electromagnetism and weak interactions was developed by Weinberg, Salam and Glashow in 1960s. Accordingly, the electromagnetic and weak interaction are understood as two aspects of the same interaction called electro-weak interaction. The idea has its origin in the following shared characteristics.

1. Both forces affect equally all forms of matter, leptons as well as hadrons
2. Both are vector in character
3. Both possess universal coupling strengths
4. Both are renormalizable
5. Both arise from a gauge principle.

8.6.1 Weak Spin

The electroweak interaction is elegantly described by the introduction of a new quantum number T, the weak isospin, analogous to isospin of strong interaction.

Each family of left handed quarks and leptons forms a doublet of fermions which can transform into each other by absorbing or emitting a W boson. The weak isospin assigned to the fermions in a doublet is $T = 1/2$ and the third component is $T_3 = \pm 1/2$, their electric charge $Z_f \cdot e$ of the two fermions in a doublet differing by one

Table 8.1 Details of T, T_3 and Z_f for leptons and quarks

	Fermion Multiplets			T	T_3	Z_f
Lepton	$\binom{\nu_e}{e}_L$	$\binom{\nu_\mu}{\mu}_L$	$\binom{\nu_\tau}{\tau}_L$	$\frac{1}{2}$	$+\frac{1}{2}$	0
					$-\frac{1}{2}$	-1
	e_R	μ_R	τ_R	0	0	-1
Quarks	$\binom{u}{d'}_L$	$\binom{c}{s'}_L$	$\binom{t}{b'}_L$	$\frac{1}{2}$	$+\frac{1}{2}$	$+\frac{2}{3}$
					$-\frac{1}{2}$	$-\frac{1}{3}$
	u_R	c_R	t_R	0	0	$+\frac{2}{3}$
	d_R	s_R	b_R	0	0	$-\frac{1}{3}$

unit. For right-handed antifermions, the signs of T_3 and Z_f are reversed. On the other hand, right-handed fermions (and left-handed antifermions) do not couple to W bosons. They are said to be singlets ($T = T_3 = 0$). Thus the left handed leptons and the (Cabibbo-rotated) left-handed quarks of each family form two doublets and there are additionally three right-handed singlets. Table 8.1 gives the details of T, T_3 and Z_f for leptons and quarks.

8.7 The Weinberg Angle

Charged current reactions are required to conserve T_3. The W^- boson is therefore assigned the quantum number $T_3(W^-) = -1$ and the W^+ boson $T_3(W^+) = +1$. A third state must therefore exist with $T = 1$, $T_3 = 0$, coupling with the same strength g as the W^+ to the fermion doublets. This state is denoted by W^0, and together with the W^+ and the W^- constitutes a weak isospin triplet.

The W^0 cannot be identical with the Z^0, since the coupling of the latter also depends on the electric charge. The existence of an additional state B^0, a singlet of the weak spin ($T = 0$, $T_3 = 0$), is postulated. Its coupling strength need not be equal to that of the triplet ($W^\pm$, W^0). The corresponding weak charge is denoted by g'. The B^0 and W^0 couple to fermions without changing their weak isospin.

The theory begins with massless Yang-Mills particles. These are denoted by W^+, W^0, W^- and B. The W's form a triplet of a new symmetry "weak isospin" while B is an isosinglet. At the same time the two neutral particles W^0 and B mix to produce two physical particles, the photon and the Z^0. The photon is massless. Z^0 acquires a mass comparable to that of W. The basic idea of electroweak unification is to describe photon and Z^0 as mutually orthogonal linear combinations of B^0 and W^0. The mixing is analogous to the description of quark mixing in terms of the Cabibbo angle expressed as a rotation through the so-called electroweak mixing angle θ_W, also known as the Weinberg angle.

$$|\gamma\rangle = \cos\theta_W |B^0\rangle + \sin\theta_W |W^0\rangle \tag{8.14}$$

$$|Z^0\rangle = -\sin\theta_W |B^0\rangle + \cos\theta_W |W^0\rangle \tag{8.15}$$

The connection between the Weinberg angle θ_W, the weak charges g and g' and the electric charge e is given by demanding that the photon couples to the charges of the left and right handed fermions but not to the neutrinos. One then obtains

$$\tan\theta_W = \frac{g'}{g}, \qquad \sin\theta_W = \frac{g'}{\sqrt{g^2+g^2}}, \qquad \cos\theta_W = \frac{g}{\sqrt{g^2+g^2}} \tag{8.16}$$

The electromagnetic charge is given by

$$e = g\sin\theta_W \tag{8.17}$$

The Weinberg angle can be determined in a number of independent ways.

1. Neutrino-electron scattering
2. Electro-weak interference in e^+–e^- scattering
3. Width of the Z^0 boson
4. Ratio of masses $m_{W\pm}/m_{Z^0}$
5. Ratio of neutral current/charge current events.

The combined analysis gives $\sin^2\theta_W = 0.2319 \pm 0.0005$.

8.8 The Unification Condition

In the unified theory it is important that the infinities which occur in the higher order diagrams should cancel in all possible processes. This is accomplished by finding a relationship between various coupling constants. When all the diagrams of a given order are added together the divergences in integrals cancel leading to finite contribution overall. The cancellation is not accidental but is a consequence of relations between various coupling constants, associated with the γ, $W^\pm$ and Z^0 vertices in unified field theory.

The unification which is based on gauge invariance demands two equations called the unification condition and the anomaly condition to be satisfied. The unification condition which explicitly relates the weak and electromagnetic constants is

$$e = g\sin\theta_W = g'\cos\theta_W \tag{8.18}$$

The relation encapsulates the unification condition and the fact that the coupling to $W^\pm$ and Z^0 is the same as in the purely electromagnetic case, the apparent difference between weak and electromagnetic interactions being associated with the heavy W and Z propagators.

One can also express the mass of the heavy bosons in terms of the Weinberg angle.

$$M_W = \left(\frac{\sqrt{2}e^2}{8G\sin^2\theta_W}\right)^{\frac{1}{2}}$$

which after inserting values becomes

$$M_W = \frac{37.4}{\sin\theta_W}\ \text{GeV} \tag{8.19}$$

Further

$$M_{Z^0} = \frac{75}{\sin 2\theta_W} \tag{8.20}$$

The anomaly condition relates the electric charges Q_l and Q_a of the lepton l and quarks 'a' and is

$$\sum_l Q_l + 3 \sum_a Q_a = 0 \tag{8.21}$$

The sums extend over all leptons l and all quark flavours, $a = u, d, s, \ldots$ and the factor three arises due to quark colours. The anomaly condition is satisfied with 6 known leptons and the 6 quarks.

8.9 $W^{\pm}$ and Z^0 Bosons

Although the discovery of neutral currents was significant for the electroweak theory the experimental observation of $W^{\pm}$ and Z^0 bosons was crucial and decisive test for the Weinberg-Salam-Glashow theory.

The weak interaction is associated with elementary spin-1 bosons which are the force carriers between quarks *and/or* leptons. These are the charged bosons W^+ and W^- and the neutral Z^0. However, unlike the photons (spin-1) and gluons (spin-1) for the electromagnetic and strong fields, these bosons which were discovered at CERN in 1983 (Arnisom et al., Bagnaia et al., Banner et al.) are very massive particles and consequently of very short range. Their masses are measured to be

$$M_W = 80.6\ \text{GeV}/c^2 \quad \text{and} \quad M_Z = 91.2\ \text{GeV}/c^2$$

The ranges calculated from the uncertainty principle are

$$R_W \simeq R_Z \simeq 2 \times 10^{-3}\ \text{fm}$$

for the weak interactions. These are very small even when compared with the size of the nucleon. At low energies, the weak interaction can be approximated to zero-range interaction. At high energies, however, the zero-range approximation is no longer valid.

The $W^{\pm}$ and Z^0 bosons were first discovered using a proton-antiproton collider at CERN. At that time the proton and antiproton beams had maximum energy of 270 GeV each providing a total centre of mass energy of 540 GeV, which is much above the rest energy of W or Z^0.

The $W^{\pm}$ and Z^0 bosons are highly unstable particles which were first produced in the reactions

$$p^- + p \to W^+ + X^-$$
$$p^- + p \to W^- + X^+$$
$$p^- + p \to Z^0 + X^0$$

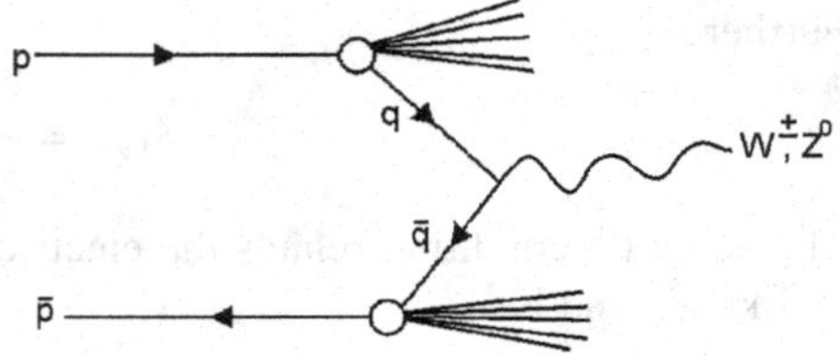

Fig. 8.10 The production mechanism of the bosons involving the quark-antiquark annihilation processes

where $X^{\pm}$ and X^0 are hadronic states allowed by the conservation laws. The heavy bosons were detected via their subsequent decays

$$W^+ \rightarrow l^+ + \nu_l\left(e^+\nu_e, \mu^+\nu_\mu, \tau^+\nu_\tau\right)$$
$$W^- \rightarrow l^- + \overline{\nu}_l\left(e^-\overline{\nu_e}, \mu^-\overline{\nu}_\mu, \tau^-\overline{\nu}_\tau\right)$$
$$Z^0 \rightarrow l^+ + l^-\left(e^+e^-, \mu^+\mu^-, \tau^+\tau^-, \nu_x\overline{\nu}_x\right)$$

where the charged leptons $l^{\pm}$ were either muons, electrons or tauons. The $W^{\pm}$ and Z^0 bosons may also decay into quark-antiquark pairs

$$W^- \rightarrow q\overline{q}(d\overline{u}, s\overline{u})$$
$$W^+ \rightarrow \overline{q}q'(\overline{d}u, \overline{s}u)$$
$$Z^0 \rightarrow q\overline{q}(u\overline{u}, d\overline{d}, s\overline{s})$$

The production mechanism of these bosons involves the quark-antiquark annihilation processes and is shown in Fig. 8.10. The processes are

$$u + \overline{d} \rightarrow W^+, \qquad d + \overline{u} \rightarrow W^-$$

and

$$u + \overline{u} \rightarrow Z^0, \qquad d + \overline{d} \rightarrow Z^0$$

for charged and neutral vector bosons, respectively. In the centre of mass of the $q\overline{q}$ system the total energy of the $q\overline{q}$ pair must be at least 81 GeV or 91 GeV corresponding to the production of either $W^{\pm}$ or a Z^0 boson at rest. However, the energy of the $p^- p$ system has to be considerably higher for a reasonable reaction rate to occur, because even quark (or antiquark) has only a fraction of the parent proton (or antiproton) energy.

8.9.1 Detection of $W^{\pm}$ and Z^0 Bosons

The proton and antiproton beams each of energy 270 GeVwere brought together in the intersection region of the collider. A very large elaborate detecting system called UAI was arranged in this region, Fig. 8.11 is the schematic diagram of the cross-section showing the end-on view, along the beam direction. As one moves away from the intersection region, the components are

1. A central tracking detector CD used to observe charged particles and to measure their momenta from the curvature of the tracks in an applied magnetic field.

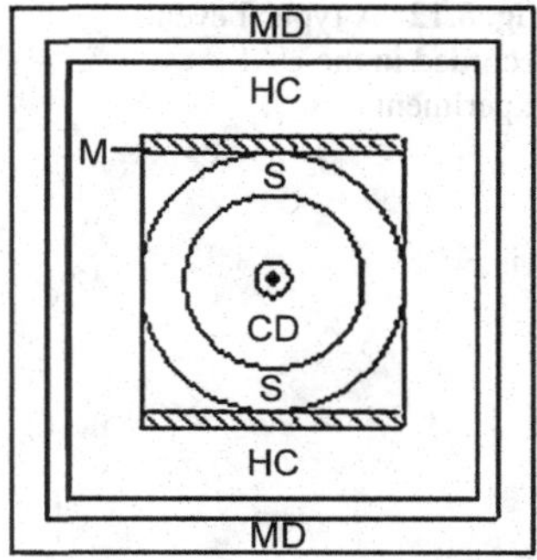

Fig. 8.11 Schematic diagram of the cross-section of UAI showing the end-on view, along the beam direction

2. Electromagnetic shower counters surrounding the central track detector used to detect both electrons and photons. These electrons are already detected in the central detector although the photons are not.
3. Coils M used to produce magnetic field in the central detector.
4. Hadron calorimeters HC which are much larger to detect the hadrons jets outside.
5. Muon detectors MD used to detect muons which are the only charged particles which can escape from the hadron calorimeters.

The calorimeters are segmented in intervals of polar and azimuthal angles. Thus, the only particles which are not detected are the neutrinos.

The main problem in the detection of the heavy bosons is the smallness of production cross-section, of the order of 1 nb (10^{-33} cm^2) for W and 0.1 nb for Z^0. This is to be compared with a value of σ (total) $= 40$ mb $= 4 \times 10^7$ nb for the total $p^-{-}p$ cross-section. The extraction of such rare events against a heavy background (at the level of 10^{-8}–10^{-9} of the total) is possible because of the circumstance that the electrons, muons and neutrinos from the decays have large momenta (because the $W^{\pm}$ or Z^0 bosons are very heavy) and are often emitted at wide angles to the beam direction, so that the leptons have large transverse momenta; $P_T \leq M_W/2 \simeq 40$ GeV/c. Leptons arising from other sources as from hadron decays have much smaller P_T values.

8.9.2 The Z^0 Events

The Z^0 was detected by observing its decay into an e^+e^- pair in the central detector in which the tracks were almost straight in spite of the magnetic field because of their large momenta. The electromagnetic shower detectors identified them as electrons. Various trigger devices were used to eliminate unwanted events. Cuts were imposed to require two well-defined isolated showers in the electromagnetic calorimeter with $E_T > 25$ GeV and with matching tracks in the central detector. Finally the invariant mass for Z^0 boson was found out. A typical event recorded in the UAI experiment is shown in Fig. 8.12. (a) shows all reconstructed vertex associated

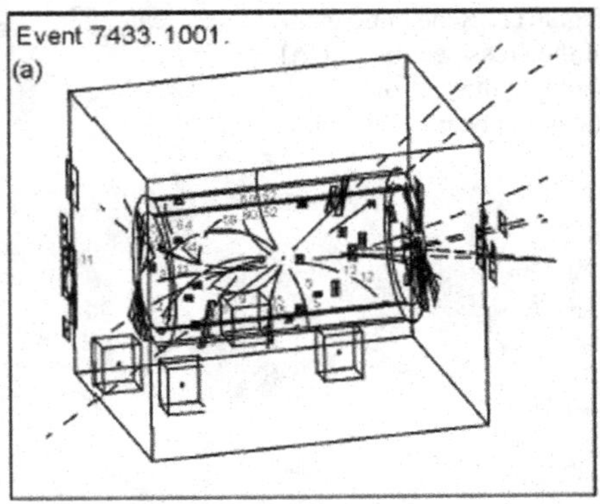

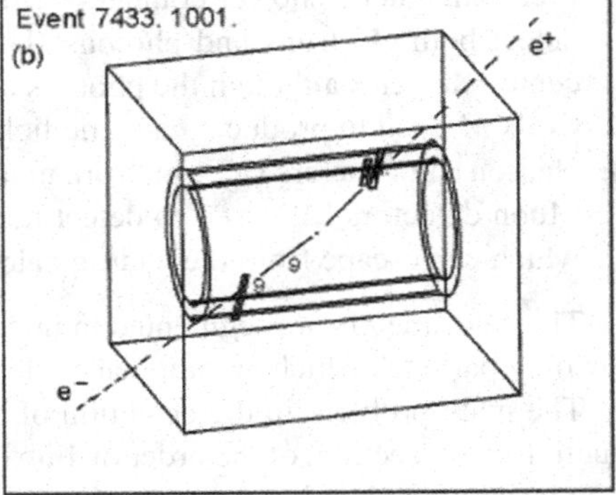

Fig. 8.12 A typical event recorded in the UAI experiment

tracks while (b) shows the same event but including only tracks with $P_{\tau} > 2$ GeV/c and calorimeter hits with $E_T > 2$ GeV. Only the e^+e^- pair survives the cuts.

Figure 8.13 shows the energy deposited in the elements of the electromagnetic calorimeter. Later the experiment was repeated at CERN and Fermi Lab.

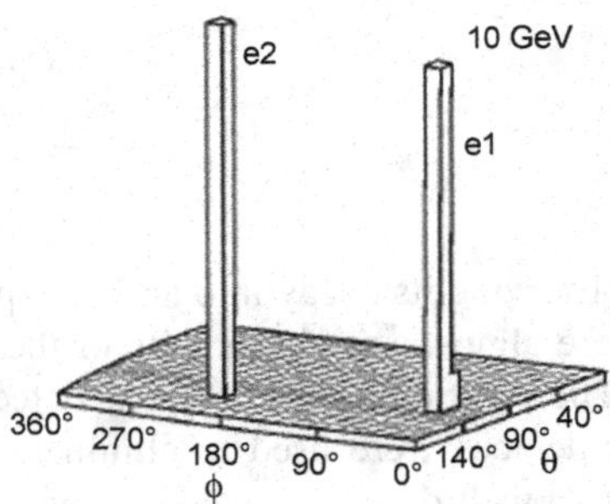

Fig. 8.13 "Lego diagram" of one of the first events of the reaction $q\overline{q} \rightarrow Z^0 \rightarrow e^+e^-$, in which the Z^0 boson was detected at CERN. The transverse energies of the electron and positron detected in the calorimeter elements are plotted as a function of the polar and azimuthal angles [1]

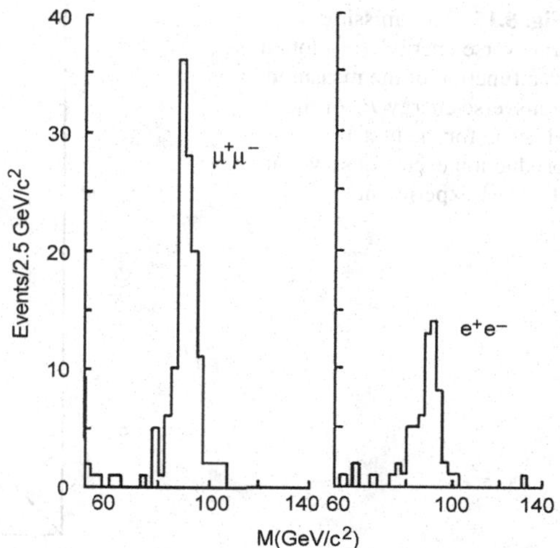

Fig. 8.14 Z^0 peaks observed in the e^+e^- and $\mu^+\mu^-$ mass distribution from a recent Fermilab experiment by the CDF collaboration [2]

Figure 8.13 shows the output from the electromagnetic shower counters due to the energy deposited for a typical Z^0 candidate. The direction in space is also recorded.

Figure 8.14 shows the e^+e^- and $\mu^+\mu^-$ mass distributions obtained in a recent experiment at Fermi lab with much greater statistics. A clear peak is seen in both the distributions, the mass being

$$M_z = 91.16 \pm 0.03 \text{ GeV}/c^2$$

and a width

$$T_Z = 2.53 \pm 0.03 \text{ GeV}$$

the corresponding lifetime being 2.6×10^{-25} sec.

8.9.3 $W^{\pm}$ Events

A typical $W^{\pm}$ event is recognized by its decay, $W \to \mu + \nu$ or $W \to e + \nu$. The $W^{\pm}$ event resembles a typical Z^0 event except that one of the electron or muon tracks is replaced by an invisible neutrino. It therefore becomes necessary to estimate the missing invisible energy associated with neutrino. Longitudinal energy balance is not so useful as it is grossly affected by the loss of particles which are undetected along the beam direction. Also, the spectator quarks tend to be projected longitudinally. For this reason focus is directed toward the transverse momentum and energy balance. The transverse momenta of the observed particles is summed up and the missing transverse momentum attributed to neutrino is deduced.

The UAI experiment yielded a sample of 43 events which were suitable for analysis. The vector momentum of the electron and that calculated for the neutrino from

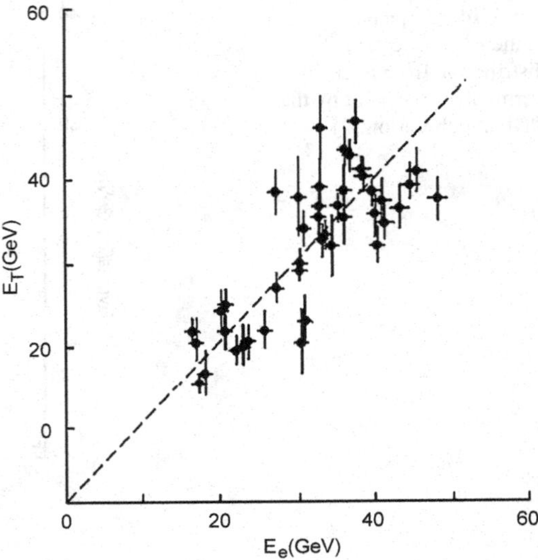

Fig. 8.15 The 'missing transverse energy' E_T plotted as a function of the measured transverse energy E_e of the electron for the first $W^{\pm}$ production events observed in the UA1 experiment [3]

the missing energy are found essentially back to back. Figure 8.15 shows good balance between the transverse energies of electron and neutrino. These features are precisely those expected from the decay of a very heavy slow particle decaying to $e\nu$.

Since the unobserved W does have some momentum, it is not possible to determine the mass of the W for individual events. One way to get round this difficulty is to measure the 'transverse mass', m_T^W where

$$m_T^W = \left[2E_T^e E_T^\nu (1 - \cos\theta_{e\nu})\right]^{1/2} \tag{8.22}$$

E_T^e and E_T^ν are the transverse energies of the electron and neutrino respectively and $\theta_{e\nu}$ is the angle between the electron and neutrino directions. If W has zero longitudinal momentum then $m_T^W = m_W$, otherwise $m_T^W < m_W$. Under some plausible assumptions it is possible to plot the expected distribution of m_T^W, as a function of m_W, which permits the mass of W to be determined. A value of $m_W = 80.3$ GeV/c^2 was obtained.

The decay width is found to be

$$\Gamma_W = 2.25 \pm 0.14 \text{ GeV}$$

corresponding to a lifetime of 2.9×10^{-25} sec. The branching ratio for each of the decay modes $e\nu_e$ and $\mu\nu_\mu$ is ~10 %.

8.10 Electroweak Reactions

In any process γ can be exchanged and Z^0 can be exchanged as well. This follows from the fact that to each of the basic electromagnetic vertices corresponds the Z^0 vertex, Fig. 8.16.

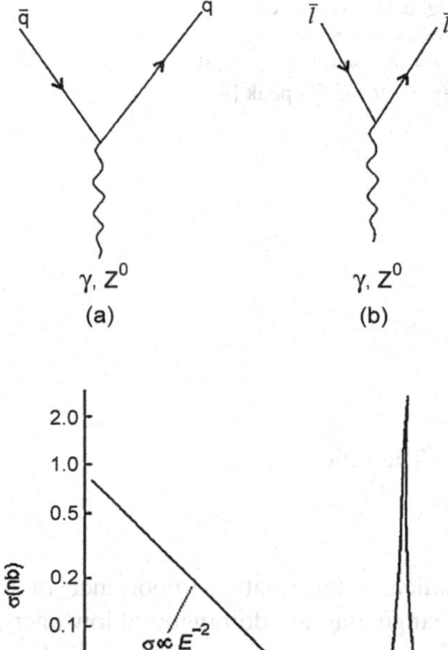

Fig. 8.16 Each of the basic electromagnetic vertices has a corresponding Z^0 vertex

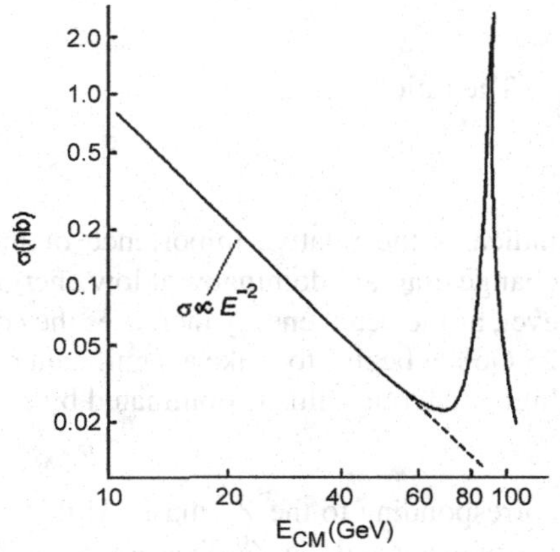

Fig. 8.17 Total cross-section for the reaction $e^+ + e^- \rightarrow \mu^+ + \mu^-$ as a function of total centre-of-mass energy (8.26). The *dashed line* shows the extrapolation of the low-energy behavior (8.23) in region of the Z^0 peak

When energy and momentum transfer are small compared with m_Z the reaction would be basically electromagnetic, the Z^0 contribution being negligible.

At high energy and large momentum transfer Z^0 contribution becomes important. In that case we will be dealing with the electroweak interaction. As an example consider the annihilation reaction $e^+ + e^- \rightarrow \mu^+ + \mu^-$. The variation of cross-section is shown in Fig. 8.17.

If we neglect the lepton masses compared with the beam energy E in the overall centre of mass system, then the second order diagram for one-photon exchange contribution, Fig. 8.4 gives as in (8.23).

$$\sigma_\gamma \approx \frac{\alpha^2}{E^2} \tag{8.23}$$

where α is the fine-structure constant.

For the contribution of the Z^0-exchange diagram

$$\sigma_Z \approx G_Z^2 E^2 \tag{8.24}$$

where G_Z is the effective zero-range coupling constant. G_Z is for neutral current in the same way the Fermi constant G_F is for the charge current.

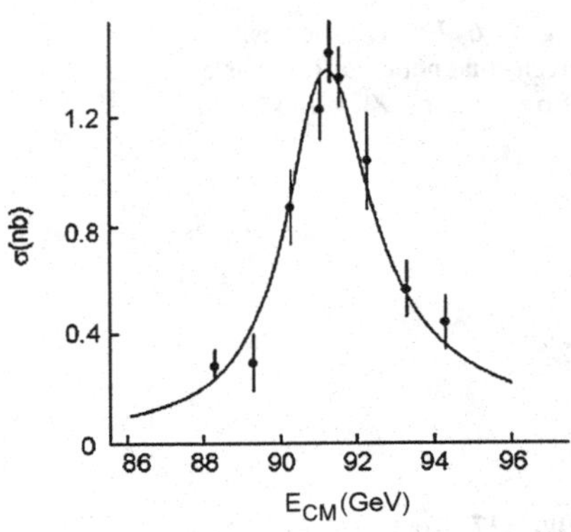

Fig. 8.18 Measured cross-section for $e^+ + e^- \rightarrow \mu^+ + \mu^-$ in the region of the Z^0 peak [4]

The ratio

$$\frac{\sigma_Z}{\sigma_\gamma} \approx \frac{G_Z^2 E^4}{\alpha^2} \approx \frac{E^4}{M_Z^2} \tag{8.25}$$

indicates the relative importance of the two diagrams. Thus the one-photon exchange diagram dominates at low energies and the cross-section falls as E^{-2}. However, as the beam energy increases the contribution of Z^0 rapidly increases. At about 25 GeV it begins to make a significant contribution to the total cross-section. At still higher energies, this is dominated by a very large peak at a centre of mass energy

$$E_{CM} = 2E = M_Z \tag{8.26}$$

corresponding to the Z^0 mass. At this energy the low-energy approximation (8.24) is irrelevant and the Z^0 diagram in Fig. 8.18 corresponds to the physical production of Z^0.

$$e^+ + e^- \rightarrow Z^0 \tag{8.27}$$

followed by the subsequent decay

$$Z^0 \rightarrow \mu^+ + \mu^- \tag{8.28}$$

This feature is exactly like the formation and decay of strong resonances discussed in Chap. 5, the only difference is that here we are dealing with weak rather than strong resonance, Fig. 8.18.

The peak occurs at $m_Z = 91$ GeV on the Breit-Wigner curve with a width equal to the total decay rate

$$\Gamma_Z = \tau_Z^{-1} = 2.53 \text{ GeV} \tag{8.29}$$

where τ_Z is the lifetime of the Z^0 boson.

8.11 Decays of the Z^0 Boson

The Z^0 bosons decay not only by $\mu^+\mu^-$ mode as in (8.28) but also to other final states allowed by the conservation laws. The colliders, Stanford Linear collider

(SLC) and the 'large electron-positron' (LEP) collider at CERN, which produce Z^0 bosons copiously have provided wealth of information regarding the decay modes. The processes observed are

$$e^+ + e^- \rightarrow l^+ + l^- \quad (l = e, \mu, \tau) \tag{8.30}$$

and

$$e^+ + e^- \rightarrow q + q^- \rightarrow \text{hadrons} \tag{8.31}$$

For each of the individual channels the Z^0 peak is observed. The cross-section is described by the Breit-Wigner formula

$$\sigma\left(e^+e^- \rightarrow X\right) = \frac{12\pi M_Z^2}{E_{CM}^2}\left[\frac{\Gamma(Z^0 \rightarrow e^+e^-)\Gamma(Z^0 \rightarrow X)}{(E_{CM}^2 - M_Z^2)^2 + M_Z^2 T_Z^2}\right] \tag{8.32}$$

where $T(Z^0 \rightarrow X)$ is the decay rate of the Z^0 to the observed final state X and Γ_Z is the total decay rate (8.29).

The positions and the widths of the peaks in (8.32) are determined by the Z^0 mass and total width T_Z, their heights are proportional to the products of branching ratios

$$B\left(Z^0 \rightarrow e^+e^-\right)B\left(Z^0 \rightarrow X\right) = \frac{\Gamma(Z^0 \rightarrow e^+e^-)}{T_Z}\frac{\Gamma(Z^0 \rightarrow X)}{T_Z} \tag{8.33}$$

By fitting the observed data, one obtains not only precise values, $M_Z = 91.16 \pm 0.03$ GeV/c^2 and $T_Z = 2.534 \pm 0.027$ GeV for the mass and total decay rate, but also the various decay rates $T(Z^0 \rightarrow X)$.

Analysis of the experiments at LEP and SLC yield the following branching ratios:

$$\begin{aligned} Z^0 \rightarrow e^+ + e^- \quad & 3.366 \pm 0.008\ \% \\ \mu^+ + \mu^- \quad & 3.367 \pm 0.013\ \% \\ \tau^+ + \tau^- \quad & 3.360 \pm 0.015\ \% \\ \nu_e, \mu, \tau + \overline{\nu}, \mu, \tau \quad & 20.01 \pm 0.16\ \% \\ \text{hadrons} \quad & 69.90 \pm 0.15\ \% \end{aligned} \tag{8.34}$$

Thus, the probability for a decay into charged leptons is significantly different from the decay probability into neutrinos. The coupling of the Z^0 boson apparently depends on the electric charge.

In the unified theory, the only Z^0 decays that arise from the basic vertices of Figs. 8.5 and 8.6 are

$$Z^0 \rightarrow \nu_l + \overline{\nu}_l \quad (l = e, \mu, \tau, \ldots)$$

$$Z^0 \rightarrow l^+ + l^- \quad (l = e, \mu, \tau)$$

and

$$Z^0 \rightarrow q + \overline{q} \tag{8.35}$$

corresponding to the diagrams of Figs. 8.19(a), (b) and (c), respectively. In (8.35), the quark pair is not seen directly but fragments into two or more jets of hadrons which are observed in the final state. The decay rate to neutrino pairs obviously cannot be measured directly but can be calculated from the diagram of Fig. 8.19(a).

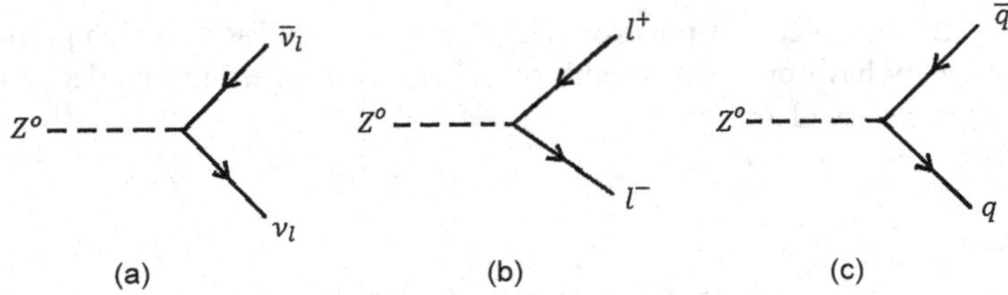

Fig. 8.19 Feynman diagrams for the Z^0 decays 14.194 (**a**), (**b**), (**c**)

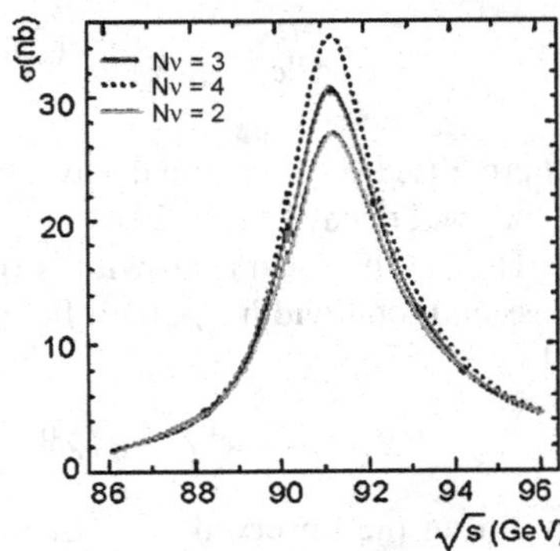

Fig. 8.20 Cross-section of the reaction $e^+e^- \to$ hadrons close to the Z^0 resonance. The data shown are the results of the OPAL experiment at CERN

8.12 Types of Neutrinos

We are familiar with three types of neutrinos, ν_e, ν_μ and ν_τ. The question is, do any more exist? The answer is provided by the analysis of the partial decay modes of Z^0. If N_ν is the number of neutrino types in the sequence $\nu_e, \nu_\mu, \nu_\tau, \ldots$ and assuming that there are only three charged leptons e, μ and τ, the balance equation for the decay rate can be written as

$$\Gamma_Z = \Gamma\left(Z^0 \to \text{hadrons}\right) + 3\Gamma\left(Z^0 \to l^+l^-\right) + N_\nu \Gamma\left(Z^0 \to \nu_l\overline{\nu}_l\right) \tag{8.36}$$

Experimental values are: $\Gamma_Z = 2.534$ GeV, $\Gamma(Z^0 \to \text{hadrons}) = 1.797$ GeV, $\Gamma(Z^0 \to l^+l^-) = 0.084$ GeV and the theoretical value is, $\Gamma(Z^0 \to \nu_l\overline{\nu}_l) = 0.166$ GeV. Using the above values in (8.36), we find $N_\nu = 2.92$ or 3. According to (8.32) the measured width of the resonance yields the total cross-section. The larger types of light leptons exist, the smaller the fraction of the total cross-section that remains for the hadron production. The graphs in Fig. 8.20 are the predicted curves based on the measured width of the resonance, assuming that 2, 3 or 4 massless neutrinos exist.

If a fourth type of light neutrino were to couple to the Z^0 in the same way then the total width would be larger by 166 MeV. On the other hand with only two types of neutrinos the total width would be smaller by 166 MeV We thus conclude that exactly three types of light neutrinos exist, meaning that the total number of generations of quarks and leptons is three and three only. This is an important result

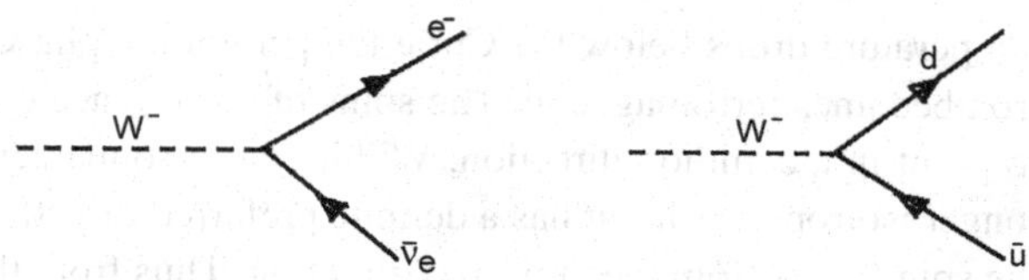

Fig. 8.21 Possible leptonic and hadronic decays for W

both in particle physics and cosmology. Note that the standard model does not have any restriction on the number of neutrinos so long the unification condition (8.21) is satisfied.

8.13 Decay of $W^\pm$ Bosons

The $W^\pm$-boson can decay into lepton or quark pairs (Fig. 8.21)

$$\begin{aligned} W^- &\to l^-\overline{\nu}_l(e^-\overline{\nu}_e, \mu^-\overline{\nu}_\mu, \tau^-\overline{\nu}_\tau) \\ &\to q\overline{q}'(d\overline{u}, s\overline{u}, \ldots) \end{aligned}$$

with corresponding decays for W^+ with particles replaced by antiparticles.

For each of the leptonic modes

$$\Gamma(W \to l\overline{\nu}_l) = \frac{GM_W^2}{G\pi\sqrt{2}} = 250\,\text{MeV}$$

For the quark-antiquark pairs a multiplying factor of 3 arises from the colour and some effects due to non-negligible quark masses may also enter. The leptonic branching ratio for the W's turns out to be

$$\frac{\Gamma(W \to l\nu_l)}{\Gamma(W \to \text{all})} = 8.5\,\%$$

in the case of three generations of quarks and leptons.

8.14 Spontaneous Symmetry Breaking

In the electroweak theory the mediating particles are $W^\pm$, Z^0 and γ. The mixture of states described by the Weinberg rotation should only occur for states with similar energies (masses). However, the photon is massless and the W and Z bosons are very massive. How this is possible is explained by invoking for the concept of spontaneous symmetry breaking, a mechanism which is familiar in phase transitions in other disciplines such as bar magnetism and super-conductivity. Let us consider the analogy of ferromagnetism. Above the curie temperature, iron is paramagnetic and the spins of valence electrons are isotropically distributed. No force is required to alter spin orientations. In so far as the spatial rotations are concerned the fields that carry the magnetic interaction may be considered massless. When the temperature drops below the Curie temperature, a phase transition takes place and iron becomes ferromagnetic. The spins of the valence electrons turn spontaneously to point in a common direction. Within the ferromagnetic material the space is no longer isotropic, rather it has a definite preferred direction. Force is required to turn the spins away from the preferred direction. Thus from the point of view of rotation the carriers of the magnetic interaction now possess a mass. This process is called spontaneous symmetry breaking.

In the case of electroweak interactions Higgs and independently Englert and Brout postulated Higgs fields. At very high temperatures (or energies) the Z^0 and $W^{\pm}$ bosons are massless like the photon. Below the energy of the phase transition, the bosons are endowed with masses by the Higgs fields. The Higgs fields are scalar.

In the theory of electroweak unification there are four Higgs fields, one for each boson. In the course of cooling of the system, three Higgs bosons, the quanta of the Higgs field are absorbed by the Z^0 and by the $W^{\pm}$. This generates their masses. Since the photon remains massless there must still be a free Higgs boson.

The standard Model does not, however, predict the mass of the Higgs boson, a circumstance which makes observation of this particle specially difficult.

The Higgs boson is expected to be electrically neutral, otherwise its coupling to the photon would generate a photon mass. It turns out that the coupling of the Higgs boson to a pair of fermions or bosons is proportional to the mass of the fermions or bosons involved. Therefore, the coupling is largest to the heaviest decay products and such decay channels will dominate. For Higgs mass greater than $2M_{Z^0}$, the ratios of coupling for the decays to Z^0Z^0, W^+W^-, $\tau + \tau^-$, pp^-, $\mu^+\mu^-$, e^+e^-, will be

$$1.00 : 0.88 : 0.02 : 0.01 : 0.001 : 5.5 \times 10^{-6}$$

The Higgs boson will not decay into $\gamma\gamma$ or gluons as they are massless.

The width of the Higgs decay peak will depend on the number of kinematically accessible channels. For $m_H < m_Z$ it will be relatively small. For Higgs mass of the order of 1 TeV/c^2, the Higgs width would be comparable to its mass and Higgs boson would not be detected as a particle or resonance but only through indirect effects.

Very low masses for the Higgs boson in the range few MeV/c^2 to a few GeV/c^2 are excluded as Higgs boson has not been observed in the decays of nuclei, kaons, B or γ particles.

If m_H is less than $2m_\mu$ so that the only accessible channel is decay to e^+e^- then the Higgs boson will have a lifetime of the order of a nanosecond or greater permitting it to travel observable distances in detectors.

Searches to higher mass will continue at the Fermi lab and CERN. In fact the construction of the Large Hadron collider at CERN was the main motivation for the search for Higgs boson.

8.15 The Standard Model of Particle Physics

The theory of quarks and the description of their interactions embodied in chromodynamics as well as the theory of weak and electromagnetic interactions developed over the period 1961–68 by Glashow, Salam and Weinberg has come to be known as the 'Standard Model'.

8.16 Experimental Tests of the Standard Model

8.16.1 Neutral Current

The standard model predicts the existence of neutral currents. In 1973, the existence of ν_μ interactions without a charged lepton in the final state were first demonstrated in a bubble chamber experiment at CERN by Hasert et al. These were called neutral current events and attributed to reactions of the form

$$\nu_\mu N \rightarrow \nu_\mu X$$

$$\overline{\nu} N \rightarrow \overline{\nu} X$$

where X is any hadronic state. In these reactions the identity of the neutrino or antineutrino is unchanged.

8.16.2 Weak-Electromagnetic Interference in, $e^+e^- \rightarrow \mu^+\mu^-$, Reaction

Since muons suffer only weak and electromagnetic interactions they can be identified by their penetration through large amount of material such as iron without producing electromagnetic showers.

The process $e^+e^- \rightarrow \mu^+\mu^-$ has been studied at the PETRA storage ring at DESY using a large detector system. The process may be described by weak and electromagnetic amplitudes A_{em} and A_{wk}. The cross-section is proportional to the square of the sum of the amplitudes

$$\sigma \propto |A_{em} + A_{wk}|^2 = |A_{em}|^2 + |A_{wk}|^2 + 2\,\mathrm{Re}\big(A_{em} A_{wk}^*\big) \tag{8.37}$$

At a total energy, $\sqrt{s} = 34$ GeV, the approximate values of the three terms are 0.1 nb, 1.5×10^{-4} nb and 8×10^{-3} nb, respectively. Thus the interference term is significant. The interference term shows up most readily by the forward-backward asymmetry for the ratio μ^+/μ^- measured by a parameter which is a function of s.

$$A(s) = \frac{N_F - N_B}{N_F + N_B} \tag{8.38}$$

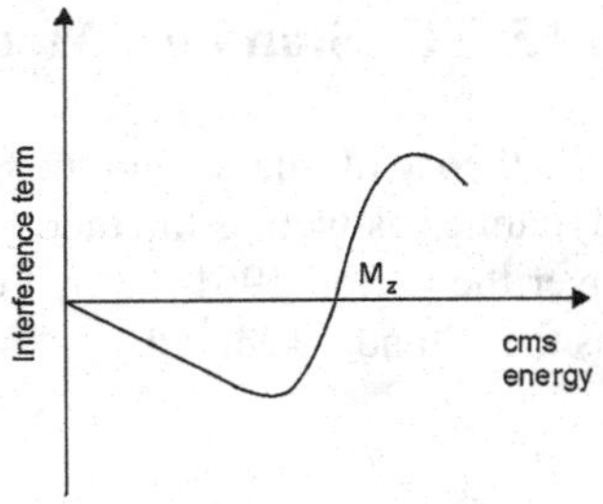

Fig. 8.22 The interference term in $e^+e^- \rightarrow \mu^+\mu^-$ as a function of centre-of-mass energy

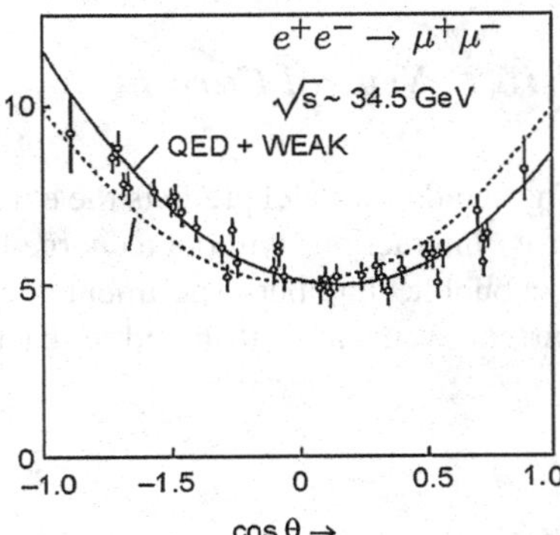

Fig. 8.23 Interference between the weak and electro-magnetic amplitudes

where N_F and N_B are the number of events with a positive lepton in the forward and backward (with respect to the incident positron) hemispheres respectively.

The variation of the interference term as a function of cms energy is shown in Fig. 8.22. The interference term passes through zero near the Z^0 mass and changes the sign. The asymmetries in $e^+e^- \rightarrow \mu^+\mu^-$ from various detectors at PETRA storage ring at the DESY laboratory at Hamburg show evidence for interference between the weak and electromagnetic amplitudes, Fig. 8.23. Similar results are obtained for $e^+e^- \rightarrow \tau^+\tau^-$ [5].

8.16.3 Parity Non-conservation in Inelastic Electron Scattering

Electro-weak theory predicts that the interference between weak and electromagnetic amplitudes should give rise to very small parity violating effects in electromagnetic processes. Such an effect was first observed in the experiment carried out using a polarised electron beam at SLAC to study inelastic scattering off deuterium

$$\vec{e} + d \rightarrow e' + x$$

First an electron beam is rendered longitudinally polarised by means of linearly polarised laser light. As this electron beam passes through the magnets of the beam-

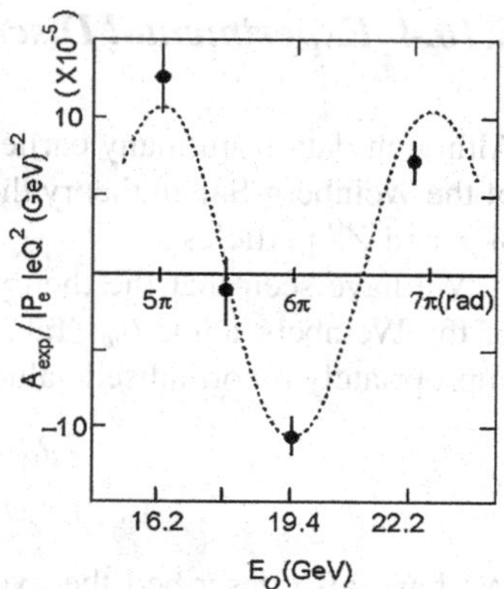

Fig. 8.24 The asymmetry in the SLAC *e-d* scattering experiment (note scale) as a function of the energy E_0 or, equivalently, the angle of precession of the electron spin

transport system the electron spin processes, the angle of precession being given by

$$\theta_{precession} = \frac{E_0(\text{GeV})\pi}{3.237} \text{ radians} \tag{8.39}$$

which is directly proportional to the electron energy E_0. The polarisation of the beam was measured by the use of the Moller scattering (elastic electron-electron scattering) polarimeter. Electrons scattered from the deuterium target were analyzed by a spectrometer consisting of bending and focusing magnets and detected first by a nitrogen-filled Cerenkov counter and then by a lead-glass shower counter. The asymmetries could be measured to a precision of about 10^{-5}. The asymmetry measured was

$$A_{\text{exp}} = \frac{N_+ - N_-}{N_+ + N_-} \tag{8.40}$$

where N_+ and N_- are the (normalised beam intensity) scattered intensities for opposite helicities if for a fully-polarised beam A is the true asymmetry arising from the weak-electromagnetic interference

$$A_{\text{exp}} = |P_e| A \cos\left[\left(E_0(\text{GeV})/3.237\right)\pi\right]$$

where P_e is the actual electron polarisation.

A is expected to be proportional to the momentum-transfer squared Q^2 to the recoiling hadronic system. The behaviour of $(A_{\text{exp}}/Q^2|P_e|)$ as a function of E_0 or equivalently the angle of precession of the electron spin is shown in the data of SLAC *e-d* scattering experiment. Although the maximum asymmetry is tiny ($\sim 1 \times 10^{-4}$) the experiment is sufficiently sensitive to show the $\cos(E_0)$ dependence arising from the precession, Fig. 8.24.

The result, for the first time demonstrated a non-conserving effect in electromagnetic interactions, as expected for Weinberg-Salam theory and was fitted by $\sin^2\theta_w = 0.224 \pm 0.020$ in complete agreement with the value of $\sin^2\theta_w$ obtained in the apparently very different neutrino-scattering experiments.

8.16.4 Experimental Discovery of Heavy Bosons

Although data from many earlier experiments provided strong evidence in support of the Weinberg-Salam theory the crucial test remained the actual observation of the $W^\pm$ and Z^0 particles.

We have seen that the theory predicts the masses of the $W^\pm$ and Z^0 in terms of the Weinberg angle θ_w. Taking into account radiative corrections and using an appropriately renormalised value of θ_w the predicted values are

$$M_{W^\pm} = 83^{+3.0}_{-2.8}\ \text{GeV}/c^2$$
$$M_{Z^0} = 93.8^{+2.5}_{-2.4}\ \text{GeV}/c^2$$

We have also described the experiments at CERN targeted to the observation of heavy bosons (see Sects. 8.9.1, 8.9.2, 8.9.3) in some detail. The experimental values are

$$M_W = 82.1 \pm 1.7\ \text{GeV}/c^2$$
$$M_Z = 93.0 \pm 1.7\ \text{GeV}/c^2$$

Thus there is in complete agreement within the errors between measurement and the theoretical predictions. Furthermore the measurement of the Z^0 width establishes the fact that the number of different kinds of neutrinos is just three.

It is remarkable that the authors of the standard model were awarded the Nobel prize five years before the discovery of heavy bosons (1983), and the actual discovery must have given a sigh of relief to the Nobel prize award committee!

8.17 Grand Unification Theories (GUTS)

In Sect. 8.6 we have seen that Glashow, Weinberg and Salam (GWS) succeeded in unifying the weak and electromagnetic forces at high energies. The theory starts with four massless mediators, as it develops, three of them, $W^\pm$, and Z^0 acquire mass by the Higgs mechanism, while the fourth one, photon, remains massless.

Although a reaction mediated by Z^0 is quite different from one mediated by the γ, they are all manifestations of a single interaction known as electroweak interaction. The relative weakness of the weak forces in attributed to the enormous mass of the intermediate vector bosons, its intrinsic strength is in fact somewhat greater than that of the electromagnetic force.

The next step was to combine the strong force in the form of chromodynamics with the electroweak force. This unification has been called the grand unification or for brevity GUTS. We recall that the strong coupling constant α_s decreases at short distance that is for very high-energy collisions, so does the weak force but at a slower rule. On the other hand, the electromagnetic coupling constant α_e, which is the smallest of the three increases. Is it then possible that all the three constants, converge at a common point at some exceedingly high energy (Fig. 8.25)?

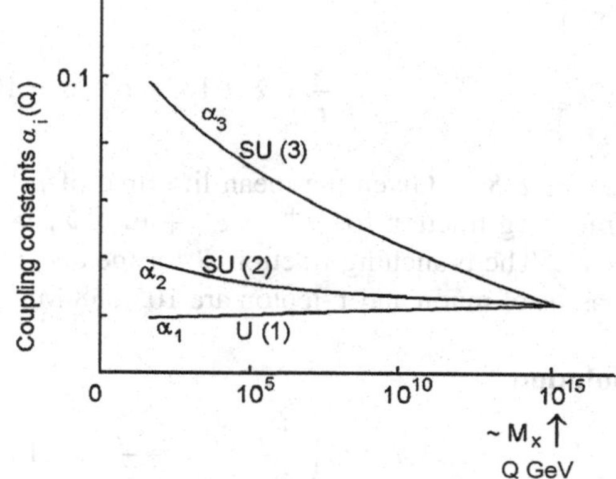

Fig. 8.25 Variation of the running coupling constants for the electroweak and strong interactions as a function of Q up to the (speculative) 'grand unification' mass at $\sim 10^{15}$ GeV

Using the form of the running coupling constants it is possible to estimate the energy at which the grand unification occurs. The predicted energy for the guts turns out to be 10^{15} GeV or the temperature of $\sim 10^{28}$ K. This energy is colossal and can never be achieved is the laboratory when we recall that the maximum energy that has been obtained is only one TeV or 10^3 GeV.

Another prediction of the GUTS is that the proton may be unstable but its life-time is at least 10^{20} times the age of the universe. According to the GUTS the baryon number is not strictly conserved nor the lepton number. The GUTS envisage interactions of the type

$$p \to e^+ + \pi^0 \quad \text{or} \quad p \to \overline{\nu}_\mu + \pi^+$$

in which baryon number and lepton number change. To this end several underground experiments have been conducted but so far the results have been negative. In the framework of the grand unified theories the whole lot of elementary particle physics will be governed by a single force.

The ultimate aim would be to bring the gravitation into the fold of unification. But this ambitious target appears to be fairly distant.

Example 8.1 Estimate the thickness of iron through which a beam of neutrinos with energy 100 GeV must travel if 1 in 2×10^9 of them is to interact. Assume that at high energies, the neutrino-nucleon total cross-section is given approximately by $\sigma_\nu \approx 10^{-38} E_\nu$ cm^2 where E_ν is in GeV. The density of iron is $\rho = 7.9$ g cm^{-3}.

Solution

$$\sigma_\nu = 10^{-38} E_\nu = (10^{-38})(100) = 1.0 \times 10^{-36}\ \text{cm}^2$$

$$\text{Number of iron atoms/cm}^3 = \frac{N_{av}\rho}{A} = \frac{6 \times 10^{23} \times 7.9}{A}$$

$$\text{Number of nucleons/cm}^3,\ n = 6 \times 10^{23} \times 7.9 = 4.74 \times 10^{24}$$

$$\text{Mean free path } \lambda = \frac{1}{n\sigma_\nu} = \frac{1}{4.74 \times 10^{24} \times 10^{-36}} = \frac{10^{12}}{4.74}\ \text{cm} = 2 \times 10^{11}\ \text{cm}$$

given

$$\frac{\lambda}{t} = 2 \times 10^9 \quad \text{or} \quad t = 100\ \text{cm} = 1\ \text{m}.$$

Example 8.2 Given the mean life time of μ^+ meson is 2.197×10^{-6} sec and its branching fraction for $\mu^+ \to e^+ + \nu_e + \overline{\nu}_\mu$ is 100 %. Estimate the mean life time of τ^+. The branching fraction B for the decay $\tau^+ \to e^+ + \nu_e + \overline{\nu}_\tau$ is 17.7 %. The masses of muon and τ-lepton are 105.658 MeV and 1784 MeV respectively.

Solution

$$t_\tau = B \times t_\mu \left(\frac{m_\mu}{m_\tau}\right)^5 = \frac{17.7}{100} \times 2.197 \times 10^{-6} \times \left(\frac{105.658}{1784}\right)^5$$
$$= 0.28 \times 10^{-12}\ \text{sec}.$$

This is to be compared with the measured lifetime of 0.3×10^{-12} sec.

Example 8.3 Which of the following reactions are allowed and which forbidden as under weak interactions?

(a) $\nu_\mu + p \to \mu^+ + n$
(b) $\nu_e + p \to e^- + \pi^+ + p$
(c) $K^+ \to \pi^0 + \mu^+ + \nu_\mu$
(d) $\wedge \to \pi^+ + e^- + \overline{\nu}_e$

Solution Weak interactions conserve, B, Q, L_e, L_μ and L_τ. They need not conserve the quark number S, C and $\tilde{B}$. Thus (a) violates L_μ and (d) violates B, while (b) and (c) conserve all the quantities and are allowed.

Example 8.4 Which of the following decays are allowed in lowest-order weak interaction?

(a) $K^+ \to \pi^+ + \pi^+ + e^- + \overline{\nu}_e$
(b) $\Xi^0 \to p + \pi^- + \pi^0$
(c) $\Omega^- \to \Xi^0 + e^- + \overline{\nu}_e$

Solution

(a) forbidden by the $\Delta S = \Delta Q$ rule
(b) forbidden by the $\Delta S = 0, \pm 1$ rule
(c) allowed and has been observed

Example 8.5 Classify the following semileptonic decays of the $D^+(1869) = cd^-$ meson as Cabibbo-allowed, Cabibbo-suppressed or forbidden in lowest-order weak interactions.

(a) $D^+ \to K^+ + \pi^- + e^+ + \nu_e$

(b) $D^+ \to \pi^+ + \pi^- + e^+ + \nu_e$

Solution

(a) Cabibbo-allowed
(b) Cabibbo-suppressed

Example 8.6 Which of the following processes are allowed in electromagnetic interactions, and which are allowed in weak interactions via the exchange of a single $W^\pm$ or Z^0? (a) $K^+ \to \pi^0 + e^+ + \nu_e$; (b) $\Sigma^0 \to n + \nu_e + \overline{\nu}_e$.

Solution

(a) It is allowed by W exchange and is therefore a weak interaction.
(b) It is forbidden as electromagnetic interaction because $\Delta S \neq 0$, and also forbidden as weak interaction because there is no strangeness-changing weak neutral current.

Example 8.7 One of the bound states of positronium has a lifetime given in natural units by $\tau = 2/m\alpha^5$ where m is the mass of the electron and α is the fine structure constant. Using dimensional arguments introduce the factors $\hbar$ and c and determine τ in seconds.

Solution Set $\tau = \frac{2\hbar^x c^y}{m\alpha^5}$
The dimensional equation becomes

$$[\mathrm{T}][\mathrm{M}] = \left[\mathrm{ML}^2\mathrm{T}^{-1}\right]^x \left[\mathrm{LT}^{-1}\right]^y = [\mathrm{M}]^x [\mathrm{L}]^{2x+y} [\mathrm{T}]^{-x-y}$$

Equating the powers of T, M and L,

$$-x - y = 1$$
$$x = 1$$
$$2x + y = 0$$

We find

$$x = 1, \qquad y = -2$$
$$\tau = \frac{2\hbar c^{-2}}{m\alpha^5} = \frac{2 \times (6.58 \times 10^{-22}\ \mathrm{MeV})(137)^5}{0.511\ \mathrm{MeV}}$$
$$= 1.243 \times 10^{-10}\ \mathrm{sec.}$$

Example 8.8 Classify the following semileptonic decays of the $D^+(1869) = cd^-$ meson as Cabibbo-allowed, Cabibbo-suppressed or forbidden in lowest-order weak interactions. (a) $D^+ \to K^- + \pi^+ + e^+ + \nu_e$; (b) $D^+ \to \pi^+ + \pi^+ + e^- + \overline{\nu}_e$.

Solution For the decay of charmed particles, the selection rules are (i) $\Delta C = \Delta S = \Delta Q = \pm 1$ for Cabibbo-allowed and (ii) $\Delta C = \Delta Q = \pm 1$, $\Delta S = 0$ for

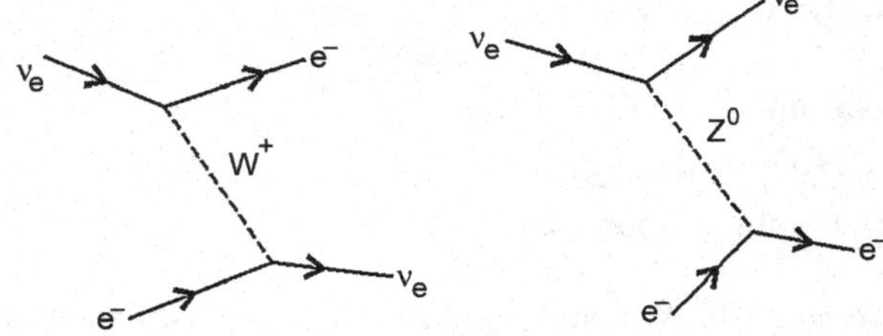

Fig. 8.26 Lowest order Feynman diagrams for the process $\nu_e + e^- \to \nu_e + e^-$

Cabibbo-suppressed decays. Using these rules (a) Cabibbo-allowed; (b) Cabibbo-forbidden.

Example 8.9 Which of the following processes are allowed in electromagnetic interactions, and which are allowed in weak interactions via the exchange of a single $W^\pm$ or Z^0? (a) $K^+ \to \pi^+ + e^+ + e^-$; (b) $\Sigma^0 \to \wedge + e^+ + e^-$.

Solution

(a) It is forbidden as electromagnetic interaction because $\Delta S \neq 0$, and also forbidden as weak interaction because there is no strangeness changing current.
(b) It is an allowed electromagnetic process.

Example 8.10 Show that both charged and neutral current interactions contribute to the process $\nu_e + e^- \to \nu_e + e^-$ by drawing the lowest order Feynman diagrams.

Solution Figure 8.26.

Example 8.11 Which of the following decays are allowed in lowest-order weak interactions?

(a) $K^+ \to \pi^+ + \pi^+ + e^- + \overline{\nu}_e$
(b) $K^- \to \pi^+ + \pi^- + e^- + \overline{\nu}_e$
(c) $\Xi^0 \to \Sigma^- + e^+ + \nu_e$
(d) $\Xi^0 \to P + \pi^- + \pi^0$
(e) $\Omega^- \to \Xi^0 + e^- + \overline{\nu}_e$
(f) $\Omega^- \to \Xi^- + \pi^+ + \pi^-$

Solution All the decays satisfy the conservation laws for weak interactions. But (a) & (c) are forbidden by the $\Delta S = \Delta Q$ rule for semileptonic decays. (d) is forbidden by the $\Delta S = 0, \pm 1$ rule. Decay (b), (e) & (f) are allowed and all have been observed experimentally.

Example 8.12 Which of the decays are allowed and which are forbidden? (a) $K^0 \to \pi^- e^+ \nu_e$; (b) $K^0 \to \pi^+ e^- \overline{\nu_e}$; (c) $\overline{K^0} \to \pi^+ e^- \overline{\nu_e}$; (d) $\overline{K^0} \to \pi^- e^+ \nu_e$.

Solution (a) & (c) are forbidden by the $\Delta S = \Delta Q$ rule. (b) & (d) are forbidden because of $\Delta S = -\Delta Q$ rule.

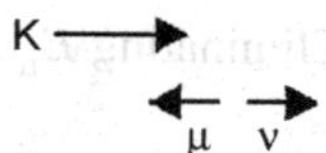

Fig. 8.27 Decay configuration of a neutrino emission

Example 8.13 Find the ratio of the decays $\frac{\Xi^- \to \Lambda + \pi^-}{\Xi^0 \to \Lambda + \pi^0}$.

Solution Introduce the spurion, a hypothetical particle of spin 1/2 and isospin 1/2 and convert the weak decay into a strong interaction.

$$\begin{array}{ccccc} & \text{Sp} + & \Xi^- \to & \Lambda + & \pi^- \\ I & \frac{1}{2} & \frac{1}{2} & 0 & 1 \\ I_3 & -\frac{1}{2} & -\frac{1}{2} & 0 & -1 \end{array}$$

The reaction must occur in pure $I = 1$ state. Looking up the Clebsch-Gordon coefficients for $\frac{1}{2} \times \frac{1}{2}$,

$$\left|\Lambda, \pi^-\right\rangle = a_1 |1, -1 >\rangle$$

For the second decay, we associate again a spurion and consider the reaction

$$\begin{array}{ccccc} & \text{Sp} + & \Xi^0 \to & \Lambda + & \pi^0 \\ I & \frac{1}{2} & \frac{1}{2} & 0 & 1 \\ I_3 & -\frac{1}{2} & -\frac{1}{2} & 0 & 0 \end{array}$$

The initial state is a mixture of $I = 0$ and $I = 1$ states. The final state can exist only in $I = 1$.

$$\left|\Lambda, \pi^0\right\rangle = \frac{1}{\sqrt{2}}\left[a_1|1, 0\rangle + a_2|0, 0\rangle\right]$$

$$\frac{\lambda(\Xi^- \to \Lambda + \pi^-)}{\lambda(\Xi^0 \to \Lambda + \pi^0)} = \frac{1}{(1/\sqrt{2})^2} = 2$$

Example 8.14 Consider the decay, $K^+ \to \mu^+ + \nu_\mu$ in flight. Sketch the decay configuration when the neutrino is emitted with maximum momentum (P_{max}). Calculate P_{max}, given $T_k = 1000$ MeV, $m_k = 494$ MeV/c^2, $m_\mu = 106$ MeV/c^2.

Solution Figure 8.27

$$\Upsilon_\nu = \Upsilon_k \Upsilon_\nu^* \left(1 + \beta_k \beta_\nu^*\right)$$

Energy released, $Q = m_K - m_\mu = 494 - 106 = 388$ MeV.

In the CMS, i.e. in the rest system of kaon, energy and momentum conservation give

$$E_\nu^* + E_\mu^* = m_k = 494$$

$$p_\nu^* = E_\nu^* = p_\mu^* = \sqrt{E_\mu^{*2} - m_\mu^2}$$

Eliminating E_μ^*, we find $E_\nu^* = 235.6$ MeV.

$$E_\nu = E_\nu^*(1 + \beta_k) = 235.6(1 + 0.944) = 458 \text{ MeV}$$

where β_k is found from Υ_k, which in turn is found from $\Upsilon_k = 1 + \frac{T_k}{m_k}$.

As $m_\nu = 0$, $cP_{max}(\nu) = E_{max}(\nu) = 458$ MeV/c.

Example 8.15 How can the neutral K-mesons, K^0 and $\overline{K^0}$ be distinguished?

Solution Because of the CPT theorem, they cannot be distinguished by their decay modes. But they can be identified through their distinct strong interaction process. For example,

$$K^0 + p \rightarrow K^+ + n \quad \text{allowed}$$

$$\left.\begin{array}{l} K^0 + p \nrightarrow \Sigma^+ + \pi^0 \\ K^0 + n \nrightarrow K^- + p \end{array}\right\} \text{ forbidden because of strangeness conservation}$$

On the other hand,

$$\left.\begin{array}{l} \overline{K^0} + p \rightarrow \Sigma^+ + \pi^0 \\ \overline{K^0} + n \rightarrow K^- + p \end{array}\right\} \text{ allowed}$$

But

$$\overline{K^0} + p \nrightarrow K^+ + n \quad \text{forbidden because of strangeness conservation}$$

Example 8.16 Antineutrinos of 2.3 MeV from the fission product decay in a reactor have a total cross-section with protons ($\overline{\nu}_e + p \rightarrow e^+ + n$) of 6×10^{-48} m^2. Calculate the mean free path of these antineutrinos in water. Assume the antineutrinos are able to interact only with the free protons.

Solution The number of free protons per cm^3 of water,

$$n = \frac{\rho N_{av}}{M} \times 2 = \frac{1 \times 6 \times 10^{23} \times 2}{18}$$
$$= 0.667 \times 10^{23}$$

The mean free path is given by

$$\lambda = \frac{1}{n\sigma} = \frac{1}{(0.667 \times 10^{23})(6 \times 10^{-44})}$$
$$= 2.5 \times 10^{20} \text{ cm} = 2.5 \times 10^{15} \text{ km}$$

Example 8.17 Assuming that the entire energy output from the PPI chain escapes from the sun, calculate the flux of neutrinos at the earth. Take the earth-sun distance as $r = 1.5 \times 10^8$ km. Assume the total energy output of the sun $L_\odot = 3.83 \times 10^{26}$ J sec^{-1}, and that each α-particle produced implies the generation of 26.72 MeV.

Solution

$$L_{\odot} = 3.83 \times 10^{26}\ \mathrm{J\,sec^{-1}}$$
$$= \frac{3.83 \times 10^{26}}{1.6 \times 10^{-13}} = 2.39 \times 10^{39}\ \mathrm{MeV\,sec^{-1}}$$

Number of α's produced, $N_\alpha = \frac{2.39 \times 10^{39}}{26.72} = 8.94 \times 10^{37}\ \mathrm{sec^{-1}}$. As the number of neutrinos produced is double the number of α's

$$N_\nu = 2N_\alpha = 1.788 \times 10^{38}\ \mathrm{sec^{-1}}$$

$$\text{Neutrino flux on earth} = \frac{N_\nu}{4\pi r^2}$$
$$= \frac{1.788 \times 10^{38}}{4\pi(1.5 \times 10^{11})^2} = 6.33 \times 10^{14}\ \mathrm{m^{-2}\,sec^{-1}}.$$

8.18 Questions

1. What was the τ–θ problem?
2. What is a Dalitz plot? State its use.
3. In what way is Fermi's theory of beta decay modified following the discovery of parity violation in weak interactions?
4. Write down the form of Hamiltonian for weak interactions.
5. Why do K^0 oscillations occur?
6. What was the need for charm?
7. What is the importance for establishing neutrino oscillations?
8. What is the solar neutrino problem?
9. Distinguish between charge and neutral currents.
10. What are Feynman diagrams?
11. What is Cabibbo angle?
12. Write down the charges of the six quarks.
13. Draw a rough diagram to indicate how the coupling constants behave with the increasing energy.
14. Give the salient features of the standard model of Weinberg Salam and Glashow.
15. How was electroweak theory confirmed?
16. What is the role of the weak mixing angle θ_W?
17. What is Higgs mechanism?
18. Indicate whether the following are electromagnetic, weak or electroweak processes?

    (a) $e^+ + e^- \rightarrow W^+ + W^-$
    (b) $Z^0 \rightarrow \mu^+ + \mu^-$

    [Ans (a) electromagnetic; (b) electroweak]
19. Give two examples of (a) Semileptonic decays (b) Non-leptonic decays.
20. What are the consequences of TCP theorem.

21. What is a Spurion?
22. Give two examples for the decay of $W^{\pm}$ and Z^0 particles.
23. Give the mass and lifetime of $W^{\pm}$ and Z^0 particles.
24. Which interaction is responsible for the decay $K^0 \to \pi^+\pi^-$, and why?
25. In a decay involving the weak interaction, state

    (a) which quark quantum numbers are conserved
    (b) which of these are not conserved and
    (c) whether there are any other quantum numbers which are conserved.

26. Explain whether it is possible for the following processes to occur:

$$\nu_\mu + p \to \mu^+ + n$$
$$\Lambda \to \pi^+ + e^- + \overline{\nu_e}$$

27. Give an outline of the standard model of particle physics, referring to the fundamental particles, the forces between them and the nature of those forces. What elements of the standard model have yet to be confirmed experimentally?
28. What part does Higgs boson play in the standard model?
29. What is meant by the statement that the electromagnetic and weak forces are unified at energies $> 10^{14}$ eV?
30. A number of μ^- decay at rest with their spins all pointing in a given direction. What is the most likely angle between the direction of the highest energy e^- from the decay and the direction in which the μ^- spins were pointing and why?
31. Experiments designed to detect neutrinos are generally very elaborate and constructed deep underground (or under water). Why?
32. How could a ν_μ interaction be distinguished from a ν_e interaction in the experiment?
33. Explain why the decay $\tau^- \to e^-\pi^0$ does not occur.
34. Muon-neutrinos have been detected by the inverse muon decay. Was tau neutrino detected in this way? If not, how was it discovered?
35. Assign the lepton generation subscript $\nu_e, \overline{\nu_e}, \nu_\mu, \overline{\nu_\mu}, \nu_\tau, \overline{\nu_\tau}$.

    (a) $\pi^+ \to \pi^0 e^+ \nu$
    (b) $\mu^+ \to e^+ \nu\nu$
    (c) $\Sigma^- \to n\mu^-\nu$
    (d) $D^0 \to K^- e^+ \nu$
    (e) $\tau^- \to \pi^-\pi^0\nu$
    (f) $\mu^+ \to e^+\nu\nu$

36. A particle decays weakly as follows:

$$X \to \pi^0 + \mu^+$$

Determine the following properties of X:

    (a) Lepton number
    (b) baryon number
    (c) charge

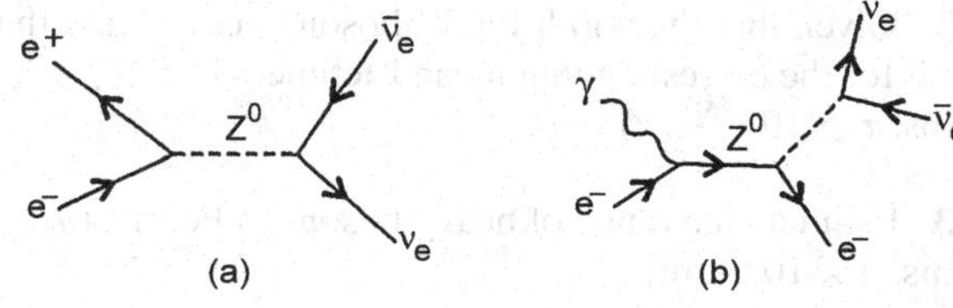

Fig. 8.28 Feynman diagram for the processes (**a**) $e^+ + e^- \to \nu_e + \overline{\nu}_e$. (**b**) $\gamma + e^- \to \nu_e + \overline{\nu}_e + e^-$

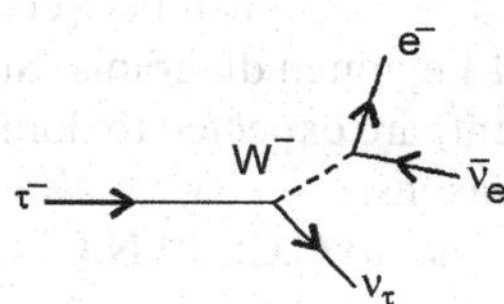

Fig. 8.29 Feynman for the decay $\tau^- \to e^- + \overline{\nu}_e + \nu_\tau$

(d) boson or fermion
(e) lower limit on its mass in MeV/c^2

[Ans. (a) $L_\mu = -1$, $L_e = 0$; (b) $B = 0$; (c) $Q = e$; (d) fermion; (e) $m_X \geq 240.6$ MeV/c^2]

37. The observation of the process, $\overline{\nu}_\mu e^- \to \overline{\nu}_\mu e^-$, signifies the presence of a neutral current interaction. Similarly, why does the process, $\overline{\nu}_e e^- \to \overline{\nu}_e e^-$, not indicate the presence of such an interaction?
[Ans. The second process can be propagated between charged lepton currents]
38. The α-decay of an excited 2^- state in ^{16}O to the ground 0^+ state of ^{12}C is found to have a width $\Gamma_\alpha \approx 1 \times 10^{-10}$ eV. Explain why this decay indicates a parity-violating potential?
[Ans: α would be emitted with $l = 2$ (even parity) while parity change is required for a $2^- \to 0^+$ transition]
39. Draw the Feynman diagram for the processes (a) $e^+ + e^- \to \nu_e + \overline{\nu}_e$; (b) $\gamma + e^- \to \nu_e + \overline{\nu}_e + e^-$.
[Ans: Fig. 8.28]
40. Draw the Feynman for the decay $\tau^- \to e^- + \overline{\nu_e} + \nu_\tau$.
[Ans: Fig. 8.29]
41. The lifetime of tauon is quoted as 3×10^{-13} *sec*, and that of the muon as 2×10^{-6} sec.
Explain what this statement means. Explain why is tauon so much less stable than the muon?

8.19 Problems

8.1 Assuming that the cross-section for a solar neutrino to interact with a nucleus is $\sim 10^{-20} b$, show that if such neutrinos are incident on the earth, then the earth would be practically transparent.

8.2 Given that the width for W boson decay is less than <6.5 MeV, estimate the limit for the corresponding mean lifetime.
[Ans. $\tau \geq 10^{-25}$ sec]

8.3 Estimate the range of heavy bosons in Fermis (m_W or $m_Z \sim 100$ GeV).
[Ans. 2×10^{-3} fm]

8.4 Distinguish between charge-current and neutral-current interactions by means of Feynman diagrams. State which of the two cross-sections (neutral or charge current) are expected to dominate at (a) $Q^2 = 10\,\text{GeV}^2$; (b) $Q^2 = 10,000\,\text{GeV}^2$, giving reasons.
[Ans. (a) C.C; (b) N.C]

8.5 Calculate the centre-of-mass energy of a proton with energy 100 GeV colliding with a liquid hydrogen. Would such a collision have enough energy to create real W particles.
[Ans. 13.8 GeV. No]

8.6 A K^0 meson decays in flight according to $K^0 \to \pi^+\pi^-$. If the π^- is produced at rest, what energy does the π^+ have? [Mass of K^0 is 498 MeV/c^2, that of $\pi^\pm$ is 140 MeV/c^2.]
[Ans. 605 MeV]

8.7 Calculate the maximum energy of positron in the decay of muon at rest according to the scheme: $\mu^+ \to e^+\overline{\nu}_\mu \nu_e$. (Masses of muon and positron are 105.7 and 0.511 MeV/c^2.)
[Ans: 52.35 MeV]

8.8 Λ can decay through nonleptonic modes $\Lambda \to p + \pi^-$ and $\Lambda \to n + \pi^0$. Introduce the spurion and determine the branching ratios for these modes.
[Ans. 2 : 1]

8.9 Analyze the following decays according to their quark content.

(a) $\Xi^- \to \Lambda + \pi^-$
(b) $K^+ \to \pi^+\pi^0$
(c) $\Sigma^+ \to p + \pi^0$
(d) $\Omega^- \to \Lambda + K^-$
(e) $\Lambda_c^+ \to p + \overline{K^0}$
(f) $\Lambda \to p + \pi^-$

8.10 Strangeness-changing decays do not conserve isospin but appear to obey $\Delta I = 1/2$ rule. Use this rule to compute the ratio

$$\frac{K_s \to \pi^+ + \pi^-}{K_s \to \pi^0 + \pi^0}$$

[Ans. 2 : 1]

8.11 Using the rule $|\Delta I| = 1/2$ for weak decay, calculate the branching ratio for the Σ^+ hyperon decay,

$$\frac{\Sigma^+ \to p + \pi^0}{\Sigma^+ \to n + \pi^+}$$

[Ans. 2 : 1]

8.12 Find the kinetic energy of each product particle in the following two-body decays when the original particle decays at rest.

(a) $\pi^+ \to \mu^+ + \nu_\mu$
(b) $K^+ \to \pi^+ + \pi^0$
(c) $\Lambda \to p + \pi^-$
(d) $\Omega^- \to \Lambda + K^-$

[Ans: (a) 4.0 MeV, 32.5 MeV; (b) 108.5 MeV, 110.5 MeV; (c) 4.8 MeV, 32.5 MeV; (d) 19.4 MeV, 43.8 MeV]

8.13 Analyze the following decays for possible violations of the basic conservation laws. In each case state which conservation laws, if any are violated and through which interaction the decay is most likely to proceed, if at all.

(a) $\Lambda^0 \to p + K^-$
(b) $K^+ \to \pi^+ + e^+ + e^-$
(c) $K^+ \to \pi^+ + \pi^+ + \pi^- + \pi^0$
(d) $K^+ \to \pi^+ + \mu^- + e^+$
(e) $\Sigma^+ \to n + e^+ + \nu_e$

8.14 How does the Cabibbo theory explain the branching ratio of the decays,

$$D^+ \to \overline{K^0}\mu^+\nu_\mu \quad \text{and} \quad D^+ \to \pi^0\mu^+\nu_\mu$$

8.15 Draw Feynman diagram for the following decays at the quark level.

(a) $\Lambda \to p + e^- + \overline{\nu_e}$
(b) $K^+ \to \pi^+ + \pi^- + \pi^+$
(c) $\Xi^- \to \Sigma^0 + \pi^-$
(d) $\tau^+ \to \pi^+ + \overline{\nu_\tau}$
(e) $K^0 \to \pi^- + e^+ + \nu_e$
(f) $D^+ \to \overline{K^0} + \mu^+ + \nu_\mu$

[Ans: Fig. 8.30]

8.20 Open Questions

The standard model in spite of possessing a self-contained picture of the fundamental building blocks of matter and of their interactions, has to address a number of questions which are still open.

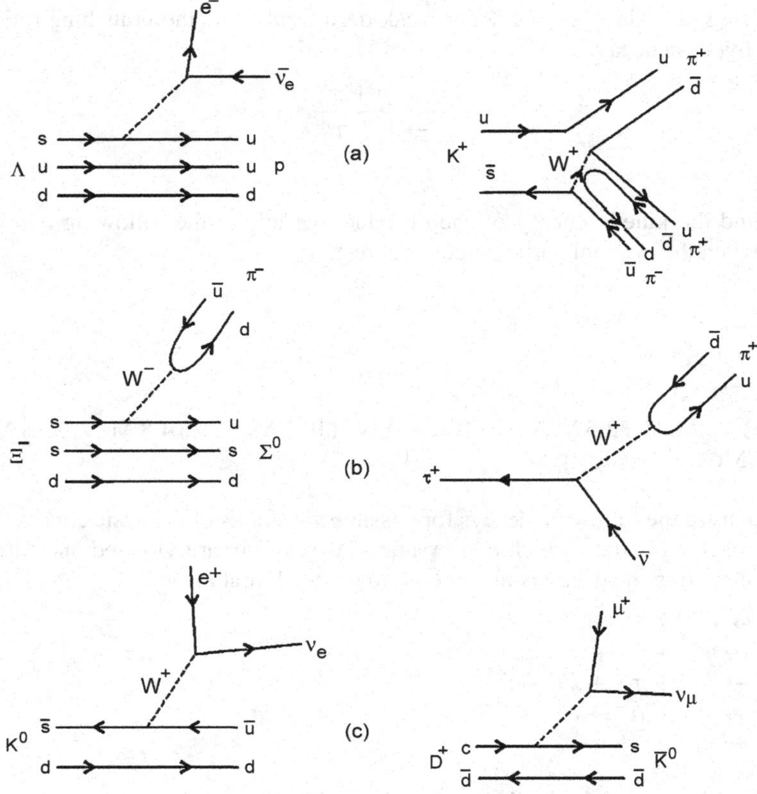

Fig. 8.30 Feynman diagram for different decays at the quark level

What is the nature of dark matter?
Why do exactly three families of leptons and quarks exist?
What is the origin of masses of various fermions and W and Z bosons?
Does the Higgs boson exist?
Why are there just four types of interactions?
What determines the magnitude of the coupling constants of the different interactions?
Are the quarks and leptons really point particles?
What is the real origin of CP violation?
Is it possible to unify the strong and electroweak interaction?
Will it be possible to include gravitation in a complete unification?

Future experiments will be able to answer these questions and as it has happened in the past quite a few surprises might be in store.

References

1. D.H. Perkins Academic (1987)
2. M. Campbell, in *Proceedings of the 1989 International Symposium on Lepton and Photon Interactions at High Energies*, Stanford University (1989), p. 274
3. M. Spiro, in *Proceedings of the 1983 International Symposium on Lepton and Photon Interactions at High Energies*, Cornell (1983), p. 1
4. The Opal Collaboration, Phys. Lett. B **240**, 497 (1990)
5. A. Bohm (1983)

Appendix A

A.1 Relativistic Kinematics

If the space coordinates x, y, z and the time coordinate t are measured in the laboratory frame Σ and the corresponding coordinates x^*, y^*, z^* and t^* in a different frame Σ moving say along the x-axis with velocity βc, then

$$\sum_{i=1}^{4} x_i^2 = \sum_{i=1}^{4} x_i^{*2} = \text{constant} \tag{A.1}$$

where we have written

$$x = x_1, \qquad y = x_2, \qquad z = x_3, \qquad ict = x_4$$

The x_i are the components of a Lorentz invariant four-vector.

One of the most useful four-vectors is that formed by the energy and the three components of the momentum. Thus, $(p_x, p_y, p_z, iE) = (p, iE)$ is a Lorentz-invariant four-vector and

$$\left(\sum E\right)^2 - \left|\sum p\right|^2 = \left(\sum E^*\right)^2 \tag{A.2}$$

since $\sum p^* = 0$ in the center of mass.

A.2 Expressions for Production Threshold

Let a particle of mass m_1, moving with velocity βc with kinetic energy T_1 hit a stationary nucleon of mass m_2 and barely produce particles of mass m_3 and m_4. Then using (A.2) in natural units

$$\begin{aligned}
&(T_1 + m_1 + m_2)^2 - p_1^2 = (m_3 + m_4)^2 \quad \text{or} \\
&(T_1 + m_1 + m_2)^2 - \left(T_1^2 + 2m_1 T_1\right) = (m_3 + m_4)^2 \\
&T_1 = \frac{1}{2m_2}\left[(m_3 + m_4)^2 - (m_1 + m_2)^2\right]
\end{aligned} \tag{A.3}$$

A. Kamal, *Particle Physics*, Graduate Texts in Physics,
DOI 10.1007/978-3-642-38661-9, © Springer-Verlag Berlin Heidelberg 2014

A.3 Expressions for γ_c and γ^*

Let a particle of mass m_1, velocity β and Lorentz factor $\gamma = (1 - \beta^2)^{-1/2}$ be incident on another particle of mass m_2 at rest. Let the CMS have Lorentz factor γ_c. Let m_1 be moving with velocity β^* in the CMS. m_2 will be moving with velocity β_c, in a direction opposite to that of m_1. By definition, the total momentum in the CMS before and after the collision is zero.

$$m_1\gamma^*\beta^* = m_2\gamma_c\beta_c \tag{A.4}$$

Squaring (A.4) and expressing β in terms of γ,

$$m_1^2(\gamma^{*2} - 1) = m_2^2(\gamma_c^2 - 1) \tag{A.5}$$

Using the invariance (A.2)

$$(m_1\gamma + m_2)^2 - m_1^2(\gamma^2 - 1) = (m_1\gamma^* + m_2\gamma_c)^2 \tag{A.6}$$

Combining (A.5) and (A.6) and calling $\nu = \frac{m_2}{m_1}$,

$$\gamma_c = \frac{\gamma + \nu}{\sqrt{1 + 2\gamma\nu + \nu^2}} \tag{A.7}$$

$$\gamma^* = \frac{\gamma + \frac{1}{\nu}}{\sqrt{1 + \frac{2\gamma}{\nu} + \frac{1}{\nu^2}}} \tag{A.8}$$

For the special case, $m_1 = m_2$, as in p–p collision

$$\gamma_c = \gamma^* = \sqrt{\frac{\gamma + 1}{2}} \tag{A.9}$$

For $\gamma \gg 1$,

$$\gamma_c \simeq \sqrt{\frac{\gamma}{2}} \tag{A.10}$$

A.4 Relation Between Lab and CMS Angles

The Lorentz transformations are:

$$E^* = Vc(E - \beta_c p\cos\theta) \tag{A.11}$$

$$p^*\cos\theta^* = \gamma_c(p\cos\theta - \beta_c E) \tag{A.12}$$

$$p^*\sin\theta^* = p\sin\theta \tag{A.13}$$

$$E = \gamma_c(E^* + \beta_c p^*\cos\theta^*) \tag{A.14}$$

$$p\cos\theta = \gamma_c(p^*\cos\theta^* + \beta_c E^*) \tag{A.15}$$

Combining (A.13) and (A.15)

$$\tan\theta = \frac{1}{\gamma_c}\frac{p^*\sin\theta^*}{p^*\cos\theta^* + \beta_c E^*} = \frac{1}{\gamma_c}\frac{\sin\theta^*}{(\cos\theta^* + \frac{\beta_c}{\beta^*})}$$

$$\left(\because \frac{p^*}{E^*} = \beta^*\right) \tag{A.16}$$

The inverse transformation of angles is

$$\tan\theta^* = \frac{\sin\theta}{r_c(\cos\theta - \frac{\beta_c}{\beta})} \tag{A.17}$$

Appendix B

B.1 Composition of Angular Momenta and the Clebsch Gordon Coefficients

The combination of two angular momenta $|I_1 m_1\rangle$ and $|I_2 m_2\rangle$ to form a total angular momentum $|JM\rangle$ must obey the following selection rules:

$$|j_1 - j_2| \leq J \leq j_1 + j_2 \tag{B.1}$$

$$M = m_1 + m_2 \tag{B.2}$$

$$J \geq |M| \tag{B.3}$$

The coupled states may be expanded with the aid of the Clebsch Gordon coefficients (CGC) $(j_1 j_2 m_1 m_2 | JM)$ in the $|j_1 j_2 JM\rangle$ basis:

$$M = m_1 + m_2$$

The weights of various allowed j-values contributing to the two-particle state are given by

$$\varphi_1(j_1 m_1)\varphi_2(j_2 m_2) = \sum_j c_j \psi(j, m) \tag{B.4}$$

The C_j's are known as Clebsch-Gordan coefficients (or Wigner coefficients).

The probability that the combination of two angular momenta $|j_1 m_1\rangle$ and $|j_2 m_2\rangle$ produces a system with total angular momentum $|JM\rangle$ is thus the square of the corresponding CGC's.

Equation (B.4) may also be applied to isospin. We list below three combinations of j_1 and j_2 which have been used in the book. The sign convention in the tables follows that of Condon and Shortley [1].

A. Kamal, *Particle Physics*, Graduate Texts in Physics,
DOI 10.1007/978-3-642-38661-9,

$\frac{1}{2}\times\frac{1}{2}$					
	J	1	1	0	1
	M	+1	0	0	−1
m_1	m_2				
$+\frac{1}{2}$	$+\frac{1}{2}$	1			
$+\frac{1}{2}$	$-\frac{1}{2}$		$\sqrt{\frac{1}{2}}$	$\sqrt{\frac{1}{2}}$	
$-\frac{1}{2}$	$+\frac{1}{2}$		$\sqrt{\frac{1}{2}}$	$-\sqrt{\frac{1}{2}}$	
$-\frac{1}{2}$	$-\frac{1}{2}$				1

$1\times\frac{1}{2}$							
	J	$\frac{3}{2}$	$\frac{3}{2}$	$\frac{1}{2}$	$\frac{3}{2}$	$\frac{1}{2}$	$\frac{3}{2}$
	M	$+\frac{3}{2}$	$+\frac{1}{2}$	$+\frac{1}{2}$	$-\frac{1}{2}$	$-\frac{1}{2}$	$-\frac{3}{2}$
m_1	m_2						
+1	$+\frac{1}{2}$	1					
+1	$-\frac{1}{2}$		$\sqrt{\frac{1}{3}}$	$\sqrt{\frac{2}{3}}$			
0	$+\frac{1}{2}$		$\sqrt{\frac{2}{3}}$	$-\sqrt{\frac{1}{3}}$			
0	$-\frac{1}{2}$				$\sqrt{\frac{2}{3}}$	$\sqrt{\frac{1}{3}}$	1
−1	$+\frac{1}{2}$				$\sqrt{\frac{1}{3}}$	$-\sqrt{\frac{2}{3}}$	
−1	$-\frac{1}{2}$						1

1×1										
	J	2	2	1	2	1	0	2	1	2
	M	+2	+1	+1	0	0	0	−1	−1	−2
m_1	m_2									
+1	+1	1								
+1	0		$\sqrt{\frac{1}{2}}$	$\sqrt{\frac{1}{2}}$						
0	+1		$\sqrt{\frac{1}{2}}$	$-\sqrt{\frac{1}{2}}$						
+1	−1				$\sqrt{\frac{1}{6}}$	$\sqrt{\frac{1}{2}}$	$\sqrt{\frac{1}{3}}$			
0	0				$\sqrt{\frac{2}{3}}$	0	$-\sqrt{\frac{1}{3}}$			
−1	+1				$\sqrt{\frac{1}{6}}$	$-\sqrt{\frac{1}{2}}$	$\sqrt{\frac{1}{3}}$			
0	−1							$\sqrt{\frac{1}{2}}$	$\sqrt{\frac{1}{2}}$	
								$\sqrt{\frac{1}{2}}$	$-\sqrt{\frac{1}{2}}$	
−1	−1									1

References

1. Condon, Shortley (1951)

Appendix C

C.1 Special Functions

C.1.1 Bessel Functions

The Bessel functions are solutions of a differential equation of the form

$$\nabla^2\varphi + B^2\varphi = 0 \tag{C.1}$$

where cylindrical coordinates are used.

Bessel's ordinary differential equation is

$$x^2\frac{d^2y}{dx^2} + x\frac{dy}{dx} + (x^2 - n^2)y = 0 \tag{C.2}$$

where the constant n is the order of the equation. The two solutions, obtained by series methods, are

$$\begin{aligned} &J_n(x) \quad \text{first kind} \\ &Y_n(x) \quad \text{second kind} \end{aligned}$$

$$J_n(x) = \sum_{K=0}^{\infty} \frac{(-1)^K (\frac{x}{2})^{n+2K}}{K!(n+K)!} \tag{C.3}$$

It converges for all finite values of x.

A second solution of Bessels equation is obtained as $J_n(x)\ln x$ plus a new series in x called $Y_n(x)$. Note that $Y_n(x)$ diverges for $x = 0$ and therefore cannot be a solution in physical problems.

The most frequently encountered orders are the zero ($n = 0$) and first ($n = 1$). Figure C.1 shows $J_0(x)$ and $J_1(x)$; Fig. C.2 shows $Y_0(x)$ and $Y_1(x)$.

Bessel functions when n is half an odd integer.

$$J_{1/2}(x) = \sqrt{\frac{2}{\pi x}}\sin x \tag{C.4}$$

A. Kamal, *Particle Physics*, Graduate Texts in Physics,
DOI 10.1007/978-3-642-38661-9, © Springer-Verlag Berlin Heidelberg 2014

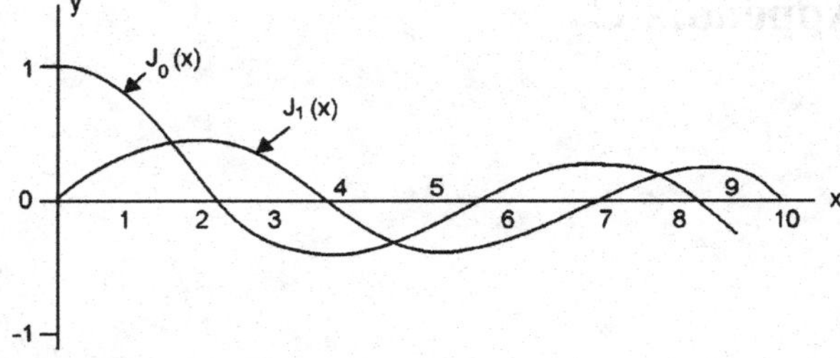

Fig. C.1 First kind of solutions $J_0(x)$ and $J_1(x)$ obtained by series method

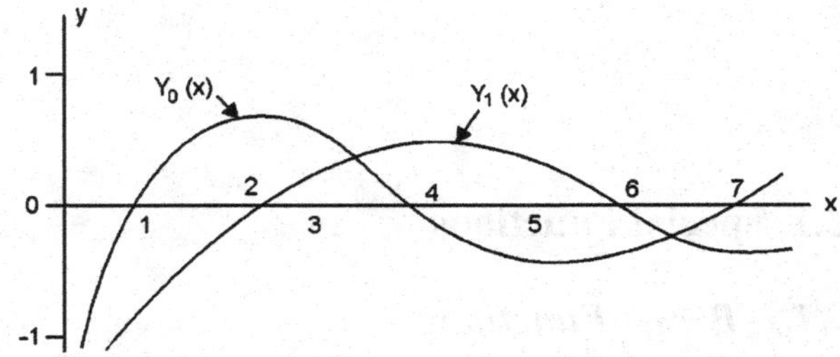

Fig. C.2 Second kind of solutions $Y_0(x)$ and $Y_1(x)$ obtained by series method

$$J_{-1/2}(x) = \sqrt{\frac{2}{\pi x}} \cos x \tag{C.5}$$

$$J_{3/2}(x) = \sqrt{\frac{2}{\pi x}} \left(\frac{\sin x}{x} - \cos x \right) \tag{C.6}$$

$$J_{-3/2}(x) = -\sqrt{\frac{2}{\pi x}} \left(\frac{\cos x}{x} + \sin x \right) \tag{C.7}$$

C.1.2 Asymptotic Expressions

$$J_{l+\frac{1}{2}}(x) = \sqrt{\frac{2}{\pi}} \frac{x^{l+\frac{1}{2}}}{(2l+1)!!} \quad x \ll 1 \tag{C.8}$$

$$J_{l+\frac{1}{2}}(x) = \sqrt{\frac{2}{\pi x}} \sin\left(x - \frac{\pi l}{2} \right) \quad x \gg 1 \tag{C.9}$$

C.1.3 Spherical Bessel Functions

The spherical Bessel functions are the solutions of the equation

$$x^2 \frac{d^2 y}{dx^2} + 2x \frac{dy}{dx} + \left[x^2 - l(l+1) \right] y = 0 \tag{C.10}$$

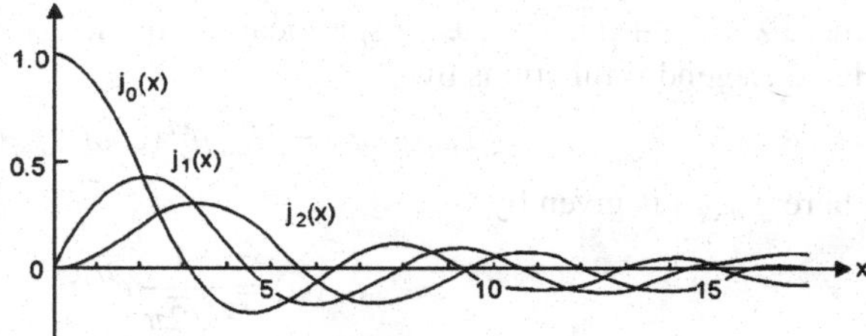

Fig. C.3 Curves for the first three j's of an ordinary Bessel function

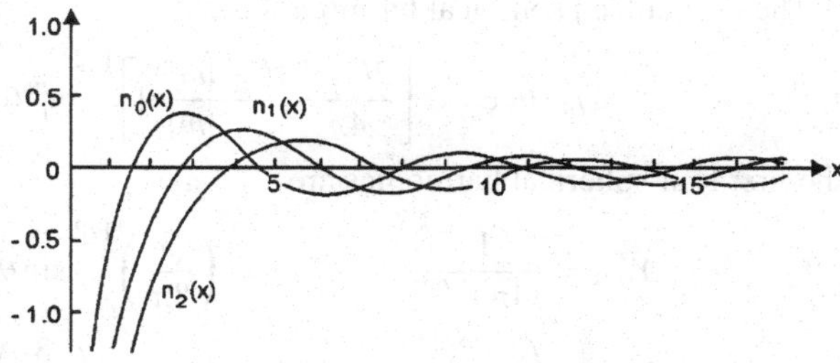

Fig. C.4 Spherical Bessel and Neumann functions for $j = 0.12$

The spherical Bessel function $j_1(x)$ which is regular at $x = 0$ is defined by

$$j_l(x) = \sqrt{\frac{\pi}{2x}}\, j_{l+\frac{1}{2}}(x) \tag{C.11}$$

where J is an ordinary Bessel function of half-odd integral order. Explicit expressions for the first three j's are

$$j_0(x) = \frac{\sin x}{x} \tag{C.12}$$

$$j_1(x) = \frac{\sin x}{x^2} - \frac{\cos x}{x} \tag{C.13}$$

$$j_2(x) = \left(\frac{3}{x^3} - \frac{1}{x}\right)\sin x - \frac{3}{x^2}\cos x \tag{C.14}$$

and the corresponding curves are shown in Fig. C.3.

Spherical Bessel and Neumann functions for different values of J are shown in Fig. C.4.

C.1.4 Spherical Harmonics

The angular part $Y_{lm}(\theta, \varphi)$ of the complete wave function, which is a solution of equation

$$\frac{1}{\sin\theta}\frac{\partial}{\partial\theta}\left(\sin\theta\frac{\partial Y}{\partial\theta}\right) + \frac{1}{\sin^2\theta}\frac{\partial^2 Y}{\partial\varphi^2} + \lambda Y = 0 \tag{C.15}$$

where $\lambda = l(l+1)$, is called a spherical harmonic. $Y_{lm}(\theta, \varphi)$ is related to the associated Legendre functions by

$$Y_{lm}(\theta, \varphi) = N_{lm} p_l^m(\cos\theta) f_m(\varphi) \tag{C.16}$$

where $f_m(\varphi)$ is given by

$$f_m(\varphi) = \frac{1}{\sqrt{2}\pi} e^{im\varphi} \tag{C.17}$$

and N_{em} is the normalization constant.

The normalized spherical harmonics are

$$Y_{lm}(\theta, \varphi) = \left[\frac{2l+1}{4\pi} \frac{(l-|m|)}{(l+|m|)}! \right]^{1/2} p_l^m(\cos\theta) e^{im\varphi} \tag{C.18}$$

The first four spherical harmonics are

$$\begin{aligned} Y_{0,0} &= \frac{1}{(4\pi)^{1/2}}, & Y_{1,1} &= \left(\frac{3}{8\pi}\right)^{1/2} \sin\theta e^{i\varphi} \\ Y_{1,0} &= \left(\frac{3}{4\pi}\right)^{1/2} \cos\theta, & Y_{1,-1} &= \left(\frac{3}{8\pi}\right)^{1/2} \sin\theta e^{-i\varphi} \end{aligned} \tag{C.19}$$

C.1.5 Associated Legendre Functions

Consider the equation

$$\frac{d}{dx}\left[(1-x^2)\frac{d}{dx} p_l^m(x)\right] + \left[l(l+1) - \frac{m^2}{1-x^2}\right] p_l^m(x) = 0 \tag{C.20}$$

$P_l^m(x)$ is called the associated Legendre function.

C.1.6 Ortho-normal Properties

$$\int_{-1}^{1} P_l^m(x) P_{l'}^m(x) dx = \frac{2}{2l+1} \frac{(l+m)!}{(l-m)!} \delta_{l'l} \tag{C.21}$$

where $\delta_{l'l} = 1$ if $l = l'$ and $= 0$ if $l \neq l'$.

Also,

$$P_l^{-m}(x) = (-1)^m \frac{(l-m)!}{(l+m)!} P_l^m(x) \tag{C.22}$$

Parity operation on spherical harmonics.

$$\begin{aligned} P Y_{lm}(\theta, \varphi) &\sim P P_l^m(\cos\theta) e^{im\varphi} \\ &= (-1)^{l+m} P_l^m(\cos\theta)(-1)^m e^{im\varphi} \\ &= (-1)^{2m}(-1)^l Y_{lm}(\theta, \varphi) = (-1)^l Y_{lm}(\theta, \varphi) \end{aligned} \tag{C.23}$$

Appendix D

D.1 Table of Physical Constants

Constant	Symbol	Value
Speed of light	c	2.998×10^{8} m s^{-1}
permeability of vacuum	μ_0	$4\pi \times 10^{-7}$ H m^{-1}
permittivity of vacuum	$\epsilon_0 = 1/\mu_0 c^2$	8.854×10^{-12} F m^{-1}
elementary charge	e	1.602×10^{-19} C
gravitational constant	G	6.673×10^{-11} N m^2 kg^{-2}
Atomic mass unit	u	1.661×10^{-27} kg
Mass		
of electron	m_e	9.109×10^{-31} kg
of proton	m_p	1.673×10^{-27} kg
of neutron	m_n	1.675×10^{-27} kg
energy equivalence of mass		
of electron	$m_e c^2$	0.511 MeV
of proton	$m_p c^2$	938.272 MeV
of neutron	$m_n c^2$	939.566 MeV
Planck constant	h	6.626×10^{-34} J s
$h/2\pi$	$\hbar$	1.055×10^{-34} J s
fine structure constant	$\alpha = e^2/4\pi\epsilon_0\hbar c$	7.297×10^{-3}
Bohr magneton	$\mu_B = e\hbar/2m_e$	9.274×10^{-24} J T^{-1}
nuclear magneton	$\mu_N = eh/2m_p$	5.051×10^{-27} J T^{-1}

A. Kamal, *Particle Physics*, Graduate Texts in Physics,
DOI 10.1007/978-3-642-38661-9,

Constant	Symbol	Value
magnetic moment of		
electron	μ_e	9.285×10^{-24} J T^{-1}
of proton	μ_p	1.411×10^{-26} J T^{-1} $= 2.793\mu_N$
of neutron	μ_n	-0.966×10^{-26} J T^{-1} $= -1.913\mu_N$
electronvolt	eV	1.602×10^{-19} J
	keV $= 10^3$ eV	1.602×10^{-16} J
	MeV $= 10^6$ eV	1.602×10^{-13} J
	GeV $= 10^9$ eV	1.602×10^{-10} J
	TeV $= 10^{12}$ eV	1.602×10^{-7} J
barn	b	10^{-28} m^2
	mb $= 10^{-3}$ b	10^{-31} m^2
	μb $= 10^{-6}$ b	10^{-34} m^2
	nb $= 10^{-9}$ b	10^{-37} m^2

Appendix E

E.1 Dalitz Plots

E.1.1 Three Similar Particles

A convenient way of representing the three-body decay is due to Dalitz [1]. The original Dalitz plot was introduced for the decay

$$K^+ \rightarrow \pi^+ + \pi^+ + \pi^-, \qquad Q = 75 \text{ MeV} = \text{total kinetic energy}.$$

In this case the three particles have equal mass and are nearly non-relativistic. The plot is made of the kinetic energies of the pions along three axes at 120°, as in Fig. E.1.

All the points within the equilateral triangle have the property that the sum of the perpendicular distances from the three sides is a constant equal to the height of the triangle. One therefore draws an equilateral triangle of height Q and for each observed decay a point is marked which has perpendicular distances of the three sides equal to the kinetic energies of the three pions. All decays have points which lie within a region slightly smaller than the inscribed circle (due to relativistic effects), because of momentum conservation. The interval dE_1dE_2 has an area which is independent of its position within the circle and it follows that the density points will be uniform if the matrix element is independent of the energy division.

Due to the symmetry of the decay all the experimental points can be plotted on a half circle. The plot gives information on the spin of K-meson. As it is for a weak decay no information is obtained for the parity of K-meson. On the other hand for a three-body decay by strong interaction, as for the ω, Dalitz plot gives information for both spin and parity, as discussed later in section Phase Space.

A statistical study of many such decays in which one measures the energies of pions and the angles between the trajectories leads to important conclusions.

The three particles emitted may have any momentum and energy permitted by the conservation laws. The largest energy carried by one of the particles, say π^- is

A. Kamal, *Particle Physics*, Graduate Texts in Physics,
DOI 10.1007/978-3-642-38661-9, © Springer-Verlag Berlin Heidelberg 2014

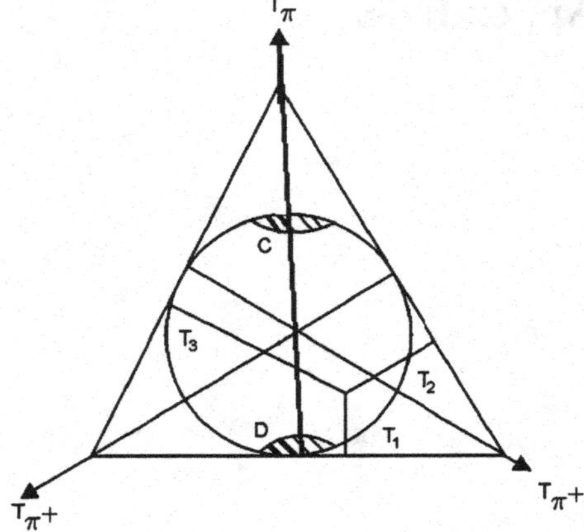

Fig. E.1 Plot of kinetic energies of the pions along three axes at 120°

when the other two pions travel in the opposite direction. It is easy to show that the maximum energy of π^- is $E_0 = (2/3)Q$, where $Q = (m_K - 3m_\pi)c^2$ obviously in the rest frame of K-meson, $E_1 + E_2 + E_3 = m_K C^2$.

The decay rate is governed by the golden rule

$$W = \frac{2\pi}{h}|M_{if}|^2\delta(P_i - P_f)\frac{dN}{dE} \tag{E.1}$$

where the matrix element may be a function of the decay pions, the four-momenta delta function ensures energy and momentum conservation, and the last factor density of final states may be calculated separately.

Kinematic Limits The conservation of energy can be satisfied for all points within the triangle. However, the conservation of momentum imposes severe limit on the allowed region. For the non-relativistic case this limit is the inscribed circle which can be proved as follows.

$$\vec{p}_1 + \vec{p}_2 + \vec{p}_3 = 0 \tag{E.2}$$

Letting $\vec{p}_1$ and $\vec{p}_2$ in the same direction,

$$(p_1 + p_2)^2 = p_3^2 \tag{E.3}$$

But

$$p_1^2 = 2E_1 m, \qquad p_2^2 = 2E_2 m, \qquad p_3^2 = 2E_3 m \tag{E.4}$$

Using (E.4) in (E.3) and simplifying

$$E_1^2 + E_2^2 + E_3^2 - 2(E_1E_2 + E_2E_3 + E_3E_1) = 0 \tag{E.5}$$

which is an inscribed circle

$$x^2 + \left(y - \frac{E_1 + E_2 + E_3}{3}\right)^2 = \left(\frac{E_1 + E_2 + E_3}{3}\right)^2 \tag{E.6}$$

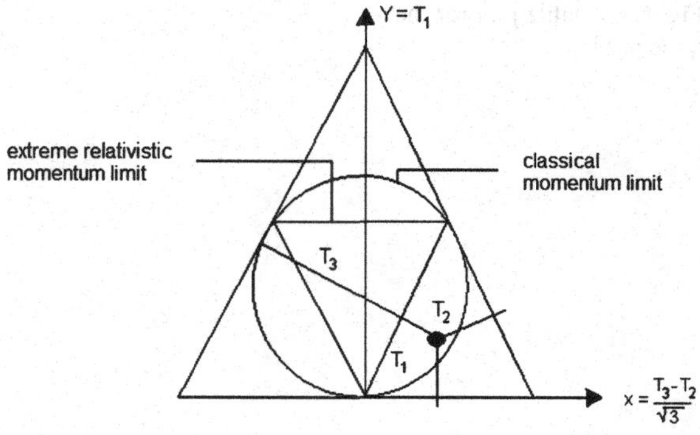

Fig. E.2 Relativistic and non-relativistic limits imposed by conservation of momentum

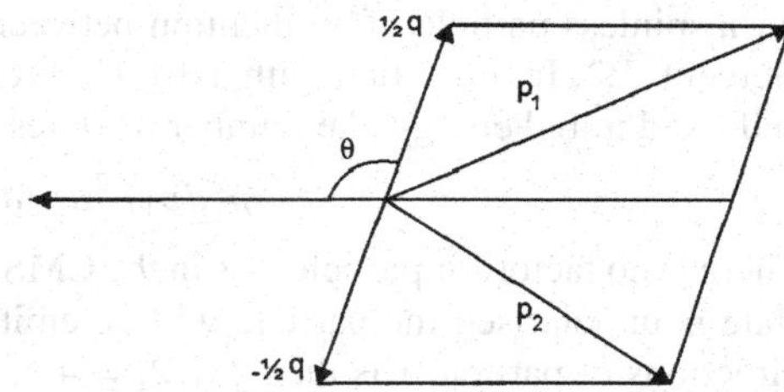

Fig. E.3 Relative momentum of the three particles

if we put $x = \frac{(E_3 - E_2)}{\sqrt{3}}$ and $y = E_1$, Fig. E.2.

In the relativistic case we get a triangle as shown in Fig. E.2. For intermediate cases we get a shape between these extremes.

We may note other features. Let two particles have relative momentum $\vec{q}$, and with respect to which the third particle has momentum $\vec{P}$, then we have the configuration as in Fig. E.3. $\vec{P} \cdot \vec{P}$ and $-\vec{P}$ are then the momenta in the overall CMS. θ is the angle between $\vec{p}$ and $\vec{q}$ so that $\cos\theta = (\vec{p} \cdot \overline{q})/pq$, Fig. E.3.

It may be verified that

$$\begin{aligned} P^2 &\propto PN \\ q^2 &\propto PQ \\ \cos\theta &= \frac{GP}{GH} \end{aligned} \tag{E.7}$$

where P, N, Q, G, H are as in Fig. E.4.

Phase-Space We shall now show that the energy density is proportional to dE_1, dE_2 for all the accessible regions of phase space, E_3 being determined once E_1 and E_2 are given.

The number of quantum states available in phase, per unit normalized volume, is

$$\frac{p^2 dp d\Omega}{\hbar^3}$$

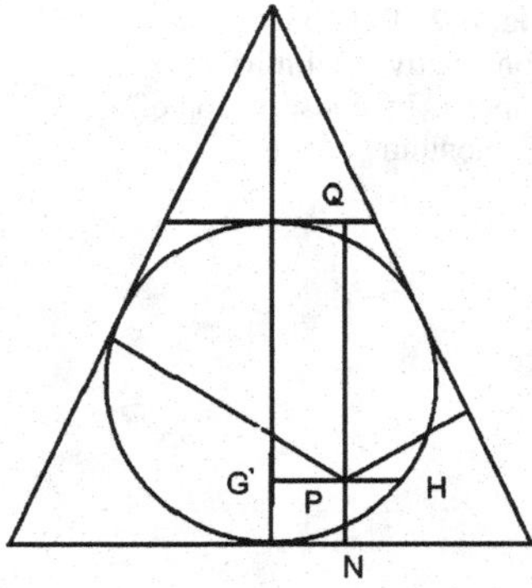

Fig. E.4 Dalitz plot for the three particles

for a spinless particle of momentum between p and $p + dp$ within the solid angle element $d\Omega$. In a reaction with a 3-body break-up into particles labeled 1, 2 and 3, and fixed initial energy, the number of states will be proportional to

$$p_1^2 dp_1 p_2^2 dp_2 d\Omega_1 d\Omega_2$$

There is no factor for particle 3 as in the CMS, $\vec{p_3} = -(\vec{p_1} + \vec{p_2})$ is fixed. If the initial state is unpolarised the particle will be emitted isotropically. The integral over all directions of particle 1 is then $\int d\Omega_1 = 4\pi$, while $d\Omega_2 = 2\pi d(\cos\theta_{12})$, where θ_{12} is the angle between particle 1 and 2. Thus, the number of states in the element $dp_1 dp_2 d(\cos\theta_{12})$ is

$$dN = \text{const}\, p_1^2 dp_1 p_2^2 dp_2 d(\cos\theta_{12})$$

The phase space expression can be made relativistically invariant by including a factor of m/E or $1/E$ for each final state particle, where E in the total energy of the particle. The form is then

$$dN = \text{const}\, \frac{p_1^2 dp_1 p_2^2 dp_2 d(\cos\theta_{12})}{E_1 E_2 E_3}$$

Using the relations

$$E_1^2 = p_1^2 + m_1^2, \qquad E_2^2 = p_2^2 + m_2^2$$
$$E_3^2 = p_3^2 + m_3^2 = p_1^2 + p_2^2 + 2p_1 p_2 \cos\theta + m_3^2$$
$$E_1 dE_1 = p_1 dp_1, \qquad E_2 dE_2 = p_2 dp_2$$
$$(E_3 dE_3)_{p_1, p_2 fixed} = p_1 p_2 d(\cos\theta),$$

one finds

$$dN = \text{const}\, \frac{E_1 dE_1 E_2 dE_2 E_3 dE_3}{E_1 E_2 E_3} = \text{const}\, dE_1 dE_2 dE_3$$

The density of final states is obtained by dividing by dE_f, where $E_f = E_1 + E_2 + E_3$ is the total energy and, for E_1 and E_2 fixed, $dE_f = dE_3$. Thus

$$\rho = \frac{dN}{dE_f} = \text{const}\, dE_1 dE_2 \tag{E.8}$$

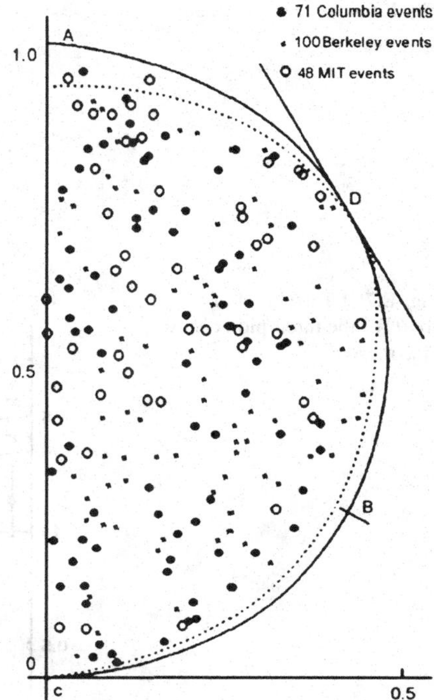

Fig. E.5 Observed distribution of events in $K \to 3\pi$, showing uniform population and thus $J_K = 0$. All events are plotted in one semicircle. The *dashed curve* is the boundary using relativistic kinematics and departs slightly from a circle [2]

when the matrix element M for the interaction is included.

$$\rho = \left|M(E_1, E_2)\right|^2 dE_1 dE_2 \tag{E.9}$$

Thus if one plots the events using E_1 and E_2 as coordinates, a diagram (Dalitz plot) is obtained in witch the accessible region in contained in a closed curve and the points representing the events are uniformly distributed within the curve if the matrix element in constant. Any departure from uniformity is to be attributed to the dependence of the matrix element on the momenta of the pions. Figure E.5 shows the observed distribution of events in $K \to 3\pi$, showing uniform population.

The three pion system may be treated in terms of a dipion, which here consists of two pions of like charge, plus an added third pion. Let the relative orbital angular momentum in the di-pion be l, and that of the third pion relative to the di-pion be L (see Fig. E.6) the parity of the three-pion system is then

$$(-1)^3(-1)^l(-1)^L = -(-1)^L$$

since the dipion must have even l for reasons of Bose symmetry. If $l > 0$ (i.e. $2, 4, \ldots$) the matrix element and Dalitz-plot density should vanish in the region C in Fig. E.1. If $L > 0$ (i.e. $1, 2, 3, \ldots$), the plot should be depleted in the region D, where the negative pion has very small energy relative to the dipion. Since the actual plot (Fig. E.5) is uniform throughout the expected circle neither l nor L can he non-zero hence it is concluded that $J_K = 0$.

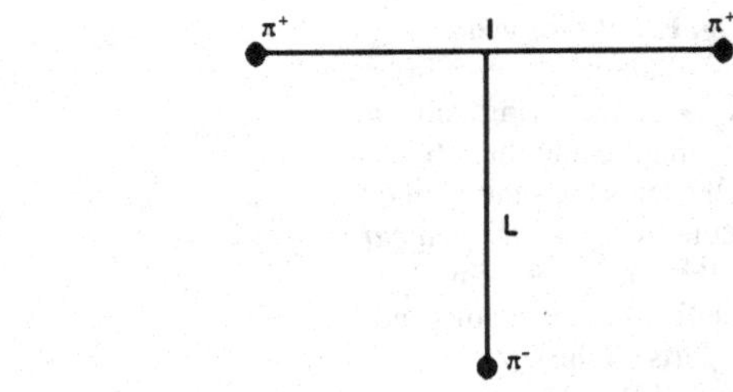

Fig. E.6 Three pion system

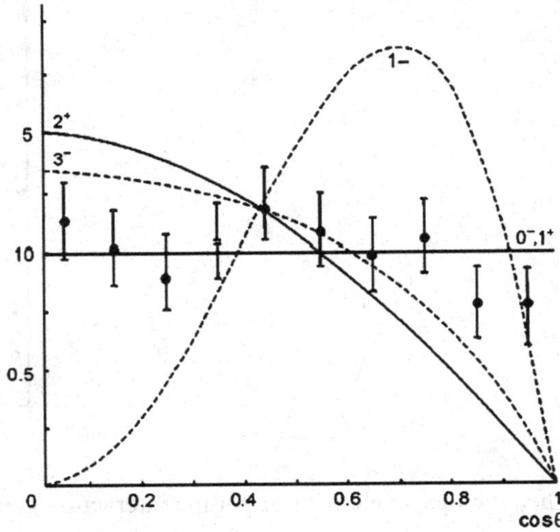

Fig. E.7 Distribution of $\cos\theta$ for the three-pion decay of k-meson

It must be stressed that the information on parity of K-meson cannot be extracted from the above analysis as parity is not conserved in weak decays and nothing can be said about the initial state. The Kaon parity may be determined by the study of hyper-nuclei (see AAK2,4).

The calculated distribution of $\cos\theta$ for the three-pion decay of K-meson is shown in Fig. E.7 for various spin-parity assignments. The approximately isotropic distribution agrees only with $J^\pi = 0^-$ or 1^+ [3].

E.1.2 Dalitz Plots Involving Three Dissimilar Particles

When the decay products have different masses there is no virtue in using the triangular plot. One uses other plots that give similar information (see Sect. 5.1.1, AAK2) for the Dalitz plot for the detection of $\Sigma(1385)$ resonance in the K^-p interactions as in Fig. 5.3, AAK2. Here we shall consider another example for the reaction

$$K^+ + p \to K^0 + p + \pi^+$$

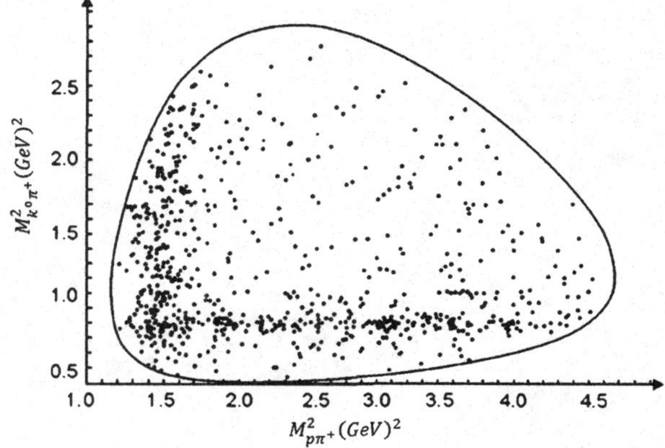

Fig. E.8 A Dalitz plot for the $K^+p \Rightarrow K^0p\pi^+$ events. $pK^+ = 3.0$ GeV/c, 747 events [4]

For each event the invariant mass of the π^+–p system and the K^0–π^+ system is plotted one against the other we obtain the points as in Fig. E.8. In the absence of resonant systems the points would have been uniformly distributed throughout the indicated region. Actually two preferred energies for the two systems, one around 1235 MeV (corresponding to Δ^{++} resonance) and the other corresponding to K^{*+} resonance at 892 MeV are observed.

References

1. Dalitz (1953)
2. Orear et al. (1956)
3. Baldo-Ceolin et al., Nuovo Cimento **6**, 84 (1957)
4. M. Ferrro Luzzi et al., Nuovo Cimento **36**, 1101 (1965)

Appendix F

F.1 Properties of Selected Particles and Resonances

F.1.1 Gauge Bosons ($J = 1$)

Particle	Mass (MeV/c^2)	Width (GeV)	Decay modes and branching fractions
γ	$<3 \times 10^{-33}$	stable	–
W	$(80.49 \pm 0.67) \times 10^3$	<6.5	$e\nu_e$, 10 %; $\mu\nu_\mu$, 12 %; $\tau\nu_\tau$, 10 %
Z	$(91.49 \pm 1.39) \times 10^3$	<5.6	e^+e^-, 4.6 %

F.1.2 Leptons ($j = \frac{1}{2}$)

Particle	Mass (MeV/c^2)	Mean life (s)	Decay modes and branching fractions
ν_e	$<1.7 \times 10^{-5}$	stable	–
e^-	$0.51099906 \pm 0.00000016$	$>2 \times 10^{22}$ years	–
ν_μ	<0.25	stable	–
μ^-	105.65839 ± 0.00006	$(2.19703 \pm 0.00004) \times 10^{-6}$	$e^-\tilde{\nu}_e\nu_\mu$, 100%
ν_τ	<35	–	–
τ^-	1784.2 ± 3.2	$(3.3 \pm 0.4) \times 10^{-3}$	$\mu^-\tilde{\nu}\nu_\tau$, (17.8 ± 0.4) % $e^-\tilde{\nu}\nu_\tau$, (17.5 ± 0.4) % $\pi^-\nu_\tau$, (10.8 ± 0.6) % $\rho^-\nu_\tau$, (22.3 ± 1.1) %

A. Kamal, *Particle Physics*, Graduate Texts in Physics,
DOI 10.1007/978-3-642-38661-9, © Springer-Verlag Berlin Heidelberg 2014

F.1.3 Baryons ($J^P = \frac{1}{2}^+$)

Particle	I	S	Mass (MeV/c^2)	Mean life (s)	Decay modes and branching fractions
P	$\frac{1}{2}$	0	938.28	stable	–
n		0	939.57	899.7 ± 8.9	$pe^-\nu_e$, 100 %
Λ	0	−1	1115.60	2.631×10^{-10}	$p\pi^-$, 64.2 %; $n\pi^0$, 35.8 %; $pe^-\tilde{\nu}_e$, 8.3×10^{-4}
Σ^+	1	−1	1189.37	0.800×10^{-10}	$p\pi^0$, 51.64 %; $n\pi^+$; 48.36 %; $\Lambda e^+\nu_e$, 2×10^{-5}
Σ^0		−1	1192.46	7.4×10^{-20}	$\Lambda\gamma$, ∼100 %
Σ^-		−1	1197.34	1.479×10^{-10}	$n\pi^-$, ∼100 %; $ne^-\tilde{\nu}_e$, 1.02×10^{-3}; $\Lambda e^-\tilde{\nu}_e$, 5.73×10^{-5}
Ξ^0	$\frac{1}{2}$	−2	1314.9	2.90×10^{-10}	$\Lambda\pi^0$, ∼100 %
Ξ^-		−2	1321.32	1.64×10^{-10}	$\Lambda\pi$, ∼100 %; $\Lambda e^-\tilde{\nu}_e$, 5.5×10^{-4}
Charmed baryon Λ_c^+	0	0	2284.9	1.8×10^{-13}	$pK^-\pi^+$, 2.2 %; and others

F.1.4 Baryon Resonances ($J^P = \frac{3}{2}^+$)

Particle	I	S	Mass (MeV/c^2)	Width (MeV)	Decay modes and branching fractions
$\Delta(1232)$	3/2	0	1232.0	115	(N)π, 99.4 %; (N)γ, 0.6 %
$\Sigma^+(1385)$	1	−1	1382.8	36	$\Lambda\pi$, 88 %; $\Sigma\pi$, 12 %
$\Sigma^0(1385)$		−1	1383.7	36	
$\Sigma^-(1385)$		−1	1387.2	39	
$\Xi^0(1530)$	1/2	−2	1531.8	9.1	$\Xi\pi$, 100 %
$\Xi^-(1530)$		−2	1535.0	9.9	
Ω^-	0	−3	1672.4	0.822×10^{-10} sec (lifetime)	ΛK^-, 67.8 %; $\Xi^0\pi^-$, 23.6 %; $\Xi^-\pi^0$, 8.6 %; $\Xi^0 e^-\tilde{\nu}_e$, 5.6×10^{-3}

F.1.5 Mesons ($J^P = 0^-$)

Particle	I	S	Mass (MeV/c^2)	Mean life or width	Decay modes and branching fractions
π^+	1	0	139.57	2.60×10^{-8} sec	$\mu^+\nu_\mu$, ∼100 %; $e^+\nu_e$, 1.23×10^{-4}
π^0		0	134.97	0.84×10^{-16} sec	$\gamma\gamma$, 98.8 %

π^-		0	139.57	2.60×10^{-8} sec	$\mu^-\tilde{\nu}_\mu$, ~100 %; $e^-\tilde{\nu}_e$, 1.23×10^{-4}
η^0	0	0	548.88	1.08 keV 1.237×10^{-8} sec	$\pi^+\pi^0\pi^-$, 23.7 %; $3\pi^0$, 31.90 %; $\gamma\gamma$, 38.9 %
K^+	1/2	+1	493.65	K_s^0 0.8928×10^{-10} sec	$\mu^+\nu_\mu$, 63.5 %; $\pi^+\pi^0$, 21.2 %; $\pi^+\pi^-\pi^-$, 5.6 %; $\pi^0 e^+\nu_e$, 4.8 %
K^0		+1	497.67	K_L^0 5.183×10^{-8} sec	$\pi^+\pi^-$, 68.6 %; $\pi^0\pi^0$, 31.4 %; $\pi^0\pi^0\pi^0$, 21.7 %; $\pi^+\pi^0\pi^-$, 12.4 %; $\pi^\pm\mu^\mp\nu_\mu$, 27.1 %; $\pi^\pm e^\mp\nu_e$, 38.7 %
$\eta'(958)$	0	0	957.57	0.24 MeV	$\eta\pi\pi$, 65.2 %; $\rho^0\gamma$, 30.0 %
$\eta_c(2980)$	0	0	2979.6	10.3 MeV	$\eta'\pi\pi$, 4.1 %; $K\tilde{K}\pi$, 5.5 %
D^+	1/2	0	1869.3	10.7×10^{-13} sec	e^+ any, 19.2 %; $K^-\pi^+\pi^+$, 7.8 %; $K^-\pi^+\pi^+\pi^0$, 3.7 %; $\tilde{K}^0\pi^+\pi^0$, 8.3 %; $\tilde{K}^0\pi^+\pi^+\pi^-$, 7.0 %
D^0		0	1864.5	4.3×10^{-13} sec	$e+$ any, 8 %; K^- any, 43 %; $K^-\pi^+\pi^0$, 12.5 %; $K^-\pi^+\pi^+\pi^-$, 7.9 %; $K^0\pi^+\pi^-$, 5.6 %
D_s^+	0	−1	1969.3	4.4×10^{-13} sec	$\phi\pi^+$; $\phi\pi^+\pi^+\pi^-$
Bottom mesons					
B^+	1/2	0	5277.6	1.4×10^{-12} sec	$\tilde{D}^0\pi^+$, 0.5 %; $D^{*-}(2010)\pi^+\pi^+\pi^0$, 4.3 %
B^0		0	5275.2		ψ any, 1 %; $\tilde{D}^0\pi^+\pi^-$, <3.9 %; $D^{*-}(2010)\pi^+$, 0.3 %

F.1.6 Mesons ($J^P = 1^-$)

Particle	I	S	Mass (MeV/c^2)	Width (MeV)	Decay modes and branching fractions
$\rho(770)$	1	0	770.0	153.0	$\pi\pi$, ~100 %; e^+e^-, 0.0044 %
$\omega(783)$	0	0	782.0	8.5	$\pi^+\pi^-\pi^-$, 89.3 %; $\pi^0\gamma$, 8.0 %; e^+e^-, 0.0071 %
$\phi(1020)$	0	0	1019.4	4.41	K^+K^-, 49.5 %; K_s^0 and K_L^0, 34.4 %; $\pi^+\pi^-\pi^0$, 1.9 %; e^+e^-, 0.031 %
$K^{*+}(892)$	1/2	+1	892.1	51.3	$K\pi$, ~100 %
$K^{*0}(892)$		+1			
$J/\psi(3097)$	0	0	3096.9	0.068	e^+e^-, 6.9 %; hadrons + radiative, 86.2 %; $\mu^+\mu^-$, 6.9 %
$\gamma(9460)$	0	0	9460.3	0.052	e^+e^-, 2.5 %; $\tau^+\tau^-$, 3.0 %; $\mu^+\mu^-$, 2.6 %
$D^{*+}(2010)$	1/2	0	2010.1	<2.0	$D^0\pi^+$, 49 %; $D^+\pi^0$, 34 %
$D^{*0}(2010)$		0	2007.1	<5	$D^0\pi^0$, 52 %; $D^0\gamma$, 48 %

Appendix G

G.1 Introductory Astrophysics

Particle astrophysics or Astro particle physics is a new branch of particle physics that studies elementary particles of astronomical origin, and their relation to astrophysics and cosmology.

The basic questions particle astrophysics attempts to answer are:

- Does the proton decay?
- What is the role of neutrinos in cosmic revolution?
- What is the origin of cosmic rays?
- What is the nature of gravitation?
- Can we detect gravitational waves?
- What is the ultimate faith of a black hole?
- What is the nature of dark matter?

The most active topics in astro particle physics being pursued are:

- Gamma-ray astronomy
- Neutrinos and neutrino astronomy
- Magnetic monopoles
- Axions

Interaction	Relative strength	Exchange particle
Strong	1	Given, spin 1, $m = 0$, quarks and binding nuclei
Electromagnetic	10^{-2}	photon between charged particles, spin 1 zero mass
Weak	10^{-7}	Weak bosons, $W^{\pm}$, Z^0, spin 1, $m_w = 80$ GeV and $m_{z^0} = 91$ GeV

For W and Z bosons, effective range of interaction $r_0 = \hbar/M_c \approx 0.0025$ fm.
Photon carries two types of charges $+$ and $-$.

A. Kamal, *Particle Physics*, Graduate Texts in Physics,
DOI 10.1007/978-3-642-38661-9, © Springer-Verlag Berlin Heidelberg 2014

Coupling of photon is

$$e^2 = 4\pi\alpha\hbar c$$

where e is the electric charge, $\hbar = c = 1$ and $\alpha = 1/137$.

Strong Colour Interactions Coupling of the quark to gluon $\rightarrow P_s^2 = 4\pi\alpha_s$, $\alpha_s \approx 1$.

Number of strong charges = 6 (R, B, G, +3 anti colour)

Photon does not carry charge but gluon does, one colour and one anti colour.

Weak Interactions

$$W^{\pm},\ Z^0$$

$W^{\pm}$ is called the *charged current weak interaction* and Z^0 is the *neutral current weak interaction*.

Electroweak Interaction Two aspects of the electromagnetic and weak interactions.

$$M_{WZ} \sim \frac{e}{\sqrt{G_F}} = \sqrt{\frac{4\pi\alpha}{G_F}} \sim 100\ \text{GeV}$$

$$G_F = 1.17 \times 10^{-5}\ \text{GeV}^{-2}$$

Electroweak means $\rightarrow$ couplings of W^- and Z bosons to the fermions is same as that of photon. $g_W = e$.

Gravitational Interactions Dimensionless

$$\frac{GM^2}{4\pi\hbar c} = 5.3 \times 10^{-40}$$

$$\left(Mc^2 = 1\ \text{GeV}\right)$$

Compare with

$$\frac{e^2}{4\pi\hbar c} = \frac{1}{137.036}$$

The Quark-Gluon Plasma (QGP) Quarks are not to be found as free particle but as three quark and antiquarks bond states, called hadrons. The quark confinement is a low energy phase but at sufficiently high temperature and density, quarks and gluons are expected to undergo a phase transition and transform into a plasma analogous to the plasma of positive ions and electrons at high temperatures when electrons and positive ions form a plasma. If such a QGP is reproduced at high level density and temperature, the state of affairs in the very early stages (the first 25 μsec) of the big bang would have been realized before the temperature fell as the expansion took place and QGP froze into hadrons.

To this end heavy ion colliders mainly at the RHIC heavy ion collider at the Brookhaven national laboratory, and the large Hadron collider at CERN, have operated (1980–1990) by smashing relativistic lead ions against each other, gold-ions against each other. The critical quantity is the energy density of the nuclear matter during the very brief (10^{-23} sec) period of the collision. In lead-lead collisions at 0.16 GeV per nucleon in each of the colliding beams, a threefold enhancement has been observed in the frequency of strange particles and antiparticles (from creation of $s\bar{s}$ pairs) as compared with proton-lead collisions at a similar energy per nucleon.

Although the results have yet to be independently verified as of February 2010, claims have been made for the creation of QGP at approximate temperature of 4 trillion degrees Celsius.

New Particles Standard Model of particle physics explains accurately in detail all experiments, at accelerators and reactors, but does not describe building blocks on large scales ($\sim$5 % of energy density only explained). Study of large scale cosmic structures, galaxies, galaxies clusters and superstructures, bulk of matter in universe is invisible (non-luminous) dark matter.

Nature and origin of dark matter so far unknown, possible candidates like fermions and bosons could be super symmetric particles. So far there is no direct experimental evidence for individual dark matter.

Dark energy exceeds any other form, following Big Bang expansion had slowed down because of pull of gravity, but now accelerating because of repulsion of dark energy.

There is no experimental evidence for grand unification of electro-weak with gravitation.

Quarks confinement is expected to disappear at high temperature ($kT >$ 0.3 GeV) and hadrons would undergo phase transition to a quark-gluon plasma.

Hubble Expansion Milky way contains $\sim 10^{11}$ stars.

Among the various types of galaxies observed, the most common are elliptical and spiral. Milkyway is an elliptical one.

Negative gravitational energy GM^2/R and the mass energy Mc^2 of the universe are about equal being $\cong 10^{70}$ J, so that total energy is zero.

In 1929, Hubble used a 100 inch Mount Wilson telescope to observe the Red shift of the spectral lines from various galaxies, due to Doppler effect. Shift depending on the apparent brightness of galaxy and hence on distance.

He also measured the recession velocity v, of a galaxy.

$$\lambda' = \lambda\sqrt{\frac{1+\beta}{1-\beta}} = \lambda(1+z)$$

where $\beta = \frac{v}{c}$ and red shift $z = \Delta\lambda/\lambda$.

Linear relationship between v and distance r.

$$v = H_0 r$$

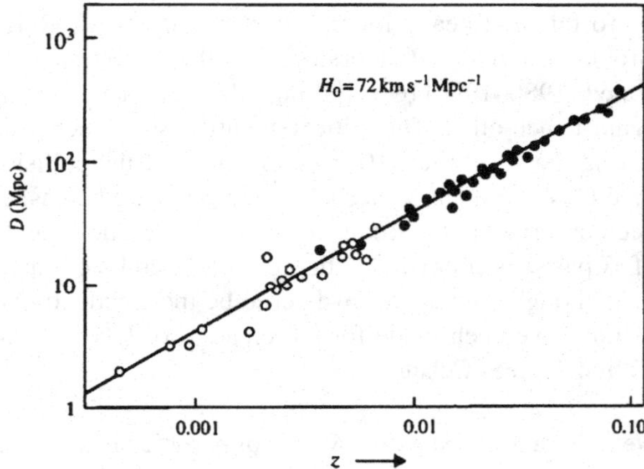

Fig. G.1 Log-log plot of distance versus red shift

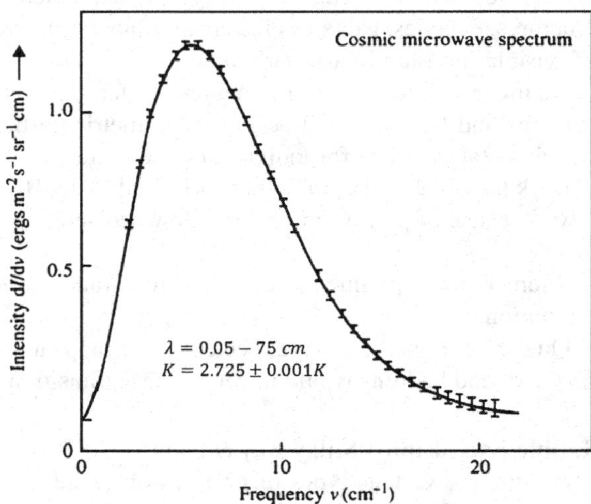

Fig. G.2 Cosmic microwave spectrum

H = Hubble constant. $H_0 = 70 \pm 10\ \mathrm{km\,s^{-1}\,Mpc^{-1}}$. 1 Mpc = Mega parsec = 3.09×10^{19} km.

If expansion is accelerating or decelerating H will be a function of time.

Subscript 0 to $H \rightarrow$ value measured today.

Interpretation of red shift applicable only for small red shifts $z < 0.003$, for nearby galaxies ($z = v/c$). Empirical relation has linear dependence of the red-shift on the distance of galaxy, as shown in Fig. G.1. Distance is estimated from apparent brightness or luminosity.

Cosmic Microwave Radiation Spectrum Cosmic microwave radiation was discovered when scientists were searching for sources of radio waves and landed with a background of microwaves that were isotropic. Figure G.2 shows the spectral dis-

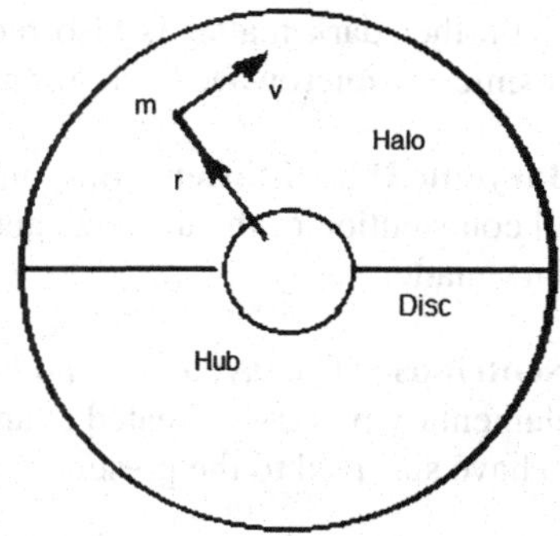

Fig. G.3 An end-on view of a spiral galaxy, consisting of a central hub, a disc, and a possible halo of dark matter

tribution of radiation as recorded by the COBE satellite and compared with Max Planck's black body radiation distribution.

The origins of cosmic microwave radiation traces back to Big Bang, when a photon fireball was cooled by expansion to a few degrees K. This showcases that in early history, universe radiation dominated and now it is matter dominated.

Note: For unification the three running constants are not meeting at the same point.

Dark Matter Indirect evidence: Missing mass from spiral galaxies (non-luminous matter).

A spiral galaxy such as our own has most of luminous material concentrated in central hub + thin disc, shown in Fig. G.3.

Let m be the mass at a distance r from galactic centre moving with tangential velocity v.

Equating gravitational and centrifugal forces

$$\frac{mv^2}{r} = \frac{mM(<r)G}{r^2}$$

where $M(<r)$ is the mass inside r.

For a star inside hub,

$$M < r \propto r^3$$

$$\therefore \quad V \propto r$$

For outside hub, $M \sim$ constant

$$\therefore \quad V \propto r^{-1/2}$$

Therefore, velocity should increase at small r and decrease at large r. However for many spiral galaxies the rotation curves are quite flat at large r values, leading to suggestion bulk of galactic mass (80–90 %) is in form of dark matter in a halo.

Also, the mismatch between the motional energy and gravitational energy requires presence of dark matter.

Effects of galaxy clusters can also be estimated directly by their effects on the images of more distant quasars.

Further dark matter is also required to account for the level of fluctuations of cosmic ray microwave background.

Baryonic Dark Matter Baryonic matter makes only a small contribution to overall contribution of the universe and a less than 25 % to the estimated total density of dark matter.

Neutrinos The favoured hypothesis is that non-baryonic dark matter is made up of elementary particles, created at an early hot stage of the universe and stable enough to have survived to the present day.

$$\gamma \leftrightarrow e^+ + e^- \leftrightarrow \nu_i + \overline{\nu}_i$$

where $i = e, \mu, \tau$.

The total energy density would be equal to the critical density if the sum of the masses of three flavours has the value

$$\sum_{e,\mu,\tau} m_\nu c^2 = 47 \text{ eV} \tag{G.1}$$

Thus relic neutrinos with mass range could make significant contribution to dark matter.

However, evidence from neutrino oscillations suggests very much smaller masses then indicated by C.

Axions The axion is a very light pseudoscalar particle (spin-parity 0^-) postulated in connection with the absence of CP violation in strong interactions quantum chromodynamics (QCD), and these would be T-violating or CP violating, as they are the weak interactions. There is no laboratory evidence for such a particle but it is a possible candidate for dark matter. The characteristics of axions—the mass m_a and the vanishingly small coupling—depend on just one parameter, which is the unknown scale of the symmetry-breaking interaction.

WIMPS The most popular hypothesis, currently, for dark matter particles (WIMPs), moving with non-relativistic velocities at the time of freeze-out and thus constituting cold dark matter.

For several reasons it is considered that super symmetric particles could be the most likely WIMP candidates. So far, there is no laboratory experiment to prove their existence.

Cosmic Rays Cosmic rays actually consist of high energy particles incident on the earth from outer space plus the secondary particles which they produce as they traverse the atmosphere.

Composition

Charged Primary Particles Protons (86 %), α-particles (11 %), nuclei of heavier elements up to uranium (1 %), and electrons (2 %).

Neutral Particles γ-rays, neutrinos and antineutrinos. Some of them can be identified as coming from point sources as γ-rays from crab nebula and active galactic nuclei, neutrinos from the sun and from supernovae.

Chemical Composition It bears a remarkable resemblance to the solar system abundances, except the big differences in those of Li, Be and B. Their comparative abundance in cosmic rays is due to spallation of carbon and oxygen nuclei as they traverse the interstellar hydrogen.

The Energy Density This is about 1 eV $\text{cm}^{-}1$ in deep space outside the influence of solar system magnetic fields, comparable with energy density in star light of 0.6 eV cm^{-3}, or cosmic microwave background radiation of 0.26 eV, or galactic magnetic of 0.25 eV cm^{-3}.

Energy Spectrum The energy spectrum of cosmic rays, above a few GeV, upto the so-called "knee" at 10^{14} eV (100 TeV) follows a simple power law

$$N(E)dE = \text{const}\, E^{-2.7} dE \tag{G.2}$$

Above the "knee" the spectrum becomes steeper with an index of about -3.0 before apparently off again above 10^{18} eV.

Acceleration of Cosmic Rays Many years ago the mechanism for the acceleration of cosmic rays to such colossal energies as 10^{20} eV and the form of energy spectrum were considered. The shock-wave acceleration from supernovae shells appears capable of accounting for the energies of cosmic ray nuclei of charge $Z|e|$ upto about 100Z TeV (10^{14} ZeV), but not beyond. There are many possible sources of shocks, but Type II supernovae shells seem to be good candidates, with shock velocities of order $10^7\ \text{m}\,\text{s}^{-1}$. A Type II supernova typically ejects a shell of material of about $10M_O$ (2×10^{31} kg), with velocity of order $10^7\ \text{m}\,\text{s}^{-1}$ into the inter stellar medium approximately once per century in our galaxy. Acceleration could occur due to shock fronts. A relativistic particle travelling in the positive x-direction, which traverses a shock-front $-u_1$ in the negative x-direction. Suppose that the particle is backscattered by the field in the ionized gas behind the front. Thus, the particle travels back across the shock-front, to be scattered by magnetized cloud, upstream of the front. If these scatter the particle backwards again (i.e. in the direction of positive x) the particle can re-cross the front and repeat the cycle of acceleration once more. Application of the Lorentz transformations shows that the fractional energy gain is of the order of the shock-front velocity.

Geomagnetic Effects East-West effect, i.e.; that at all latitudes, more (positively charged particles) arrive from the west than from the east, because of the lower momentum cut-off. The effect arises because all positively charged particles are deflected in a clockwise spiral, as viewed from above the N-pole. The detailed theory is originally due to Stormer. In particular in NW Europe, with $\lambda \sim 50°$N the vertical cut-off momentum would be 1.1 GeV c^{-1}. At the magnetic equator the vertical cut-off is 14.9 GeV c^{-1}, from the eastern horizon it is 59.6 GeV c^{-1} while that for particles from the western horizon it is only 10.2 GeV c^{-1}.

Secondary Cosmic Radiation Nuclear interaction mean free path (MFP) is $\sim$100 gm cm^{-2} for a proton compared with a total atmospheric depth of $X =$ 1030 gm cm^{-2}, the pions will be created mostly in the stratosphere Nuclear absorption will only become important for energies of order 100 GeV or more and at energies practically all charged pions decay in flight rather than interact.

At high (TeV) energies, on the contrary, majority of the pions undergo nuclear interaction before they have a chance to decay.

Muons produced from the decay of charged pions will decay in the flight ($E_\mu \leq$ 1 GeV). However, muons of energy 3 GeV has $\sim$20 km decay length and even with ionization energy loss of 2 MeV gm^{-1} cm^2 of air traversed are able to reach the sea level and go underground. Muons are said to be the **hard components** of cosmic radiation.

Neutral pions undergo electromagnetic decay $\pi^0 \to 2\gamma$ with an extremely short lifetime of 8×10^{-17} sec. This develops electron-photon cascades mostly in the high atmosphere. The cascade is easily absorbed. This constitutes the soft component of cosmic rays.

Quasars (Quasi-Stellar Radio Sources) Brightest optical and radio sources in the sky, far exceeding total light output from their host galaxies. Quasars invariably have large red shifts (upto $z \approx 5$) and correspond to most distant events in the development of the universe—billion of years ago when galaxies were formed.

Quasars are believed to be associated with very massive black holes of typically 10^6–$10^8 M_\odot$.

Pulsars Although the early theory of neutron stars was developed shortly after Chadwick discovered the neutron in 1932, major experiments had to wait the discovery of Pulsars by Hewish et al. [1]. Pulsars are rapidly rotating neutron stars which emit radiation at short and extremely regular intervals much like a rotating lighthouse beam which crosses the line of sight of an observer with regular frequency. Over 1000 pulsars are known with rotational periods ranging from 1.5 ms to 8.5 s. Only about 1 % of pulsars can be associated with past supernova remnant, since over millions of years the neutron stars have drifted from the remnant nebula. For a few young pulsars like that in the Crab the nebula is still associated with most famous example of a pulsar which has a period of 33 ms, and is remnant of the AD 1054 supernova recorded by the Chinese. Besides the radio pulsars, some 200 X-ray pulsars are known.

Stellar Stability At high densities as in stellar cores at an advanced stage, a new form of pressure called electronic degeneracy pressure becomes important apart from gas pressure and radiative pressure. At absolute zero temperature the gas falls into quantum states of the lowest possible energy and the gas is said to be degenerate. Because of Pauli's principle each quantum state can be occupied by one electron only. If the degeneracy pressure wins over the inward gravitational pressure to prevent contraction then the maximum mass of a stellar core which is stable against collapse is given by

$$M_{ch} = 1.4 M_{\odot} \tag{G.3}$$

The quantity M_{ch} is called the Chandrasekhar mass after the physicist who first discussed the stability of white dwarfs and obtained the above limit in 1931.

White Dwarf Stars Stars of relatively low mass such as our own after passing through the hydrogen- and helium-burning phases will form cores of carbon and oxygen. Higher temperature of the core enables helium to be burned in a spherical shell surrounding the core and the stellar envelope will expand by a huge amount and eventually escape to form a planetary nebula surrounding the star. The star providing its mass is less than the Chandrasekhar mass and is saved from a catastrophic collapse because of the electron degeneracy pressure in the core. Such a star bereft of its envelope and slowly cooling off is known as a **white dwarf**. All main sequence stars (sun, Sirius and alpha Centauri A and B are all main sequence stars) will end up eventually as white dwarfs. However these stars are limited to a fairly narrow mass range. The upper limit is determined by $M_{ch} = 1.4 M_{\odot}$ but there is also a lower limit of $0.25 M_{\odot}$.

The typical radius of a white dwarf is of the order of 1 % of the solar radius corresponding to the fact that, the average density is of the order of 10^6 times the mean solar density. For a white dwarf of about one solar mass the central density is calculated $\approx 10^{11}$ kg m^{-3}. Since white dwarfs as the name implies emits white light they have surface temperatures of the same order as that of the sun so that with almost 100 times smaller radius their luminosities are of the order of 10^{-3} of the solar luminosity. This guarantees that even with no nuclear energy source white dwarfs can continue shining for billions of years.

Neutron Stars Neutron star is a type of stellar remnant that can result from the gravitational collapse of a massive star during a super-novae event. The rump left behind after a supernova explosion is a neutron star, which contains neutrons, protons, electrons and heavier but with neutrons predominating.

Neutrons play a similar role in supporting a neutron star as degenerate electrons do in supporting a white dwarf. The limit to which the degenerate neutron gas can do this is analogous to the Chandrasekhar limit for electron-electron degeneracy in white dwarfs.

The limit now becomes

$$M_{max} \approx 5 M_{\odot} \tag{G.4}$$

The fate of a neutron star that undergoes gravitational collapse is a black hole.

Let R_{schw} be the Schwarzschild radius associated with a black hole for an object of mass M given by the formula

$$R_{schw} = \frac{2GM}{c^2} \tag{G.5}$$

As an example let R_{schw} radius of a star of mass $M = 5M_{\odot}$ is 10 km. Equation (G.5) implies that when the physical radius of a collapsed star falls inside R_{schw} there are no light paths to outside world. Photons from the star cannot escape its gravitational field and it becomes black to an outside observer. However an observer with R_{schw} radius would record lots of activity but would not be able to communicate with the outside world.

Black holes are inevitable consequences of Einstein's general theory of relativity.

Hawking Radiation from Black Holes In 1974 Hawking proved that in very strong gravitational field the black holes are able to emit (thermal) radiation.

Hawking temperature for a black hole of mass M is given by

$$kT_H = \frac{\hbar c^3}{8\pi GM}$$

Thus, for $M = 5M_{\odot}$, $T_H \approx 8 \times 10^{-8}$ K.

Note that as the black hole loses energy and mass, it gets hotter and thus a black hole will eventually evaporate and disappear.

The lifetime is given by

$$\tau_{BH} = \text{const} \times \frac{G^2 M^3}{\hbar c^4} \approx 10^{66} \left(\frac{M}{M_0}\right)^3 \text{ yr.}$$

Thus the time for a black hole of a typical astronomical mass to evaporate is far longer than the age of the universe.

Neutrinos from SN1987A SN1987A was a supernova in the large Magellanic cloud some 60 kpc from the Milkyway. The light from the supernova reached the earth on February 23, 1987. As it was the first supernova discovered in 1987, it was labeled "1987A". It occurred approximately 51.4 kilo parsecs from Earth, approximately 170000 light years, close enough that it could be seen from the naked eye from the southern hemisphere. It is the famous supernova from which interactions of the emitted neutrinos were observed simultaneously in the Kamiokande and IMB water Cerenkov detectors, originally designed to detect proton decay. The neutrino pulse was actually detected about seven hours before the optical signal became detectable.

The important reactions that could lead to detection of supernova neutrinos in a water detector are as follows

$$\overline{\nu_e} + p \rightarrow n + e^+ \tag{G.6}$$

$$\nu + e^- \rightarrow \nu + e^- \tag{G.7}$$

$$\overline{\nu} + e^- \rightarrow \overline{\nu} + e^- \tag{G.8}$$

The first supernova recorded by Chinese and Muslim astronomers in 185 A.D.

The integrated neutrino luminosity calculated from the event rates was

$$L \approx 3 \times 10^{46}\ \text{J} \approx 2 \times 10^{59}\ \text{MeV} \tag{G.9}$$

with an uncertainty of a factor of two.

Equation (G.9) is in excellent agreement with the calculated gravitational energy by 1.8×10^{59}. Altogether approximately 10^{58} neutrinos were emitted and even at the earth, some 170000 light years distant the flux survived a journey without attenuation implies their stability since the neutrino pulse looked less than 10 sec, the transit time of neutrinos of different energies was the same within one part in 5×10^{11}. The time of arrival on Earth, t_E will be given in terms of the emission time from the supernova, t_{SN}, its distance and the neutrino mass m and energy E by

$$t_E = t_{SN} + \left(\frac{L}{C}\right)\left[1 + \frac{m^2c^4}{2E^2}\right]$$

for $m^2 \ll E^2$. For two events with different energies E_1 and E_2 the time difference will be given by

$$\Delta t = |\Delta t_E - \Delta t_{SN}| = \left(\frac{Lm^2c^4}{2c}\right)\left[\frac{1}{E_1^2} - \frac{1}{E_2^2}\right] \tag{G.10}$$

For typical values $E_1 = 10$ MeV, $E_2 = 20$ MeV and $\Delta t < 10$ sec, (G.10) gives $m < 20$ eV, a poorer limit than obtained from tritium beta decay.

References

1. Hewish et al. (1968)

Appendix H

Landmarks in chronological order for the development of Nuclear Physics (♣), Particle Physics (♡), Particle Astrophysics (♦), award not given in physics (♠), award given in physics not connected with ♣, ♡, ♦, (†) are indicated by corresponding abridged Nobel Prize citations.

♣ 1901: W.C. Rontgen—citation "for discovery of X-rays"

† 1902:

♡ 1903: A.H. Becquerel, Pierre Curie and Marie Curie—citation "for spontaneous radioactivity"

† 1904:

♣ 1905: P.E.A. Lenard—citation "for work on Cathode rays"
♣ 1906: Sir J.J. Thompson—citation "for Conduction of electricity through gases"
♦ 1907: A.A. Michelson—citation "for precision interferometer especially for the negative result for the existence of ether"

† 1908:
† 1909:
† 1910:
† 1911:
† 1912:
† 1913:

♣ 1914: M.V. Laue—citation "for discovery of the diffraction of X-rays by crystals"

♣ 1915: Sir W.H. Bragg and Sir W.L. Bragg—citation "for analysis of crystal structure by means of X-rays"

♠ 1916:

♣ 1917: C.G. Barkla—citation "for discovery of the characteristic Rontgen radiation of the elements"
♡ 1918: Max Planck—citation "for discovery of energy quanta"

A. Kamal, *Particle Physics*, Graduate Texts in Physics,
DOI 10.1007/978-3-642-38661-9, © Springer-Verlag Berlin Heidelberg 2014

♣ 1919: Johannes Stark—citation “for discovery of Doppler Effect in canal rays and splitting of spectrum lines in electric fields”

† 1920:

♣ 1921: Albert Einstein—citation “for theoretical physics and especially for the law of photoelectric effect”

♣ 1922: Neil Bohr—citation “for structure of atoms and radiation emanating from them”

♣ 1923: R.A. Milikan—citation “for elementary charge of electricity and photoelectricity”

♣ 1924: K.M.G. Siegbahn—citation “for discoveries in X-ray spectroscopy”

† 1925:

† 1926:

♣ 1927: A.H. Compton—citation “for Compton effect” and C.T.R. Wilson—citation “for his method of making the paths of electricity visible by condensation of vapor”

♣ 1928: Sir Owen Williams Richardson—citation “for thermionic phenomenon”

♣ 1929: Prince L.V. De Broglie—citation “for his discovery of the wave nature of electrons”

† 1930:

♠ 1931:

♣ 1932: Werner Heisenberg—citation “for the creation of quantum mechanics”

♣ 1933: Erwin Schrodinger and P.A.M. Dirac—citation “for new productive forms of atomic theory”

♠ 1934:

♡ 1935: Sir James Chadwick—citation “for the discovery of neutrons”

♡ 1936: C.D. Anderson—citation “for the discovery of positron”
♦ V.F. Hess—citation “for discovery of cosmic radiation”

♣ 1937: C.J. Davisson and Sir G.B. Thompson—citation “for experimental discovery of diffraction of electrons by crystals”

♣ 1938: Enrico Fermi—citation “for existence of new radioactive elements produced by neutron irradiation and discovery of nuclear reactions by slow neutrons”

♡ 1939: E.O. Lawrence—citation “for cyclotron and results on artificial radioactive elements”

♠ 1940:

♠ 1941:

♠ 1942:

♣ 1943: Otto Stern—citation “for development of molecular ray method and discovery of magnetic moment of proton”

♣ 1944: Isaac Rabi—citation “for resonance method for recording magnetic properties of atomic nuclei”

♣ 1945: Wolfgang Pauli—citation “for discovery of exclusion principle (Pauli’s principle)”

† 1946:

† 1947:

♣ 1948: Lord P.M.S. Blackett—citation “for development of the Wilson cloud chamber method, and discoveries in nuclear physics and cosmic radiation”

♡ 1949: Hideki Yukawa—citation “for his prediction of the existence of mesons on the basis of theoretical work on nuclear forces”

♡ 1950: C.F. Powell—citation “for development of photographic method and discoveries regarding mesons with this method”

♣ 1951: Sir J.D. Cockcroft and E.T.S. Walton—citation “for transmutation of atomic nuclei by artificially accelerated atomic particles”

♣ 1952: Felix Bloch and E.M. Purcell—citation “for nuclear magnetic precision measurements (MRI)”

† 1953:

♡ 1954: Walter Bothe—citation “for coincidence method”

♣ 1955: W.E. Lamb—citation “for fine structure of hydrogen spectrum and Polykarp Kusch “for precise magnetic moment of electron”

† 1956:

♡ 1957: C.N. Yang and T.D. Lee—citation “for violation of parity in weak interactions (matter- anti matter asymmetry)”

♡ 1958: P.A. Cerenkov, I.M. Frank and I.Y. Tamm—citation “for discovery and interpretation of the Cerenkov effect”

♡ 1959: E.G. Segré and Owen Chamberlain—citation “for discovery of antiproton”

♡ 1960: D.A. Glaser—citation “for invention of bubble chamber”

♣ 1961: Robert Hofstader—citation “for scattering in atomic nuclei and structure of nuclei and R.L. Mösbauer “for recoilless absorption of gamma radiation”

† 1962:

♣ 1963: M.G. Mayer and J.H.D. Jenson—citation “for nuclear shell structure” and E.P. Wigner “for symmetry principles in particles and nuclei”

† 1964:

♡ 1965: S.I. Tomenaga, Julian Schwinger and R.P. Feynman—citation “for quantum electrodynamics related to elementary particles”

† 1966:

♦ 1967: H.A. Bethe—citation “for energy production in stars”

♡ 1968: L.W. Alvarez—citation “for discovery of a large number of resonance states, large hydrogen bubble chamber and data analysis”

♡ 1969: Murray Gell-Mann—citation “for classification of elementary particles and their interactions”

† 1970:
† 1971:
† 1972:
† 1973:

♦ 1974: Sir Martin Ryle—citation "for aperture synthesis technique and Antony Hewish for radio astrophysics (pulsars)"

♣ 1975: Aage Bohr, Ben Mottelson and James Rainwater—citation "for discovery between collective motion and particle motion in atomic nuclei"

♡ 1976: Burton Richter Samuel Ting—citation "for discovery of a heavy elementary particle of a new kind (ψ/J)"

♠ 1977:

♦ 1978: A.A. Penzias and R.W. Wilson—citation "for discovery of cosmic microwave background radiation"

♡ 1979: S.L. Glashow, Abdus Salam and Steven Weinberg—citation "for unified weak and electromagnetic interaction between elementary particles including prediction of weak neutral current"

♡ 1980: J.W. Cronin and V.L. Fitch—citation "for CP violation in neutral K-meson"

† 1981:
† 1982:

♦ 1983: S. Chandrasekhar—citation "for theoretical studies of structure and evolution of stars and W.A. Fowler "for theoretical and experimental studies of nuclear reactions leading to chemical elements in the universe"

♡ 1984: Carlo Rubbia and Simon Vander Meer—citation "for stochastic cooling which led to the discovery of W and Z bosons"

† 1985:
† 1986:
† 1987:

♡ 1988: L.M. Lederman, Melvin Schawartz and Jack Steinberger—citation "for neutrino beam method and discovery of muon neutrino"

† 1989:

♡ 1990: J.L. Friedman, H.W. Kendall and R.E. Taylor—citation "for deep inelastic scattering of electrons and bound neutrons—important for quark model"

† 1991:

♡ 1992: Georges Charpak—citation "for invention of multiwire proportional chamber"

♦ 1993: R. A. Hulse and J.H. Taylor Jr.—citation "for study of gravitation by pulsar"

♡ 1994: B.N. Bruckhouse and C.G. Shull—citation "for development of neutron scattering technique"

♡ 1995: M.L. Perl and Froderin Reines—citation "for lepton physics"

♡ 1999: G. Hooft and M.J.G. Veltman—citation "for quantum structure of electroweak interactions"

♡ 2004: D.J. Gross, H.D. Politzer and F. Wilczek—citation "for discovery of asymptotic freedom in the theory of strong interaction"

♡ 2008: Y.C. Nambu, Makoto Kabajashi and Toshide Masakawa—citation "for mechanism of spontaneous broken symmetry—quantum mechanical explanation of CKM matrix"

♦ 2011: Saul Perlmutter, B.P. Schimdle and A.G. Reiss—citation "for discovery of accelerating expansion of universe through observation of distant supernovae"

Some Interesting Facts

- Father and son combination: Neil Bohr and Aage Bohr, Sir W.L. Bragg, Sir W.L. Bragg.
- Youngest Nobel Laureate: (age 25) William Laurence Bragg.
- Only Laureate to win the Nobel prize twice in Physics: 1952 and 1956—John Bardeen.
- Marie Curie received Nobel prize twice in different subjects—physics in 1903 and chemistry in 1911.

References

1. C. Albajar et al. (UA1 Collab.), Search for B0 anti-B0 oscillations at the CERN proton–anti-proton collider. 2. Phys. Lett. B **186**, 247 (1987)
2. Alitti et al. (UA2 Collab), An improved determination of the ratio of W and Z masses at the CERN $\bar{p}p$ collider. Phys. Lett. B **276**, 354 (1992)
3. R. Ansari et al. (UA2 Collab.), Measurement of the standard model parameters from a study of W and Z bosons. Phys. Lett. B **186**, 440 (1987)
4. G. Armison et al. (UA1 Collab.), Experimental observation of isolated large transverse energy electrons with associated missing energy at $\sqrt{s} = 540$-GeV. Phys. Lett. B **122**, 103 (1983)
5. J.N. Bahcal, Solar neutrino experiments. Rev. Mod. Phys. **50**, 881 (1978)
6. J.N. Bahcal, *Neutrino Astrophysics* (Cambridge University Press, Cambridge, 1989)
7. C. Baltay et al., Antibaryon production in antiproton proton reactions at 3.7 BeV/c. Phys. Rev. B **140**, 1027 (1965)
8. J.D. Barrow, The baryon asymmetry of the universe. Surv. High Energy Phys. **1**, 182 (1980)
9. W. Bartel et al. (JADE Collab.), Tau lepton production and decay at PETRA energies. Phys. Lett. B **161**, 188 (1985)
10. H.J. Behread (CELLO-Collab.), A measurement of the muon pair production in e^+e^- annihilation at 38.3 GeV $\leq \sqrt{s} \leq$ 46.8 GeV. Phys. Lett. B **191**, 209 (1987)
11. H.A. Bethe, *Elementary Nuclear Theory* (Wiley, New York, 1947)
12. A.A. Bethe, J. Ashkin, Passage of radiations through matter, in *Experimental Nuclear Physics*, vol. 1, ed. by E. Segre (Wiley, New York, 1953), p. 166
13. H.A. Bethe, P. Morrison, *Elementary Nuclear Theory* (Wiley, New York, 1952)
14. J.B. Birks, *Scintillation Counters* (McGraw-Hill, New York, 1953)
15. J. Blatt, V.F. Weisskopf, *Theoretical Nuclear Physics* (Wiley, New York, 1952)
16. E. Bleur, G.J. Goldsmith, *Nucleonics* (Holt, Rinehart and Winston, New York, 1963)
17. M.H. Blewett, Characteristic of typical accelerators. Annu. Rev. Nucl. Sci. **17**, 427 (1967)
18. R.J. Blin-Stoyle, *Nuclear and Particle Physics* (Chapman and Hall, London, 1991)
19. A. Bohr, B.R. Mottelson, *Nuclear Structure* (Benjamin, New York, 1969)
20. H. Bradner, Bubble chambers. Annu. Rev. Nucl. Sci. **10**, 109 (1960)
21. D.M. Brink, *Nuclear Forces* (Pergamon, Oxford, 1965)
22. H.N. Brown et al., Observation of production of a e^-e^+ pair. Phys. Rev. Lett. **8**, 255 (1962)
23. W.E. Burcham, *Elements of Nuclear Physics* (Longman, Harlow, 1979)
24. W.E. Burcham, M. Jobes, *Nuclear and Particle Physics* (Longman, Harlow, 1995)
25. J. Button et al., Evidence for the reaction, $p^- p \to \Sigma^0 \Lambda$. Phys. Rev. Lett. **4**, 530 (1960)
26. N. Cabibbo, Unitary symmetry and leptonic decays. Phys. Rev. Lett. **10**, 531 (1963)
27. C.G. Callan Jr., D.J. Gross, High-energy electroproduction and the constitution of the electric current. Phys. Rev. Lett. **22**, 156 (1969)

A. Kamal, *Particle Physics*, Graduate Texts in Physics,
DOI 10.1007/978-3-642-38661-9, © Springer-Verlag Berlin Heidelberg 2014

28. G. Charpak, F. Sauli, High resolution electronic particle detectors. Annu. Rev. Nucl. Part. Sci. **34**, 285 (1984)
29. F.E. Close, *An Introduction to Quarks and Partons* (Academic Press, New York, 1979)
30. F.E. Close, The quark parton model. Rep. Prog. Phys. **42**, 1285 (1979)
31. B.L. Cohen, *Concepts of Nuclear Physics* (McGraw-Hill, New York, 1971)
32. E.D. Commins, *Weak Interactions* (McGraw-Hill, New York, 1973)
33. E.D. Courant, Accelerators for high intensities and high energies. Annu. Rev. Nucl. Sci. **18**, 435 (1968)
34. C.L. Cowan Jr., F. Reines et al., Detection of the free neutrino: a confirmation. Science **124**, 103 (1956)
35. R.H. Dalitz, On the analysis of τ-meson data and the nature of the τ-meson. Philos. Mag. **44**, 1068 (1953)
36. R.H. Dalitz, Strange particle resonant states. Annu. Rev. Nucl. Sci. **13**, 339 (1963)
37. M. Danysz, J. Pniewsky, Delayed disintegration of a heavy nuclear fragment: I. Philos. Mag. **44**, 348 (1953)
38. A. Das, T. Ferbel, *Introduction to Nuclear and Particle Physics* (Wiley, New York, 1993)
39. G.G. Eichholz, J.W. Poston, *Principles of Nuclear radiation detection* (Ann Arbor Science, Ann Arbor, 1979)
40. J. Ellis et al., Physics of intermediate vector bosons. Annu. Rev. Nucl. Part. Sci. **32**, 443 (1982)
41. L.R.B. Elton, *Introductory Nuclear Theory* (Pitmans, London, 1959)
42. H.A. Enge, *Introduction to Nuclear Physics* (Addison-Wesley, Reading, 1966)
43. R.D. Evans, *The Atomic Nucleus* (McGraw-Hill, New York, 1955)
44. C.W. Fabjan, H.G. Fischer, Particle detectors. Rep. Prog. Phys. **43**, 1003 (1980)
45. C.W. Fabjan, T. Ludlam, Calorimetry in high energy physics. Annu. Rev. Nucl. Part. Sci. **32**, 335 (1982)
46. G. Feinberg, L. Lederman, Physics of muons and muon neutrinos. Annu. Rev. Nucl. Sci. **13**, 431 (1963)
47. E. Fermi, *Nuclear Physics*, 5th edn. (University of Chicago Press, Chicago, 1953)
48. R.P. Feynman, *Photon-Hadron Interactions* (Benjamin, New York, 1972)
49. Firestone et al., Observation of the anti-omega. Phys. Rev. Lett. **26**, 410 (1971)
50. W.R. Fraser, *Elementary Particles* (Prentice Hall, Englewood Cliffs, 1966)
51. J.I. Friedman, H.W. Kendall, Deep inelastic electron scattering. Annu. Rev. Nucl. Sci. **22**, 203 (1972)
52. S. Gasiorowicz, *Elementary Particle Physics* (Wiley, New York, 1966)
53. M. Gell-Mann, Y. Ne'eman, *The Eight Fold Way* (Benjamin, New York, 1964)
54. M. Gell-Mann, A. Pais, Behavior of neutral particles under charge conjugation. Phys. Rev. **97**, 1387 (1955)
55. D. Glaser, The bubble chamber, in *Encyclopaedia of Physics*, ed. by S. Fluegge (Springer, Berlin, 1955), p. 45
56. S.L. Glashow, Towards a unified theory, threats in tapestry. Rev. Mod. Phys. **52**, 539 (1980)
57. M. Goldhaber, L. Grodzins, A. Sunyar, Helicity of neutrinos. Phys. Rev. **109**, 1015 (1958)
58. M. Goppert-Mayer, J.H.D. Jensen, *Elementary Theory of Nuclear Shell Structure* (Wiley, New York, 1955)
59. K. Gottfried, V.F. Weisskopf, *Concepts of Particle Physics*, vol. 1 (Clarendon, Oxford, 1984)
60. K. Gottfried, V.F. Weisskopf, *Concepts of Particle Physics*, vol. 2 (Clarendon, Oxford, 1986)
61. A.E.S. Green, *Nuclear Physics* (McGraw-Hill, New York, 1955)
62. O.W. Greenberg, Quarks. Annu. Rev. Nucl. Sci. **28**, 327 (1978)
63. D. Griffiths, *Introduction to Elementary Particles* (Wiley, New York, 1987)
64. C. Grupen, *Particle Detectors* (Cambridge University Press, Cambridge, 1996)
65. D. Halliday, *Introduction to Nuclear Physics* (Wiley, New York, 1955)
66. T. Hamada, L.D. Johnson, A potential model representation of two nucleon data below 315 MeV. Nucl. Phys. **34**, 382 (1962)

67. W.D. Hamilton, Parity violation in electromagnetic and strong interaction processes. Prog. Part. Nucl. Phys. **10**, 1 (1969)
68. H. Harairi, Quarks and leptons. Phys. Rep. **42**, 235 (1978)
69. E.M. Henley, Parity and time reversal invariance in nuclear physics. Annu. Rev. Nucl. Sci. **19**, 367 (1969)
70. P.E. Hodgson, E. Gadioli, E. Gadioli Erba, *Introductory Nuclear Physics* (Clarendon, Oxford, 1997)
71. R. Hofstadter, Nuclear and nucleon scattering of high energy electrons. Annu. Rev. Nucl. Sci. **7**, 231 (1957)
72. I.S. Hughes, *Elementary Particles*, 2nd edn. (Cambridge University Press, Cambridge, 1987)
73. G. Hutchinson, Cerenkov detectors. Prog. Part. Nucl. Phys. **8**, 195 (1960)
74. J.D. Jackson, *The Physics of Elementary Particles* (Princeton University, Princeton, 1958)
75. L.W. Jones, A review of quark search experiments. Rev. Mod. Phys. **49**, 717 (1977)
76. G. Kallen, *Elementary Particle Physics* (Addison-Wesley, Reading, 1964)
77. N. Kemmer, J.C. Polkinghorne, D. Pursey, Invariance in elementary particle physics. Rep. Prog. Phys. **22**, 368 (1959)
78. K. Klein-Knecht, *Detectors for Particle Radiation* (Cambridge University Press, Cambridge, 1999)
79. M. Kobayashi, T. Maskawa, CP violation in the renormalizable theory of weak interaction. Prog. Theor. Phys. **49**, 652 (1973)
80. J.J. Kokkedee, *The Quark Model* (Benjamin, New York, 1969)
81. E.J. Konipinski, Experimental clarification of the laws of β-radioactivity. Annu. Rev. Nucl. Sci. **9**, 99 (1959)
82. K. Krane, *Introductory Nuclear Physics* (Wiley, New York, 1987)
83. R.E. Lapp, H.L. Andrews, *Nuclear Radiation Physics* (Prentice Hall, New York, 1972)
84. C.M.G. Lattes, H. Muirhead, G.P.S. Occhialini, C.F. Powell, Processes involving charge mesons. Nature **159**, 694 (1947)
85. J.D. Lawson, M. Tigner, The physics of particle accelerators. Annu. Rev. Nucl. Part. Sci. **34**, 99 (1984)
86. L. Lederman, Neutrino physics, in *Pure and Applied Physics*, vol. 25, ed. by Burhop (Academic Press, New York, 1967)
87. L.M. Lederman, Lepton production in hadron collisions. Phys. Rep. **26**, 149 (1976)
88. L. Lederman, The upsilon particle. Sci. Am. **239**, 60 (1978)
89. T.D. Lee, C.S. Wu, Weak interactions. Annu. Rev. Nucl. Sci. **15**, 381 (1965)
90. M.S. Livingston, J.P. Blewett, *Particle Accelerators* (McGraw-Hill, New York, 1962)
91. M.S. Livingstone, *High Energy Accelerators* (Interscience, New York, 1954)
92. P. Marmier, E. Sheldon, *Physics of Nuclei and Particles*, vols. I and II (Academic Press, San Diego, 1969/1970)
93. B.R. Martin, G. Shaw, *Particle Physics* (Wiley, New York, 1992)
94. I.E. McCarthy, *Nuclear Reactions* (Pergamon, Elmsford, 1970)
95. E.M. McMillan, Particle accelerators, in *Experimental Nuclear Physics*, ed. by Segre (Wiley, New York, 1959), p. 3
96. W. Meyarhof, *Elements of Nuclear Physics* (McGraw-Hill, New York, 1967)
97. M. Moe, P. Vogel, Double beta-Decay. Annu. Rev. Nucl. Part. Sci. **44**, 247 (1994)
98. L. Montanet et al., Review of particle properties. Phys. Rev. D **50**, 1173 (1994)
99. A. Muirhead, *The Physics of Elementary Particles* (Pergamon, London, 1965)
100. R.L. Murray, *Nuclear Reactor Physics* (Wiley, New York, 1960)
101. P.E. Nemirovskii, *Contemporary Models of the Atomic Nucleus* (Pergamon, Elmsford, 1963) Student editions
102. L.B. Okun, *Weak Interactions of Elementary Particles* (Pergamon, London, 1965)
103. J. Orear, G. Harris, S. Taylor, Spin and parity analysis of bevatron τ mesons. Phys. Rev. **102**, 1676 (1956)
104. P.J. Ouseph, *Introduction to Nuclear Radiation Detectors* (Plenum, New York, 1933)
105. C. Pelligrini, Colliding beam accelerators. Annu. Rev. Nucl. Sci. **22**, 1 (1972)

106. D.H. Perkins, Inelastic lepton nucleon scattering. Rep. Prog. Phys. **40**, 409 (1977)
107. D.H. Perkins, *Introduction to High Energy Physics* (Addison-Wesley, Reading, 1987)
108. M.L. Perl, The tau lepton. Annu. Rev. Nucl. Part. Sci. **30**, 299 (1980)
109. B. Povh, K. Rith, C. Scholz, F. Zetsehe, *Particles and Nuclei* (Springer, Berlin, 1995)
110. C.F. Powell, P.H. Fowler, D.H. Perkins, *The Study of Elementary Particles by the Photographic Method* (Pergamon, Elmsford, 1959)
111. M.A. Preston, *Physics of the Nucleus*, 2 edn. (Addison-Wesley, Wokingham, 1963)
112. W.J. Price, *Nuclear Radiation Detectors* (McGraw-Hill, New York, 1964)
113. D. Prowse, M. Baldo-Ceolin, Anti-lambda hyperon. Phys. Rev. Lett. **1**, 179 (1958)
114. N.F. Ramsey, Electric dipole moments of particles. Annu. Rev. Nucl. Part. Sci. **32**, 211 (1982)
115. F. Reins, Neutrino interactions. Annu. Rev. Nucl. Sci. **10**, 1 (1960)
116. G.D. Rochester, C.C. Butler, Evidence for the existence of new unstable elementary particles. Nature **160**, 855 (1947)
117. M.N. Rosenbluth, High energy elastic scattering of electrons on protons. Phys. Rev. **79**, 615 (1950)
118. B. Rossi, *High Energy Particles* (Prentice Hall, Englewood Cliffs, 1952)
119. E.G. Rowe, E.J. Squires, Present status of C-, P-, and T-invariance. Rep. Prog. Phys. **32**, 273 (1969)
120. R.R. Roy, B.P. Nigam, *Nuclear Physics* (Wiley Eastern, New Delhi, 1993)
121. C. Rubbia, Experimental observation of the intermediate vector bosons W^+, W^- and Z^0. Rev. Mod. Phys. **57**, 699 (1985)
122. R.G. Sachs, *Nuclear Theory* (Addison-Wesley, Reading, 1953)
123. J.J. Sakurai, *Invariance Principles and Elementary Particles* (Princeton University Press, Princeton, 1964)
124. A. Salam, Gauge unification of fundamental forces. Rev. Mod. Phys. **52**, 525 (1980)
125. G.R. Satchler, *Introduction to Nuclear Reactions*, 2nd edn. (MacMillan, London, 1990)
126. L.I. Schiff, *Quantum Mechanics* (McGraw-Hill, New York, 1949)
127. E. Segre, *Experimental Nuclear Physics, 3 Vols.* (Wiley, New York, 1963)
128. E. Segre, *Nuclei and Particles* (Benjamin, Elmsford, 1977)
129. J.R. Stanford, The Fermi national accelerator laboratory. Annu. Rev. Nucl. Part. Sci. **26**, 151 (1976)
130. R. Stephenson, *Introduction to Nuclear Engineering* (McGraw-Hill, New York, 1954)
131. G. 't Hooft, Gauge theories of the forces between elementary particles. Sci. Am. **243**, 90 (1980)
132. J.M. Taylor, *Semiconductor Particle Detectors* (Butterworths, Stoneham, 1989)
133. R.D. Tripp, Spin and parity determination of elementary particles. Annu. Rev. Nucl. Sci. **15**, 325 (1965)
134. S. Van der Meer, Stochastic cooling and accumulation of antiprotons. Rev. Mod. Phys. **57**, 699 (1985)
135. S. Weinberg, Unified theories of elementary particle interactions. Sci. Am. **231**, 50 (1974)
136. S. Weinberg, Conceptual foundations of the unified theory of weak and electromagnetic interactions. Rev. Mod. Phys. **52**, 515 (1980)
137. G.B. West, Electron scattering from atoms, nuclei and nucleons. Phys. Rep. C **18**, 264 (1975)
138. G.C. Wick, Invariance principles of nuclear physics. Annu. Rev. Nucl. Sci. **8**, 1 (1958)
139. W.S. Williams, *An Introduction to Elementary Particles* (Academic Press, New York, 1971)
140. S.S.M. Wong, *Introductory Nuclear Physics* (Prentice Hall, New York, 1990)
141. C.S. Wu et al., Experimental test of parity conservation in beta decay. Phys. Rev. **105**, 1413 (1957)
142. H. Yukawa, On the interaction of elementary particles. Proc. Phys. Math. Soc. Jpn. **17**, 48 (1935)

Index

A. Kamal, *Particle Physics*, Graduate Texts in Physics,
DOI 10.1007/978-3-642-38661-9, © Springer-Verlag Berlin Heidelberg 2014

Printed in the United States
By Bookmasters